VOLUME ONE HUNDRED AND THIRTY THREE

CURRENT TOPICS IN DEVELOPMENTAL BIOLOGY

Vertebrate Skeletal Development

CURRENT TOPICS IN DEVELOPMENTAL BIOLOGY

"A meeting-ground for critical review and discussion of developmental processes"
A.A. Moscona and Alberto Monroy (Volume 1, 1966)

VOLUME ONE HUNDRED AND THIRTY THREE

CURRENT TOPICS IN DEVELOPMENTAL BIOLOGY

Vertebrate Skeletal Development

Edited by

BJORN R. OLSEN
Developmental Biology, Harvard School of Dental Medicine,
Cell Biology, Harvard Medical School,
Boston, MA, United States

ELSEVIER

ACADEMIC PRESS
An imprint of Elsevier

Academic Press is an imprint of Elsevier
50 Hampshire Street, 5th Floor, Cambridge, MA 02139, United States
525 B Street, Suite 1650, San Diego, CA 92101, United States
The Boulevard, Langford Lane, Kidlington, Oxford OX5 1GB, United Kingdom
125 London Wall, London, EC2Y 5AS, United Kingdom

First edition 2019

Notices

Knowledge and best practice in this field are constantly changing. As new research and experience broaden our understanding, changes in research methods, professional practices, or medical treatment may become necessary.

Practitioners and researchers must always rely on their own experience and knowledge in evaluating and using any information, methods, compounds, or experiments described herein. In using such information or methods they should be mindful of their own safety and the safety of others, including parties for whom they have a professional responsibility.

To the fullest extent of the law, neither the Publisher nor the authors, contributors, or editors, assume any liability for any injury and/or damage to persons or property as a matter of products liability, negligence or otherwise, or from any use or operation of any methods, products, instructions, or ideas contained in the material herein.

ISBN: 978-0-12-810487-3
ISSN: 0070-2153

Publisher: Zoe Kruze
Acquisition Editor: Fiona Pattison
Editorial Project Manager: Shellie Bryant
Production Project Manager: Denny Mansingh
Cover Designer: Greg Harris

Typeset by SPi Global, India

Contents

Contributors

Bashar Alkhatib
Department of Cell Developmental and Integrative Biology, University of Alabama at Birmingham, Birmingham, AL, United States

Deepak H. Balani
Endocrine Unit, Massachusetts General Hospital and Harvard Medical School, Boston, MA, United States

Frank Beier
Department of Physiology and Pharmacology, Schulich School of Medicine and Dentistry, and Western University Bone and Joint Institute, University of Western Ontario, London, ON, Canada

Yang Chai
Center for Craniofacial Molecular Biology, University of Southern California, Los Angeles, CA, United States

Danny Chan
School of Biomedical Sciences, The University of Hong Kong, Pok Fu Lam, Hong Kong

Kathryn S.E. Cheah
School of Biomedical Sciences, The University of Hong Kong, Pok Fu Lam, Hong Kong

Qian Cong
Department of Developmental Biology, Harvard School of Dental Medicine, Boston, MA, United States

Rebekah S. Decker
Genomics Institute of the Novartis Research Foundation, San Diego, CA, United States

Christine Hartmann
Institute of Musculoskeletal Medicine, Department Bone and Skeletal Research, Medical Faculty of the Westphalian Wilhelms University of Münster, Münster, Germany

Karl E. Kadler
Wellcome Centre for Cell-Matrix Research, Faculty of Biology, Medicine & Health, University of Manchester, Manchester Academic Health Science Centre, Manchester, United Kingdom

Eiki Koyama
Translational Research Program in Pediatric Orthopaedics, Division of Orthopaedic Surgery, The Children's Hospital of Philadelphia, Philadelphia, PA, United States

Henry M. Kronenberg
Endocrine Unit, Massachusetts General Hospital and Harvard Medical School, Boston, MA, United States

Véronique Lefebvre
The Children's Hospital of Philadelphia, Philadelphia, PA, United States

Karen M. Lyons
Department of Orthopaedic Surgery, Geffen School of Medicine, UCLA, Los Angeles, CA, United States

Pierre J. Marie
UMR-1132 Inserm (Institut national de la Santé et de la Recherche Médicale) and University Paris Diderot, Sorbonne Paris Cité, Hôpital Lariboisière, Paris, France

Noriaki Ono
University of Michigan School of Dentistry, Ann Arbor, MI, United States

David M. Ornitz
Department of Developmental Biology, Washington University School of Medicine, St. Louis, MO, United States

Maurizio Pacifici
Translational Research Program in Pediatric Orthopaedics, Division of Orthopaedic Surgery, The Children's Hospital of Philadelphia, Philadelphia, PA, United States

Carina Prein
Department of Physiology and Pharmacology, Schulich School of Medicine and Dentistry, and Western University Bone and Joint Institute, University of Western Ontario, London, ON, Canada

Vicki Rosen
Department of Developmental Biology, Harvard School of Dental Medicine, Boston, MA, United States

Danielle Rux
Translational Research Program in Pediatric Orthopaedics, Division of Orthopaedic Surgery, The Children's Hospital of Philadelphia, Philadelphia, PA, United States

Rosa Serra
Department of Cell Developmental and Integrative Biology, University of Alabama at Birmingham, Birmingham, AL, United States

Stefan Teufel
Institute of Musculoskeletal Medicine, Department Bone and Skeletal Research, Medical Faculty of the Westphalian Wilhelms University of Münster, Münster, Germany

Sade Williams
Department of Cell Developmental and Integrative Biology, University of Alabama at Birmingham, Birmingham, AL, United States

Ruoshi Xu
State Key Laboratory of Oral Diseases, National Clinical Research Center for Oral Diseases, Department of Cariology and Endodontology, West China Hospital of Stomatology, Sichuan University, Chengdu, China

Yingzi Yang
Department of Developmental Biology, Harvard School of Dental Medicine, Boston, MA, United States

Ching-Yan Chloé Yeung
Institute of Sports Medicine Copenhagen, Bispebjerg Hospital; Center for Healthy Aging, University of Copenhagen, Copenhagen, Denmark

Raymond K.H. Yip
School of Biomedical Sciences, The University of Hong Kong, Pok Fu Lam, Hong Kong

Yuan Yuan
Center for Craniofacial Molecular Biology, University of Southern California, Los Angeles, CA, United States

Preface

This volume of *Current Topics in Developmental Biology* covers outstanding research on mechanisms responsible for the formation, growth, and maintenance of the vertebrate skeleton in health and disease. In chapters written by authors who have made substantial contributions to answering critical questions regarding skeletal development and function, the current state of the field is documented and future steps that need to be taken for an even deeper understanding of this crucial organ system are suggested.

The functional vertebrate musculoskeletal organ system consists of several types of tissues, such as bone, cartilage, soft connective tissues, tendons and ligaments, muscles, bone marrow, nerves, and blood vessels. Therefore, many different types of cells participate in building, maintaining, and repairing this robust system that allows humans and other vertebrate animals to walk, run, work, and play. The reviews in this volume, however, are primarily focused on the cellular and molecular processes responsible for generating and regulating the cartilage, bones and teeth, tendons and ligaments, growth plates, and joints in the skeleton. Chapter 1, written by Ono, Balani, and Kronenberg, provides a timely discussion of stem and progenitor cells in skeletal development. The authors examine the relevance of the traditional definition of skeletal stem and progenitor bone-forming cells when compared with more unambiguous in vivo-tracing experiments in mice. These more recent approaches are leading to a better identification of skeletal stem cells in the perichondrium of fetal cartilage models of future bones, in the marrow space and periosteal regions of postnatal bones, and the resting zone of growth plates. Discussion of data from studies of endochondral ossification at different stages—from growth plate formation to establishment and maintenance of bone marrow stroma in adult bones—is followed by reviews of data on the contributions of periosteal and craniofacial suture cells to the formation of chondrocytes and osteoblasts. The chapter concludes with an excellent review of the action of a parathyroid hormone-related polypeptide on osteoblastic precursor cells and the effects of Wnt/β-catenin signaling on bone formation. Finally, the authors remind readers that although major advances in understanding the cellular aspects of bone development have been made, the relationships between different types of skeletal stem cells are only beginning to be understood; much fundamental work is clearly needed.

The cellular activities responsible for the formation and maintenance of cartilage and bone tissues depend on chemical and mechanical signaling between cells and their extracellular matrix environment. In Chapter 2, Prein and Beier describe recent advances in the understanding of how extracellular matrix-binding cell-surface receptors are linked to intracellular signaling pathways controlling chondrocytic activities at several developmental stages—when cartilage models of future bones are formed and when cartilage is replaced by bone during endochondral ossification. The authors start with a review of integrins—heterodimeric transmembrane receptors encoded by at least 18 α- and 8 β-subunit genes in the human genome—and their roles in activating several pathways for the control of cell proliferation, differentiation, migration, survival, and interactions within the intracellular cytoskeleton. This is followed by a discussion of intracellular integrin partners, such as focal adhesion kinase, Rho GTPases, MAP kinases, and their contributions to the differentiation of chondrocytes. The chapter also provides data on nonintegrin chondrocytic receptors, including CD44, the principal receptor for hyaluronan; Syndecans-transmembrane proteoglycans containing heparan sulfate side chains and serving as coreceptors for growth factors, cytokines, and extracellular matrix components; and the receptor tyrosine kinases Discoidin Domain Receptors DDR1 and DDR2, activated by binding to collagens and inducing expression of matrix metalloproteinase 1 (MMP1), involved in remodeling of the extracellular matrix.

While Chapters 1 and 2 provide a background of current understanding of the mechanisms by which stem and progenitor cells and the extracellular matrix generate cartilage and bone, Chapter 3 by Williams, Alkhatib, and Serra provides an excellent up-to-date review of complex signaling mechanisms governing the development of the axial skeleton and discusses advances in tissue engineering that may help generation of novel strategies for improving treatment of human diseases associated with dysregulated mechanisms in the development and postnatal functions of the spine. The description of different steps in the development of somite-derived structures—somitogenesis; specification of sclerotomes; resegmentation; development of vertebrae, annulus fibrosis, tendons, and ligaments—is based on a large number of important studies in the field. The authors also provide a detailed review of the role of the notochord in the development of the nucleus pulposus, the extracellular matrix composition and function of the notochordal sheath, efforts aimed at identifying markers for notochordal and nucleus pulposus cells, mechanisms responsible for transition of notochordal tissue to nucleus pulposus, as well as more recent studies aimed at identifying genes critical for the maintenance of the nucleus pulposus. Finally, the authors point out that since there is currently no high-quality source of notochordal cells that may be used for regeneration of the nucleus pulposus in patients with intervertebral disc disorders, studies of stem cell differentiation to nucleus pulposus-like cells are needed. Recent studies showing that human iPSCs can be converted to nucleus pulposus-like cells in culture, combined with the identification of genes that are critical for the maintenance of nucleus pulposus, support the conclusion of the authors that "mimicking embryonic development" by tissue engineering may provide a strategy for treatment of patients with intervertebral disc disorders.

The potential use of stem cells and tissue engineering for the treatment of skeletal disorders is also relevant to the discussion in Chapter 4. Based on up-to-date data, Yuan and Chai provide an excellent overview of the development of jaw bones by cranial neural crest cells and formation of teeth, with epithelial cells being responsible for the formation of enamel and mesenchymal cells giving rise to dentin-producing odontoblasts, dental pulp cells, cementoblasts, and cells in the periodontal ligament. This is followed by a systematic review of cellular contributions to the formation of the maxilla and mandible, a discussion of the critical genes and proteins involved and the molecular identity of the two jaw bones. After describing the fate and role of Meckel's cartilage and associated regulatory mechanisms, steps and details of mandibular bone osteogenesis are reviewed. The authors have also included interesting sections on hemifacial microsomia and the use of quantitative analysis, and dynamic imaging and anatomical landmarks, to analyze defects in the mandible and maxilla. Finally, the chapter covers several aspects of tooth development, including early interactions of odontogenic ectoderm and ectomesenchyme; signaling pathways regulating formation of dentin and enamel; development of the roots of teeth; interactions between teeth and jaw bones; identification and properties of dental stem cells. It ends with a discussion of stem cells and regenerative therapies in cases of mandibular hypoplasia and injury or lesions in teeth.

The development of a functional vertebrate skeleton requires biological devices, such as joints, to allow movement and flexibility. In Chapter 5, Rux, Decker, Koyama, and Pacifici discuss early events in the development and morphogenesis of joints in the appendicular skeleton as well as evolutionary influences on these processes. Several references to reviews that cover other important aspects of the development of joints are cited. The starting point for the discussion of the early events in this case is the "interzone," an area of interconnected and closely bound mesenchymal cells, flanked by cartilage-forming cells within the developing cartilage models of endochondral bones. Based on data from several important experiments the authors describe the functions and fates of interzone cells. The interzone cells, expressing the TGFβ superfamily member GDF5, can generate many of the cell types and tissues needed to build the synovial types of joint. This is followed by a review of the mechanisms that regulate the postnatal growth and morphogenesis of the articular cartilage in joints like the knee joint. Based on elegant studies of such postnatal growth in mice, it appears that the remarkably rapid thickening of postnatal articular cartilage is to a large extent due to an increase in the size and stacking of the chondrocytes within the cartilage, somewhat similar to what happens in growth plate cartilage of appendicular long bones. Finally, the authors end the chapter with an interesting and thought-provoking review that highlights important aspects of the organization, structure, architecture, and function in the appendicular skeleton during the evolution of fishes and the transition of these vertebrates from water to land. In addition, they provide a "map" for future experiments aimed at understanding how interzone cells can generate the multiple tissues and the different architectures of different joints and mechanisms responsible for the remarkable structure and mechanical functionality of articular cartilage.

At the end of Chapter 5, the authors point out recent discoveries of genes/proteins and signaling pathways, including BMP signaling, that may have substantial roles in the development and function of articular cartilage. In Chapter 6, Lyons and Rosen discuss in detail the important role of BMP signaling in joint development. Following an up-to-date overview of BMP and TGFβ signaling pathways and how they are regulated, the authors describe experimental evidence in support of the conclusion that inhibition of BMP must occur before a functional interzone can form and that inhibition of BMP is important for the development of different joint tissues. They also present experimental evidence that GDF5 indeed has a role in joint development and is important for the formation of the different tissue components in the joint. Interestingly, GDF5 appears to have a dual function in the developing cartilage and joint, in that it promotes chondrogenesis by cells when they are committed to the chondrogenic fate, but it inhibits chondrocyte differentiation of cells in the interzone. To explain these different effects, the authors speculate that effects of GDF5 are mediated by TAK1, and they discuss evidence suggesting that GDF5 in the interzone may help to prevent inhibition of BMP signaling outside the zone. Finally, the review provides analyses of the complex roles of TGFβ signaling in skeletal development and formation of joints, and it addresses the question of how TGFβ signaling may interfere with BMP signaling in interzone cells. This is followed by a brief summary of what is currently known and the important message that much remains to be discovered before there is a solid basis for the development of effective clinical strategies aimed at joint tissue repair and regeneration.

In Chapters 7–10, investigators who have made substantial contributions to the knowledge of transcription factors, growth factors, and signaling pathways in skeletal development provide detailed analyses of experimental evidence and summaries of areas that need more work. In Chapter 7, Lefebvre discusses the studies that led to the discovery of transcription factors in the SOX family, their structural and functional similarities and differences, and their relationship to the superfamily of high-mobility-group domain containing proteins. Their importance in normal human development is underscored by the finding that mutations in 10 SOX genes, including SOX genes involved in development of the skeleton, cause syndromic developmental disorders. In mice, *Sox11* deficiency results in death at birth, associated with underdevelopment of both intramembranous and endochondral bones. The combined inactivation of SOX4, SOX11, and SOX12 in mesodermal progenitor cells impairs formation of the skull in mice. Following a detailed and informative summary of the control of skeletal progenitor cells by SOX genes, the author discusses the essential roles of the genes in chondrogenesis. Without SOX9 or SOX5/SOX6 genes notochordal cells die before they can transition into nucleus pulposus cells in intervertebral discs; all three genes are required for the formation of cartilage growth plates, and SOX9 is required for maintaining the chondrocyte fate of hypertrophic chondrocytes. Thus, SOX genes are crucial for skeletal development and growth. As pointed out by the author, however, the understanding of how SOX genes and proteins are regulated is still in its infancy. Several posttranslational modifications of SOX proteins have been studied in vitro, but only a few have been examined in vivo. Major pieces of the SOX puzzle have been identified, but critical details are yet to be discovered.

In Chapter 8, Ornitz and Marie review critical roles of fibroblast growth factors (FGFs) and their receptor-mediated (FGFR) signaling pathways in skeletal development and growth. They provide a detailed description of how FGF/FGFR expression is required at all stages of appendicular development, from preparing mesenchymal cells for the condensation process that is followed by chondrocytic differentiation, to growth and remodeling, maintenance, and repair of mature bones. During these processes, signaling through FGFR1, 2, 3, and 4 is regulated in location- and time-dependent manners by interactions with different FGFs, other signaling pathways, and transcription factors such as SOX9. A detailed and up-to-date review of the crucial role of FGFR3-mediated signaling at all stages of chondrogenesis includes a description of its stimulation of immature chondrocyte proliferation, inhibitory effects on proliferation and maturation of chondrocytes in growth plate cartilage, its suppression of other signaling pathways in cartilage, and its effect on the length and transport activities of the primary cilium in chondrocytes. In an up-to-date section on FGFR signaling in osteoblasts, the authors provide information on its control of proliferation, differentiation, and survival, including increased expression and acetylation/stabilization of RUNX2, the key transcription factor for osteoblastic differentiation. Interestingly, the osteoblastic effects are also controlled by cell surface and secreted heparan sulfate proteoglycans, acting as FGF coreceptors, as well as interactions with BMP, TGFβ, and Wnt/β-catenin signaling pathways. In the last part of the chapter, the consequences of mutations in the FGF/FGFR signaling pathways are described. Covering chondrodysplasia, CATSHL, and craniosynostosis syndromes in humans, the authors discuss current ideas and efforts to develop FGFR signaling pathway-based therapeutics. They conclude, however, that development of novel therapeutic approaches aimed at attenuating dysregulated skeletal phenotypes requires a better understanding of mechanisms underlying FGFR signaling.

In Chapter 9, Teufel and Hartmann provide a detailed and informative discussion of how members of a large number of cysteine-rich glycoproteins, Wnts, signal through a variety of different intracellular pathways to control differentiation of mesenchymal cells to chondrocytes or osteoblasts, regulate osteocyte and osteoclast functions, and affect development, maintenance, and degeneration of joints. During the early steps of bone formation in the limbs, Wnt-signaling inhibits chondrocyte differentiation with the consequence that mesenchymal cells in the core of the growing limb bud are kept in an undifferentiated state. Within the different layers of chondrocytes in the growth plates of endochondral bones, Wnts regulate the transition of cells from one layer to the next and induce expression of Indian hedgehog in prehypertrophic chondrocytes, thus promoting maturation of growth plate chondrocytes. During development and maintenance of bones, Wnt-signaling regulates both bone resorption and bone formation. Wnt16, required for the increased rate of bone formation induced by mechanical loading, also acts as an inhibitor of osteoclast formation. Postnatal loss of Wnt/β-catenin signaling in preosteoblasts shifts their differentiation toward adipocytes. Polymorphisms in the human Wnt3a gene are associated with variations in bone mineral density. The authors also discuss the substantial and important roles of Wnt-signaling and signaling inhibitors expressed by osteocytes, the effects of Wnt-signaling in osteoclastogenesis, and in the development, maintenance, and degeneration of joints. Several studies have reported upregulated Wnt-signaling in osteoarthritic (OA) patients and drugs modulating Wnt-signaling are being investigated for treatment of OA. Finally, the dysregulated processes in skeletal disorders caused by mutations in Wnt-signaling genes reflect the importance of the pathways involved. Although these discoveries have led to new strategies for development of drugs to treat skeletal disorders, the authors point out the need for better understanding of the ligands associated with many of the Wnt-signaling pathways, and they propose investigations aimed at identifying fine-tuning mechanisms that may be used to control the different pathways involved.

In Chapter 10, Cong, Xu, and Yang describe the contributions of $G\alpha_S$ signaling to skeletal development, homeostasis, and diseases. In the introductory section they provide information regarding the importance of G-protein-coupled receptors, the largest family of cellular surface receptor families, for the control of developmental and physiological processes. Heterotrimeric G-protein complexes, consisting of $G\alpha$, $G\beta$, and $G\gamma$ subunits, transduce the signaling of the receptors. $G\alpha_S$, encoded by the GNAS gene in humans, is a stimulatory subunit that regulates several steps in skeletal development, including chondrocyte proliferation and hypertrophy, osteoblastic differentiation, and mineralization. The authors provide a concise, yet detailed, review of skeletal diseases caused by activating and inactivating mutations in GNAS. The activating mutations cause hematopoietic and adipocytic tissues in bone marrow spaces to be replaced by fibrous tissue and are associated with abnormalities in bone architecture and variable dysfunctions in other organ systems. Heterozygous inactivating mutations are seen in progressive osseous heteroplasia, Albright's hereditary osteodystrophy, pseudo-pseudohypoparathyroidism, and osteoma cutis. Detailed and up-to-date experimental data, discussed in the following sections, indicate that $G\alpha_S$ facilitates commitment of mesenchymal cells to osteoblasts, helps the production of high-quality bone, and is able to affect bone formation via augmentation of Hedgehog (Hh)- and Wnt-signaling pathways. In the last part of the chapter, the authors review mouse models of GNAS mutations, generated to improve the understanding of pathogenetic mechanisms and develop a basis for novel therapeutic approaches aimed at treating patients with fibrous dysplasia, progressive osseous heteroplasia, and heterotopic ossification. The mouse models also provide tools to allow a better understanding of the role of $G\alpha_S$ signaling in both GPCR-independent and -dependent pathways, in addition to testing whether drugs targeting $G\alpha_S$ effector pathways, such as Wnt- and Hh-signaling pathways, may be used to treat patients with disorders caused by $G\alpha_S$ or GPCR mutations.

It has been known for some time that many homeostatic cell and tissue processes, regulated by different signaling mechanisms, exhibit a circadian rhythm. Tissues that are rich in extracellular matrix components, such as cartilage and tendons, are peripheral circadian clocks. In Chapter 11, Yeung and Kadler provide an excellent review of circadian clock mechanisms that regulate proper development of tendons and give them their remarkable ability to transmit forces from muscles to bones during joint movements over a lifetime of exposure to mechanical stress. An introductory section summarizes embryonic processes resulting in formation of an organized assembly of tenocytes containing intracellular compartments with thin collagen fibrils of uniform diameters, and the release of the fibrils into the extracellular space as tenocytes grow and reorganize into a network of cells after birth. This is followed by a detailed and up-to-date review of the mammalian clock system, beginning with the "master clock" in the brain, followed by descriptions of peripheral clocks, their tissue specificity and entrainment, and a discussion of clock-dependent regulation of tendon homeostasis. Transcriptome data indicate that about 4.5% of the tendon transcriptome are rhythmic with a 24-h period and a comparison with other musculoskeletal tissues indicates that this circadian regulation of gene expression in tendons is tissue specific. Interestingly, transcription of collagen type I genes is not rhythmic in murine tendons, but collagen type I production is regulated by the tendon clock at a posttranscriptional level. The tendon clock also regulates posttranslational modifications of procollagen type I, the transport of collagens from the endoplasmic reticulum (ER) to the Golgi, and it prevents calcification of tendons via regulation of BMP signaling.

These findings are followed by a discussion of the remodeling of tendon extracellular matrices. Tendons from animals at different time points of the 24-h cycle exhibit different viscoelastic properties. This is consistent with diurnal variations in biochemical properties of the extracellular matrix and the demonstration that mice with a compromised tendon clock develop tendons with abnormal collagen fibril structures and poor mechanical properties. Finally, in sections on aging of the tendon clock, methods for tendon clock entrainment, and around-the-clock tendon care, the authors raise several important issues that require more research, such as potential effects of the central clock on tendon clocks, the potential impact of exercise and age-dependent changes in the extracellular matrix, the effects of the rest–activity cycle, and intertissue synchronization of clocks (tendon to bone or muscle).

In Chapter 12, Yip, Chan, and Cheah provide an engaging, informative, up-to-date overview of many of the mutations that disrupt the patterning, architecture, and growth of the skeleton. Starting with mutations that affect patterning of the appendicular skeleton, they systematically review the genetic disruptions affecting integrated control of cartilaginous and osteoblastic skeletal elements and discuss mutations responsible for ciliopathies and planar cell polarity defects in growth plates. These sections are followed by a review of skeletal genetic disorders associated with endoplasmic reticulum stress, mutations in noncoding regulatory DNA elements affecting skeletal development, an up-to-date discussion of disorders caused by changes in the 3D genome, and how mechanistic insights from skeletal disorders can impact the path to therapies. As examples of disorders for which mechanistic insights have led to therapies in human clinical trials, the authors are discussing achondroplasia, the most common type of short-limbed dwarfism in humans—caused by activating mutations in *FGFR3*—and Schmid metaphyseal chondrodysplasia—caused by mutations in collagen type X in hypertrophic chondrocytes, affecting its folding and resulting in activation of ER stress responses. A large number of in vitro and mouse studies have led to the identification of the mechanisms by which FGFR3 negatively affects growth plate activity. Based on these data, testing of a drug that counteracts these effects has been through clinical trials and is now in Phase 3. In the case of Schmid metaphyseal chondrodysplasia, a drug to reduce the level of ER stress has been tested in laboratory experiments and clinical trials have been initiated.

Throughout the sections of the chapter, the authors link the contributions of causative mutations in skeletal dysplasias to the understanding of regulatory mechanisms responsible for the development and maintenance of the vertebrate skeleton. They also include several examples demonstrating that since mutations in human syndromes are not generated for testing specific mechanistic hypotheses, they can provide essential, unexpected, and new information about dysregulated mechanisms. A good example is Liebenberg syndrome, a rare dominant condition characterized by the forelimbs acquiring hindlimb characteristics. Genetic screening showed that the patients carried heterozygous deletions and translocations, causing a switch in the chromatin architecture, upstream of the gene encoding PITX1; a protein expressed by mesenchymal cells in mouse hindlimb buds and required for bud outgrowth. As a result, the transcription of PITX1 in Liebenberg syndrome patients is activated in the forelimbs by an enhancer that normally is only positioned close to the PITX1 gene in the hindlimbs. Preaxial polydactyly, caused by mutations affecting Sonic Hedgehog (Shh) expression, is another example of how mutations in noncoding DNA regions can cause severe dysregulation of skeletal development. An enhancer, located about 1 Mb away from the Shh promoter, is regulating the anterior area-restricted expression of Shh in the early limb bud. Mutations in the enhancer cause expression of Shh beyond the restricted area, with polydactyly as the result.

In a "Future directions and perspectives" section, innovative ideas and suggestions for future work are discussed. Based on recent discoveries that hypertrophic chondrocytes may differentiate into osteoblasts, the authors propose experiments aimed at determining whether a chondrocyte-to-osteoblast transition may occur in cases of ectopic bone formation. Considering the recent advances in understanding the 3D genome architecture and how it is regulated, they stress the need for "integrating knowledge of gene regulatory networks, genetic variants, phenotypes, and proteosomes into a system-level understanding of the molecular basis of skeletal dysplasias."

Bjorn R. Olsen
Developmental Biology, Harvard School of Dental Medicine,
Cell Biology, Harvard Medical School,
Boston, MA, United States

CHAPTER ONE

Stem and progenitor cells in skeletal development

Noriaki Ono[a,*], Deepak H. Balani[b], Henry M. Kronenberg[b]
[a]University of Michigan School of Dentistry, Ann Arbor, MI, United States
[b]Endocrine Unit, Massachusetts General Hospital and Harvard Medical School, Boston, MA, United States
*Corresponding author: e-mail address: noriono@umich.edu

Contents

Abstract

Accumulating evidence supports the idea that stem and progenitor cells play important roles in skeletal development. Over the last decade, the definition of skeletal stem and progenitor cells has evolved from cells simply defined by their in vitro behaviors to cells fully defined by a combination of sophisticated approaches, including serial transplantation assays and in vivo lineage-tracing experiments. These approaches have led to better identification of the characteristics of skeletal stem cells residing in multiple sites, including the perichondrium of the fetal bone, the resting zone of the postnatal growth

Current Topics in Developmental Biology, Volume 133
ISSN 0070-2153
https://doi.org/10.1016/bs.ctdb.2019.01.006

plate, the bone marrow space and the periosteum in adulthood. These diverse groups of skeletal stem cells appear to closely collaborate and achieve a number of important biological functions of bones, including not only bone development and growth, but also bone maintenance and repair. Although these are important findings, we are only beginning to understand the diversity and the nature of skeletal stem and progenitor cells, and how they actually behave in vivo.

1. Introduction

Deliberate coordination of cell differentiation is essential to skeletal development. The skeletal system is comprised of closely connected but functionally distinct tissues such as bones, cartilages and tendons that connect the former two with muscles. Bones, as a central component of the skeletal system, are characterized by strong and rigid structures owing to mineralized matrix, but their functions are not limited to protection of vital organs or levering effects allowing body movement. Bones host and nurture hematopoietic cells within their marrow space; at the same time, they secrete hormones that regulate carbohydrate and mineral ion metabolism, provide large stores of calcium and phosphate available for regulation of systemic mineral ion homeostasis, as well as fertility and brain function. Bones, therefore, have many functions, which are achieved by the coexistence of multiple distinct types of highly active differentiated cells within their structure.

The currently prevailing view is that stem and progenitor cells stand at the pinnacle of the skeletal lineage and provide a significant source of these differentiated cells. Stem cells are characterized by two important functions: self-renewal, which is the ability to replicate themselves while maintaining their properties, and multipotency, which is the ability to give rise to multiple types of differentiated cell types. Progenitor cells are their downstream offspring with similar but potentially more limited capabilities. Bones undergo a number of biologically important steps throughout their life cycle, such as morphogenesis and development, explosive growth and functional maturation, maintenance and repair of proper architecture and function. There is a constant demand for differentiated cells at each step so that bones can become bigger and stronger, while maintaining their strength and functions throughout life. Thus, the significance of stem and progenitor cells playing major roles in these processes has been emphasized.

Stem and progenitor cells play distinctive roles in supporting growth and repair of bones in stage-specific and tissue-specific manners. In skeletal development, bones start as relatively simple primordial structures termed mesenchymal condensations, which then increase their complexity over time and differentiate into each component of the skeletal system. While stem cells in mesenchymal condensations act as multipotent stem cells that can give rise to the entire spectrum of the skeletal cell lineage, tissue-specific stem cells with more dedicated functions at later stages might be even more important to achieve proper tissue growth and regeneration. How stem and progenitor cells alter their properties over various stages of skeletal development are not well understood. This is largely due to the technical and conceptual limitation that these particular cell types cannot be easily identified within each skeletal tissue, since they are embedded within highly complex three-dimensional structures. In addition, complexity and plasticity of the skeletal cell lineage and lack of stage-specific markers contribute to hampering our understanding of these important cell populations. The notion that one or a few types of omnipotent skeletal stem cells can orchestrate the entire process of skeletal development and regeneration might be too simplistic. The current notion rather supports the hypothesis that multiple distinct types of skeletal stem and progenitor cells collaborate and cooperatively establish the network of the skeletal system. In the first chapter, we discuss recent advances in the concept of stem and progenitor cells in skeletal development.

2. Colony-forming unit fibroblasts (CFU-Fs) and mesenchymal/skeletal stem cells (MSCs/SSCs): A traditional definition for skeletal stem and progenitor cells

Most of the work on stem and progenitor cells in skeletal tissues has been strongly motivated by the goal of regenerative medicine, which is to identify cells capable of restoring functions to human bones. The bulk of existing knowledge on stem and progenitor cells of the skeletal lineage has been built on experiments based on human and rodent bone marrow cells. Traditionally, culture of bone marrow cells and subsequent heterotopic transplantation of *in vitro* expanded cells into immunodeficient mice has been used as the gold standard to identify these putative stem cells (Bianco, 2014). The first discovery that bone marrow may include stem cells capable of making bones was almost serendipitously made in 1960s. When whole human bone marrow cells were subcutaneously transplanted into immunodeficient mice, they formed ossicles that included blood cells inside (Friedenstein, Piatetzky-Shapiro, & Petrakova, 1966). Later, colony-forming unit fibroblasts

(CFU-Fs), which are defined as cells capable of adhering to a plastic culture dish and establishing colonies, were identified as cells responsible for heterotopic ossicle formation (Castro-Malaspina et al., 1980). These hybrid ossicles contained osteoblasts and stromal cells of the donor origin and blood cells of the recipient origin. Bone marrow cells from rodents showed similar properties. Therefore, CFU-Fs residing in human and rodent bone marrow are capable of reconstituting bone marrow in a new environment.

CFU-Fs are highly heterogeneous in gene expressions and functions, despite universally possessing similar colony-forming capabilities in cultured conditions. According to the current definition, mesenchymal stem cells (MSCs), or more accurately "skeletal" stem cells, represent a subset of CFU-Fs (Bianco et al., 2013). In fact, only a subset of CFU-Fs possesses self-renewability and multipotency, as not all individual CFU-Fs have the capability to give rise to so-called trilineage cells, i.e., osteoblasts, chondrocytes and adipocytes *in vitro*, or to form ossicles upon transplantation (Sacchetti et al., 2007). Interestingly, individually-cloned CFU-Fs demonstrate much reduced efficiency for ossicle formation upon transplantation (Sacchetti et al., 2007), suggesting that heterotopic interactions among multiple and heterogeneous CFU-Fs might facilitate reconstituting a hematopoietic microenvironment. Locations and properties of CFU-Fs and MSCs cannot be easily clarified in the native environment because these cells can be identified only retrospectively after exposure to culture conditions.

In human bone marrow, use of cell surface markers and cell sorting technologies led to the hypothesis that these putative stem and progenitor cells reside in a perivascular location and assume the morphology closely resembling adventitial reticular cells in bone marrow sinusoids (Bianco, 2014). Initially, the STRO-1 antibody was used to identify and enrich clonal human bone marrow mesenchymal cells (Shi & Gronthos, 2003). Further studies showed that all CFU-Fs were recovered in the $CD146^+$ fraction of human bone marrow cells. $CD146^+$ cells meet the criteria of "skeletal stem cells (SSCs)," as they can be serially transplanted and generate $CD146^+$ cells on secondary transplantation. These $CD146^+$ cells correspond to adventitial reticular cells lining bone marrow sinusoids *in vivo*. Further studies revealed that CD51 (αV integrin)$^+$ platelet-derived growth factor receptor-α (PDGFRα)$^+$ cells represent a small subset of $CD146^+$ cells with even more enriched colony-forming activities (Pinho et al., 2013). Therefore, these studies established the idea that cell surface markers that are typically expressed by perivascular stromal cells can be used to enrich human CFU-Fs in bone marrow.

Genetically modified mice, especially those engineered to express easily assayable proteins, such as green fluorescent protein (GFP), have been particularly useful in identifying putative stem and progenitor cell populations, when combined with cell surface markers and cell sorting technologies. Many of these markers are expressed in proximity to bone marrow vasculature, such as PDGFRα, stem cell antigen-1 (Sca1), chemokine (C-X-C motif) ligand 12 (CXCL12), nestin and α-smooth muscle actin (αSMA). PDGFRα$^+$Sca1$^+$ nonhematopoietic cells (PαS cells) reside in a perivascular space *in vivo* and are enriched for CFU-Fs (Morikawa et al., 2009). These cells, if uncultured and cotransplanted with HSCs, can engraft into irradiated recipients and become osteoblasts, stromal cells, adipocytes and, more importantly, PαS cells themselves, suggesting their self-renewal capability *in vivo*, while cultured PαS cells do not have such capability. Nestin is an intermediate filament protein and a marker for neural stem cells. *Nestin*-GFP is highly expressed in pericytes of bone marrow arterioles (small arteries), and *Nestin*-GFP$^+$ cells in bone marrow include all CFU-F activities and form self-renewable "mesenspheres" that can go through serial heterotopic transplantations (Méndez-Ferrer et al., 2010). The CD51$^+$PDGFRα$^+$ fraction of *Nestin*-GFP$^+$ cells is further enriched for CFU-F activities (Pinho et al., 2013). Alpha-smooth muscle actin (αSMA) is a marker for pericytes of bone marrow arteries. Pericytes associated with vasculature in bone marrow are marked by αSMA-GFP/mCherry transgenic expression and exhibit trilineage differentiation potential *in vitro* (Grcevic et al., 2012). Further, connective tissue growth factor (CTGF) is expressed by peri-trabecular stromal cells, and *CTGF*-GFP$^+$ cells appear to give rise to clonal cells in vitro that are further transplantable (Wang et al., 2015). It is increasingly evident that bone marrow is not the sole location from which CFU-Fs and MSCs can be isolated. Other important locations in bones, such as the periosteum and the growth plate, also house clonogenic cell populations that may have distinct functions *in vivo* (we will mention these points later in the chapter). Therefore, comprehensive approaches combining cell surface and transgenic markers have proven to be particularly useful in gaining insight into CFU-Fs and MSCs.

3. How relevant are CFU-Fs and MSCs/SSCs to skeletal development?

As described above, analyses focusing on CFU-Fs and MSCs have produced a wealth of insight into stem and progenitor cells in skeletal tissues. However, the potential limitation of these studies is that they require multiple maneuvers to evaluate their properties, i.e., cell isolation from their native environment, cell culture in exogenous conditions and transplantation into recipient mice. It is tempting to think that this knowledge accumulated from transplantation assays can be extrapolated into cell lineage development of native skeletal tissues. An appealing hypothesis would be that these skeletal stem and progenitor cells defined by transplantation assays could actively participate in bone formation in their native environment. Yet, whether these cells (CFU-Fs and MSCs) represent precursors of osteoblasts in normal development or during bone remodeling, or perhaps represent cells that respond to injury, is an important question that needs to be addressed through more rigorous analyses.

The emerging idea is that stem cells defined by transplantation assays may have limited relevance to physiological conditions. In the field of hematopoietic stem cell research, transplantation assays have been used as the gold standard to evaluate stem cell properties. However, an elegant study utilizing a cellular barcoding system based on a Sleeping Beauty transposase reveals that these traditionally defined hematopoietic stem cells have minimal contribution to native hematopoiesis; rather lineage-restricted progenitors are the main drivers for blood cell production during most of adulthood (Sun et al., 2014). Whether the same principle applies to MSCs is unknown, current and further lineage-tracing experiments are needed to fully clarify the possible role of bone marrow MSCs in normal skeletal homeostasis and repair.

4. In vivo lineage-tracing experiments in mice: An unambiguous approach to reveal cell fates

The gold standard for investigating fates of particular cells *in vivo* in their native environment is lineage-tracing experiments using transgenic mice. This approach is important for studying the process of development because cell lineages can be investigated without any perturbation. Typically, lineage-tracing strategies utilize the *cre-loxP* technology to permanently mark cells of interest using a bigenic system. *Cre* recombinase is expressed in a promoter/enhancer-specified manner in the first transgene, and acts on the reporter locus in the second transgene. In the reporter locus, multiple sequences containing polyA sequences, as well as translation termination codons in all three reading frames ("STOP" sequences) prevent continued transcription and translation of reporter genes. These "STOP" sequences are flanked by *loxP* sites; when *cre* recombinase acts on the *Rosa26* locus and removes these sequences, the reporter gene becomes expressed under the direction of a ubiquitously active promoter. In a modified version, the *cre* recombinase is covalently bound to the ligand-binding domain of the estrogen receptor (*creERt*) that has been mutated so that 4-hydroxy-tamoxifen (4-OHT), but not estradiol, can bind and alter its tertiary structure. Translocation of the *creERt* complex to the nucleus depends on the presence of 4-OHT, which is an active form of tamoxifen produced after being metabolized in the liver. Therefore, in the *creERt* system, administration of tamoxifen can temporarily activate *cre-loxP* recombination only for 24–48 h until 4-OHT is cleared away from cells. Recombination in the reporter locus is irreversible. Therefore, the reporter gene is continually expressed in the targeted cells and their descendants as long as they survive, even after the promoter that drove expression of *cre* recombinase is no longer active.

Several different versions of the modified *Rosa26* reporter allele have been developed, including *R26R*-LacZ (encoding β-galactosidase), *R26R*-YFP (encoding yellow fluorescent protein), *R26R*-tdTomato (encoding tandem dimer of red fluorescent protein, DsRed), and *R26R*-Confetti. The last allele encodes four different fluorescent proteins (nuclear GFP, YFP, tdTomato and CFP [cyan fluorescent protein]), in which one of them becomes stochastically expressed after *cre-loxP* recombination. It should be noted that each version has its own advantages and disadvantages. For example, although the *R26R*-Confetti allele provides four different colors that may be instrumental for *in vivo* clonal analysis, this allele is relatively insensitive to *cre-loxP* recombination because of the complexity of the transgene design requiring both excision and inversion of *loxP*-flanked sequences (N. Ono, Unpublished Observations).

The lineage-tracing technology has been extensively applied to investigate progenitor-descendant relationships in an unperturbed native environment. To achieve this purpose, it is important to ensure that promoters driving the transgene expression are active in a narrow array of desirable cell types. It is also highly desirable that given promoters have minimal "off-site" activities, especially in their putative descendant cells. Although this genetic approach has successfully uncovered important information on stem and progenitor cells in skeletal tissues, heterogeneity of cell populations of interest marked by a promoter-based approach may complicate overall interpretation of presented data. This is particularly true in studies using a "constitutively active" version of *cre* recombinase. The difference between "constitutively active" *cre* and "inducible" *creERt* is significantly impactful. Because *cre* lacks a temporal factor and induces recombination whenever the given promoter becomes active, the possible relationships between different cell types cannot be easily ascertained. More sophisticated inducible genetic tools that can specifically mark putative stem and progenitor cells in particular skeletal tissues are highly desirable to better characterize these important cell populations.

5. Endochondral bone development/Phase 1: Formation of the growth plate

To understand potential roles of stem and progenitor cells for skeletal development, it is essential to understand the fundamental processes of skeletal development. Here, we review these processes in conjunction with currently available findings from *in vivo* lineage-tracing or fate-mapping studies. Bones are formed through two common mechanisms: intramembranous and endochondral bone formation. The former intramembranous bone formation is a relatively simple process in which immature mesenchymal cells directly differentiate into osteoblasts that lay down the mineralized matrix. The latter endochondral bone formation is a complex process in which initial cartilaginous templates are replaced later by bones. In both mechanisms, a primordial structure called mesenchymal condensations set the frame for the future domain of bones. In this process, mesenchymal cells in a specific domain of the embryonic tissue align together to form cell clusters that exclude blood vessels.

Most bones in mammals are formed through endochondral bone formation. In this process, mesenchymal cells in condensations further differentiate into two distinct but closely related cell types, chondrocytes and perichondrial cells. Chondrocytes develop in the vasculature-free central portion of condensations, whereas perichondrial cells develop in the highly vascularized outer layer of condensations. This process is initiated when condensing mesenchymal cells express the transcription factor Sox9, a master regulator of chondrogenesis. Indeed, Sox9 is absolutely required for these mesenchymal cells to remain within condensations. Sox9 can directly bind to regulatory elements of major cartilage matrix genes, including those encoding collagens (such as type II, IX and XI collagen) and proteoglycans (e.g., aggrecan).

As a result, cartilaginous extracellular matrix is established, and mesenchymal cells in condensations become programmed as chondrocytes or perichondrial cells in their surrounding region.

Subsequently, chondrocytes form the fetal growth plate. This is triggered when chondrocytes in the center of the cartilage stop proliferation, drastically change their cell morphology and differentiate into hypertrophic chondrocytes. Simultaneously, chondrocytes in the flanking region of the cartilage also differentiate and establish the highly hierarchical structure of the growth plate (Kozhemyakina, Lassar, & Zelzer, 2015). The fetal growth plate is primarily composed of three distinct types of chondrocytes: round, flat and hypertrophic chondrocytes. Especially, flat chondrocytes are stacked up on each other like pancakes, and continue to proliferate to generate columnar chondrocytes. These chondrocytes eventually stop proliferating, change their cell morphology again, and differentiate into prehypertrophic and hypertrophic chondrocytes.

Lines of evidence clearly indicate that stem and progenitor cells during these early stages are established locally within mesenchymal condensations and their subsequent structures. In the limb, bones originate from the lateral plate mesoderm during development. The transcription factor Prrx1 is expressed in these mesodermal cells. In fact, *Prrx1-cre*, in which *cre* recombinase is expressed under the direction of a 2.4 kb *Prrx1* promoter, marks essentially all limb mesenchymal cells at later stages, including chondrocytes, perichondrial cells, osteoblasts and stromal cells, but not muscle cells (Logan et al., 2002). As mentioned earlier, Sox9 is expressed in condensing mesenchyme and potently stimulates expression of type II collagen (Col2). Both *Sox9-cre* and *Col2-cre* mark essentially all chondrocytes, perichondrial cells and osteoblasts in a similar way to *Prrx1-cre* (Akiyama et al., 2005). For the technical reasons mentioned in the preceding section, it remains as the subject of future studies if, when, and where early stem and progenitor cells are established during skeletal development.

6. Endochondral bone development/Phase 2: Formation of the perichondrium and osteoblast precursors

Hypertrophic chondrocytes are the master regulators of endochondral bone formation. These cells abundantly express type X collagen (Col10), whose expression is also tightly regulated by Sox9 and its other transcription factors (Dy et al., 2012; He, Ohba, Hojo, & McMahon, 2016). More importantly, hypertrophic chondrocytes secrete paracrine factors, including vascular endothelial growth factor (VEGF) that induce invasion of blood vessels from the perichondrium, and Indian hedgehog (Ihh) that regulates proliferation and differentiation of neighboring chondrocytes and perichondrial cells. Ihh executes a number of important functions through its interactions with its receptor, patched 1 (Ptch1). Ptch1 action blocks the activation of a G protein coupled receptor, smoothened (Smo), and thereby blocks the action of the transcription factors of the Gli family. This inhibitory action of Ptch1 is canceled when Ihh binds to Ptch1. Ihh facilitates formation of columnar chondrocytes by inducing differentiation of these cells from round chondrocytes (Kobayashi et al., 2005). Ihh also promotes proliferation of flat chondrocytes and their differentiation into hypertrophic chondrocytes (Mak, Kronenberg, Chuang, Mackem, & Yang, 2008). Ihh also acts further on periarticular chondrocytes at the end of the cartilage and promotes production of parathyroid hormone-related peptide (PTHrP) (St-Jacques, Hammerschmidt, & McMahon, 1999). PTHrP maintains flat chondrocytes in the proliferating pool and delays their differentiation into hypertrophic chondrocytes through its receptor, the PTH/PTHrP receptor, therefore indirectly delaying Ihh production (Schipani et al., 1997). These series of interactions establish the PTHrP-Ihh feedback loop that is essential to maintaining the growth plate structure (Lanske et al., 1996; Vortkamp et al., 1996).

Cells in the perichondrium adjacent to hypertrophic chondrocytes, especially those on the innermost layer of the perichondrium, become committed to the osteoblast lineage largely due to the actions of Ihh released from these cells (Ono, Ono, Mizoguchi, et al., 2014). This portion of the perichondrium is described as the osteogenic perichondrium, and is the first location in which osteoblast precursors appear in endochondral bone formation. These cells enter the program regulated by transcription factors Runx2 and osterix (Osx), in which Osx acts at a level genetically downstream of Runx2 (Nakashima et al., 2002). These cells further differentiate into mature osteoblasts and abundantly produce bone matrix proteins including type I collagen (Col1), osteopontin, and osteocalcin to make the bone collar, which replaces a layer of the perichondrium. The bone collar eventually becomes the cortical bone, while other perichondrial cells become cells in the periosteum composed of multiple layers of mesenchymal cells with diverse functions (we will mention the periosteum in later sections).

7. Endochondral bone development/Phase 3: Formation of the primary ossification center and the bone marrow cavity

The nascent bone marrow cavity is formed within the cartilage template by attracting blood vessels from the perichondrium. Hypertrophic chondrocytes orchestrate this process by producing VEGF, one of the most potent mediators of angiogenesis (Maes, 2017). At the same time, mesenchymal cells coinvade into the nascent bone marrow cavity and give rise to the mesenchymal stromal compartment of the marrow space. There are at least three routes for these incoming mesenchymal cells.

The first route is through the adjacent perichondrium. Cells committed to the osteoblast lineage on the innermost layer of the osteogenic perichondrium translocate into the nascent bone marrow cavity in a pericyte-like fashion before they become mature osteoblasts. This was demonstrated by lineage-tracing experiments using an *Osx-creERt* line and a *Col1(3.2kb)-creERt* line (Maes et al., 2010); the former labels precursors for osteoblasts, while the latter labels mature matrix-producing osteoblasts in the fetal perichondrium. Importantly, only Osx^+ cells can translocate into the nascent marrow cavity; therefore, such migrating capability is unique to relatively undifferentiated populations of osteoblasts. Interestingly, Osx^+ cells from the fetal perichondrium can proliferate only for a limited period, and eventually disappear from the marrow space (Mizoguchi et al., 2014; Ono, Ono, Mizoguchi, et al., 2014). Thus, the osteogenic perichondrium in the fetal growth plate appears to represent only a transient source of the marrow stromal compartment.

The second route is through the hypertrophic layer of the growth plate. Some hypertrophic chondrocytes do not die from apoptosis but transform into cells that eventually become osteoblasts in an area right beneath the hypertrophic layer (Yang, Tsang, Tang, Chan, & Cheah, 2014). This idea that hypertrophic chondrocytes can behave as precursors of osteoblasts has a growing body of evidence based on lineage-tracing experiments. Cells marked by *Col10-cre* or *Col10-creERt* can differentiate into osteoblasts (Yang et al., 2014; Zhou, von der Mark, et al., 2014). How cells with a large size can turn into very compact mesenchymal cells in the marrow space, and whether this pathway represents a major source of osteoblasts need to be addressed by further experimentation. In addition, heterogeneity of cell populations marked by a *Col10* promoter/enhancer needs to be carefully dissected. The hypertrophic layer represents another transient source of the marrow stromal compartment.

The third route is through the "borderland" between the growth plate and the perichondrium. Borderline chondrocytes are a special group of chondrocytes immediate beneath the perichondrium, which line up parallel to perichondrial cells but perpendicular to other growth plate chondrocytes (Bianco, Cancedda, Riminucci, & Cancedda, 1998). Our lineage-tracing experiments using *PTHrP-creERt* (which will be mentioned in more detail later) demonstrate that borderline chondrocytes in the neonatal stage can translocate into the marrow space (N.O., Unpublished Data). $PTHrP^+$ borderline chondrocytes can proliferate only for a limited period, and eventually disappear from the marrow space. Therefore, the borderland represents another transient source of the marrow stromal compartment.

These mesenchymal cells of heterogeneous origins contribute to osteoblasts and stromal cells within a highly vascularized environment called the primary spongiosa. Whether different cellular origins denote functional differences of these mesenchymal cells is unknown. Interestingly, *Runx2* and *Osx* are also expressed by prehypertrophic and hypertrophic chondrocytes. The partly overlapping gene programs of hypertrophic chondrocytes and osteoblastic cells suggest that mesenchymal cells arriving at the marrow cavity might already have some epigenetic signatures tuned to the osteoblast lineage. Supporting this notion, perivascular marrow stromal cells possess different transcriptome signatures depending on different developmental origins (Sacchetti et al., 2016), suggesting that mesenchymal cells may retain part of their original identities in their perivascular destinations.

As the growth plate continues to grow further, the marrow cavity and the primary ossification center continue to enlarge. In this process, osteoblasts and stromal cells need to be replenished constantly. This requires beyond three transient routes for migrating mesenchymal cells that we described above. Stem and progenitor cells of particular skeletal components are likely to provide the source of continuous generation of these mesenchymal cells. There are two possible mechanisms.

The first mechanism is that growth plate chondrocytes and perichondrial cells continue to feed into the growing marrow cavity. This is likely to happen because continuous mitotic activities in the flat proliferating layer of the growth plate give rise to an ample number of hypertrophic and borderline chondrocytes in the end, some of which may transform into these mesenchymal cells. Moreover, vigorous cell proliferation in the perichondrium provides a sufficient number of cells that can eventually feed into the marrow space. Evidence for this mechanism is supported by lineage-tracing experiments using multiple transgenes active in chondrocytes and their precursors. Cells within and in proximity of the growth plate marked by *Col2-creERt*, *Sox9-creERt* and *Acan-creERt* continue to generate growth plate chondrocytes, osteoblasts and stromal cells (Ono, Ono, Nagasawa, & Kronenberg, 2014), unambiguously demonstrating that these transgenes can mark the self-renewing population that may represent stem and progenitor cells of skeletal development. Also, cells marked by a Notch-responsive transgene, *Hes1-creERt*, can mark early cells prior to the commitment to the osteoblast lineage in the fetal perichondrium, and these cells can generate osteoblasts and stromal cells for a longer period and to a greater extent than fetal perichondrial *Osx-creERt* osteoblast precursors do (N.O., Unpublished Data), suggesting that this transgene is expressed in putative stem and progenitor cells of the perichondrium.

The second mechanism is that precursors for osteoblasts and stromal cells are re-established within the marrow space and replicate themselves for an extended period. This is also likely to happen because a group of bone marrow stromal cells is believed to behave as stem and progenitor cells with the important properties of self-renewability and multilineage differentiation potential, including those called "mesenchymal stem cells (MSCs)." Bone marrow stromal cells are generally positioned at a perivascular location surrounding arterioles (small arteries) and sinusoids (capillary vessels), and universally express CXCL12 (also known as stromal cell-derived factor 1, SDF-1) (Ara et al., 2003) and stem cell factor (SCF, also known as KIT ligand) (Keller, Ortiz, & Ruscetti, 1995) to attract and retain hematopoietic progenitor cells in the marrow cavity. It is generally believed that bone marrow mesenchymal precursors are located in proximity to blood vessels (we will discuss their potential cell fates in the next section). It is possible that aforementioned transgenes (*Col2-creERt*, *Sox9-creERt* and *Acan-creERt*) are also active in these newly established marrow precursor cells or mark precursors of these marrow stem cells, although more details need to be clarified. When *Sox9-creERt* is used to mark skeletal precursors shortly after birth (Ono, Ono, Nagasawa, et al., 2014) or when mice are 6 weeks old (Balani, Ono, & Kronenberg, 2017), a chase over time shows that these cells slowly and relatively modestly become CXCL12-expressing cells, which are likely to include the perivascular cells described above, in addition to rapidly becoming type I collagen (Col1a1)-expressing osteoblasts.

8. Endochondral bone development/Phase 4: Formation of the postnatal growth plate and continued growth of the marrow space

The postnatal growth plate continues to be active well into early adulthood, playing roles as an important driver for active bone growth. The formation of this structure is closely linked with the formation of the secondary ossification center within the fetal growth plate, which differentiate the existing round layer into two tissues with distinct functions: the postnatal growth plate and the articular cartilage, a transient and a permanent cartilage, respectively. The postnatal growth plate is formed as a disk-like tissue between the primary and secondary ossification centers with characteristic columnar chondrocytes. It is similar to the fetal growth plate, but there are certain differences. Most notably, chondrocytes at the top of the postnatal growth plate are slowly dividing, termed as resting or reserve chondrocytes. As described in the preceding section, PTHrP is expressed in the periarticular layer in the fetal growth plate. In the postnatal growth plate, PTHrP is also expressed in the resting zone. This resting zone has been suggested to contain stem-like cells that give rise to clones of proliferating chondrocytes and produce cytokines that orient columns parallel to the long axis of the bone, as demonstrated by surgical autograft tissue transplantation experiments in rabbits (Abad et al., 2002). Our lineage-tracing experiments using a *PTHrP-creERt* line confirmed and expanded this finding further: $PTHrP^{+}$ resting chondrocytes continue to form columnar chondrocytes long term, which undergo hypertrophy and became osteoblasts and marrow stromal cells beneath the growth plate (Mizuhashi et al., 2018); interestingly, these cells do not become marrow adipocytes (we will discuss more about this point later). $PTHrP^{+}$ chondrocytes possess colony-forming abilities, behave as "skeletal stem cells (SSCs)" and overlap with transplantable "mouse SSCs (mSSCs)" identified by a combination of cell surface markers, including αV integrin (CD51), Thy (CD90), endoglin (CD105) and OX2 (CD200) that are found in the growth plate (Chan et al., 2015). Similar stem cells have been also found in the human growth plate, termed "human SSCs (hSSCs)" identified by expression of $PDPN^{+}$ $CD146^{-}$ $CD73^{+}$ $CD146^{+}$ (Chan et al., 2018). These studies present an emerging idea that the resting zone of the postnatal growth plate is one of the niches in which stem and progenitor cells for bone growth are housed. How these resting chondrocytes maintain themselves long term within the postnatal growth plate is largely unknown; however, actions of Ihh released from the hypertrophic layer appear to be important in maintaining stem cells and their niche.

The postnatal growth plate is flanked by a highly vascularized marrow space enriched with trabecular bones in the primary and secondary ossification centers. The perivascular space of these vasculature-rich epiphyseal and metaphyseal regions is considered to contain abundant stem and progenitor cells pertinent to active bone growth. Gremlin 1 (Grem1), a bone morphogenetic protein (BMP) antagonist, is expressed in marrow stromal cells of the perivascular space. Lineage-tracing experiments using a *Grem1-creERt* demonstrate that these cells can become osteoblasts, chondrocytes and reticular stromal cells, but not marrow adipocytes, and are therefore identified as osteo-chondro-reticular (OCR) stem cells (Worthley et al., 2015). These vasculature-rich regions are adjacent to the prehypertrophic/hypertrophic layer that abundantly releases Ihh; therefore, these stem and progenitor cells might be highly responsive to Hedgehog signaling. Transcription factor Gli1 is an authentic Hedgehog-responsive gene that mediates a number of biological actions of Hedgehog signaling. Gli1 marks "metaphyseal mesenchymal progenitors (MMPs)," as lineage-tracing experiments using a *Gli1-creERt* demonstrate that these cells, unlike *Grem1-creERt*, contribute to osteoblasts, adipocytes and stromal cells (Shi et al., 2017). Interestingly, a group of these stem and progenitor cells might be transiently committed to the osteoblast lineage that can be defined by Osx expression. Lineage-tracing experiments using an *Osx-creERt* line demonstrate that these cells in the perinatal stage can contribute to osteoblasts and stromal cells that persist in the marrow long term (Mizoguchi et al., 2014); interestingly, Osx^+ cells in adulthood do not possess such capabilities. As mentioned earlier, pericytes associated with bone marrow arteries show specific elongated cell morphology and express α-smooth muscle actin (α-SMA). Lineage-tracing experiments using an *αSMA-creERt* demonstrate that these pericytes rather represent transient precursors for osteoblasts and, possibly, stromal cells (Grcevic et al., 2012). These lines of studies illustrate the variable fates of cells capable of generating multiple mesenchymal lineages. The studies also support the notion that distinct types of stem and progenitor cells for bone growth exist to support explosive growth of bone and its marrow space in early life.

9. Endochondral bone development/Phase 5: Establishment and maintenance of the adult bone marrow stroma

Bones need constant maintenance to sustain their structures and function throughout the lifespan even after bone growth slows and stops. As the growth plate activity slows down toward adulthood, the contribution of chondrocytes and perichondrial cells to the marrow space is likely to become negligible. It is probably at this post-growth stage that bone marrow stromal cells make a significant contribution to diverse skeletal cell types. Because matrix-producing osteoblasts are relatively short-lived, a continuous supply of new osteoblasts is necessary to maintain the bone structure. The current notion is that bone marrow stromal cells can differentiate into osteoblasts during normal adult skeletal homeostasis.

The first endeavor to mark and trace the fate of bone marrow stromal cells was made in 2010 using a *Nestin-creERt* line (Méndez-Ferrer et al., 2010). Although these cells certainly become osteoblasts and chondrocytes, this transgene simultaneously marks a large number of endothelial cells (Ono, Ono, Mizoguchi, et al., 2014), complicating the interpretation of overall outcomes. Leptin receptor (LepR) is expressed by bone marrow stromal cells around sinusoids and arterioles (Zhou, Yue, Murphy, Peyer, & Morrison, 2014). Fate-mapping studies using a *LepR-cre* knock-in allele demonstrate that these cells contribute to a large fraction of CFU-Fs, osteoblasts and marrow adipocytes in adult bones, particularly only after 2 months of age in mice (Zhou, Yue, et al., 2014). CXCL12 is abundantly expressed by marrow stromal cells surrounding sinusoids, which are termed CXCL12-abundant reticular (CAR) cells. The work from Nagasawa's group, using *Ebf3-creERt* to mark CAR cells in a tamoxifen-dependent manner, shows that CAR cells can become osteoblasts in adulthood (Seike, Omatsu, Watanabe, Kondoh, & Nagasawa, 2018). Our lineage-tracing experiments using a *Cxcl12-creERt* line demonstrate that these cells remain dormant without generating new reticular cells during growth of the marrow space, and become trabecular osteoblasts, but not cortical osteoblasts (N.O., Unpublished Data). The question whether there is any particular stem and progenitor cell populations maintaining the adult bone marrow stroma has not been completely answered. It is also possible that more differentiated cells (such as "bone-lining cells") play more important roles.

10. Periosteum and craniofacial sutures

Outside the marrow space encased by the cortical bone, the periosteum is formed as a result of activities of perichondrial cells. This tissue also contains clonogenic cell populations (CFU-Fs) and possesses highly regenerative capabilities. Lineage-tracing experiments demonstrate that perichondrial cells marked by *Prrx1-creERt* and *αSMA-creERt* lines can respond to injury and robustly generate chondrocytes and osteoblasts during fracture healing (Duchamp de Lageneste et al., 2018; Kawanami, Matsushita, Chan, & Murakami, 2009; Matthews et al., 2014). The suture of craniofacial bones is somewhat similar to the perichondrium of endochondral bones, and considered as the niche where stem cells reside. Lineage-tracing experiments demonstrate that suture mesenchymal cells marked by *Gli1-creERt* and *Prrx1-creERt* can contribute to normal turnover and regeneration of calvarial bones (Wilk et al., 2017; Zhao et al., 2015). The significance of these cells populations to normal skeletal development needs to be clarified by further experimentation. Recent work from Greenblatt's group confirms that the periosteum is the source of stem cells that can form osteoblasts and contribute heavily to callus formation after fracture, but do not support hematopoiesis (Debnath et al., 2018).

11. Sox9$^+$ osteoblast precursors

Earlier, we mentioned that *Sox9-creERt* can be "chased" into reticular cells expressing high levels of CXCL12, first described by Nagasawa's group, that strongly support hematopoiesis. At early times after administration of tamoxifen to mice shortly after birth or 6 weeks later, *Sox9-creERt* marks, in addition to growth plate chondrocytes, three groups of cells in bone. The largest group are metaphyseal cells immediately adjacent to the growth plate, separating the growth plate from the marrow contents; a second group marks cells in the perichondrium, probably those noted in (Debnath et al., 2018); a third group is found immediately below the osteoblast/lining cells layer at the surface of cortical and trabecular bone. Cells from these three locations can be chased into groups of osteoblasts lining trabecular bone, and osteoblasts at the endosteal and periosteal surfaces. It is potentially relevant to stem cells found in many other tissues expressing Sox9, such as hair follicle stem cells (Nowak, Polak, Pasolli, & Fuchs, 2008), breast (Guo et al., 2012), liver (Furuyama et al., 2011), intestine (Formeister et al., 2009) and pancreatic stem cells (Kawaguchi, 2013).

12. Parathyroid hormone (PTH) action on skeletal precursors

Parathyroid hormone (PTH) is a major regulator of mammalian calcium homeostasis that acts on bone (Hock, 2001). When teriparatide (hPTH (1–34)) is administered by daily subcutaneous injection to humans or rodents, bone mass increases as a result of an increase in the number of bone-forming osteoblasts and an increase in bone formation rate (Esbrit & Alcaraz, 2013). The cause of this teriparatide-mediated increase in the number of mature osteoblasts has been extensively studied. The *Sox9-creERt* lineage tracing experiment showed that, besides acting on the post-mitotic mature osteoblasts and osteocytes, teriparatide-mediated increase in bone mass also involves action on very early cells of the osteoblast lineage. Teriparatide suppresses apoptosis of early skeletal precursors, thus contributes to the overall increase in the number of osteoblast precursors seen in the first 7 days after the start of teriparatide administration (Balani et al., 2017). The early cells of the osteoblast lineage express *Pth1r* (*PTH/PTHrP receptor*) mRNA. When the PTH receptor is knocked out specifically from *Sox9-creERt*-expressing cells, teriparatide failed to increase the number of osteoblast precursors, suggesting that the actions of teriparatide on early skeletal precursors require direct signaling via PTH/PTHrP receptors (Balani et al., 2017). When daily subcutaneous administration of PTH was continued for a longer duration, the number of reporter-marked *Sox9-creERt* cells continued to increase and differentiated into osteoblasts with a higher rate of differentiation compared to controls. In control mice, a portion of *Sox9-creERt* descendant cells are adipocytes, in addition to stromal cells and osteoblasts. However, no adipocytes that descended from *Sox9-creERt* cells are observed at the end of 4 weeks of teriparatide administration. As expected, when teriparatide administration is then halted and mice are followed for another 4 weeks, most of the increase in bone mass caused by teriparatide administration is reversed. Strikingly, however, the bone marrow of the teriparatide-withdrawn mice shows a dramatic increase in adipocytes descended from the *Sox9-creERt+* cells, unlike what happens in the controls. This observation is consistent with previous work done by other groups showing that PTH suppresses adipocytic differentiation of human stromal cells *in vitro* (Yu et al., 2012). However, this experimental design cannot determine whether Sox9-expressing cells or, instead, one or more groups of descendant cells were the source of the new marrow adipocytes. Strikingly, when an analogous experiment is performed using *Ocn-creERt*, reporter mice that primarily label mature osteoblasts and their descendants, several adipocytes seen after teriparatide withdrawal are descendants of *Ocn-creERt+* cells (Balani et al., 2017). The conversion of *Sox9-creERt+* or *Ocn-creERt+* cells to adipocytes was not seen as long as teriparatide was administered once daily, suggesting that the molecular cues that direct adipocytic differentiation of early precursors is suppressed as long as PTH is administered.

In the marrow adipocyte pool, the overall contribution of cells that differentiated from *Sox9-creERt* or *Ocn-creERt* cells ranged between 10% and 30%. In these studies, tamoxifen was administered to 6-week-old mice, at the same time as the initiation of teriparatide treatment. Interestingly, when the withdrawal experiment was performed in *Sox9-creERt* mice, in which Tomato+ cells were labeled soon after birth at day P3 and teriparatide administration began at 6-weeks of age, Tomato+ cells contributed >98% of the marrow adipocytes formed after PTH-withdrawal *in vivo*. These findings suggest that a larger fraction of the adipocyte precursors express Sox9 at P3 than at 6 weeks of age, perhaps because Sox9 descendants, no longer expressing Sox9, accumulate with age and can serve as adipocyte precursors.

13. Wnt/β-catenin signaling and cell fate decision

Wnt/β-catenin signaling plays a crucial role in controlling osteoblast and adipocyte differentiation. Removal of β-catenin from bone marrow stromal cells (BMSCs) *in vitro* blocks osteoblastic differentiation and causes these cells to more readily differentiate into the adipocytic lineage (Christodoulides, Lagathu, Sethi, & Vidal-Puig, 2009). Wnt protein supplementation to culture media increases osteoblast differentiation and suppresses adipogenesis in the bone marrow-derived ST2 stromal cell line. β-Catenin activation plays a key role in the canonical Wnt pathway and regulates Wnt target gene transcription. When this pathway is activated by any of the variety of Wnt ligands (Wnt10b, Wnt1, Wnt3, for example), activated β-catenin is transported into the nucleus and initiates targeted gene transcription (MacDonald, Tamai, & He, 2009). Stabilization of β-catenin in BMSCs promotes osteoblast differentiation and inhibits their adipogenic differentiation. In contrast, overexpression of β-catenin in pre-adipocytes suppresses their differentiation into mature adipocytes (Ross et al., 2000). Conditional deletion of β-catenin *in vivo* from osterix-expressing pre-osteoblasts leads to a substantial fraction of these cells differentiating into adipocytes (Song et al., 2012). Teriparatide administration, both *in vitro* and *in vivo*, increases Wnt/β-catenin signaling (Lee & Partridge, 2009). Teriparatide administration also suppresses Wnt inhibitors such as DKK-1 (Guo et al., 2010) and sclerostin (Keller & Kneissel, 2005); this, in turn, leads to an increase in Wnt/β-catenin signaling. Our report (Balani et al., 2017) suggests that teriparatide withdrawal causes a sudden decrease in Wnt/β-catenin signaling in skeletal stem cells, based on measurements of the non-phosphorylated (active) form of β-catenin accumulated in vivo. Teriparatide-withdrawn mice showed a dramatic decrease in the active β-catenin in *Sox9-creERt+* cells compared to the levels in control mice subjected to continued one daily PTH administration. The molecular mechanism of this decrease in activation of β-catenin is not known, but we hypothesize that there must be mechanisms of skeletal homeostasis that regulate the increase and decrease of bone mass to optimize the stresses on individual osteoblasts/osteocytes under normal conditions. We further speculate that the increase in bone mass in response to one daily teriparatide administration leads to a higher bone mass than the mouse otherwise "needs," and that the bones sense this high-bone mass state. That is, sudden cessation of teriparatide administration may activate homeostatic mechanisms that decrease bone mass until bone mass falls and reaches its basal level. When the number of osteoblasts/osteocytes is artificially high after teriparatide administration, they may each receive insufficient stress signals from gravity and muscle pull to activate canonical Wnt signaling. This could lead to a decrease in activation of β-catenin, and then to lower bone mass and an increase in adipocyte generation from accumulated precursors.

14. Conclusions and perspectives

Bone cells have multiple missions throughout life. Early in life, growth is relatively explosive, while in adulthood, bones must maintain multiple cellular configurations, from the rapidly turning over trabecular bone to the more slowly remodeling cortical bone and repair the bone when cracks or fractures occur. The missions during adulthood include regulation of mineral ion homeostasis and hematopoiesis, and provision of levers for muscle-generated movement and protection of inner organs. Considering all these tasks, it is not surprising that multiple sites are used to generate skeletal stem cells, from the bone marrow and perichondrium of fetal bone, the resting zone of the postnatal growth plate, the marrow cavity in adulthood, and the periosteum. These stem cells have distinct properties and missions that are regulated both by local factors and by hormones. Here we have delineated how both the tools of regenerative medicine and of lineage tracing can be used to identify skeletal stem cells. Although we are off to a good start, we are only beginning to understand the relationships between the different skeletal stem cells. Very importantly, it is likely that most stem cell groups identified *in vivo* are quite heterogeneous, since they are primarily identified by the expression of one particular promoter driving *cre* expression. Finding ways to effectively combine the powerful cell sorting tools of regenerative medicine with the cell lineage tools needed to determine what actually happens *in vivo* in normal or injured bone remains an enormous challenge for the future.

Acknowledgments

The authors acknowledge NIH Grants R01DE026666 (N.O.), P01 DK011794 (H.M.K.), an Early Investigator Fellowship from the Swiss National Science Foundation (D.H.B.) and the Pfizer Aspire Young Investigator Grant (D.H.B.).

References

Abad, V., Meyers, J. L., Weise, M., Gafni, R. I., Barnes, K. M., Nilsson, O., et al. (2002). The role of the resting zone in growth plate chondrogenesis. *Endocrinology*, *143*, 1851–1857.

Akiyama, H., Kim, J. E., Nakashima, K., Balmes, G., Iwai, N., Deng, J. M., et al. (2005). Osteo-chondroprogenitor cells are derived from Sox9 expressing precursors. *Proceedings of the National Academy of Sciences of the United States of America*, *102*, 14665–14670.

Ara, T., Itoi, M., Kawabata, K., Egawa, T., Tokoyoda, K., Sugiyama, T., et al. (2003). A role of CXC chemokine ligand 12/stromal cell-derived factor-1/pre-B cell growth stimulating factor and its receptor CXCR4 in fetal and adult T cell development in vivo. *Journal of Immunology*, *170*, 4649–4655.

Balani, D. H., Ono, N., & Kronenberg, H. M. (2017). Parathyroid hormone regulates fates of murine osteoblast precursors in vivo. *The Journal of Clinical Investigation*, *127*, 3327–3338.

Bianco, P. (2014). "Mesenchymal" stem cells. *Annual Review of Cell and Developmental Biology*, *30*, 677–704.

Bianco, P., Cancedda, F. D., Riminucci, M., & Cancedda, R. (1998). Bone formation via cartilage models: The "borderline" chondrocyte. *Matrix Biology*, *17*, 185–192.

Bianco, P., Cao, X., Frenette, P. S., Mao, J. J., Robey, P. G., Simmons, P. J., et al. (2013). The meaning, the sense and the significance: Translating the science of mesenchymal stem cells into medicine. *Nature Medicine*, *19*, 35–42.

Castro-Malaspina, H., Gay, R. E., Resnick, G., Kapoor, N., Meyers, P., Chiarieri, D., et al. (1980). Characterization of human bone marrow fibroblast colony-forming cells (CFU-F) and their progeny. *Blood*, *56*, 289–301.

Chan, C. K. F., Gulati, G. S., Sinha, R., Tompkins, J. V., Lopez, M., Carter, A. C., et al. (2018). Identification of the human skeletal stem cell. *Cell*, *175*, 43–56.e21.

Chan, C. K., Seo, E. Y., Chen, J. Y., Lo, D., McArdle, A., Sinha, R., et al. (2015). Identification and specification of the mouse skeletal stem cell. *Cell*, *160*, 285–298.

Christodoulides, C., Lagathu, C., Sethi, J. K., & Vidal-Puig, A. (2009). Adipogenesis and WNT signalling. *Trends in Endocrinology and Metabolism*, *20*, 16–24.

Debnath, S., Yallowitz, A. R., McCormick, J., Lalani, S., Zhang, T., Xu, R., et al. (2018). Discovery of a periosteal stem cell mediating intramembranous bone formation. *Nature*, *562*, 133–139.

Duchamp de Lageneste, O., Julien, A., Abou-Khalil, R., Frangi, G., Carvalho, C., Cagnard, N., et al. (2018). Periosteum contains skeletal stem cells with high bone regenerative potential controlled by Periostin. *Nature Communications*, *9*, 773.

Dy, P., Wang, W., Bhattaram, P., Wang, Q., Wang, L., Ballock, R. T., et al. (2012). Sox9 directs hypertrophic maturation and blocks osteoblast differentiation of growth plate chondrocytes. *Developmental Cell*, *22*, 597–609.

Esbrit, P., & Alcaraz, M. J. (2013). Current perspectives on parathyroid hormone (PTH) and PTH-related protein (PTHrP) as bone anabolic therapies. *Biochemical Pharmacology*, *85*, 1417–1423.
Formeister, E. J., Sionas, A. L., Lorance, D. K., Barkley, C. L., Lee, G. H., & Magness, S. T. (2009). Distinct SOX9 levels differentially mark stem/progenitor populations and enteroendocrine cells of the small intestine epithelium. *American Journal of Physiology. Gastrointestinal and Liver Physiology*, *296*, G1108–G1118.
Friedenstein, A. J., Piatetzky-Shapiro, I. I., & Petrakova, K. V. (1966). Osteogenesis in transplants of bone marrow cells. *Journal of Embryology and Experimental Morphology*, *16*, 381–390.
Furuyama, K., Kawaguchi, Y., Akiyama, H., Horiguchi, M., Kodama, S., Kuhara, T., et al. (2011). Continuous cell supply from a Sox9-expressing progenitor zone in adult liver, exocrine pancreas and intestine. *Nature Genetics*, *43*, 34–41.
Grcevic, D., Pejda, S., Matthews, B. G., Repic, D., Wang, L., Li, H., et al. (2012). In vivo fate mapping identifies mesenchymal progenitor cells. *Stem Cells*, *30*, 187–196.
Guo, W., Keckesova, Z., Donaher, J. L., Shibue, T., Tischler, V., Reinhardt, F., et al. (2012). Slug and Sox9 cooperatively determine the mammary stem cell state. *Cell*, *148*, 1015–1028.
Guo, J., Liu, M., Yang, D., Bouxsein, M. L., Saito, H., Galvin, R. J., et al. (2010). Suppression of Wnt signaling by Dkk1 attenuates PTH-mediated stromal cell response and new bone formation. *Cell Metabolism*, *11*, 161–171.
He, X., Ohba, S., Hojo, H., & McMahon, A. P. (2016). AP-1 family members act with Sox9 to promote chondrocyte hypertrophy. *Development*, *143*, 3012–3023.
Hock, J. M. (2001). Anabolic actions of PTH in the skeletons of animals. *Journal of Musculoskeletal & Neuronal Interactions*, *2*, 33–47.
Kawaguchi, Y. (2013). Sox9 and programming of liver and pancreatic progenitors. *The Journal of Clinical Investigation*, *123*, 1881–1886.
Kawanami, A., Matsushita, T., Chan, Y. Y., & Murakami, S. (2009). Mice expressing GFP and CreER in osteochondro progenitor cells in the periosteum. *Biochemical and Biophysical Research Communications*, *386*, 477–482.
Keller, H., & Kneissel, M. (2005). SOST is a target gene for PTH in bone. *Bone*, *37*, 148–158.
Keller, J. R., Ortiz, M., & Ruscetti, F. W. (1995). Steel factor (c-kit ligand) promotes the survival of hematopoietic stem/progenitor cells in the absence of cell division. *Blood*, *86*, 1757–1764.
Kobayashi, T., Soegiarto, D. W., Yang, Y., Lanske, B., Schipani, E., McMahon, A. P., et al. (2005). Indian hedgehog stimulates periarticular chondrocyte differentiation to regulate growth plate length independently of PTHrP. *The Journal of Clinical Investigation*, *115*, 1734–1742.
Kozhemyakina, E., Lassar, A. B., & Zelzer, E. (2015). A pathway to bone: Signaling molecules and transcription factors involved in chondrocyte development and maturation. *Development*, *142*, 817–831.
Lanske, B., Karaplis, A. C., Lee, K., Luz, A., Vortkamp, A., Pirro, A., et al. (1996). PTH/PTHrP receptor in early development and Indian hedgehog-regulated bone growth. *Science*, *273*, 663–666.
Lee, M., & Partridge, N. C. (2009). Parathyroid hormone signaling in bone and kidney. *Current Opinion in Nephrology and Hypertension*, *18*, 298–302.
Logan, M., Martin, J. F., Nagy, A., Lobe, C., Olson, E. N., & Tabin, C. J. (2002). Expression of Cre recombinase in the developing mouse limb bud driven by a Prxl enhancer. *Genesis*, *33*, 77–80.
MacDonald, B. T., Tamai, K., & He, X. (2009). Wnt/beta-catenin signaling: Components, mechanisms, and diseases. *Developmental Cell*, *17*, 9–26.

Maes, C. (2017). Signaling pathways effecting crosstalk between cartilage and adjacent tissues: Seminars in cell and developmental biology: The biology and pathology of cartilage. *Seminars in Cell & Developmental Biology*, *62*, 16–33.

Maes, C., Kobayashi, T., Selig, M. K., Torrekens, S., Roth, S. I., Mackem, S., et al. (2010). Osteoblast precursors, but not mature osteoblasts, move into developing and fractured bones along with invading blood vessels. *Developmental Cell*, *19*, 329–344.

Mak, K. K., Kronenberg, H. M., Chuang, P. T., Mackem, S., & Yang, Y. (2008). Indian hedgehog signals independently of PTHrP to promote chondrocyte hypertrophy. *Development*, *135*, 1947–1956.

Matthews, B. G., Grcevic, D., Wang, L., Hagiwara, Y., Roguljic, H., Joshi, P., et al. (2014). Analysis of αSMA-labeled progenitor cell commitment identifies notch signaling as an important pathway in fracture healing. *Journal of Bone and Mineral Research*, *29*, 1283–1294.

Méndez-Ferrer, S., Michurina, T. V., Ferraro, F., Mazloom, A. R., Macarthur, B. D., Lira, S. A., et al. (2010). Mesenchymal and haematopoietic stem cells form a unique bone marrow niche. *Nature*, *466*, 829–834.

Mizoguchi, T., Pinho, S., Ahmed, J., Kunisaki, Y., Hanoun, M., Mendelson, A., et al. (2014). Osterix marks distinct waves of primitive and definitive stromal progenitors during bone marrow development. *Developmental Cell*, *29*, 340–349.

Mizuhashi, K., Ono, W., Matsushita, Y., Sakagami, N., Takahashi, A., Saunders, T. L., et al. (2018). Resting zone of the growth plate houses a unique class of skeletal stem cells. *Nature*, *563*, 254–258.

Morikawa, S., Mabuchi, Y., Kubota, Y., Nagai, Y., Niibe, K., Hiratsu, E., et al. (2009). Prospective identification, isolation, and systemic transplantation of multipotent mesenchymal stem cells in murine bone marrow. *The Journal of Experimental Medicine*, *206*, 2483–2496.

Nakashima, K., Zhou, X., Kunkel, G., Zhang, Z., Deng, J. M., Behringer, R. R., et al. (2002). The novel zinc finger-containing transcription factor osterix is required for osteoblast differentiation and bone formation. *Cell*, *108*, 17–29.

Nowak, J. A., Polak, L., Pasolli, H. A., & Fuchs, E. (2008). Hair follicle stem cells are specified and function in early skin morphogenesis. *Cell Stem Cell*, *3*, 33–43.

Ono, N., Ono, W., Mizoguchi, T., Nagasawa, T., Frenette, P. S., & Kronenberg, H. M. (2014). Vasculature-associated cells expressing nestin in developing bones encompass early cells in the osteoblast and endothelial lineage. *Developmental Cell*, *29*, 330–339.

Ono, N., Ono, W., Nagasawa, T., & Kronenberg, H. M. (2014). A subset of chondrogenic cells provides early mesenchymal progenitors in growing bones. *Nature Cell Biology*, *16*, 1157–1167.

Pinho, S., Lacombe, J., Hanoun, M., Mizoguchi, T., Bruns, I., Kunisaki, Y., et al. (2013). PDGFRα and CD51 mark human nestin+ sphere-forming mesenchymal stem cells capable of hematopoietic progenitor cell expansion. *The Journal of Experimental Medicine*, *210*, 1351–1367.

Ross, S. E., Hemati, N., Longo, K. A., Bennett, C. N., Lucas, P. C., Erickson, R. L., et al. (2000). Inhibition of adipogenesis by Wnt signaling. *Science*, *289*, 950–953.

Sacchetti, B., Funari, A., Michienzi, S., Di Cesare, S., Piersanti, S., Saggio, I., et al. (2007). Self-renewing osteoprogenitors in bone marrow sinusoids can organize a hematopoietic microenvironment. *Cell*, *131*, 324–336.

Sacchetti, B., Funari, A., Remoli, C., Giannicola, G., Kogler, G., Liedtke, S., et al. (2016). No identical "mesenchymal stem cells" at different times and sites: Human committed progenitors of distinct origin and differentiation potential are incorporated as adventitial cells in microvessels. *Stem Cell Reports*, *6*, 897–913.

Schipani, E., Lanske, B., Hunzelman, J., Luz, A., Kovacs, C. S., Lee, K., et al. (1997). Targeted expression of constitutively active receptors for parathyroid hormone and parathyroid hormone-related peptide delays endochondral bone formation and rescues mice that lack parathyroid hormone-related peptide. *Proceedings of the National Academy of Sciences of the United States of America, 94*, 13689–13694.

Seike, M., Omatsu, Y., Watanabe, H., Kondoh, G., & Nagasawa, T. (2018). Stem cell niche-specific Ebf3 maintains the bone marrow cavity. *Genes & Development, 32*, 359–372.

Shi, S., & Gronthos, S. (2003). Perivascular niche of postnatal mesenchymal stem cells in human bone marrow and dental pulp. *Journal of Bone and Mineral Research, 18*, 696–704.

Shi, Y., He, G., Lee, W. C., McKenzie, J. A., Silva, M. J., & Long, F. (2017). Gli1 identifies osteogenic progenitors for bone formation and fracture repair. *Nature Communications, 8*, 2043.

Song, L., Liu, M., Ono, N., Bringhurst, F. R., Kronenberg, H. M., & Guo, J. (2012). Loss of wnt/β-catenin signaling causes cell fate shift of preosteoblasts from osteoblasts to adipocytes. *Journal of Bone and Mineral Research, 27*, 2344–2358.

St-Jacques, B., Hammerschmidt, M., & McMahon, A. P. (1999). Indian hedgehog signaling regulates proliferation and differentiation of chondrocytes and is essential for bone formation. *Genes & Development, 13*, 2072–2086.

Sun, J., Ramos, A., Chapman, B., Johnnidis, J. B., Le, L., Ho, Y. J., et al. (2014). Clonal dynamics of native haematopoiesis. *Nature, 514*, 322–327.

Vortkamp, A., Lee, K., Lanske, B., Segre, G. V., Kronenberg, H. M., & Tabin, C. J. (1996). Regulation of rate of cartilage differentiation by Indian hedgehog and PTH-related protein. *Science, 273*, 613–622.

Wang, W., Strecker, S., Liu, Y., Wang, L., Assanah, F., Smith, S., et al. (2015). Connective tissue growth factor reporter mice label a subpopulation of mesenchymal progenitor cells that reside in the trabecular bone region. *Bone, 71*, 76–88.

Wilk, K., Yeh, S. A., Mortensen, L. J., Ghaffarigarakani, S., Lombardo, C. M., Bassir, S. H., et al. (2017). Postnatal calvarial skeletal stem cells expressing PRX1 reside exclusively in the calvarial sutures and are required for bone regeneration. *Stem Cell Reports, 8*, 933–946.

Worthley, D. L., Churchill, M., Compton, J. T., Tailor, Y., Rao, M., Si, Y., et al. (2015). Gremlin 1 identifies a skeletal stem cell with bone, cartilage, and reticular stromal potential. *Cell, 160*, 269–284.

Yang, L., Tsang, K. Y., Tang, H. C., Chan, D., & Cheah, K. S. (2014). Hypertrophic chondrocytes can become osteoblasts and osteocytes in endochondral bone formation. *Proceedings of the National Academy of Sciences of the United States of America, 111*, 12097–12102.

Yu, B., Zhao, X., Yang, C., Crane, J., Xian, L., Lu, W., et al. (2012). Parathyroid hormone induces differentiation of mesenchymal stromal/stem cells by enhancing bone morphogenetic protein signaling. *Journal of Bone and Mineral Research, 27*, 2001–2014.

Zhao, H., Feng, J., Ho, T. V., Grimes, W., Urata, M., & Chai, Y. (2015). The suture provides a niche for mesenchymal stem cells of craniofacial bones. *Nature Cell Biology, 17*, 386–396.

Zhou, X., von der Mark, K., Henry, S., Norton, W., Adams, H., & de Crombrugghe, B. (2014). Chondrocytes transdifferentiate into osteoblasts in endochondral bone during development, postnatal growth and fracture healing in mice. *PLoS Genetics, 10*. e1004820.

Zhou, B. O., Yue, R., Murphy, M. M., Peyer, J. G., & Morrison, S. J. (2014). Leptin-receptor-expressing mesenchymal stromal cells represent the main source of bone formed by adult bone marrow. *Cell Stem Cell, 15*, 154–168.

CHAPTER TWO

ECM signaling in cartilage development and endochondral ossification

Carina Prein, Frank Beier*

Department of Physiology and Pharmacology, Schulich School of Medicine and Dentistry, and Western University Bone and Joint Institute, University of Western Ontario, London, ON, Canada
*Corresponding author: e-mail address: fbeier@uwo.ca

Contents

Abstract

During cartilage development chondrocytes undergo a multi-step process characterized by consecutive changes in cell morphology and gene expression. Cell proliferation, polarity, differentiation, and migration are influenced by chemical and mechanical signaling between the extracellular matrix (ECM) and the cell. Several structurally diverse transmembrane receptors such as integrins, discoidin domain receptor 2 (DDR 2), and CD44 mediate the crosstalk between cells and their ECM. However, the contribution of cell-matrix interactions during early chondrogenesis and further cartilage development through cell receptors and their signal transduction pathways is still not fully understood. Determination of receptor signaling pathways and the function of downstream targets will aid in a better understanding of musculoskeletal pathologies such as chondrodysplasia, and the development of new approaches for the treatment of cartilage disorders. We will summarize recent findings, linking cell receptors and their potential signaling pathways to the control of chondrocyte behavior during early chondrogenesis and endochondral ossification.

Current Topics in Developmental Biology, Volume 133
ISSN 0070-2153
https://doi.org/10.1016/bs.ctdb.2018.11.003

1. Introduction

The supramolecular structure and composition of cartilage extracellular matrix (ECM) components influence the functionality of many connective tissues. The ECM of cartilage consists of collagen and proteoglycans, the two main components, as well as non-collagenous glycoproteins. This ECM is secreted by chondrocytes and is organized into specific macromolecular assemblies. Together, they establish an environment that provides mechanical and structural support to the tissue, and consequently determines tissue-specific function such as compressive strength and hydraulic permeability in the case of articular cartilage. In addition, ECM provides spatial context for cell-matrix interactions and controls cell behavior through specific signaling pathways. The crosstalk between cells and their surrounding ECM plays an important role not only in adult cartilage homeostasis, but also during cartilage development and the establishment of the skeletal framework during endochondral ossification. Specific chemical and mechanical signals are transmitted within the ECM and influence many aspects of cell behavior such as matrix synthesis, migration, proliferation, survival, and differentiation (Geiger, Bershadsky, Pankov, & Yamada, 2001). Integrins and proteoglycans are the main ECM adhesion receptors transducing external ECM signals into the cell, determining cell fate during early development and tissue homeostasis *via* various intracellular signaling pathways, and interactions with the chondrocyte cytoskeleton.

2. Chondrogenesis and endochondral ossification

During chondrogenesis and endochondral ossification, chondrocytes form the cartilaginous template for the future bone. In this process they undergo a highly regulated differentiation program accompanied by compositional changes in their surrounding matrix (Erlebacher, Filvaroff, Gitelman, & Derynck, 1995). Chondrogenesis and endochondral ossification can be subdivided into consecutive but overlapping stages (Fig. 1). First mesenchymal cells migrate toward and condense at the sites of future skeletal elements. This condensation results in an increased cell density and increased number of cell-cell contacts accompanied by the expression of molecules involved in cell-cell adhesion such as N-cadherin (Delise & Tuan, 2002; Olsen, Reginato, & Wang, 2000). Committed cell aggregates are separated from the surrounding tissues by a thin layer of elongated cells, called perichondrium. Inside the

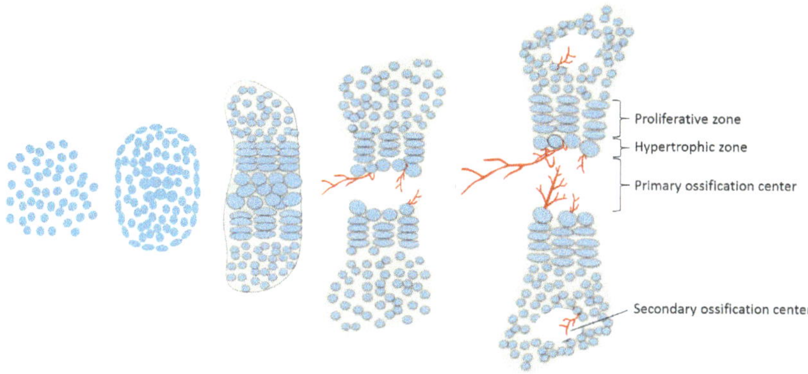

Fig. 1 During chondrogenesis, mesenchymal cells condense at the sites of future skeletal bones. Subsequently, condensed cells start to differentiate into chondrocytes. Chondrocytes start to proliferate and deposit a cartilage-specific matrix shaping the cartilaginous template. Chondrocytes within the center of the anlage undergo hypertrophy which triggers molecular processes that attract blood vessels, osteoclasts, and osteoblasts that will replace cartilage with bone in the area of the primary ossification center. The secondary ossification center within the epiphyses recapitulates this process.

condensations, the mesenchymal cells rapidly differentiate into chondrocytes forming the hyaline cartilage model. Chondrocytes at this stage are marked by the expression of Sox5, 6, and 9, which regulate the expression of genes encoding ECM molecules collagen type II and the proteoglycan aggrecan (Lefebvre, 2002; Lefebvre & de Crombrugghe, 1998; Lefebvre & Smits, 2005). The collagen meshwork provides tensile strength to the tissue, while aggrecan maintains the high water content and hydraulic permeability of the tissue due to its negatively charged glycosaminoglycan side chains (Han, Grodzinsky, & Ortiz, 2011). Subsequently, some chondrocytes will differentiate further in a multistep process characterized by successive changes in their morphology and gene expression. First, chondrocytes begin to form longitudinally oriented columns. These cells are embedded into an extensive ECM and make up the proliferative zone of the growth plate cartilage. Within the core of the diaphysis, chondrocytes stop dividing, enlarge their cell volume, and differentiate into prehypertrophic and then hypertrophic chondrocytes under the control of exogenous factors such as thyroid hormones, Indian hedgehog, and parathyroid hormone-related peptide (Kronenberg & Chung, 2001). Chondrocytes of the hypertrophic growth plate zone stop to express collagen type II and secrete collagen type X, as well as bone morphogenetic proteins (BMPs), the vascular endothelial growth factor A (VEGF), and receptor activator of nuclear factor kappa-B ligand (RANKL). This switch in protein expression is mediated by different transcription factors such as Runx2/3, MEF2C, and FoxA2/3 (Karsenty & Wagner, 2002; Sun & Beier, 2014). VEGF is an angiogenesis promoting factor initiating the invasion of vessels into the mid-diaphysis, allowing the migration of osteoblast precursors cells derived from the perichondrium into the ossification front, and stimulating their differentiation into osteoblasts within the perichondrium as well as the primary ossification center (Gerber et al., 1999; Maes et al., 2010). Furthermore, blood vessel invasion brings undifferentiated mesenchymal cells, which will also differentiate into osteoblasts (Rabie, Leung, Chayanupatkul, & Hagg, 2002).

Thus, the perichondrium around the central diaphysis gradually becomes periosteum, laying down the bone collar by intramembranous ossification. The fate of hypertrophic chondrocytes at the cartilage-bone border is controversial. Either they die due to programmed cell death/apoptosis, or they transdifferentiate into osteoblasts and actively participate in bone formation (Komori, 2016). Osteoclasts/chondroclasts partially remove the mineralized matrix, whereas osteoblasts start to construct trabecular bone on the cartilage remnants. The extensive ECM secretion, cell proliferation and differentiation, and growth (hypertrophy) lead to a longitudinal spreading of the primary ossification front in both directions and thus the longitudinal growth of future bones. At the epiphyses, the same process occurs after birth, resulting in secondary ossification centers. At this point, just a small layer of cartilage remains at the end of long bones that becomes the permanent articular cartilage. The cartilage between primary and secondary ossification centers continues to control growth of bones.

Chondrocytes display distinct behaviors reflected by their morphology as well as their gene and protein expression patterns in the various growth plate zones. Consequently, different growth plate zones form a highly individualized ECM with distinct composition. Skeletal disorders, affecting growth plate (e.g., chondrodysplasia) or articular cartilage (e.g., osteoarthritis), are manifested by changes in chondrocyte morphology and metabolic activity, as well as ECM structure and composition. Cell proliferation, differentiation, and survival all crucially depend on the continuous interaction between cells and their surrounding ECM. Cells and their surrounding matrix should be considered as one unit operating in a constant feedback loop; cell survival depends on the composition and structure of the microenvironment, which provides a substrate for cell anchorage, and transmits environmental signals and nutrients, while the ECM is controlled by the balance of synthesis and degradation orchestrated by embedded cells. This close cell-matrix interaction largely depends on cell surface receptors such as integrins. The binding of these receptors to sites on ECM molecules transduces chemical and mechanical environmental information to the cell interior, which will affect tissue-specific cell behavior and fate (Geiger et al., 2001; Song & Park, 2014).

3. Roles of integrins in chondrogenesis and further chondrocyte maturation

Integrins are heterodimeric transmembrane receptors composed of α and β subunits that mediate cell-matrix and cell-cell interactions from embryonic development to mature tissue formation (Legate, Wickström, & Fässler, 2009). In the human genome at least 18 α and 8 β subunits have been described. Each subunit is composed of a large extracellular domain, a single transmembrane α-helix, and a short cytoplasmic domain (Wegener & Campbell, 2008). In mammals these subunits form at least 24 dimeric integrin receptors which can bind ECM components such as collagens, laminin, and fibronectin (Hynes, 2002). The ligand specificity of the integrin heterodimer depends on the composition of α and β subunits. By connecting the ECM to the intracellular actin cytoskeleton, integrins transmit extracellular chemical as well as mechanical signals into the cell. Upon ligand binding, a large number of molecules is recruited to the short cytoplasmic domain of the integrins in order to form so-called focal adhesion (FAs) sites (Harburger & Calderwood, 2009). FAs are molecular platforms composed of adaptor, cytoskeletal, and signaling molecules, which are involved in the regulation of various signaling pathways that control cell behaviors such as survival, proliferation, differentiation, migration, and adhesion (Harburger & Calderwood, 2009). Furthermore, FAs link the ECM to the intracellular cytoskeleton by the recruitment of various scaffolding proteins. The focal adhesion complex includes signaling molecules such as focal adhesion kinase (FAK) and Src tyrosine kinases. Signals will be further propagated through pathways such as MAP and Akt kinases or Rho family GTPases (Song & Park, 2014).

In cartilage several, mainly β1-containing integrin heterodimers have been found on the surface of chondrocytes, including receptors for collagen II (α1β1, α2β1, and α10β1), laminin (α6β1) and fibronectin (α5β1, αvβ3, and αvβ5) (Loeser, 2000, 2014). The integrin subunit β1 is ubiquitously expressed and can associate with more than half of all α subunits. Therefore, it is not surprising that the genetic deletion (knockout) of β1 integrin severely compromises mouse development. Constitutive β1 knockout mouse embryos die shortly after or at implantation, implying a crucial role of β1 integrin during early development (Fässler et al., 1995). Cartilage-specific inactivation of the β1 integrin gene using the Col2a1-Cre and Prx1-Cre transgenes resulted in chondrodysplasia during embryogenesis and lethality shortly after birth (Aszodi, Hunziker, Brakebusch, & Fässler, 2003; Logan et al., 2002; Raducanu, Hunziker, Drosse, & Aszodi, 2009; Sakai et al., 2001). However, the early differentiation of chondrocytes does not seem to be affected by the lack of β1 integrin since cartilage tissues are generated in the absence of the β1 subunit. Interestingly, the mutant growth plate is devoid of all levels of cell polarity. The normally highly organized, flattened proliferative chondrocytes are roundish and never elongate along the mediolateral axis or organize into longitudinal columns. Chondrocytes of the developing cartilage show a reduced proliferation rate, while the number of apoptotic cells is increased. Chondrocytes are often binucleated and display actin organization abnormalities. Furthermore, ultrastructural alterations of the collagen network of the mutant growth plate are observed. These findings nicely demonstrate the close connection between ECM and cell fate. The absence of β1 integrin results in a disruption of cell-matrix signaling, which ultimately disturbs cell cytokinesis due to impaired adhesion-generated pulling forces at the cleavage furrow (Aszodi et al., 2003). Similar but milder growth plate abnormalities were observed using a cartilage-specific knockout model for integrin-linked kinase (ILK) (Grashoff, Aszódi, Sakai, Hunziker, & Fässler, 2003; Terpstra et al., 2003).

In vitro experiments confirmed the relation between β1 integrin activity and chondrocyte proliferation rate. The blocking of α5β1 interactions of growth plate chondrocytes resulted in a decreased proliferation rate (Motomi et al., 1997). The investigation of another integrin subunit partially recapitulated growth plate abnormalities seen in the β1 knockout models. Together with the β1 subunit, the subunit α10 works as a receptor for collagen type II. α10 integrin knockout mice, generated by using the Col2a1-Cre and Prx1-Cre transgenes, showed a chondrodysplasia including a mild change in chondrocyte shape and growth plate disorganization (Bengtsson et al., 2005). Not surprisingly, these studies show that the loss of all β1-containing integrins is more severe than the lack of a single α subunit. However, they clearly demonstrate the crucial impact of β1 integrin on normal chondrocyte behavior and cartilage development. Other integrin subtypes expressed in chondrocytes do not seem to critically influence cartilage development (Bouvard et al., 2001). These studies on β1 integrin strongly support the view that cells sense their environment *via* integrins.

This opens up new opportunities for therapeutic approaches in musculoskeletal diseases. However, morphological evaluation of mouse models is not sufficient in order to understand the multiple tasks of integrins during chondrogenesis and further steps in cartilage development. Integrin-mediated adhesion stimulates a large number of signaling molecules, including Rho family of GTPases, tyrosine phosphatases, and cAMP-dependent protein kinase. Thus, further investigations of signaling defects in integrin-deficient mice are required for the unraveling of the highly complex signaling machinery triggered by integrins. Some of the molecules involved in integrin signaling and recent findings related to their involvement during chondrogenesis and endochondral ossification will be discussed in Section 4.

4. Integrin downstream partners: Connecting ECM to the cell cytoskeleton

It is well established that the actin cytoskeleton plays a critical role during chondrogenesis (Song & Park, 2014). Forcing cells to adopt a spherical cell shape *via* pharmacological inhibition of actin polymerization stimulates dedifferentiated cells to re-express chondrogenic matrix molecules and promotes the commitment of mesenchymal cells to the chondrogenic lineage (Benya, Brown, & Padilla, 1988; Benya & Padilla, 1993; Brown & Benya, 1988; Zanetti & Solursh, 1984). Furthermore, mutations in actin modifying proteins have been linked to human skeletal disorders (Loty, Forest, Boulekbache, & Sautier, 1995; Robertson et al., 2003).

Consequently, it was concluded that cellular receptor-mediated signaling pathways, which regulate cell shape, also influence cell differentiation. Thus, researchers began to focus on intracellular integrin-binding partners and their roles in signaling pathways that affect cell behavior during chondrogenesis and subsequent chondrocyte differentiation.

4.1 Focal adhesion kinase

The activation of the protein tyrosine kinase FAK is one of the earliest events upon integrin activation. FAK is a major signaling molecule involved in regulation of cell proliferation, survival, and motility. FAK activation and subsequent auto-phosphorylation lead to recruitment of Src, another tyrosine kinase, and to the formation of a dual FAK-Src signaling complex (Mitra, Hanson, & Schlaepfer, 2005). FAK-deficient mouse embryos display a general defect in mesoderm development and die at embryonic day 8.5 (Furuta et al., 1995; Ilic et al., 1995). Thus, researchers tried to unravel the role of FAK at later embryonic stages by utilizing high density micromass cultures of *Fak*$^{+/+}$ and *Fak*$^{-/-}$ fibroblasts, cultured for 6 days (Pala et al., 2008). When compared to wild type cells, mutant cells deficient in FAK expression exhibited increased cellular staining for *Peanut Agglutinin* (PNA), known to bind to condensed mesenchymal chondroprogenitor cells (Stringa & Tuan, 1996). Furthermore, L-Sox5 and Col2a1 mRNA expression was significantly higher than in controls, while Sox9 expression was unaltered. These findings indicate that loss of FAK expression promotes chondrogenesis *in vitro*. An explanation for this finding might be that loss of FAK leads to a situation where cell-cell interactions, which are necessary during early chondrogenesis, are favored. Furthermore, FAK is required for cell migration, but migration is not required for condensation of cells in these high density micromass cultures. However, the presence of FAK might be important during later stages of chondrocyte differentiation due to its role in the activation of other downstream signaling partners of integrin (Bang et al., 2000; DeMali, Wennerberg, & Burridge, 2003).

4.2 Rho GTPases: Family members with different functions

Rho GTPases connect signals from the ECM to the actin cytoskeleton of cells and act upstream and downstream of integrin ligand binding (Woods, Wang, & Beier, 2007). The three most studied GTPase family members RhoA, Rac1 and Cdc42 have been shown to be involved in chondrocyte differentiation and they very likely act as regulators of cytoskeletal control during chondrogenesis. RhoA signaling stimulates stress fiber formation through its main effector Rho-associated protein kinase ROCK (ROCKI and ROCKII). ROCK signaling results in phosphorylation of the actin depolymerizing protein cofilin and thereby stabilizes actin filaments. *In vitro*, the formation of stress fibers is associated with chondrocyte de-differentiation into a fibroblast-like phenotype (Riento & Ridley, 2003). In this context, the Beier lab and others showed that overexpression of RhoA inhibits early chondrogenesis as well as hypertrophic differentiation of the ATDC5 chondrocyte cell line (Kumar & Lassar, 2009; Wang et al., 2004; Woods, Wang & Beier, 2005). In contrast, pharmacological inhibition of RhoA/ROCK signaling or disruption of the cytoskeleton promotes glycosaminoglycan production as well as Sox9 expression and induces a spherical, chondrocyte-like cell shape, in some but not all cellular contexts (Wang et al., 2004; Woods & Beier, 2006; Woods et al., 2005). Furthermore, RhoA/ROCK signaling supports chondrocyte proliferation, most likely due to an increased expression of cyclin D1, and inhibits hypertrophic differentiation (Beier et al., 2001; Wang et al., 2004). Overall, these data suggest that RhoA suppresses chondrogenesis in most contexts, consistent with what is also seen in a 3D collagen matrix culture (Lu, Doulabi, Huang, Bank, & Helder, 2008). However, its regulatory influence on actin dynamics might play a role during endochondral ossification and growth plate establishment. Indeed, in embryonic tibia organ culture, the use of cytochalasin D, an inhibitor of actin polymerization, caused loss of normal growth plate architecture. The inhibition of Rho/ROCK signaling had similar but milder effects on growth plate development (Woods, James, Wang, Dupuis, & Beier, 2009). Importantly, no *in vivo* models with genetic interruption of RhoA (or RhoB, RhoC) or ROCK1/2 in cartilage have been reported.

Interestingly, in contrast to RhoA, other members of the Rho GTPases family such as Rac1 promote early chondrogenesis through the stimulation of N-cadherin as well as Sox5, 6, and 9 expression. Cdc42 also promotes chondrogenesis through the control of Sox9 expression (Woods et al., 2007). Together, Rac1 and Cdc42 stimulate chondrocyte hypertrophy, most likely through activation of p38 MAP kinase, and inhibit chondrocyte proliferation when overexpressed in cell culture (Bobick & Kulyk, 2008; Wang & Beier, 2005).

In vivo analysis of the role of Rac1 using cartilage-specific knockout mice, generated by using the collagen II Cre driver line, showed that Rac 1-deficient mice display a high rate of perinatal lethality. Surviving mice exhibited kyphosis, shortened long bones and disorganized, hypocellular growth plates with reduced chondrocyte proliferation, increased apoptosis, and decreased expression of cyclin D1 (Wang et al., 2007). Surprisingly, the decreased hypertrophic differentiation seen *in vitro* was not observed *in vivo*. Instead, a premature exit from the cell cycle, based on reduced cyclin D1 expression, was detected. Rac1-deficient chondrocytes isolated from growth plate cartilage exhibited reduced adhesion to collagen type II and fibronectin, as well as changes in cell shape and actin organization, indicating a disrupted signaling between cells and the ECM. Follow-up studies demonstrated that Rac1 controls chondrocyte proliferation through a pathway involving iNOS, nitric oxide, and the transcription factor ATF3, directly suppressing the cyclin D1 promoter (James, Woods, Underhill, & Beier, 2006; Wang et al., 2011). Mice with Rac1 conditionally knocked out in mouse limb bud mesenchyme, using the Prx1-Cre driver line, showed a syndactyly phenotype due to loss of apoptosis in interdigital areas, in addition to reduced body length and weight, as well as disorganized growth plates, similar to what is seen in conditional Rac1 knockout mice using the collagen II Cre driver line (Suzuki et al., 2009).

Interestingly, recent studies, based on overexpression of an activated Rac1 under control of either the collagen II or Prx1-Cre drivers, resulted in more subtle phenotypes, partially resembling those of conditional knockout lines (Suzuki, Bush, Bryce, Kamijo, & Beier, 2017). Moreover, crosses of conditional knockout and overexpressing lines partially rescued the defects observed in individual mutants. These data suggest that the exact level of Rac1 activity is essential in the control of cartilage development and endochondral bone growth.

Recently generated conditional Cdc42 knockout mice, using both the Prx1 and collagen II Cre driver lines, showed similar phenotypes of syndactyly, reduced skeletal growth and growth plate disorganization, including loss of columnar arrangements. The growth plate hypertrophic zone was enlarged, while the levels of ECM hypertrophic markers, such as collagen X and MMP13, were reduced (Aizawa et al., 2012; Suzuki et al., 2015).

In order to investigate the function of both Rac1 and Cdc42 during chondrogenesis and endochondral ossification researchers generated double conditional knockout mice (Ikehata et al., 2018). Not surprisingly, these mice died shortly after birth with abnormal skeletal formation as well as disorganized growth plates. These changes appeared more dramatic than defects in single conditional Rac1 and Cdc42 knockout mice, indicating that Rac1 and Cdc42 have cooperating roles during skeletal development.

Taken together, these studies demonstrate that Rho GTPases are crucial for multiple steps in chondrocyte differentiation and endochondral ossification. However, there is still a need for defining upstream and downstream pathways for Rho GTPases as well as their interactions with other receptors in order to unravel their detailed connections to cellular responses and ECM signaling. Furthermore, microarray data demonstrate changes in the expression profiles of other members of the Rho family during cartilage differentiation *in vitro* and *in vivo*, and this needs to be further addressed (James, Appleton, Ulici, Underhill, & Beier, 2005; James et al., 2010).

4.3 MAP kinase cascade

Mitogen-activated protein kinases (MAPK) are downstream targets for FAK and many other pathways. MAPK pathways are organized into cascades that originate with the phosphorylation of upstream kinases resulting in the activation of MAPK. Three major MAPK families have been identified: ERK1/ERK2 (extracellular signal regulated kinase), p38MAPK, and JNK (Jun N-terminal kinases) (Han, Lee, Bibbs, & Ulevitch, 1994; Kyriakis et al., 1994; Marais & Marshall, 1996). During cartilage development MAPK pathways are involved in the early mesenchymal condensations as well as the differentiation into chondrocytes. The role of MAPK in chondrogenesis was first studied using different cell culture systems and selective MAPK inhibitors. One study showed that the activity and phosphorylation of p38 are increased during chondrogenesis, while ERK phosphorylation is decreased (Oh et al., 2000). Furthermore, the authors of this study demonstrated that inhibition of p38 activity blocks chondrogenesis, while inhibition of ERK had the opposite effect. JNK phosphorylation is not affected during chondrogenesis, indicating that this MAPK has a minor role in this process (Nakamura et al., 1999). In agreement with these findings, inhibition of chondrogenesis by epidermal growth factor (EGF) resulted in activation of ERK phosphorylation and p38 inhibition (Yoon et al., 2000). Inhibition of chondrogenesis *via* rapamycin or retinoic acid resulted in reduced activity of p38 (Oh et al., 2001; Weston, Chandraratna, Torchia, & Underhill, 2002). Other *in vitro* studies suggested a close connection between p38 and growth factors that are highly involved in chondrocyte differentiation and function. This includes TGF-β, involved in the regulation of collagen II and aggrecan expression in response to these growth factors (Nakamura et al., 1999; Tuli et al., 2003; Yosimichi et al., 2001; Zuzarte-Luis et al., 2004). In a micromass culture system, p38 signaling has been shown to positively regulate hypertrophic chondrocyte differentiation, whereas treatment with p38 inhibitors resulted in delay of hypertrophic differentiation, accompanied by reduced expression of marker genes, including collagen X (Stanton, Sabari, Sampaio, Underhill, & Beier, 2004). Furthermore, ERK1/2 has been implicated in the control of collagen X transcription *via* c-RAF kinase (Beier, Taylor, & LuValle, 1999).

These *in vitro* studies reveal the crucial impact of MAPKs in various phases of chondrogenesis and endochondral ossification, whereas p38 and ERK show opposing roles. In order to confirm these findings and particularly to verify involvement MAPKs in chondrogenesis and chondrocyte differentiation *in vivo*, mice with targeted deficiency in p38alpha, the predominant p38 member in chondrocytes, were generated using the Collagen II Cre strain (Hutchison, 2013). Knockout mice were dwarfed and exhibited decreased growth plate width, disrupted columnar cell formation in the proliferative zone, as well as delayed hypertrophic chondrocyte differentiation, accompanied by altered expression of Sox9 and Runx2. Another study showed direct upregulation of Sox9 transcriptional activity using mice with constitutively activated MKK6, a MAPK kinase that specifically activates p38 (Zhang, Murakami, Coustry, Wang, & de Crombrugghe, 2006). Interestingly, the mice showed a phenotype comparable to that found in mice with targeted disruption in p38alpha with additional delay of primary and secondary ossification. In addition, the expression of Indian hedgehog and PTH/PTH-related peptide receptor, which are known to positively regulate chondrocyte proliferation, was decreased. These findings indicate that the p38 MAPK pathway regulates expression of various proteins and tightly controls cartilage development.

A strong skeletal phenotype was also observed in conditional ERK1$^{-/-}$; ERK2$^{fl/fl}$ mice using the Col2a1-Cre transgene. Embryonic day 16.5 embryos showed absence of primary ossification centers in the axial skeleton and a widening of growth plate hypertrophic zones in long bones. At embryonic day 18.5 loss of growth plate columnar structure was observed (Matsushita et al., 2009). Long bones of these knockout mice were significantly longer and the epiphyses significantly wider than in controls, indicating that ERK1 and ERK2 negatively regulate the growth of the cartilaginous elements during endochondral ossification (Sebastian et al., 2011).

Taken together, these studies show that MAPK signaling in chondrocytes affects master transcription factors of cartilage such as Sox9, and therefore chondrocyte differentiation and the establishment of the cartilage ECM.

5. Other non-integrin cell receptors

5.1 CD44

The principal cell receptor for hyaluronan, CD44, is another well-studied cell membrane receptor in many cell types, including chondrocytes. The CD44 domain is structurally similar to the hyaluronan binding site of aggrecan and cartilage link protein. Hyaluronan is a long, linear glycosaminoglycan chain, which can bind up to ~100 aggrecan monomers. The aggrecan/hyaluronan aggregates, stabilized by link protein, are the major space filling components of the cartilage ECM and are embedded into the tensile strength network formed by several types of collagen (Warren et al., 2000). Thus, the connection between cells and this aggrecan/hyaluronan molecular complex *via* CD44 provides strong communications between cells and the matrix.

During the early stage of limb bud formation, cells are separated from each other due to a large volume of ECM mainly composed of hyaluronan

(Kosher, Savage, & Walker, 1981; Thorogood & Hinchliffe, 1975). During this period, it is assumed that hyaluronan promotes cell migration and division and maintains the space between cells. With the initiation of condensation, the hyaluronan content gets depleted *via* CD44, which is known to participate in hyaluronan endocytosis and degradation (Culty, Nguyen, & Underhill, 1992). Consequently, the intracellular spaces in the regions of future cartilage development decrease. Indeed, an *in vitro* study by Ishida et al. demonstrated that the adhesion of chondrocytes to hyaluronan *via* CD44 induces cell proliferation as well as expression of TGF-β and c-myc mRNA, which are related to cell proliferation and chondrocyte maturation as well (Osamu, Yoshiya, Isao, Masaharu, & Sumiya, 1997). However, although CD44 is already active before cartilage matrix assembly begins, later on the expression of CD44 in the developing mouse long bones seems to be restricted to the zone of erosion and ossification, as seen by immunohistochemistry in mouse long bones at postnatal day 19 (Knudson & Toole, 1987; Pavasant, Shizari, & Underhill, 1994). CD44 appears to be expressed in cultured chondrocytes isolated from articular cartilage; however, it was not detected in growth plate chondrocytes *in vivo*. In addition, CD44 knockout mice showed no histomorphometric skeletal defects (de Vries, Schoenmaker, Beertsen, van der Neut, & Everts, 2005). Thus, the most recent studies of CD44 focus on its role in articular cartilage homeostasis and disease, which is beyond the scope of this review.

5.2 Syndecan

Syndecans belong to the family of small type I transmembrane proteoglycans that serve as co-receptors for growth factors as well as cytokines and ECM components. They contain heparan sulfate side chains and elicit signal transduction through their cytoplasmic tails (Oh & Couchman, 2004; Shimazu et al., 1996). Syndecans include four members, which all have been implicated in regulation of cytoskeletal organization and are expressed in tissue-specific and developmentally regulated patterns (Kirsch, Koyama, Liu, Golub, & Pacifici, 2002; Oh & Couchman, 2004). Syndecan 1 is mainly expressed in epithelial cells, while syndecan 2 can be detected in mesenchymal cells. Syndecan 3 is mostly associated with bone development, and syndecan 4 is ubiquitously expressed (Kim, Goldberger, Gallo, & Bernfield, 1994). Expression of syndecan 1, 3, and 4 was also reported in rat chondrocytes, as well as in progenitor cells during mandibular condyle development (Molteni, Modrowski, Hott, & Marie, 1999). Syndecan 3, a typical integral membrane proteoglycan, exhibits a cytoplasmic domain which is believed to interact with the cytoskeleton and thus may play an important role during chondrocyte development. Indeed, syndecan 3 expression was found to be increased at the onset of chondrogenesis in condensed chick limb mesenchyme (Gould, Upholt, & Kosher, 1992). Syndecan 3 mRNA was found in proliferating immature cartilage, while its transcription decreased in mature hypertrophic cartilage, suggesting an important role during chondrocyte differentiation. Indeed, the injection of polyclonal syndecan 3 neutralizing antibodies into the progress zone of early chick embryos *in ovo* resulted in inhibition of local cell proliferation, resulting in severe skeletal abnormalities and reduced growth (Dealy, Seghatoleslami, Ferrari, & Kosher, 1997). In addition, immunohistochemistry of chick embryo tibia (P18) confirmed the spatially restricted presence of syndecan 3 in the growth plate top zone, close to the developing articular cartilage (Shimazu et al., 1996). Other data show that syndecan 3 gene expression is changing with increasing embryonic age of the developing skeletal elements, whereas it is continuously associated with proliferating chondrocytes (Kirsch et al., 2002). All these studies indicate that syndecan 3 plays a regulatory role during chondrocyte proliferation.

The treatment of cultured chondrocytes with FGF-2 leads to enhanced cell proliferation and syndecan 3 gene expression, while treatment with heparinases counteracts this effect. When syndecan 3 antibodies, similar to those used in the limb interference studies described above, were used to specifically test the role of syndecan 3 core protein, the stimulation of chondrocyte proliferation by FGF-2 was counteracted. Thus, researchers concluded that the chondrocyte response to FGF-2 might be directly mediated by heparan sulfate proteoglycans such as syndecan 3 and not by its glycosaminoglycan side chains (Rapraeger, 1995; Shimazu et al., 1996). However, the underlying mechanism by which syndecan 3 or its antibodies actually operate still remains unknown. Furthermore, a recent study suggests that syndecan 3 influences or even mediates the activity of Indian hedgehog during chick limb development, supporting the idea that there is close interaction between syndecan and growth factor mediated proliferation (Shimo et al., 2004). However, more studies need to be done to unravel the scope of syndecan 3 during chondrogenesis and endochondral ossification. Recent studies focus on the recapitulation of findings by using mammalian models instead of avian systems (Koyama et al., 2007).

Syndecan 4 is mainly associated with osteoarthritis (Echtermeyer et al., 2001). Just recently, researchers started to focus on its function during endochondral ossification using syndecan 4 knockout mice generated using the lacZ-knockin strategy. Knockout mice showed no obvious postnatal abnormalities, although syndecan 4 promoter activity was found in all phases of chondrocyte differentiation during endochondral ossification (Bertrand et al., 2013). Interestingly, the studies revealed a marked increase of syndecan 2 in the absence of syndecan 4, indicating a compensatory mechanism between these two syndecans.

5.3 Discoidin domain receptors

Discoidin domain receptors (DDRs) are non-integrin receptors that bind directly to ECM molecules. The discoidin domain receptor 1 (DDR1) and DDR2 are receptor tyrosine kinase receptors, that are activated by collagen (Vogel, Gish, Alves, & Pawson, 1997). In comparison to other receptor tyrosine kinases, which are activated by small diffusible growth factors, the activation of DDRs is slow but persistent. DDRs are single-span transmembrane proteins. The extracellular region contains an N-terminal discoidin homology domain, which includes the collagen binding site (Carafoli et al., 2012; Ichikawa et al., 2007). DDRs are widely expressed in mammalian tissues and bind various collagens with distinct specificity.

The activation of DDR2 *in vitro* induces the expression of matrix-metalloproteinase 1, which is known to be involved in ECM remodeling, for example, during morphogenesis (Vogel et al., 1997; Werb, 1997). The analysis of DDR2 expression in the tibial growth plate of 1-week-old wild-type mice by *in situ* hybridization showed a characteristic pattern in the proliferative zone. Expression of collagens II and X, as well as alkaline phosphatase, osteocalcin (osteoblast marker), and different metalloproteinases was analyzed at different postnatal stages, but no differences in the expression pattern between wild type and DDR2 knockout mice were found. However, DDR2 knockout mice showed reduced postnatal chondrocyte proliferation and growth of long bones (Labrador et al., 2001). The number of proliferating cells in DDR2 knockout mice was markedly lower compared to wild type mice at postnatal week two and three. In contrast, the number of apoptotic cells remained constant. The development of latter lineages appeared to be normal in DDR2 knockout mice, indicating a role for DDR2 in chondrocyte proliferation, rather than differentiation. Interestingly, mutations in DDR2 were reported to lead to a rare form of growth retardation in humans (Ali et al., 2010; Bargal et al., 2009). However, more studies need to be done in order to elucidate downstream events mediated by DDR2. DDR1 also works as a receptor for collagen but is predominant in epithelial cells and not in cells of connective tissues. Thus, the knockout of DDR1 does not result in an obvious skeletal phenotype (Gross et al., 2004).

6. Conclusions and implications

The control of chondrocyte behavior during chondrogenesis and endochondral ossification is complex and involves many different types of stimulatory and inhibitory events. There is substantial progress in the field of adhesion receptor-mediated signaling, but much still remains to be learned. Apparently, many of the cell receptor-mediated pathways are very complex. Upstream regulators, downstream mediators, as well as target genes and receptor crosstalk will have to be investigated before a deeper understanding of chondrogenesis and associated diseases can be reached. In addition, integration of chemical and biomechanical signals and the organization of the cytoskeleton need to be analyzed further. New techniques that enable imaging of living organisms in real time will help to visualize many spatial and temporal cellular events that are downstream of cell surface receptors. Furthermore, many findings are based solely on *in vitro* experiments and it is not known how well ECM substrates used in cell and tissue cultures correspond to the actual *in vivo* situations regarding their compositions and mechanical properties.

Acknowledgments

We thank all members of the Beier laboratory for discussions. Work in the laboratory is supported by grants from the Canadian Institutes of Health Research and The Arthritis Society (Canada). C.P. is the recipient of a postdoctoral fellowship from the Collaborative Program in Musculoskeletal Health Research at Western University, and F.B. is the recipient of a Canada Research Chair Award.

References

Aizawa, R., Yamada, A., Suzuki, D., Iimura, T., Kassai, H., Harada, T., et al. (2012). Cdc42 is required for chondrogenesis and interdigital programmed cell death during limb development. *Mechanisms of Development*, *129*(1), 38–50. https://doi.org/10.1016/j.mod.2012.02.002.

Ali, B. R., Xu, H., Akawi, N. A., John, A., Karuvantevida, N. S., Langer, R., et al. (2010). Trafficking defects and loss of ligand binding are the underlying causes of all reported DDR2 missense mutations found in SMED-SL patients. *Human Molecular Genetics*, *19*(11), 2239–2250. https://doi.org/10.1093/hmg/ddq103.

Aszodi, A., Hunziker, E. B., Brakebusch, C., & Fässler, R. (2003). β1 integrins regulate chondrocyte rotation, G1 progression, and cytokinesis. *Genes and Development*, *17*(19), 2465–2479. https://doi.org/10.1101/gad.277003.

Bang, O.-S., Kim, E.-J., Chung, J. G., Lee, S.-R., Park, T. K., & Kang, S.-S. (2000). Association of focal adhesion kinase with fibronectin and paxillin is required for precartilage condensation of chick mesenchymal cells. *Biochemical and Biophysical Research Communications*, *278*(3), 522–529. https://doi.org/10.1006/bbrc.2000.3831.

Bargal, R., Cormier-Daire, V., Ben-Neriah, Z., Le Merrer, M., Sosna, J., Melki, J., et al. (2009). Mutations in DDR2 gene cause SMED with short limbs and abnormal calcifications. *American Journal of Human Genetics, 84*(1), 80–84. https://doi.org/10.1016/j.ajhg.2008.12.004.
Beier, F., Ali, Z., Mok, D., Taylor, A. C., Leask, T., Albanese, C., et al. (2001). TGFβ and PTHrP control chondrocyte proliferation by activating cyclin D1 expression. *Molecular Biology of the Cell, 12*(12), 3852–3863.
Beier, F., Taylor, A. C., & LuValle, P. (1999). Raf signaling stimulates and represses the human collagen X promoter through distinguishable elements. *Journal of Cellular Biochemistry, 72*(4), 549–557.
Bengtsson, T., Aszodi, A., Nicolae, C., Hunziker, E. B., Lundgren-Äkerlund, E., & FÃssler, R. (2005). Loss of α10β1 integrin expression leads to moderate dysfunction of growth plate chondrocytes. *Journal of Cell Science, 118*(5), 929–936.
Benya, P. D., Brown, P. D., & Padilla, S. R. (1988). Microfilament modification by dihydrocytochalasin B causes retinoic acid-modulated chondrocytes to reexpress the differentiated collagen phenotype without a change in shape. *The Journal of Cell Biology, 106*(1), 161–170.
Benya, P. D., & Padilla, S. R. (1993). Dihydrocytochalasin B enhances transforming growth factor-beta-induced reexpression of the differentiated chondrocyte phenotype without stimulation of collagen synthesis. *Experimental Cell Research, 204*(2), 268–277. https://doi.org/10.1006/excr.1993.1033.
Bertrand, J., Stange, R., Hidding, H., Echtermeyer, F., Nalesso, G., Godmann, L., et al. (2013). Syndecan 4 supports bone fracture repair, but not fetal skeletal development, in mice. *Arthritis and Rheumatism, 65*(3), 743–752. https://doi.org/10.1002/art.37817.
Bobick, B. E., & Kulyk, W. M. (2008). Regulation of cartilage formation and maturation by mitogen-activated protein kinase signaling. *Birth Defects Research. Part C, Embryo Today: Reviews, 84*(2), 131–154. https://doi.org/10.1002/bdrc.20126.
Bouvard, D., Brakebusch, C., Gustafsson, E., Aszódi, A., Bengtsson, T., Berna, A., et al. (2001). Functional consequences of integrin gene mutations in mice. *Circulation Research, 89*(3), 211–223.
Brown, P. D., & Benya, P. D. (1988). Alterations in chondrocyte cytoskeletal architecture during phenotypic modulation by retinoic acid and dihydrocytochalasin B-induced reexpression. *The Journal of Cell Biology, 106*(1), 171–179.
Carafoli, F., Mayer, M. C., Shiraishi, K., Pecheva, M. A., Chan, L. Y., Nan, R., et al. (2012). Structure of the discoidin domain receptor 1 extracellular region bound to an inhibitory Fab fragment reveals features important for signaling. *Structure, 20*(4), 688–697. https://doi.org/10.1016/j.str.2012.02.011.
Culty, M., Nguyen, H. A., & Underhill, C. B. (1992). The hyaluronan receptor (CD44) participates in the uptake and degradation of hyaluronan. *The Journal of Cell Biology, 116*(4), 1055–1062.
de Vries, T. J., Schoenmaker, T., Beertsen, W., van der Neut, R., & Everts, V. (2005). Effect of CD44 deficiency on in vitro and in vivo osteoclast formation. *Journal of Cellular Biochemistry, 94*(5), 954–966. https://doi.org/10.1002/jcb.20326.
Dealy, C. N., Seghatoleslami, M. R., Ferrari, D., & Kosher, R. A. (1997). FGF-stimulated outgrowth and proliferation of limb mesoderm is dependent on syndecan-3. *Developmental Biology, 184*(2), 343–350. https://doi.org/10.1006/dbio.1997.8525.
Delise, A. M., & Tuan, R. S. (2002). Analysis of N-cadherin function in limb mesenchymal chondrogenesis in vitro. *Developmental Dynamics, 225*(2), 195–204. https://doi.org/10.1002/dvdy.10151.
DeMali, K. A., Wennerberg, K., & Burridge, K. (2003). Integrin signaling to the actin cytoskeleton. *Current Opinion in Cell Biology, 15*(5), 572–582. https://doi.org/10.1016/S0955-0674(03)00109-1.

Echtermeyer, F., Streit, M., Wilcox-Adelman, S., Saoncella, S., Denhez, F., Detmar, M., et al. (2001). Delayed wound repair and impaired angiogenesis in mice lacking syndecan-4. *The Journal of Clinical Investigation*, *107*(2), R9–R14. https://doi.org/10.1172/jci10559.
Erlebacher, A., Filvaroff, E. H., Gitelman, S. E., & Derynck, R. (1995). Toward a molecular understanding of skeletal development. *Cell*, *80*(3), 371–378.
Fässler, R., Pfaff, M., Murphy, J., Noegel, A. A., Johansson, S., Timpl, R., et al. (1995). Lack of β1 integrin gene in embryonic stem cells affects morphology, adhesion, and migration but not integration into the inner cell mass of blastocysts. *Journal of Cell Biology*, *128*(5), 979–988.
Furuta, Y., Ilic, D., Kanazawa, S., Takeda, N., Yamamoto, T., & Aizawa, S. (1995). Mesodermal defect in late phase of gastrulation by a targeted mutation of focal adhesion kinase, FAK. *Oncogene*, *11*(10), 1989–1995.
Geiger, B., Bershadsky, A., Pankov, R., & Yamada, K. M. (2001). Transmembrane crosstalk between the extracellular matrix and the cytoskeleton. *Nature Reviews. Molecular Cell Biology*, *2*, 793. https://doi.org/10.1038/35099066.
Gerber, H.-P., Vu, T. H., Ryan, A. M., Kowalski, J., Werb, Z., & Ferrara, N. (1999). VEGF couples hypertrophic cartilage remodeling, ossification and angiogenesis during endochondral bone formation. *Nature Medicine*, *5*, 623. https://doi.org/10.1038/9467.
Gould, S. E., Upholt, W. B., & Kosher, R. A. (1992). Syndecan 3: A member of the syndecan family of membrane-intercalated proteoglycans that is expressed in high amounts at the onset of chicken limb cartilage differentiation. *Proceedings of the National Academy of Sciences of the United States of America*, *89*(8), 3271–3275.
Grashoff, C., Aszódi, A., Sakai, T., Hunziker, E. B., & Fässler, R. (2003). Integrin-linked kinase regulates chondrocyte shape and proliferation. *EMBO Reports*, *4*(4), 432–438. https://doi.org/10.1038/sj.embor.embor801.
Gross, O., Beirowski, B., Harvey, S. J., McFadden, C., Chen, D., Tam, S., et al. (2004). DDR1-deficient mice show localized subepithelial GBM thickening with focal loss of slit diaphragms and proteinuria. *Kidney International*, *66*(1), 102–111. https://doi.org/10.1111/j.1523-1755.2004.00712.x.
Han, L., Grodzinsky, A. J., & Ortiz, C. (2011). Nanomechanics of the cartilage extracellular matrix. *Annual Review of Materials Research*, *41*, 133–168.
Han, J., Lee, J., Bibbs, L., & Ulevitch, R. (1994). A MAP kinase targeted by endotoxin and hyperosmolarity in mammalian cells. *Science*, *265*(5173), 808–811. https://doi.org/10.1126/science.7914033.
Harburger, D. S., & Calderwood, D. A. (2009). Integrin signalling at a glance. *Journal of Cell Science*, *122*(2), 159–163. https://doi.org/10.1242/jcs.018093.
Hutchison, M. R. (2013). Mice with a conditional deletion of the neurotrophin receptor TrkB are dwarfed, and are similar to mice with a MAPK14 deletion. *PLoS One*, *8*(6), e66206. https://doi.org/10.1371/journal.pone.0066206.
Hynes, R. O. (2002). Integrins: bidirectional, allosteric signaling machines. *Cell*, *110*(6), 673–687.
Ichikawa, O., Osawa, M., Nishida, N., Goshima, N., Nomura, N., & Shimada, I. (2007). Structural basis of the collagen-binding mode of discoidin domain receptor 2. *The EMBO Journal*, *26*(18), 4168–4176. https://doi.org/10.1038/sj.emboj.7601833.
Ikehata, M., Yamada, A., Fujita, K., Yoshida, Y., Kato, T., Sakashita, A., et al. (2018). Cooperation of Rho family proteins Rac1 and Cdc42 in cartilage development and calcified tissue formation. *Biochemical and Biophysical Research Communications*, *500*(3), 525–529. https://doi.org/10.1016/j.bbrc.2018.04.032.
Ilic, D., Furuta, Y., Kanazawa, S., Takeda, N., Sobue, K., Nakatsuji, N., et al. (1995). Reduced cell motility and enhanced focal adhesion contact formation in cells from FAK-deficient mice. *Nature*, *377*(6549), 539–544. https://doi.org/10.1038/377539a0.

James, C. G., Appleton, C. T., Ulici, V., Underhill, T. M., & Beier, F. (2005). Microarray analyses of gene expression during chondrocyte differentiation identifies novel regulators of hypertrophy. *Molecular Biology of the Cell, 16*(11), 5316–5333. https://doi.org/10.1091/mbc.E05-01-0084.

James, C. G., Stanton, L. A., Agoston, H., Ulici, V., Underhill, T. M., & Beier, F. (2010). Genome-wide analyses of gene expression during mouse endochondral ossification. *PLoS One, 5*(1), e8693. https://doi.org/10.1371/journal.pone.0008693.

James, C. G., Woods, A., Underhill, T. M., & Beier, F. (2006). The transcription factor ATF3 is upregulated during chondrocyte differentiation and represses cyclin D1 and A gene transcription. *BMC Molecular Biology*, 7(1), 30. https://doi.org/10.1186/1471-2199-7-30.

Karsenty, G., & Wagner, E. F. (2002). Reaching a genetic and molecular understanding of skeletal development. *Developmental Cell, 2*(4), 389–406.

Kim, C. W., Goldberger, O. A., Gallo, R. L., & Bernfield, M. (1994). Members of the syndecan family of heparan sulfate proteoglycans are expressed in distinct cell-, tissue-, and development-specific patterns. *Molecular Biology of the Cell, 5*(7), 797–805.

Kirsch, T., Koyama, E., Liu, M., Golub, E. E., & Pacifici, M. (2002). Syndecan-3 is a selective regulator of chondrocyte proliferation. *The Journal of Biological Chemistry, 277*(44), 42171–42177. https://doi.org/10.1074/jbc.M207209200.

Knudson, C. B., & Toole, B. P. (1987). Hyaluronate-cell interactions during differentiation of chick embryo limb mesoderm. *Developmental Biology, 124*(1), 82–90. https://doi.org/10.1016/0012-1606(87)90462-3.

Komori, T. (2016). Cell death in chondrocytes, osteoblasts, and osteocytes. *International Journal of Molecular Sciences, 17*(12), 2045. https://doi.org/10.3390/ijms17122045.

Kosher, R. A., Savage, M. P., & Walker, K. H. (1981). A gradation of hyaluronate accumulation along the proximodistal axis of the embryonic chick limb bud. *Journal of Embryology and Experimental Morphology, 63*, 85–98.

Koyama, E., Young, B., Shibukawa, Y., Nagayama, M., Enomoto-Iwamoto, M., Iwamoto, M., et al. (2007). Conditional Kif3a ablation causes abnormal hedgehog signaling topography, growth plate dysfunction and ectopic cartilage formation in mouse cranial base synchondroses. *Development (Cambridge, England), 134*(11), 2159–2169. https://doi.org/10.1242/dev.001586.

Kronenberg, H. M., & Chung, U. (2001). The parathyroid hormone-related protein and Indian hedgehog feedback loop in the growth plate. *Novartis Foundation Symposium, 232*, 144–152. discussion 152-147.

Kumar, D., & Lassar, A. B. (2009). The transcriptional activity of Sox9 in chondrocytes is regulated by RhoA signaling and actin polymerization. *Molecular and Cellular Biology, 29*(15), 4262–4273. https://doi.org/10.1128/MCB.01779-08.

Kyriakis, J. M., Banerjee, P., Nikolakaki, E., Dai, T., Rubie, E. A., Ahmad, M. F., et al. (1994). The stress-activated protein kinase subfamily of c-Jun kinases. *Nature, 369*, 156. https://doi.org/10.1038/369156a0.

Labrador, J. P., Azcoitia, V., Tuckermann, J., Lin, C., Olaso, E., Mañes, S., et al. (2001). The collagen receptor DDR2 regulates proliferation and its elimination leads to dwarfism. *EMBO Reports, 2*(5), 446–452. https://doi.org/10.1093/embo-reports/kve094.

Lefebvre, V. (2002). Toward understanding the functions of the two highly related Sox5 and Sox6 genes. *Journal of Bone and Mineral Metabolism, 20*(3), 121–130.

Lefebvre, V., & de Crombrugghe, B. (1998). Toward understanding S0X9 function in chondrocyte differentiation. *Matrix Biology, 16*(9), 529–540.

Lefebvre, V., & Smits, P. (2005). Transcriptional control of chondrocyte fate and differentiation. *Birth Defects Research Part C: Embryo Today: Reviews, 75*(3), 200–212. https://doi.org/10.1002/bdrc.20048.

Legate, K. R., Wickström, S. A., & Fässler, R. (2009). Genetic and cell biological analysis of integrin outside-in signaling. *Genes and Development*, *23*(4), 397–418.

Loeser, R. F. (2000). Chondrocyte integrin expression and function. *Biorheology*, *37*(1–2), 109–116.

Loeser, R. F. (2014). Integrins and chondrocyte–matrix interactions in articular cartilage. *Matrix Biology: Journal of the International Society for Matrix Biology*, *39*, 11–16. https://doi.org/10.1016/j.matbio.2014.08.007.

Logan, M., Martin, J. F., Nagy, A., Lobe, C., Olson, E. N., & Tabin, C. J. (2002). Expression of Cre recombinase in the developing mouse limb bud driven by a Prxl enhancer. *Genesis*, *33*(2), 77–80. https://doi.org/10.1002/gene.10092.

Loty, S., Forest, N., Boulekbache, H., & Sautier, J.-M. (1995). Cytochalasin D induces changes in cell shape and promotes in vitro chondrogenesis: A morphological study. *Biology of the Cell*, *83*(2), 149–161. https://doi.org/10.1016/0248-4900(96)81303-7.

Lu, Z. F., Doulabi, B. Z., Huang, C. L., Bank, R. A., & Helder, M. N. (2008). β1 integrins regulate chondrogenesis and rock signaling in adipose stem cells. *Biochemical and Biophysical Research Communications*, *372*(4), 547–552. https://doi.org/10.1016/j.bbrc.2008.05.063.

Maes, C., Kobayashi, T., Selig, M. K., Torrekens, S., Roth, S. I., Mackem, S., et al. (2010). Osteoblast precursors, but not mature osteoblasts, move into developing and fractured bones along with invading blood vessels. *Developmental Cell*, *19*(2), 329–344. https://doi.org/10.1016/j.devcel.2010.07.010.

Marais, R., & Marshall, C. J. (1996). Control of the ERK MAP kinase cascade by Ras and Raf. *Cancer Surveys*, *27*, 101–125.

Matsushita, T., Chan, Y. Y., Kawanami, A., Balmes, G., Landreth, G. E., & Murakami, S. (2009). Extracellular signal-regulated kinase 1 (ERK1) and ERK2 play essential roles in osteoblast differentiation and in supporting osteoclastogenesis. *Molecular and Cellular Biology*, *29*(21), 5843–5857. https://doi.org/10.1128/MCB.01549-08.

Mitra, S. K., Hanson, D. A., & Schlaepfer, D. D. (2005). Focal adhesion kinase: In command and control of cell motility. *Nature Reviews. Molecular Cell Biology*, *6*(1), 56–68. https://doi.org/10.1038/nrm1549.

Molteni, A., Modrowski, D., Hott, M., & Marie, P. J. (1999). Differential expression of fibroblast growth factor receptor-1, -2, and -3 and syndecan-1, -2, and -4 in neonatal rat mandibular condyle and calvaria during osteogenic differentiation in vitro. *Bone*, *24*(4), 337–347.

Motomi, E. I., Masahiro, I., Kazuhisa, N., Yoshiki, M., David, B., Maurizio, P., et al. (1997). Involvement of α5β1 integrin in matrix interactions and proliferation of chondrocytes. *Journal of Bone and Mineral Research*, *12*(7), 1124–1132. https://doi.org/10.1359/jbmr.1997.12.7.1124.

Nakamura, K., Shirai, T., Morishita, S., Uchida, S., Saeki-Miura, K., & Makishima, F. (1999). p38 mitogen-activated protein kinase functionally contributes to chondrogenesis induced by growth/differentiation factor-5 in ATDC5 cells. *Experimental Cell Research*, *250*(2), 351–363. https://doi.org/10.1006/excr.1999.4535.

Oh, C. D., Chang, S. H., Yoon, Y. M., Lee, S. J., Lee, Y. S., Kang, S. S., et al. (2000). Opposing role of mitogen-activated protein kinase subtypes, erk-1/2 and p38, in the regulation of chondrogenesis of mesenchymes. *The Journal of Biological Chemistry*, *275*(8), 5613–5619.

Oh, E. S., & Couchman, J. R. (2004). Syndecans-2 and -4; close cousins, but not identical twins. *Molecules and Cells*, *17*(2), 181–187.

Oh, C. D., Kim, S. J., Ju, J. W., Song, W. K., Kim, J. H., Yoo, Y. J., et al. (2001). Immunosuppressant rapamycin inhibits protein kinase C alpha and p38 mitogen-activated protein kinase leading to the inhibition of chondrogenesis. *European Journal of Pharmacology*, *427*(3), 175–185.

Olsen, B. R., Reginato, A. M., & Wang, W. (2000). Bone development. *Annual Review of Cell and Developmental Biology*, *16*, 191–220.
Osamu, I., Yoshiya, T., Isao, M., Masaharu, T., & Sumiya, E. (1997). Chondrocytes are regulated by cellular adhesion through CD44 and hyaluronic acid pathway. *Journal of Bone and Mineral Research*, *12*(10), 1657–1663. https://doi.org/10.1359/jbmr.1997.12.10.1657.
Pala, D., Kapoor, M., Woods, A., Kennedy, L., Liu, S., Chen, S., et al. (2008). Focal adhesion kinase/Src suppresses early chondrogenesis: Central role of CCN2. *The Journal of Biological Chemistry*, *283*(14), 9239–9247. https://doi.org/10.1074/jbc.M705175200.
Pavasant, P., Shizari, T. M., & Underhill, C. B. (1994). Distribution of hyaluronan in the epiphysial growth plate: Turnover by CD44-expressing osteoprogenitor cells. *Journal of Cell Science*, *107*(Pt. 10), 2669–2677.
Rabie, A. B., Leung, F. Y., Chayanupatkul, A., & Hagg, U. (2002). The correlation between neovascularization and bone formation in the condyle during forward mandibular positioning. *The Angle Orthodontist*, *72*(5), 431–438. https://doi.org/10.1043/0003-3219(2002)072<0431:tcbnab>2.0.co;2.
Raducanu, A., Hunziker, E. B., Drosse, I., & Aszodi, A. (2009). β1 integrin deficiency results in multiple abnormalities of the knee joint. *Journal of Biological Chemistry*, *284*(35), 23780–23792.
Rapraeger, A. C. (1995). In the clutches of proteoglycans: How does heparan sulfate regulate FGF binding? *Chemistry & Biology*, *2*(10), 645–649.
Riento, K., & Ridley, A. J. (2003). ROCKs: Multifunctional kinases in cell behaviour. *Nature Reviews. Molecular Cell Biology*, *4*, 446. https://doi.org/10.1038/nrm1128.
Robertson, S. P., Twigg, S. R., Sutherland-Smith, A. J., Biancalana, V., Gorlin, R. J., Horn, D., et al. (2003). Localized mutations in the gene encoding the cytoskeletal protein filamin A cause diverse malformations in humans. *Nature Genetics*, *33*, 487–491. https://doi.org/10.1038/ng1119.
Sakai, K., Hiripi, L., Glumoff, V., Brandau, O., Eerola, R., Vuorio, E., et al. (2001). Stage- and tissue-specific expression of a Col2a1-Cre fusion gene in transgenic mice. *Matrix Biology*, *19*(8), 761–767.
Sebastian, A., Matsushita, T., Kawanami, A., Mackem, S., Landreth, G., & Murakami, S. (2011). Genetic inactivation of ERK1 and ERK2 in chondrocytes promotes bone growth and enlarges the spinal canal. *Journal of Orthopaedic Research: Official Publication of the Orthopaedic Research Society*, *29*(3), 375–379. https://doi.org/10.1002/jor.21262.
Shimazu, A., Nah, H.-D., Kirsch, T., Koyama, E., Leatherman, J. L., Golden, E. B., et al. (1996). Syndecan-3 and the control of chondrocyte proliferation during endochondral ossification. *Experimental Cell Research*, *229*(1), 126–136. https://doi.org/10.1006/excr.1996.0350.
Shimo, T., Gentili, C., Iwamoto, M., Wu, C., Koyama, E., & Pacifici, M. (2004). Indian hedgehog and syndecans-3 coregulate chondrocyte proliferation and function during chick limb skeletogenesis. *Developmental Dynamics*, *229*(3), 607–617. https://doi.org/10.1002/dvdy.20009.
Song, E. K., & Park, T. J. (2014). Integrin signaling in cartilage development. *Animal Cells and Systems*, *18*(6), 365–371. https://doi.org/10.1080/19768354.2014.987319.
Stanton, L. A., Sabari, S., Sampaio, A. V., Underhill, T. M., & Beier, F. (2004). p38 MAP kinase signalling is required for hypertrophic chondrocyte differentiation. *The Biochemical Journal*, *378*(Pt. 1), 53–62. https://doi.org/10.1042/bj20030874.
Stringa, E., & Tuan, R. S. (1996). Chondrogenic cell subpopulation of chick embryonic calvarium: Isolation by peanut agglutinin affinity chromatography and in vitro characterization. *Anatomy and Embryology (Berlin)*, *194*(5), 427–437.
Sun, M. M., & Beier, F. (2014). Chondrocyte hypertrophy in skeletal development, growth, and disease. *Birth Defects Research. Part C, Embryo Today: Reviews*, *102*(1), 74–82. https://doi.org/10.1002/bdrc.21062.

Suzuki, D., Bush, J. R., Bryce, D. M., Kamijo, R., & Beier, F. (2017). Rac1 dosage is crucial for normal endochondral bone growth. *Endocrinology*, *158*(10), 3386–3398. https://doi.org/10.1210/en.2016-1691.

Suzuki, W., Yamada, A., Aizawa, R., Suzuki, D., Kassai, H., Harada, T., et al. (2015). Cdc42 is critical for cartilage development during endochondral ossification. *Endocrinology*, *156*(1), 314–322. https://doi.org/10.1210/en.2014-1032.

Suzuki, D., Yamada, A., Amano, T., Yasuhara, R., Kimura, A., Sakahara, M., et al. (2009). Essential mesenchymal role of small GTPase Rac1 in interdigital programmed cell death during limb development. *Developmental Biology*, *335*(2), 396–406. https://doi.org/10.1016/j.ydbio.2009.09.014.

Terpstra, L., Prud'homme, J., Arabian, A., Takeda, S., Karsenty, G., Dedhar, S., et al. (2003). Reduced chondrocyte proliferation and chondrodysplasia in mice lacking the integrin-linked kinase in chondrocytes. *The Journal of Cell Biology*, *162*(1), 139–148. https://doi.org/10.1083/jcb.200302066.

Thorogood, P. V., & Hinchliffe, J. R. (1975). An analysis of the condensation process during chondrogenesis in the embryonic chick hind limb. *Journal of Embryology and Experimental Morphology*, *33*(3), 581–606.

Tuli, R., Tuli, S., Nandi, S., Huang, X., Manner, P. A., Hozack, W. J., et al. (2003). Transforming growth factor-beta-mediated chondrogenesis of human mesenchymal progenitor cells involves N-cadherin and mitogen-activated protein kinase and Wnt signaling cross-talk. *The Journal of Biological Chemistry*, *278*(42), 41227–41236. https://doi.org/10.1074/jbc.M305312200.

Vogel, W., Gish, G. D., Alves, F., & Pawson, T. (1997). The discoidin domain receptor tyrosine kinases are activated by collagen. *Molecular Cell*, *1*(1), 13–23. https://doi.org/10.1016/S1097-2765(00)80003-9.

Wang, G., & Beier, F. (2005). Rac1/Cdc42 and RhoA GTPases antagonistically regulate chondrocyte proliferation, hypertrophy, and apoptosis. *Journal of Bone and Mineral Research*, *20*(6), 1022–1031. https://doi.org/10.1359/JBMR.050113.

Wang, G., Woods, A., Agoston, H., Ulici, V., Glogauer, M., & Beier, F. (2007). Genetic ablation of Rac1 in cartilage results in chondrodysplasia. *Developmental Biology*, *306*(2), 612–623. https://doi.org/10.1016/j.ydbio.2007.03.520.

Wang, G., Woods, A., Sabari, S., Pagnotta, L., Stanton, L. A., & Beier, F. (2004). RhoA/ROCK signaling suppresses hypertrophic chondrocyte differentiation. *The Journal of Biological Chemistry*, *279*(13), 13205–13214. https://doi.org/10.1074/jbc.M311427200.

Wang, G., Yan, Q., Woods, A., Aubrey, L. A., Feng, Q., & Beier, F. (2011). Inducible nitric oxide synthase–nitric oxide signaling mediates the mitogenic activity of Rac1 during endochondral bone growth. *Journal of Cell Science*, *124*(20), 3405–3413. https://doi.org/10.1242/jcs.076026.

Warren, K., Brian, C., Yoshihiro, N., Wolfgang, E., Kuettner, K. E., & Knudson, C. B. (2000). Hyaluronan oligosaccharides perturb cartilage matrix homeostasis and induce chondrocytic chondrolysis. *Arthritis and Rheumatism*, *43*(5), 1165–1174. https://doi.org/10.1002/1529-0131(200005)43:5<1165::AID-ANR27>3.0.CO;2-H.

Wegener, K. L., & Campbell, I. D. (2008). Transmembrane and cytoplasmic domains in integrin activation and protein-protein interactions. *Molecular Membrane Biology*, *25*(5), 376–387.

Werb, Z. (1997). ECM and cell surface proteolysis: Regulating cellular ecology. *Cell*, *91*(4), 439–442. https://doi.org/10.1016/S0092-8674(00)80429-8.

Weston, A. D., Chandraratna, R. A. S., Torchia, J., & Underhill, T. M. (2002). Requirement for RAR-mediated gene repression in skeletal progenitor differentiation. *The Journal of Cell Biology*, *158*(1), 39–51. https://doi.org/10.1083/jcb.200112029.

Woods, A., & Beier, F. (2006). RhoA/ROCK signaling regulates chondrogenesis in a context-dependent manner. *The Journal of Biological Chemistry, 281*(19), 13134–13140. https://doi.org/10.1074/jbc.M509433200.

Woods, A., James, C. G., Wang, G., Dupuis, H., & Beier, F. (2009). Control of chondrocyte gene expression by actin dynamics: A novel role of cholesterol/Ror-α signalling in endochondral bone growth. *Journal of Cellular and Molecular Medicine, 13*(9b), 3497–3516. https://doi.org/10.1111/j.1582-4934.2009.00684.x.

Woods, A., Wang, G., & Beier, F. (2005). RhoA/ROCK signaling regulates Sox9 expression and actin organization during chondrogenesis. *The Journal of Biological Chemistry, 280*(12), 11626–11634. https://doi.org/10.1074/jbc.M409158200.

Woods, A., Wang, G., & Beier, F. (2007). Regulation of chondrocyte differentiation by the actin cytoskeleton and adhesive interactions. *Journal of Cellular Physiology, 213*(1), 1–8. https://doi.org/10.1002/jcp.21110.

Yoon, Y. M., Oh, C. D., Kim, D. Y., Lee, Y. S., Park, J. W., Huh, T. L., et al. (2000). Epidermal growth factor negatively regulates chondrogenesis of mesenchymal cells by modulating the protein kinase C-alpha, Erk-1, and p38 MAPK signaling pathways. *The Journal of Biological Chemistry, 275*(16), 12353–12359.

Yosimichi, G., Nakanishi, T., Nishida, T., Hattori, T., Takano-Yamamoto, T., & Takigawa, M. (2001). CTGF/Hcs24 induces chondrocyte differentiation through a p38 mitogen-activated protein kinase (p38MAPK), and proliferation through a p44/42 MAPK/extracellular-signal regulated kinase (ERK). *European Journal of Biochemistry, 268*(23), 6058–6065.

Zanetti, N. C., & Solursh, M. (1984). Induction of chondrogenesis in limb mesenchymal cultures by disruption of the actin cytoskeleton. *The Journal of Cell Biology, 99*(1 Pt. 1), 115–123.

Zhang, R., Murakami, S., Coustry, F., Wang, Y., & de Crombrugghe, B. (2006). Constitutive activation of MKK6 in chondrocytes of transgenic mice inhibits proliferation and delays endochondral bone formation. *Proceedings of the National Academy of Sciences of the United States of America, 103*(2), 365–370. https://doi.org/10.1073/pnas.0507979103.

Zuzarte-Luis, V., Montero, J. A., Rodriguez-Leon, J., Merino, R., Rodriguez-Rey, J. C., & Hurle, J. M. (2004). A new role for BMP5 during limb development acting through the synergic activation of Smad and MAPK pathways. *Developmental Biology, 272*(1), 39–52. https://doi.org/10.1016/j.ydbio.2004.04.015.

CHAPTER THREE

Development of the axial skeleton and intervertebral disc

Sade Williams, Bashar Alkhatib, Rosa Serra*
Department of Cell Developmental and Integrative Biology, University of Alabama at Birmingham, Birmingham, AL, United States
*Corresponding author: e-mail address: rserra@uab.edu

Contents

Abstract

Development of the axial skeleton is a complex, stepwise process that relies on intricate signaling and coordinated cellular differentiation. Disruptions to this process can result in a myriad of skeletal malformations that range in severity. The notochord and the sclerotome are embryonic tissues that give rise to the major components of the intervertebral discs and the vertebral bodies of the spinal column. Through a number of mouse models and characterization of congenital abnormalities in human patients, various growth factors, transcription factors, and other signaling proteins have been demonstrated to have critical roles in the development of the axial skeleton. Balance between opposing growth factors as well as other environmental cues allows for cell fate specification and divergence of tissue types during development. Furthermore, characterization of progenitor cells for specific cell lineages has furthered the understanding of specific spatiotemporal cues that cells need in order to initiate and complete development of distinct tissues. Identifying specific marker genes that can distinguish between the various embryonic and mature cell types is also of importance. Clinically, understanding developmental clues can aid in the generation of therapeutics

Current Topics in Developmental Biology, Volume 133
ISSN 0070-2153
https://doi.org/10.1016/bs.ctdb.2018.11.018

for musculoskeletal disease through the process of developmental engineering. Studies into potential stem cell therapies are based on knowledge of the normal processes that occur in the embryo, which can then be applied to stepwise tissue engineering strategies.

Abbreviations

AF annulus fibrosus
CEP cartilage end plate
ECM extracellular matrix
ESC embryonic stem cell
hiPSCs human induced pluripotent stem cells
iAF inner annulus fibrosus
IM intermediate mesoderm
iPSCs induced pluripotent stem cells
IVD intervertebral disc
LPM lateral plate mesoderm
MET mesenchymal to epithelial transition
MRI magnetic resonance imaging
NCCN notochordal cell containing NP tissue
NP nucleus pulposus
oAF outer annulus fibrosus
P postnatal day
PM paraxial mesoderm
PSM presomitic mesoderm
SLRPs small leucine rich proteoglycans
VB vertebral body

1. Introduction

The vertebrate axial skeleton was an evolutionary development that provided support for the body and protection of the spinal cord (Alkhatib, Ban, Williams, & Serra, 2018; Cox & Serra, 2014). The spine consists of two major components, the bony vertebrae derived from cartilage models through endochondral bone formation and fibrous connective tissues including the intervertebral discs (IVDs), ligaments and tendons. The IVD is the shock absorber of the spine and consists of two compartments, the nucleus pulposus (NP) and the annulus fibrosus (AF). The spine has a segmented structure, with the IVDs positioned between neighboring vertebral bodies (VBs). This organization is key for proper function, since the individual parts of the spinal column have discrete roles. The rigid, bony vertebrae provide support and protection. The IVDs distribute weight and other mechanical loads. The NP of the IVD consists of a gelatinous core that gives the IVD its weight and load distribution properties. The AF surrounds the NP and provides structural support and integrity (Adams & Roughley, 2006).

The NP and the AF cohesively form the mature IVD but are derived from very different embryonic structures (Paavola, Wilson, & Center, 1980; Rufai, Benjamin, & Ralphs, 1995; Theiler, 1988). The NP is derived from the notochord, while the AF is derived from the sclerotome of the somites (Christ, Huang, & Scaal, 2004, 2007; Christ, Huang, & Wilting, 2000). Somites are transient structures that determine the segmented nature of the vertebral column and further differentiate into the sclerotome, which eventually forms most of the connective tissues of the axial skeleton, including the vertebral body, AF, ligaments and tendons (Fig. 1). The signaling mechanisms involved in the development of these embryonic tissues into the terminal structures of the spine are critical for its proper development. Since the spine is derived from the notochord, somites, and sclerotome,

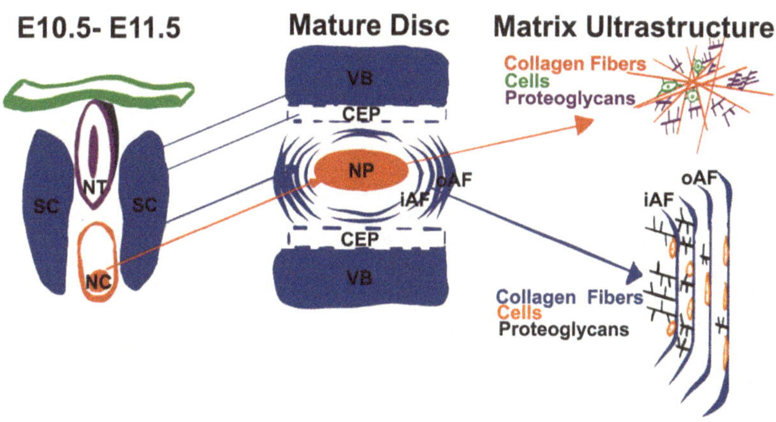

Fig. 1 Illustration of a brief summary of IVD development and matrix organization. At E10.5–11.5, the neural tube (NT) and the notochord (NC) are located in between the sclerotome (SC). The NC is positioned ventral to the NT. The nucleus pulposus (NP) is derived from the NC (*red arrow* between E10.5–11.5 and mature disc) and the inner and outer annulus fibrosus (iAF and oAF), the cartilaginous endplate (CEP), and the vertebral body (VB) are derived from the SC (*blue arrows* between E10.5–11.5 and mature disc). The matrix ultrastructure of the NP is composed of randomly aligned collagen fibers with cells entrapped. A high abundance of proteoglycans helps keep the collagen ultrastructure together and helps NP cells communicate with the extracellular matrix. The AF is composed of organized, aligned collagen fibers. Cells are aligned parallel to the collagen fibers and are more round in the iAF and more elongated in the oAF. Proteoglycans are more abundant in the iAF than the oAF. NT, neural tube; NC, notochord; SC, sclerotome; VB, vertebral body; CEP, cartilaginous endplate; NP, nucleus pulposus; iAF, inner annulus fibrosus; oAF, outer annulus fibrosus.

alterations in the signaling pathways important for the formation of any of these tissues can result in severe developmental disorders (Cox & Serra, 2014). Therefore, insight into the mechanisms of how the axial skeleton develops and the important factors involved in its development can help to address pathology of the spine. An understanding of the basic mechanisms of how the axial skeleton develops can provide a basis for treatment, repair, or regeneration strategies for injury or damage to the spine (Alkhatib, Ban, et al., 2018; Cox & Serra, 2014; Gadjanski, Spiller, & Vunjak-Novakovic, 2012; Lenas, Luyten, Doblare, Nicodemou-Lena, & Lanzara, 2011; Lenas, Moos, & Luyten, 2009a, 2009b). This concept of "developmental engineering" could be used in the future for cell and/or tissue replacement therapy. This chapter will cover the signaling mechanisms governing the major steps of development of the axial skeleton, human diseases associated with dysregulation of these signals and the current advancements in tissue engineering strategies.

2. Development of somite derived structures

2.1 Somitogenesis

Somites are transient, spherical-shaped structures that pattern and segment the embryo along the anterior to posterior axis (Christ et al., 2004, 2000; Stockdale, Nikovits, & Christ, 2000). The process of somitogenesis begins with a dynamic reorganization of the embryo due to the formation of the primitive streak, which marks the beginning of gastrulation at E6.0 in mice (Arendt & Nubler-Jung, 1999; Mikawa, Poh, Kelly, Ishii, & Reese, 2004). Gastrulation establishes the three germ layers of the embryo; the ectoderm, which forms terminal structures such as epidermis and neurons; the endoderm, which forms the digestive and respiratory tract; and the mesoderm, which forms structures in much of the connective tissues in the body (Mikawa et al., 2004; Tam & Behringer, 1997). Most of the skeleton is formed from mesoderm although many components of the head and face are derived from neural crest. Trunk mesoderm is further divided into three compartments, lateral plate mesoderm (LPM), intermediate mesoderm (IM), and paraxial mesoderm (PM). Somites are derived specifically from the PM, also known as presomitic mesoderm, and eventually differentiate into muscle, dermis, and the fibrous and cartilaginous tissues of the spine (Brand-Saberi & Christ, 2000; Pourquie, 2011). Follistatin, an Activin antagonist, has been shown to be required for the specification of the PM.

Follistatin works in conjunction with Noggin to inhibit BMP signaling from the LPM and thus permit PM formation (Stafford, Monica, & Harland, 2014).

Much research has been devoted to stimulating the *in vitro* differentiation of stem cells into PM using small molecule inhibitors that target growth factors previously shown to be involved PM differentiation. A recent study derived PM-like cells from pluripotent embryonic stem cells (ESCs) by utilizing such small molecule inhibitors. Formation of the PM is dependent upon Wnt3a and Noggin signaling (Aulehla & Pourquie, 2010; Yamaguchi, Takada, Yoshikawa, Wu, & McMahon, 1999). By using a GSK3 inhibitor to mimic Wnt3a signaling and an inhibitor of BMP type 1 receptors to replace Noggin, the ESCs began to express PM markers, Tcf15 and Meox1 (Zhao et al., 2014). Another study stimulated the differentiation of 12 human mesodermal cell lineages from induced pluripotent stem cells using extrinsic factors previously shown to be critical during mesoderm formation and differentiation (Loh et al., 2016). These studies highlight the advancements in regenerative science by using "developmental engineering" strategies to potentially repair damaged connective tissues in the spine (Gadjanski et al., 2012; Lenas et al., 2011, 2009a, 2009b). Engineered PM is important because it can be used as a starting point to engineer all of the musculoskeletal derivatives of the somite. For example, a recent study showed stimulation of chondrocyte differentiation from mouse ESCs by first generating Flk-1-/Pdgfrα-positive PM cells with Activin, Wnt, and VEGF and subsequently treating those cells with BMP4 or GDF5 to stimulate chondrogenesis (Craft et al., 2013). Table 1 contains a list of known factors essential for somitogenesis. These factors can be potential targets for future studies of developmental engineering strategies to generate PM.

The mesenchymal, rod-like structure of the PM begins to undergo somitogenesis in an anterior to posterior fashion along the embryo axis at E8.0 in mice. The spherical somites are pinched off from the PM in a fashion that is highly regulated temporally and spatially (Brand-Saberi & Christ, 2000; Pourquie, 2011). The current model for somitogenesis is the Clock and Wavefront Model where Notch and Wnt signaling act as the "clock" that provides the temporal, permissive signal, while FGF and Retinoic Acid signaling acts as the "wave" that provides spatial specificity. The wave is responsible for the anterior to posterior formation of somites due to FGF and Retinoic Acid. These signals provide the directionality and spatial regulation needed to stimulate the mesenchymal to epithelial transition (MET) of the mesenchymal cells of the PM to the epithelial outer layer of the newly formed somite (Bellairs, Curtis, & Sanders, 1978; Brand-Saberi & Christ, 2000;

Table 1 Proteins involved in important signaling pathways during somitogenesis.

Protein name[a]	Abbreviation[a]	Protein function[a]	Skeletal disorders[b,c]
Activin	ACTIVIN	TGFβ superfamily ligand	Osteogenesis imperfecta type 1: OMIM #166200
Ephrin	EPHRIN	Cell surface transmembrane ligand	n/a
EPH Receptor A4	EPHA4	Receptor tyrosine kinase	n/a
Fibroblast Growth Factor 8	FGF8	Ligand for FGF receptors	n/a
Fibronectin	FN1	Glycoprotein	Spondylometaphyseal Dysplasia, Corner Fracture Type: OMIM #184255
Follistatin	FST	Activin antagonist	Spinal muscular atrophy, type iv: OMIM #271150
Hes Family BHLH Transcription Factor 1	HES1	Basic helix loop helix transcription factor	n/a
Mesenchyme Homeobox 1	MEOX1	Homeobox transcription factor	Klippel-Feil syndrome, OMIM #613702; Diaphanospondylodysostosis, OMIM #608022
Neural Cadherin	NCAD	Calcium-dependent cell adhesion protein	n/a
Neural Cell Adhesion Molecule	NCAM	Cell adhesion	n/a
Noggin	NOG	BMP inhibitor	Multiple synostoses syndrome 1, OMIM #186500
Notch	NOTCH	Type 1 transmembrane protein	Adams-Oliver syndrome 5, OMIM #616028

Table 1 Proteins involved in important signaling pathways during somitogenesis.—cont'd

Protein name[a]	Abbreviation[a]	Protein function[a]	Skeletal disorders[b,c]
Paired Box 3	PAX3	PAX family transcription factor	Craniofacial-deafness-hand syndrome, OMIM #122880
Platelet Derived Growth Factor Receptor Alpha	PDGFRα	Cell surface tyrosine kinase receptor	n/a
Retinoic Acid	RA	Ligand for nuclear RA receptors	n/a
Transcription Factor 15 (Paraxis)	TCF15	Basic helix loop helix transcription factor	n/a
Vascular Endothelial Growth Factor	VEGF	Growth factor	n/a
Vascular Endothelial Growth Factor Receptor 2	VEGFR2 (FLK-1)	Cell surface tyrosine kinase receptor	n/a
Wnt Family Member 3A	WNT3a	Secreted ligand	Craniodiaphyseal Dysplasia, OMIM #122860

[a]Genecards.org.
[b]Omim.org.
[c]Malacards.org.

Cooke & Zeeman, 1976; Pourquie, 2011). The MET causes somites to consist of two cell types, an outer layer composed of epithelial cells and an inner cell mass composed of mesenchyme called the somitocoel (Ferrer-Vaquer, Viotti, & Hadjantonakis, 2010; Mittapalli, Huang, Patel, Christ, & Scaal, 2005). Various studies have identified key factors and signaling pathways salient for proper epithelization during somitogenesis. Adhesion molecules such as N-cadherin, fibronectin, cytoactin, and neural cell adhesion molecules are expressed in the developing somite and are important for epithelization (Crossin, Hoffman, Grumet, Thiery, & Edelman, 1986; Duband et al., 1987). Two transcription factors involved in MET are Pax3 and Paraxis.

Studies have shown that overexpression of Pax3 is sufficient to stimulate epithelialization of mesenchymal cell lines *in vitro* (Wiggan, Fadel, & Hamel, 2002), and Pax3 expression in the PSM is also required to maintain the epithelial integrity later in somites (Mansouri, Pla, Larue, & Gruss, 2001). Similarly, loss of Paraxis expression disrupts epithelialization (Burgess, Rawls, Brown, Bradley, & Olson, 1996). The tyrosine kinase EphA4 is also required for proper epithelialization of the somite. Attenuation of EphA4/Ephrin signaling results in somite boundaries, but no epithelial layer formation (Barrios et al., 2003). The mechanism of how the Clock and Wavefront Model fully translates into MET has yet to be determined, but Notch regulates the expression of transcription factor Hes1, which regulates Ephrin expression (Glazier, Zhang, Swat, Zaitlen, & Schnell, 2008).

2.2 Sclerotome specification

Shortly after MET, the somite begins to differentiate into its respective tissues: the dermatome, the myotome, and the sclerotome. The dermatome forms the dermis of the back, the myotome forms all the skeletal muscle of the body, and the sclerotome forms the connective tissues of the axial skeleton: vertebrae (VB), cartilaginous end plates, annulus fibrosus (AF), tendon and ligament (Brand-Saberi & Christ, 2000; Kalcheim & Ben-Yair, 2005). The sclerotome is a transient, embryonic tissue composed of pluripotent, mesenchymal stem cells located in the ventromedial region of the somite. The localization and specification of the sclerotome is a tightly controlled and highly dynamic process induced by Shh signaling from the floor plate of the neural tube and notochord, which induces expression of early sclerotome markers Pax1, Pax 9, and Mfh1 (Borycki, Mendham, & Emerson, 1998; Brand-Saberi & Christ, 2000; Chiang et al., 1996; Dockter, 2000; Fan & Tessier-Lavigne, 1994). The embryonic knock out of both Pax1 and Pax9 causes the complete loss of the VB and AF (Peters et al., 1999). This defect can be caused by two alternative possibilities. Pax 1/9 loss can result in a failure of the initial formation of the sclerotome. Alternatively, sclerotome formation may occur, but the VB and AF do not form due to defects in proliferation and subsequent differentiation. Nkx3.2, also known as Bapx1, is another early marker of sclerotome formation. Its expression is regulated by Shh and it initially has the same temporal and spatial expression pattern as Pax1/9; however, Nkx3.2 remains expressed throughout the sclerotome while Pax expression later becomes restricted to more fibrous tissues (Murtaugh, Zeng, Chyung, & Lassar, 2001). BMP from the LPM has the capacity to disrupt sclerotomal specification by interfering with Shh signaling. However, several BMP antagonists are expressed regionally that carefully restrict BMP activity, such as Noggin and Gremlin1 (Rider & Mulloy, 2010; Stafford, Brunet, Khokha, Economides, & Harland, 2011). Noggin and Gremlin1 cooperate to antagonize BMP signaling, allowing sclerotome differentiation in the presence of Shh (Stafford et al., 2011). Noggin and Gremlin1 are required for sclerotome differentiation in mice, since their deletion results in a lack of sclerotome cells, while the dermomyotome remains completely unaffected. Inhibition of BMP alone is not sufficient to specify sclerotome or expand sclerotome differentiation *in vivo* or *in vitro* suggesting that antagonism of BMP is a permissive factor for sclerotome differentiation (Rider & Mulloy, 2010). Sclerotome formation is coupled with an epithelial to mesenchymal transition to permit migration and further differentiation of the mesenchymal cells (Christ & Ordahl, 1995; Dockter & Ordahl, 2000). More information about the important factors that contribute to sclerotome specification can be found in Table 2.

2.3 Resegmentation

After the sclerotome has been specified, the process of resegmentation begins. Resegmentation is the formation of rostral and caudal domains within the sclerotome that separate and recombine with the adjacent domain in the neighboring sclerotomal segment. Reorganization of the rostral and caudal domains creates a spatial organization where the myotome is staggered one half segment to the developing VB to allow the nerves projecting from the spinal cord to innervate the developing muscles (Bagnall, Higgins, & Sanders, 1988; Goldstein & Kalcheim, 1992; Huang, Zhi, Brand-Saberi, & Christ, 2000; Tanaka & Uhthoff, 1981). Resegmentation is necessary for the proper formation of the Intervertebral Disc (IVD) and VB. Therefore, deviations to this process have clinical implications. Rare diseases that occur in humans called Proatlas Segmentation Anomalies OMIM: 109500 are thought to be caused by failure in the proatlas sclerotome, which will partially go on to form the C1 VB, to undergo resegmentation. These disorders result in dysplasia within the craniovertebral junction and can manifest in various ways clinically, such as cervical VB fusion and ventral brain stem compression (Muthukumar, 2016; Spittank, Goehmann, Hage, & Sacher, 2016; Umegaki et al., 2017). Klippel-Feil Syndrome (OMIM: 613702), where patients suffer from cervical vertebral fusion due to lack of IVD formation, has also been associated with a disruption in resegmentation and mutations in the *GDF6*

Table 2 Proteins involved in important signaling pathways during sclerotome specification.

Protein name[a]	Abbreviation[a]	Protein function[a]	Skeletal disorders[b,c]
Forkhead Box C2	FOXC2 (MFH1)	Forkhead family transcription factor	n/a
Gremlin	GREM1	BMP antagonist	Sclerosteosis, OMIM #269500
Noggin	NOG	BMP inhibitor	Multiple synostoses syndrome 1, OMIM #186500
NK3 Homeobox2	NXK3.2 (BAXP1)	Transcriptional repressor	Spondylo-megaepiphyseal-metaphyseal dysplasia, OMIM #613330
Paired Box 1	PAX1	PAX family transcription factor	Otofaciocervical syndrome 2, OMIM #615560; Klippel-Feil syndrome, OMIM #613702; Diaphanospondylodysostosis, OMIM #608022
Paired Box 9	PAX9	PAX family transcription factor	Tooth agenesis, OMIM #604625
Sonic Hedgehog	SHH	Secreted ligand	Laurin-Sandrow syndrome, OMIM #135750; Hypoplasia or aplasia of tibia with polydactyly, OMIM #188740; Solitary median maxillary central incisor, OMIM #147250

[a]Genecards.org.
[b]Omim.org.
[c]Malacards.org.

gene (Tassabehji et al., 2008). For more information concerning skeletal disorders and the associated proteins that cause these disorders see Table 3.

Rostral-caudal polarity is established in the sclerotome by the expression of rostral markers such as Mesp2 and Tbx18 and caudal markers such as Ripply 1 and 2, Uncx4.1, and Pax1/9 (Kawamura, Koshida, & Takada, 2008; Leitges, Neidhardt, Haenig, Herrmann, & Kispert, 2000; Morimoto et al., 2007; Neubuser, Koseki, & Balling, 1995). The sclerotome migrates ventrally to surround the notochord, splits into two separate halves along the

Table 3 Proteins involved in important signaling pathways during resegmentation.

Protein name[a]	Abbreviation[a]	Protein function[a]	Skeletal disorders[b,c]
Growth Differentiation Factor 6	GDF6	TGFβ superfamily ligand	Klippel-Feil syndrome, OMIM #613702; Multiple Synostoses syndrome 4, OMIM #617898
Mesoderm Posterior BHLH Transcription Factor 2	MESP2	Basic helix loop helix transcription factor	Spondylocostal dysostosis 2, OMIM #608681
Paired Box 1	PAX1	PAX family transcription factor	Otofaciocervical syndrome 2, OMIM #615560; Klippel-Feil syndrome, OMIM #613702; Diaphanospondylodysostosis, OMIM #608022
Paired Box 9	PAX9	PAX family transcription factor	Tooth agenesis, OMIM #604625
Ripply Transcriptional Repressor	RIPPLY1/2	Transcriptional repressor	Klippel-Feil syndrome, OMIM #613702; Spondylocostal dysostosis 6, autosomal recessive, OMIM #616566
T-box Transcription Factor TBX18	TBX18	Transcriptional repressor	n/a
UNC Homeobox	UNCX4.1	Homeobox transcription factor	n/a

[a]Genecards.org.
[b]Omim.org.
[c]Malacards.org.

rostral-caudal border called von Ebner's fissure and combines with half of the adjacent sclerotome (Bagnall et al., 1988; Christ et al., 2000). As a result, in mice, the IVD is derived from the cells in the caudal half of the sclerotome near von Ebner's fissure. The recombined sclerotomal halves form the VB, while the AF of the IVD differentiates from cells in the caudal half in mice and the rostral half in chicks near the fissure (Bruggeman et al., 2012). Some of the resegmented cells will then migrate dorsally to surround the neural tube and form the neural arches (Monsoro-Burq, 2005). A new model coined the "resegmentation-shift" model suggests that after the sclerotome resegments the cells that migrate dorsally to surround the neural tube do so in a non-linear fashion. The model suggests that as sclerotomal cells migrate dorsally to form the neural arches, there is a cranial shift in their migration pattern. This cranial shift was visualized *in vivo* by using lipophilic dyes to label the migrating sclerotome in the chick model and was only observed in the lumbo-sacral region of the spine. It is believed this cranial shift is necessary to establish the slant of the neural arch pedicle specifically in the lumbo-sacral region (Ward, Evans, & Stern, 2017).

2.4 Sclerotome derivatives

2.4.1 Vertebra

Sclerotome cells organize to form many structures within the spine, including hyaline cartilage that will form the VB and Cartilaginous End Plate (CEP) *via* chondrogenesis (Fig. 1). A list of signaling molecules involved in specification of the vertebrae is provided in Table 4. Defects in chondrogenesis result in improper formation of the VB and lead to severe spinal disorders such as kyphosis and scoliosis (Sivakamasundari et al., 2017). Vertebral chondrocyte cell fate is specified early when the sclerotome first forms (Murtaugh, Chyung, & Lassar, 1999). Shh secreted from the notochord, primes the cells to respond to the chondrogenic actions of BMP signaling. Pax1 and Pax9, which are regulated by Shh, are early sclerotome markers but they also regulate condensation of the sclerotome mesenchyme and the initiation of chondrogenesis (Sivakamasundari et al., 2017). Pax1/9 regulates the transcription factor Sox 5, which in conjunction with Sox 6 and Sox 9, is necessary for the activation of early chondrocyte differentiation markers, Col2a1 and Aggrecan (Han & Lefebvre, 2008; Lefebvre, Behringer, & de Crombrugghe, 2001). Pax 1/9 and Shh also regulate Nkx3.2, another early marker of sclerotome formation and a required factor in stimulating chondrogenic differentiation. Nkx3.2 is one of the earliest markers of prechondrogenic cells in the axial skeleton and induces the expression of Sox9, a master regulator of chondrogenesis (Tribioli & Lufkin, 1999; Zeng, Kempf, Murtaugh, Sato, & Lassar, 2002). Nkx3.2 acts as a transcriptional repressor by inhibiting repression of Sox9 and thus allowing the sclerotome to be competent to undergo chondrogenesis by responding to BMP signals, while simultaneously inhibiting anti-prechondrogenic factors in the prospective VB. The embryonic deletion

Table 4 Proteins involved in important signaling pathways to form sclerotomal derivatives.

Protein name[a]	Abbreviation[a]	Protein Function[a]	Skeletal disorders[b,c]
Vertebra and Cartilaginous End Plate			
Aggrecan	ACAN	Proteoglycan	Spondyloepimetaphyseal dysplasia, aggrecan type, OMIM #612813; Osteochondritis Dissecans, OMIM# 165800
Bone Morphogenetic Protein	BMP	TGFβ superfamily ligand	Brachydactyly, Type A2, OMIM#112600, Osteonecrosis, OMIM #608805, Short Stature, Facial Dysmorphism, and Skeletal Anomalies with or Without Cardiac Anomalies, OMIM #617877
Bone Morphogenetic Protein Receptor Type 1A	BMPR1a	Serine/Threonine kinase transmembrane receptor	n/a
Bone Morphogenetic Protein Receptor Type 1B	BMPRlb	Serine/Threonine kinase transmembrane receptor	Brachydactyly, type a1, d, OMIM # 616849, Acromesomelic dysplasia, Demirhan type, OMIM #609441
Chordin	CHRD	BMP antagonist	
Collagen Type II Alpha 1 Chain	COL2al	Extracellular matrix protein	Kniest Dysplasia, OMIM #156550; Spondyloepiphyseal dysplasia congenital type, OMIM #83900; Achondrogenesis 2, OMIM #200610; Legg-Calve-Perthes disease, OMIM #150600
Crossveinless	CV-2	BMP inhibitor	Diaphanospondylodysostosis, OMIM #608022; Ischio-Vertebral syndrome, OMIM#n/a
Dachsous Cadherin-Related 1	DCHS1	Calcium-dependent cadherin	Van Maldergem syndrome 1, OMIM #601390; Hennekam syndrome, OMIM #235510
FAT Atypical Cadherin 4	FAT4	Protocadherin	Van maldergem syndrome 2, OMIM #615546; Hennekam syndrome, OMIM #235510

Continued

Table 4 Proteins involved in important signaling pathways to form sclerotomal derivatives.—cont'd

Protein name[a]	**Abbreviation**[a]	**Protein Function**[a]	**Skeletal disorders**[b,c]
Growth and Differentiation Factor 5	GDF5	TGFβ superfamily ligand	Acromesomelic chondrodysplasia, Grebe type, OMIM #200700; Brachydactyly C, OMIM #113100; Du Pan syndrome, OMIM #228900; Multiple synostoses syndrome 2, OMIM #610017; Symphalangism, proximal IB, OMIM #615298
NK3 Homeobox 2	NKX3.2 (BAXP1)	Transcriptional repressor	Spondylo-megaepiphyseal-metaphyseal dysplasia, OMIM #613330
Notch	NOTCH	Type 1 transmembrane protein	Adams-Oliver syndrome 5, OMIM #616028
Paired Box 1	PAX1	PAX family transcription factor	Otofaciocervical syndrome 2, OMIM #615560; Klippel-feil syndrome, OMIM #613702; Diaphanospondylodysostosis, OMIM #608022
Paired Box 9	PAX9	PAX family transcription factor	Tooth agenesis, OMIM #604625
Recombination Signal Binding Protein For Immunoglobulin Kappa J	RBPJ	Transcriptional Repressor	Adams-Oliver syndrome, OMIM #100300
SMAD Family Member 1	SMAD1	Transcriptional coactivator	Buschke-Ollendorff syndrome, OMIM #166700; Osteopoikilosis, OMIM #166705
SMAD Family Member 5	SMAD 5	Transcriptional coactivator	n/a
Sonic Hedgehog	SHH	Secreted ligand	Laurin-Sandrow syndrome, OMIM #135750; Hypoplasia or aplasia of
SRY-Box 5	SOX5	SOX family transcription factor	Lamb-Shaffer syndrome, OMIM #616803
SRY-Box 6	SOX6	SOX family transcription factor	Multiple synostoses syndrome 1, OMIM #186500
SRY-Box 9	SOX9	SOX family transcription factor	Campomelic dysplasia, OMIM #114290

Transforming Growth Factor Beta	TGFβ	TGFβ superfamily ligand	Camurati-Engelmann disease, OMIM #131300; Loeys-Dietz syndrome 5, OMIM #615582; Holt-Oram syndrome, OMIM #142900
Transforming Growth Factor Beta Receptor 2	TGFβR2	Serine/Threonine kinase transmembrane receptor	Loeys-Dietz syndrome 2, OMIM #610168
Annulus Fibrosus			
Bone Morphogenetic Protein	BMP	TGFβ superfamily ligand	Brachydactyly, Type A2, OMIM#112600, Osteonecrosis, OMIM #608805, Short Stature, Facial Dysmorphism, and Skeletal Anomalies with or Without Cardiac Anomalies, OMIM #617877
Collagen Type 1 A	COL1a	Extracellular matrix protein	Caffey disease, OMIM #114000; Ehlers-Danlos syndrome, classic type, OMIM #130000; Osteogenesis imperfecta 1, OMIM #166200; Osteoporosis, OMIM #166710
Fibromodulin	FMOD	Proteoglycan	Hypochondrogenesis, OMIM #200610; Pseudoachondroplasia, OMIM #177170
Filamin B	FLNB	Cytoskeleton protein	Atelosteogenesis 1, OMIM #108720; Boomerang dysplasia, OMIM #112310; Larsen syndrome, OMIM #150250; pondylocarpotarsal synostosis syndrome; OMIM #272460
Kelch Like Family Member 14	KIhI14	Long non-coding RNA	n/a
Mohawk	MKX	IRX family-related homeobox protein	Cleft Palate, Isolated, OMIM #119540
Noggin	NOG	BMP inhibitor	Multiple synostoses syndrome 1, OMIM #186500
NK3 Homeobox2	NKX3.2 (BAXP1)	Transcriptional repressor	Spondylo-megaepiphyseal-metaphyseal dysplasia, OMIM #613330

Continued

Table 4 Proteins involved in important signaling pathways to form sclerotomal derivatives.—cont'd

Protein name[a]	Abbreviation[a]	Protein Function[a]	Skeletal disorders[b,c]
Paired Box 1	PAX1	PAX family transcription factor	Otofaciocervical syndrome 2, OMIM #615560; Klippel-Feil syndrome, OMIM #613702;
Paired Box 9	PAX9	PAX family transcription factor	Tooth agenesis, OMIM #604625
Scleraxis	scx	Basic helix loop helix transcription factor	Acrocallosal syndrome, OMIM #200990; Wilson-Turner X-Linked Mental Retardation Syndrome, OMIM #309585
SMAD Family Member 2	SMAD 2	Transcriptional coactivator	Buschke-Ollendorff syndrome, OMIM #166700; Melorheostosis, OMIM #155950
SMAD Family Member 3	SMAD 3	Transcriptional coactivator	Loeys-Dietz syndrome 3, OMIM #613795
Sonic Hedgehog	SHH	Secreted ligand	Laurin-Sandrow syndrome, OMIM #135750; Hypoplasia or aplasia of tibia with polydactyly, OMIM #188740; Solitary median maxillary central incisor, OMIM #147250
SRY-Box 6	SOX6	SOX family transcription factor	Multiple synostoses syndrome 1, OMIM #186500
SRY-Box 9	SOX9	SOX family transcription factor	Campomelic dysplasia, OMIM #114290
Syndecan 4	SDC4	Transmembrane (type 1) heparan sulfate proteoglycan	n/a
Transforming Growth Factor Beta	TGFβ	TGFβ superfamily ligand	Camurati-Engelmann disease, OMIM #131300; Loeys-Dietz syndrome 5, OMIM #615582; Holt-Oram Syndrome, OMIM #142900
Transforming Growth Factor Beta Receptor 2	TGFβR2	Serine/Threonine kinase transmembrane receptor	Loeys-Dietz syndrome 2, OMIM #610168

[a] Genecards.org.
[b] Omim.org.
[c] Malacards.org.

of Nkx3.2 causes perinatal lethality in mice and results in severe spinal dysplasia due to failure of the sclerotome to undergo chondrogenesis (Murtaugh et al., 2001).

BMP signaling is essential for chondrocyte specification and differentiation and is a potent regulator upstream of Sox 5, 6, and 9 transcription factors (Lefebvre et al., 2001; Ohba, He, Hojo, & McMahon, 2015; Yoon et al., 2005). Research has established the requirement of BMP signaling for chondrogenesis by creating an embryonic knockout of BMPR1b and conditional knockout of BMPR1a in Collagen Type 2-expressing cells in mice. This study demonstrated that progenitor cells can condense but do not further differentiate into chondrocytes (Yoon et al., 2005). *In vitro* studies have shown that mesenchymal stem cells exogenously treated with BMP ligands 2, 4, and 6 begin to undergo chondrogenic differentiation and express chondrocyte markers such as Col2a and Aggrecan (Sekiya, Larson, Vuoristo, Reger, & Prockop, 2005). Once chondrogenesis is initiated in the axial skeleton, chondrocyte differentiation and endochondral bone formation occur in the vertebra.

During development the sclerotome is organized into a patterned structure of alternating loose and dense mesenchyme. The dense mesenchyme will differentiate into the AF of the IVD, and the loose mesenchyme will differentiate into the VB. A sharp boundary exists between the two compartments due to the differential expression of certain factors. High activity levels of BMP signaling are required for differentiation of the vertebral cartilage; however, BMP mRNA is synthesized in the presumptive IVD cells (Zakin, Chang, Plouhinec, & De Robertis, 2010; Zakin, Metzinger, Chang, Coffinier, & De Robertis, 2008). To get BMP from the tissue of production, the IVD, to the responsive tissues, the VB, BMP expression must be relocalized and concentrated. The BMP interacting proteins, Crossveinless-2 (Cv-2) and Chordin (Chd), have been shown to be required for this translocation. Cv-2 mRNA is made in cartilaginous cells of the presumptive VB, and Chd mRNA is made in the presumptive IVD; however, most of the protein is localized to the vertebral compartment. Chd binds to and inactivates BMP in the developing IVD, and subsequently moves BMP to the developing VB. Once there, Chd binds to Cv-2, and is cleaved by a protease, thus releasing BMP (Zakin et al., 2010, 2008). The deletion of Cv-2 or Chd results in small vertebral bodies and a slightly expanded intervertebral space. This movement and relocation of BMP helps define the discrete boundaries within the developing spinal column. In addition, knock out of Tgfbr2 in collagen type II expressing cells indicated that Tgfbr2 is required to maintain the sharp boundary between the developing VB and AF (Baffi, Moran, & Serra, 2006). TGFβ signaling is a known antagonist of BMP signaling; therefore, it is possible that loss of TGFβ results in inappropriate BMP activity in the presumptive IVD where BMP mRNA is synthesized (Candia et al., 1997; Li et al., 2006).

The protocadherins Fat4 and Dchs1 act as a receptor-ligand pair to establish planar cell polarity and have also been shown to be required for proper formation of the VB. The double knockout of Fat4 and Dchs1 shows a reduced number of chondrogenic cells within the developing VB and malformed VB that is split along the midline (Kuta et al., 2016). These results suggest that cell adhesion and polarity are important in chondrocyte differentiation and formation of the vertebrae.

2.4.2 Annulus fibrosus

The compartment of the sclerotome that will form the AF of the IVD can be traced back to the somitocoel cells of the somite (Mittapalli et al., 2005). As described above, somites are transient structures that create the segmented organization of the embryo. They consist of an outer epithelial ball with a central core of mesenchymal cells called the somitocoel. After sclerotome formation, Shh from the notochord causes an epithelial-to-mesenchymal transition in the ventral-lateral portion of the somite, causing these cells along with the mesenchymal cells of the somitocoel to migrate ventrally to surround the notochord (Monsoro-Burq, 2005) (Fig. 1). After migration, the sclerotome reorganizes during resegmentation so that cells in the caudal boundary near von Ebner's fissure will end up in between the VB as the AF. This area has been termed the arthrotome. The importance of the somitocoel in IVD formation was shown in the chick model by surgically removing the somitocoel and replacing the space with an inert bead (Mittapalli et al., 2005). This resulted in about half of the operated embryos lacking IVDs and the fusion of adjacent articular processes. Studies have shown that although somitocoel cells can contribute to the AF, they are not committed to AF differentiation and can be driven toward another cell fate if moved to a different environment (Senthinathan, Sousa, Tannahill, & Keynes, 2012).

The developing AF is further differentiated into two compartments: the inner AF (iAF) and outer AF (oAF) (Fig. 1). iAF is fibrocartilage that exhibits characteristics of both fibrous tissues and cartilage. oAF is more fibrous and resembles ligament and tendon (van den Akker et al., 2017). Pax1/9 expression becomes enriched in the caudal region of the sclerotome where the AF will form, even though expression is initially throughout the entire sclerotome and is required for the initiation of chondrogenesis. Expression is eventually solely restricted to the developing AF and marks the boundary between the IVD and developing VB by E12.5 in mice (Sivakamasundari et al., 2017). This restricted expression pattern may be due to a feed-back loop in which Pax1/9 induces expression of Sox proteins, which in turn inhibits further expression of Pax1(Sivakamasundari et al., 2017). The transcriptome of Pax1/9-GFP cells was recently characterized using a GFP tag to isolate and sort fluorescent cells. Pax1/9-GFP expressing cells were enriched for genes involved in salient AF developmental processes, such as cell proliferation, condensation, and collagen fibrillogenesis, and had reduced expression of genes associated with cartilage development when compared to GFP-negative controls. Pax1/9 was also shown to promote both iAF and oAF expression profiles in differentiating cells; however, Pax1/9 in combination with TGFβ and BMP was primarily involved in promoting iAF (Sivakamasundari et al., 2017; Sohn, Cox, Chen, & Serra, 2010). Additionally, Pax1/9 was recently shown to regulate a long non-coding RNA, Klhl14, during IVD development. In Pax1/9 knockout mice, Klhl14 was the most downregulated transcript in the vertebral column and seemed to be regulated by Pax1/9, Sox6/9 and Nxk.2, specifically in the iAF. These data suggest a potential role for Klhl14 in AF development that has yet to be elucidated (Kraus et al., 2018).

The cells in the developing AF are fibroblast-like and elongate to organize into concentric circles around the developing NP (Hayes & Ralphs, 2011; Peacock, 1951; Rufai et al., 1995). This orientation is caused by the organization of an intracellular network of actin stress fibers and is thought to provide a template for the deposition of collagen, the protein responsible for the trademark lamellar structure of the AF (Hayes, Benjamin, & Ralphs, 1999; Hayes, Hughes, Ralphs, & Caterson, 2011; Hayes, Isaacs, Hughes, Caterson, & Ralphs, 2011). Fibromodulin is strongly expressed in the AF relative to the vertebral cartilage and plays an important role in collagen cross-linking, packing, and fibril diameter (Hayes & Ralphs, 2011; Kalamajski, Bihan, Bonna, Rubin, & Farndale, 2016; Smits & Lefebvre, 2003; Sohn et al., 2010). Recent work suggests heparan sulfate proteoglycans (HSPG) may be important in the formation of lamellar structure in the AF as well. A recent study showed that Syndecan-4, a co-receptor for integrin signaling in formation of focal adhesion complexes, is localized to the lamellar structures in the oAF of rats (Beckett, Ralphs, Caterson, & Hayes, 2015; Choi, Chung, Jung, Couchman, & Oh, 2011). Syndecan 4 is expressed very early on the cell surface of developing AF cells and becomes restricted to the oAF as the AF begins to organize. The co-receptor has an expression pattern very similar to that of Fibronectin, an extracellular matrix component known to be critical for the initiation of lamellar structures. This suggests a potential role for Syndecan 4 in promoting lamellar structure in AF as well.

TGFβ signaling is an important regulator of spinal column development and is required for proper development of the AF. Mice with targeted deletion of the Tgfbr2 gene in the sclerotome manifested severe defects in the development of the AF, which were reduced or completely missing (Baffi et al., 2006, 2004). The annulus was most affected with the expression of Fibromodulin highly reduced or altogether missing. Vertebral markers were ectopically expressed in the presumptive AF and peanut agglutinin, which normally only stains the presumptive VB, stained the length of the developing spine. Microarray analysis comparing wild-type and Tgfbr2 mutant IVDs showed that the AF from Tgfbr2 mutant mice resembled wild-type VB more so than AF (Sohn et al., 2010). Additionally, a significant number of BMP- and TGFβ-regulated genes were also regulated by Pax1/9, suggesting that Pax1/9 cooperates with BMP and TGFβ signaling pathways during axial skeleton development (Sivakamasundari et al., 2017).

A balance between TGFβ and BMP signaling is also important in maintaining cell fate and maintenance of postnatal AF. A study showed that Filamin B knockout mice demonstrate an increase in both canonical and non-canonical TGFβ signaling, as measured by increased phosphorylated Smad2/3, activated ERK and increase in target genes CTGF and P21. Similarly, both canonical and non-canonical BMP signaling were increased as measured by phosphorylated Smad1/5, activation of P38, and an increase in the target gene Msx2 (Zieba et al., 2016). A recent study showed that mice with disrupted Tgfbr2 signaling, whether through dominant-negative interference of TGFβ signaling or deletion of Tgfbr2 in Aggrecan Cre expressing cells, cannot maintain postnatal AF (Alkhatib, Liu, & Serra, 2018). Filamin B knockout mice also display Spondylocarpotarsal Synostosis Syndrome (OMIM: 272460), characterized by vertebral fusion, lordosis, and scoliosis (Mitter, Krakow, Farrington-Rock, & Meinecke, 2008). The AF cells showed characteristic hyaline chondrocyte morphology and began to overtake the NP space. Collagen deposition was disrupted, and Collagen type II expression was inappropriately localized in the oAF, which normally expresses Collagen type I, a more fibrous collagen. The oAF also expressed the hypertrophic marker Collagen type X, indicating inappropriate endochondral bone formation.

Taken together the data suggest that Filamin B is important to maintain cell types of the AF in the postnatal IVD (Mitter et al., 2008; Zieba et al., 2016).

Scleraxis (Scx) is a transcription factor that is required for tendon formation but has been shown through lineage tracing of Scx positive cells to be localized in both the developing and mature oAF (Sugimoto et al., 2013; Yoshimoto et al., 2017). However, Scx does not appear to be required for formation of the AF since IVD is formed in Scx knockout mice (Murchison et al., 2007; Schweitzer et al., 2001). When Sox9, an initiator of chondrogenesis, is deleted in Scx positive cells, mice demonstrate an expanded oAF and a decreased iAF size (Blitz, Sharir, Akiyama, & Zelzer, 2013; Sugimoto et al., 2013). The results indicate an important relationship between Sox9 and Scx positive cells in the division between iAF and oAF. These data also suggest that although Scx may not be required for development of the AF, the Scx positive population of cells does contribute to its formation. Another tendon-associated transcription factor, Mohawk (Mkx), is involved in AF development (Nakamichi et al., 2016). A recent study showed that Mkx is expressed in the oAF in embryonic and adult stages in mice and has a role in collagen fiber formation in oAF by affecting expression of collagens and small leucine-rich proteoglycans (SLRPs). Collagens and SLRPS have an established, essential role in forming the lamellar structure in oAF (Aszodi, Chan, Hunziker, Bateman, & Fassler, 1998; Furukawa et al., 2009). Mkx has also been shown to promote regeneration of oAF when expressed in mesenchymal cells from degenerated IVD, thus suggesting Mkx is a critical regulator of cell fate in mesenchymal cells (Nakamichi et al., 2016). Together the results indicate a close relationship between AF and tendon/ligament differentiation. A list of regulators of AF differentiation is provided in Table 4.

2.4.3 Tendon/ligament

Axial tendons develop from the syndetome, a compartment of the sclerotome (Brent, Schweitzer, & Tabin, 2003). Since tendons link the muscle to the vertebrae, localization during development is critical. The epithelial dermomyotome is formed from the most dorsal part of the somite and the sclerotome forms the ventral and medial parts. The myotome is immediately dorsal and adjacent to the sclerotome. It expresses master regulators of muscle development MyoD and Myf5 and secretes fibroblast growth factors (FGF) 4 and 6. Receptors for FGF are located on the sclerotome cells and FGF signaling stimulates expression of Scx, inducing their differentiation to tendon cells. The ventral side of the sclerotome sees high levels of Shh from the notochord, which prevents tendon formation and promotes formation of the vertebral bodies. Thus, the tendon forms in between the vertebra and muscle (Brent, Braun, & Tabin, 2005; Brent et al., 2003). More recently, it was shown that TGF-β is also an important regulator of tendon development and maintenance. Conditional deletion of Tgfbr2 or deletion of both Tgfb1 and Tgfb3 ligands in mice results in loss of most tendons (Pryce et al., 2009). Other signals that regulate formation of tendons and ligaments are now beginning to be elucidated (Huang, Lu, & Schweitzer, 2015).

3. Development of the nucleus pulposus from notochord

3.1 Formation and function of the notochord

The embryonic notochord is a rod-like structure that forms at the midline of vertebrates (Beddington, 1994; Salisbury, 1993; Sulik et al., 1994). It is a transient structure that serves as a preliminary axial skeleton and a signaling center during development (Salisbury, 1993). The notochord gives rise to the nucleus pulposus (NP) of the intervertebral disc (IVD). During mouse development, the dorsal organizer gives rise to the notochord during gastrulation at around E6 (Stemple, 2005). Subsequent to the dorsal organizer giving rise to the chordamesoderm, cells undergo convergent extension and migrate toward the midline to lengthen the presumptive notochord (E8.5 to E10.5 in mice) (Stemple, 2005). Chordamesodermal cells then acquire a thick sheath and become vacuolated (Stemple, 2005). Osmotic pressure acts against the notochoral sheath to give the notochord its final rod-like appearance (Stemple, 2005).

A recent study attempted to decipher the notochord's role in vertebral column segmentation (Ward, Pang, Evans, & Stern, 2018). In chick embryos, ablation of the notochord caused loss of segmentation of the vertebral bodies (VB) and IVDs. However, without the surrounding sclerotome, the notochord was not segmented. To examine whether the notochord dictates the segmentation of the sclerotome, ectopic notochords were grafted into chick embryos. The data indicated that intrinsic segmentation of the sclerotome is dominant over any segmental information that the notochord may possess. No evidence was found to suggest that the chick notochord was intrinsically segmented. Therefore, it was concluded that the segmental pattern of the VBs and IVDs in the chick is dictated by the sclerotome. Furthermore, the sclerotome must first signal to the notochord to ensure that the NP develops in register with the annulus fibrosus (AF), and the notochord is required for the maintenance of sclerotome segmentation and formation of mature VBs and IVDs.

Mechanical forces are known to play a role in the formation of the notochord and are hypothesized to also play a role in the notochord-to-NP transition. A recent study asserted that mechanical forces, rather than the alignment of the notochordal sheath, define the collagen architecture of the IVDs (Ghazanfari, Werner, Ghazanfari, Weaver, & Smit, 2018). This study compared the development of the mammal and avian IVDs. The authors postulate that upon bulging of the notochord to become the NP in mammals, a bi-axial strain is applied on the sclerotome between condensations which induces the criss-cross aligned structure of the collagen fibers within the AF. However, in birds, notochord bulging does not occur and the notochord is not squeezed (birds do not form an NP); therefore, the notochord persists and collagen fibers are concentrically organized.

Aside from mechanical forces, the matrix environment of a tissue can greatly impact cell behavior and differentiation. Cells have the ability to take cues from their environment and adapt to create a functional tissue during development and maintain that tissue into adulthood. A recent study investigated whether naïve stem cells (human induced pluripotent stem cells, hiPSCs) could differentiate into notochordal cells through direct contact with porcine NP matrix (Liu, Rahaman, & Bal, 2014). Contact and non-contact cultures both yielded functional notochordal-like cells; however, contact cultures had a significantly higher differentiation yield. Generated notochordal-like cells were highly homogenous in their expression of notochordal marker genes from both contact and non-contact cultures. Culture medium containing FGF, EGF, VEGF, and IGF-1 further supported notochordal-like cell differentiation in the presence of porcine NP matrix. Notochordal-like cells from these cultures produced aggrecan and collagen type II and had a proteoglycan-to-collagen ratio that was similar to native NP. Overall, native NP matrix could be used to differentiate hiPSCs to functional notochordal-like cells that could be implanted to treat disc degeneration.

Gene mutations have identified molecular components critical to the development and maintenance of the notochord. Of these, the *T* gene (brachyury), Danforth's short-tail (*Sd*), Sickle tail gene (*Skt*), Sonic hedgehog (*Shh*), Sox5/6, and collagen II mutations are among the best characterized. The *T* gene has been identified as a notochordal marker and its expression is restricted and maintained within the notochord (Kispert, Herrmann, Leptin, & Reuter, 1994). Homozygous mutations in the *T* gene in mice resulted in failure to establish the trunk notochord. Further mutations in the spine and allantois led to early embryonic lethality (Dobrovolskaia-Zavadskaia, 1927). Mice with a dominant-negative mutation in the *T* gene survived but demonstrated abnormal NP morphology (Stott, Kispert, & Herrmann, 1993).

While notochord formation is a vital first step, notochord maintenance is also important and is affected by Danforth's short-tail (*Sd*) mutation. The gene in which this mutation occurs is still unidentified. *Sd* mice with a homozygous mutation had a discontinuous, fragmented notochord (Paavola et al., 1980). The notochord eventually disappeared and NP morphology was severely affected. In mice with a heterozygous mutation, the NP was absent and was replaced by fibrous tissue resembling the annulus fibrosus (AF) (Semba et al., 2006). *Sd* is located on the same chromosome as the Sickle tail gene (*Skt*). In mice with mutations in the *Skt* gene, the NP did form but its position was shifted to the periphery of the disc (Semba et al., 2006). Furthermore, the boundary between the NP and AF was altered. AF collagen layers were also thin compared to wild-type mice. Polymorphisms in the *SKT* gene have been identified in Japanese and Finnish populations and are associated with lumbar disc herniation (Karasugi et al., 2009, OMIM#617367).

A Skt^{cre} mouse has been generated. Cre expression from the Skt^{cre} allele was used to activate β-galactosidase and expression in the notochord from E9.5 onward was characterized (Abe et al., 2012). When crossed to lacZ reporter (R26R) mice, reporter activity was detected at E15.5 in the NP and a portion of the AF. Mice from this study could be used to fate map notochordal cells or to develop mouse models for disc degeneration.

3.1.1 Notochord sheath

The notochordal sheath is composed of extracellular matrix (ECM) proteins such as collagens and proteoglycans (Gotz, Osmers, & Herken, 1995). It functions to contain hydrostatic pressure within the notochord (Adams, Keller, & Koehl, 1990). Previous studies have demonstrated that mutations in genes that encode proteins within the notochordal sheath lead to aberrant development and maintenance of the notochord and the prospective NP. Sox 5/6 and Shh are both expressed in the notochord while collagen II is expressed in the notochordal sheath. Sox 5/6-null mice form a rod-like notochord but it fails to be surrounded by a sheath (Smits & Lefebvre, 2003). Extracellular matrix protein genes are downregulated in notochordal cells and eventually the notochord is dismantled. This leads to formation of an irregular NP tissue. Shh is also exprcssed in the notochord and postnatally in the NP (Dahia, Mahoney, Durrani, & Wylie, 2009; DiPaola, Farmer, Manova, & Niswander, 2005). Similarly, *Shh*-null mice form a notochord but it is not maintained (Chiang et al., 1996). Early lethality of these mice precluded further study of the NP. However, when *Smoothened* (*Smo*) is conditionally deleted in *Shh*-expressing cells, the notochordal sheath is missing and the NP does not expand into the IVD region (Choi & Harfe, 2011). Notochordal cells are scattered throughout the vertebral column. Deletion of *Smo* after formation of the notochordal sheath does not affect the formation of the NP. Disruption of the ECM components of the sheath can also lead to notochord and NP aberrations. Collagen II is a major structural component of the notochordal sheath (Gotz et al., 1995; Swiderski & Solursh, 1992). In *Col2a1*-null mice, the notochord is not removed from the vertebral bodies (VBs) and the IVDs do not form (Aszodi et al., 1998). Presumably, structural weakening of the notochordal sheath leads to loss of the ability to contain osmotic pressure (Adams et al., 1990). As a result, the notochord is not removed from the VBs and does not expand in to the IVD and the NP does not form. Together, these results demonstrate an important role for the notochordal sheath in the development and maintenance of the notochord and eventually formation of the NP.

3.2 Identification of notochordal and NP markers

In order to more clearly distinguish between notochordal and NP cells, various studies have employed advanced techniques to identify marker genes for both populations. One aim is to confirm whether NP cells are in fact derived solely from the notochordal cell population. Another impetus for studies of this nature is the desire to use this knowledge to develop gene therapies to regenerate diseased NP tissue. Previously, Sonic hedgehog (Shh) was investigated in an attempt to fate map notochordal cells and discern whether NP cells are indeed derived from the notochord. In mice, Shh is highly expressed in notochordal cells and later in NP cells into adulthood. Inducible-Cre mice with a promoter under the control of Shh were utilized to track notochordal cells as they transition to NP cells (Choi, Cohn, & Harfe, 2008). Cre recombination was activated (β-galactosidase staining indicated activation of the ROSA26-lacZ locus) at E8.0 in mice and IVD tissue was examined at E13.5 and 19 months. At both ages, all NP cells were labeled with the reporter and no AF cells were labeled. A subsequent study attempted to corroborate these results by studying a different marker for notochordal cells (McCann, Tamplin, Rossant, & Seguin, 2012). Noto is a highly conserved transcription factor with restricted expression limited to the node and early notochord. Reporter activity, as judged by activation of the ROSA26-lacZ locus by the Noto-cre promoter, was observed in all mouse NP cells even into adulthood. The results of these studies indicate that NP cells are derived from notochordal cells in mice (Fig. 1). A list of genes and signaling molecules involved in notochord and NP development is provided in Table 5.

Another group of extensive studies was focused on identifying specific marker genes for both notochordal and NP cells. Previous work harnessed the power of microarray technology to determine gene expression profiles of NP, AF and articular cartilage cells from various species, including human, bovine, canine, and rodent cells (Lee et al., 2007; Minogue, Richardson, Zeef, Freemont, & Hoyland, 2010; Sakai, Nakai, Mochida, Alini, & Grad, 2009). Cumulatively, the *T* (Brachyury) gene, K8, K18, and K19 were all identified to be more highly expressed in NP than in AF cells or in articular cartilage.

Table 5 Genes and proteins involved in nucleus pulposus development.

Brachyury	**T**	**Transcription factor**	**N/A**
Collagen II Type 1A	Col2A1	Protein	Diastrophic dysplasia, OMIM#222600
CYR61, CTGF, and NOV	CCN	Protein	Childhood Progressive Pseudorheumatoid Arthropathy, OMIM#208230
Forkhead Box A	FOXA	Transcription factor	N/A
Forkhead Box O	FOXO	Transcription factor	N/A
Hypoxia Inducible Factor 1 Subunit Alpha	HIF-1α	Transcription factor	N/A
Sickle Tail Gene	Skt	Protein	Lumbar Disc Disease, OMIM#617367
Smoothened	Smo	G protein-coupled receptor	Curry-Jones syndrome, OMIM#601707
Sonic Hedgehog	Shh	Secreted ligand	Brachydactyly Type A1, OMIM#112500
Sox 5	Sox 5	Transcription factor	Multiple synostoses syndrome, OMIM#186500

The data from these studies suggested that certain marker genes clearly distinguish NP cells from other cartilaginous and fibrocartilaginous cell types. More recent studies, however, focused on delineating specific markers for notochordal and NP cells with the aim of further understanding developmental mechanisms and opening the door for therapeutic strategies. Rodrigues-Pinto et al. performed a unique, detailed transcriptomic profiling of human embryonic and fetal spine cells (Rodrigues-Pinto et al., 2018). CD24-positive notochordal cells were isolated using FACS and microarray analysis along with qPCR validation and utilized for identification of specific marker genes. *CD24, STMN2, RTN1, PRPH, CXCL12, IGF1, MAP1B, ISL1, CLDN1*, and *THBS2* were all identified as notochord-specific marker genes. Expression of these genes was confirmed in NP cells from aging and degenerated IVDs. Furthermore, ingenuity pathway analysis (analysis tool that can identify new targets or candidate biomarkers within RNA-seq or microarray data) demonstrated that genes encoding molecules involved in inhibition of vascularization (*WISP2, NOGGIN, EDN2*) and inflammation (*IL1-RN*) were master regulators of notochordal genes. Altogether, the results confirmed data from previous studies and established putative markers for the notochord and NP. However, effective markers for a specific cell type must be continually expressed even through tissue changes in adulthood. Another study performed by Richardson et al. examined NP tissue at various ages and stages of degeneration to determine whether notochordal markers continue to be expressed (Richardson et al., 2017). Gene expression and immunohistochemistry data from 116 individual tissue samples indicated that expression of the NP markers *FOXF1, PAX1, KRT8/18, CA12* and notochordal markers *TBXT, LGALS3*, and *CD24* was maintained in NP cells irrespective of age or stage of degeneration.

Determining useful markers is only one step toward the understanding of development and development of therapeutics. Profiling gene expression at different stages of development leading up to the formation of the NP is also important. To this end, a distinctive study was performed in mice to identify genes that are critical in the notochord-to-NP transition (Peck et al., 2017). Notochordal cells were sorted from *Shh*-cre;ROSA:YFP mice at E12.5 and P0. *Shh* is specifically expressed in the developing notochord. Expression of Sonic hedgehog signaling molecules along with Wnt pathway components was significantly reduced at P0 compared to E12.5. Conversely, TGF-β pathway components along with IGF pathway components were significantly increased at P0 compared to E12.5. As expected, mRNA of several ECM components was increased at P0 (*Acan, Bcan, Bgn, Dcn, Col1a1* and

Col6a1) when compared to E12.5. NP marker genes (*Tbxt*, *Krt8/18*, *Hif1α*) were comparable at both E12.5 and P0. These results indicate that signaling and biosynthesis of notochordal-derived cells can change depending on the stage of development. This could provide insights into the developmental process and identify gene targets for therapeutics.

3.3 Notochord-to-nucleus pulposus transition

Much debate has surrounded the process of how the notochord transitions to the NP. At this point, two central models exist, the "pressure" model and the "repulsion/attraction" model (Lawson & Harfe, 2015). In the "pressure model" mouse histological sections revealed that no major cell death occurred in the region where the prospective vertebrae form. Furthermore, cell proliferation was not observed in the region where the prospective NP is formed. Therefore, it was hypothesized that notochordal cells were "squeezed into" or "pushed into" the regions where the IVDs are forming. The "repulsion/attraction" model suggests that cell movement is dictated by repulsive or attractant molecules. It is hypothesized that attractant molecules lure notochordal cells to aggregate within the region where the prospective IVD will form. Conversely, it is also possible that repulsive molecules within the prospective vertebrae regions exclude notochordal cells in these regions.

Regardless of the mechanism underlying the transition from notochord to NP, several mouse models have been used to investigate genes that are important in the formation of the NP. *Foxa* genes are required for embryonic development and postnatal life in mice. *Foxa1* and *Foxa2* are expressed in the notochord (Besnard, Wert, Hull, & Whitsett, 2004; Kaestner, Hiemisch, Luckow, & Schutz, 1994; Monaghan, Kaestner, Grau, & Schutz, 1993; Sasaki & Hogan, 1993). One study performed by Maier et al. used a double knockout model to investigate whether *Foxa1* and *Foxa2* genes are required for the formation of the IVDs (Maier, Lo, & Harfe, 2013). On a *Foxa1* null background, *Foxa2* was conditionally knocked out in Shh-expressing cells and these were termed double knockouts (dKOs). Massive cell death was observed at E11.5 in the posterior somites and midline of the tail in dKO mice. At E19.5, NPs of dKO mice were compressed and small. IVD defects were more severe posteriorly than anteriorly and dKOs had shorter tails that lacked IVD and VB structures. The notochordal sheath and the notochord-toNP transition were both abnormal in dKO mice. The dKO notochordal sheath was very faint at E11.5 compared to the control when visualized using alcian blue and fast green staining. β-Galactosidase treatment of dKO mice that contained the *R26R* allele revealed that notochordal cells were dispersed throughout the vertebral column.

Expression of genes involved in notochordal development and formation of the IVDs were affected by the knockout of the *Foxa* genes (Maier et al., 2013). Evidence from a *Shh* enhancer element suggests that FOXA proteins can regulate *Shh* within the notochord. *Shh* expression and signaling were decreased in the notochord of dKO tails and floorplates. In contrast, expression of *Pax1* and *Tbx18* within the sclerotome of dKO mice was indistinguishable from control mice. This indicated that *Foxa1* and *Foxa2* expression within the notochord is not required for the formation of the sclerotome. Dorsal-ventral patterning of the neural tube, however, was greatly affected in dKO mice. Expression of markers such as *Nkx6.1*, *Nkx2.2*, and *Pax3* was irregular and indicated that *Foxa* expression within the notochord is critical for dorsal-ventral patterning of the neural tube. Altogether, the results suggest that *Foxa* genes are critical for the development and formation of the IVDs.

Shh is secreted from the notochord and NP and is required for the formation and post-natal maintenance of the IVD (Choi & Harfe, 2011; Dahia, Mahoney, & Wylie, 2012). A recent study took an interesting approach to further confirm the importance of Shh in the development of the NP. The sacrum in vertebrates is formed by collapse of the IVDs between the vertebrae and their fusion into a single bone. Bonavita et al. demonstrated that, in mice, the collapse of these sacral discs is associated with a downregulation of *Shh* signaling in the NP (Bonavita, Vincent, Pinelli, & Dahia, 2018). Furthermore, experimental postnatal reactivation of *Shh* signaling in dormant NP cells can preserve sacral discs in the mouse spine. This evidence suggests that loss of *Shh* signaling is pathological in one region of the spine but part of normal turnover in another.

The CCN family of secreted matricellular proteins serves as multifunctional signaling mediator that regulates interactions between cells, growth factors and the ECM (Perbal, 2004). Mouse knockout models have demonstrated that these proteins are involved in angiogenesis, embryonic cartilage and bone formation, as well as inflammation and fibrosis in adults (Brigstock, 2003; Moussad & Brigstock, 2000). CCN2 has been demonstrated to be expressed in the embryonic node and notochord (Tamplin, Cox, & Rossant, 2011). In the IVD, CCN2 was first discovered as an anabolic factor secreted by notochordal cells that induces NP cell proliferation and aggrecan production *in vitro* (Aguiar, Johnson, & Oegema, 1999; Erwin,

Ashman, O'Donnel, & Inman, 2006; Erwin & Inman, 2006). In a recent study, Bedore et al. conditionally knocked out *CCN2* to study the effects on the mouse IVD at various timepoints (Bedore et al., 2013). Expression of the Cre recombinase was under the control of a notochord-specific promoter (*Noto*^{cre}) which allowed for conditional deletion of *CCN2* in notochordal cells. Notochord segmentation and IVD patterning were not affected by the conditional *CCN2* knockout. At E15.5, CCN3 expression was increased in the prospective NP; however, aggrecan and collagen II expression was decreased. ECM perturbations were more pronounced in P1 mice. Levels of aggrecan and collagen II were significantly decreased and collagen I expression was increased. No appreciable differences were observed between conditional knockout mice and wild-type littermates at P28. However, at 12 months of age, conditional knockout IVDs had lost distinct NP-AF boundaries. The cellularity of the NP had significantly decreased and the IVD structure was disorganized. This was accompanied by a decrease in aggrecan levels within the NP and decreased signal intensity on a T2/T1-weighted MRI. Conditional knockout mice at 17 months of age had more pronounced intervertebral disc degeneration. IVDs were herniated and osteophytes had formed within the IVD tissue. These IVDs also had low signal intensities on T2/T1-weighted MRI scans. Overall, these results demonstrated that *CCN2* expression in the notochord is essential for the regulation of IVD development and age-associated maintenance of the NP.

Animal modeling has provided valuable information pertaining to the developmental process of the IVD. Nevertheless, recent *in vitro* experimentation has yielded vital results in the investigation of the transition process from notochord to NP. Many of these studies, with proper follow-ups, could have significant clinical implications. One such study investigated whether mechanical load would induce maturation of notochordal cell (NC)-rich porcine NP tissue (6 to 8-week old) (Purmessur et al., 2013). NC-rich NP tissue was loaded with hydrostatic pressure (0.5–2 MPa at 0.1 Hz for 2 h) and analyzed after 8 days. Three conditions were examined: Control (no pressurization), one single dose of loading, and daily loading. The percentage of cells that had transitioned from large notochordal cells to small NP-like cells was higher in daily (73.8%) and single dose (44%) than in control (28%) cultures. No cell death was observed in any of the conditions. Furthermore, loading increased metabolic activity and levels of safranin-O stained matrix, indicative of maturation of loaded tissues. These results suggested that mechanical loading is an integral component of the notochord transition to NP.

Although the study of notochordal cells and their transition to NP cells is essential for understanding the processes of aging and degeneration, regenerative medicine requires more translational ideas and methodologies. Currently, there is no consistent, high-quality source of notochordal cells that can be used for implantation to regenerate NP tissue. For this reason, the study of stem cell differentiation to NP-like cells has become paramount for the treatment of IVD degeneration. One study isolated induced pluripotent stem cells (iPSCs) from mice and sorted them using magnetic activated cell sorting (MACS) to further isolate a $CD24^+$ iPSC subpopulation (Chen et al., 2013). $CD24^+$ sorted cells exhibited an increase in expression of notochordal gene markers (*Noto*, *Shh*, *Foxa2*, and *T* (*brachyury*)) when compared with presorted cells. $CD24^+$ iPSCs were cultured on a laminin-rich culture system for 28 days and the phenotype of the cells was assessed using various biochemical methods. Post-culture, $CD24^+$ iPSCs had formed three-dimensional cell clusters that were rich in sulfated glycosaminoglycans (sGAG) and type II collagen. The cells also expressed integrin α6, vimentin and cytokeratin 5/8, markers of the immature NP cell phenotype. $CD24^-$ cells were also cultured in the laminin-rich system but under different conditions to mimic *in situ* oxygen tension. Cells were cultured in 2% O_2 and in media containing factors secreted by notochordal cell-containing NP tissue (NCCM). The results were similar to those observed with $CD24^+$ iPSC cultures. A follow-up study attempted to differentiate human iPSCs using a novel stepwise protocol (Tang et al., 2018). When colonies of human iPSCs were stimulated with BMP4/FGF2 and Wnt-3a/Activin A, the cells were found to express early mesodermal and notochordal markers. Human iPSCs were also cultured in monolayers with the addition of GDF-5 and in pellet cultures with TGF-β3 and L-Proline to stimulate matrix synthesis. This protocol eventually led to the formation of larger, proteoglycan-containing pellets and the vacuole-structure characteristic of NP-like cells. These results demonstrated that iPSCs could be used for cellular therapy to treat IVD degeneration.

Subsequent to the notochord transitioning to the NP, much debate has surrounded the true identity of the adult NP cell. It is thought that the NP cell population is a heterogenous one that consists of cells which are more prone to either anabolic or catabolic stimulation. A recent study generated 54 immortal clones of non-degenerate healthy human NP cells (van den Akker et al., 2014). Subclones were profiled using a novel set of NP markers (*CD24*, *CA12*, *PAX1*, *PTN*, *FOXF1*, and *KRT19* mRNA) and expression of these markers confirmed their NP origin. Two predominant clones were identified based on their ability to induce SOX9 and COL2A1 in a Matrigel hydrogel culture system. In a follow-up study, clones that were able to induce SOX9 and COL2A1 were termed responders (NP-R) and those that could not were termed non-responders (NP-nR) (van den Akker et al., 2016). The authors hypothesized that NP-R cells could represent undifferentiated cells and that NP-nR cells are differentiated cells. When challenged with catabolic stimuli, NP-nR clones were more responsive (higher expression of catabolic genes) than NP-R clones. These results support the hypothesis that the adult NP population is not homogenous and different subpopulations could be more susceptible to catabolic stimuli leading to IVD degeneration.

3.4 Maintenance of the nucleus pulposus

Recent studies have begun to investigate genes that are critical for the maintenance of NP tissue. Mouse modeling has provided the most evidence for genes required for homeostasis of the NP. FOXO proteins are transcription factors that are primarily involved in development, aging, and longevity (Kahn, 2015; Martins, Lithgow, & Link, 2016). One study investigated the effects of conditional deletion of all *Foxo* isoforms (1, 3, and 4) using either the *Col2a1* promoter (*Col2a1cre*) or the aggrecan promoter (*AcancreER*, deletion after skeletal maturity) to drive *cre*- or tamoxifen-inducible *cre*-dependent recombination (Alvarez-Garcia et al., 2018). In 6-month-old *Col2a1cre*-FOXO KO mice, a significant loss of cellularity was observed in the NP and cartilaginous endplate (EP). Furthermore, the boundary between the NP and AF was disrupted and AF lamellae were disorganized. The presence of hypertrophic cells was observed in the inner AF. FOXO deficiency led to severe spinal deformities with abnormal curvature of the spine and kyphosis in 6-month-old mice. To study the role of FOXO proteins in the maintenance of the mature IVD, 4-month-old *AcancreER*-FOXO KO mice were injected with tamoxifen and vertebral column tissue was analyzed at 12 months of age. *AcancreER*-FOXO KO mice had a significant reduction in NP and EP cellularity and disc height at 12 months of age. NP cell clustering, loss of NP/AF boundary demarcation, and disorganization of AF lamellae were all observed in *AcancreER*-FOXO KO mice. Studies of gene expression in both mouse models revealed that FOXO-deficient mice had impaired autophagy and reduced antioxidant defense. Human primary NP cells were also studied *in vitro* to further discern the function of FOXO proteins by gene expression analysis. FOXO directly regulates autophagy, adaptation to hypoxia, and resistance to oxidative and inflammatory stress in human primary NP cells. Overall, FOXO proteins are involved in both the maturation and maintenance of the NP.

Another study investigated the role of HIF-1α in the development and maintenance of the NP (Merceron et al., 2014). Notochordal cells were targeted using a *Foxa2-cre* line and HIF-1α was conditionally deleted in these cells. HIF-1α is stabilized within the prospective NP and is used as an adaptive response protein under a variety of stresses, including low oxygen tension within the NP. Notochord development was not affected in this study. However, at E15.5, mutant NPs appeared smaller than control NPs and lacked large vacuoles indicative of residual notochordal cells. By 1 month, NPs had completely disappeared in mutants and were replaced by fibrocartilaginous tissue that stained intensely with safranin-O. Lineage studies revealed that NP cells did not transdifferentiate into chondrocyte-like cells. Alternatively, NP cells underwent massive cell death as confirmed by TUNEL assay. Mutant IVDs were functionally inferior and had an impaired ability to absorb axial energy and distribute it. This study identified HIF-1α as a critical protein in the development and maintenance of the NP.

4. Conclusions and implications

Understanding how the axial skeleton develops will provide the basis for future repair, regeneration, and tissue engineering strategies for spine disease. In addition, defining signals involved in the embryonic development of the axial skeleton provides insights into mechanisms of spinal pathology. Using what is known about the normal developmental processes to generate new tissues *in vitro* has been termed developmental engineering (Gadjanski et al., 2012; Lenas et al., 2011, 2009a, 2009b). The focus on developmental mechanisms will provide a rational step-wise process for tissue engineering of skeletal tissues. Cartilage and bone organs have already been generated using the principles of developmental engineering (Loh et al., 2016; Oldershaw et al., 2010; Scotti et al., 2010). The spine is obviously more complex than cartilage or bone tissue alone. Nevertheless, mimicking embryonic development is likely to facilitate tissue engineering of complex tissues like the IVD, and difficult problems, such as the integration of the IVD with the surrounding end plates and ligaments, may be solved.

Acknowledgments

The authors would like to thank Ga I Ban for help with text and figure for this chapter. Spine research in RS laboratory is funded by a grant from the National Institutes of Health, National Institute of Arthritis, Musculoskeletal and Skin Diseases, R01 AR053860. B.A. was funded through NIH/NIDCR T90 DE022736. S.W. was funded by T32 AR069516, from NIH/NIAMS.

References

Abe, K., Araki, K., Tanigawa, M., Semba, K., Ando, T., Sato, M., et al. (2012). A Cre knock-in mouse line on the sickle tail locus induces recombination in the notochord and intervertebral disks. *Genesis*, *50*, 758–765.

Adams, D. S., Keller, R., & Koehl, M. A. (1990). The mechanics of notochord elongation, straightening and stiffening in the embryo of Xenopus laevis. *Development*, *110*, 115–130.

Adams, M. A., & Roughley, P. J. (2006). What is intervertebral disc degeneration, and what causes it? *Spine (Phila Pa 1976)*, *31*, 2151–2161.

Aguiar, D. J., Johnson, S. L., & Oegema, T. R. (1999). Notochordal cells interact with nucleus pulposus cells: Regulation of proteoglycan synthesis. *Experimental Cell Research*, *246*, 129–137.

Alkhatib, B., Ban, G. I., Williams, S., & Serra, R. (2018). IVD development: Nucleus pulposus development and sclerotome specification. *Current Molecular Biology Reports*, *4*(3), 132–141.

Alkhatib, B., Liu, C., & Serra, R. (2018). Loss of Tgfbr2 in Acan-Cre expressing cells results in spondylosis in mice. *JOR Spine*, *1*(2), e1025.

Alvarez-Garcia, O., Matsuzaki, T., Olmer, M., Miyata, K., Mokuda, S., Sakai, D., et al. (2018). FOXO are required for intervertebral disk homeostasis during aging and their deficiency promotes disk degeneration. *Aging Cell*, *17*, e12800.

Arendt, D., & Nubler-Jung, K. (1999). Rearranging gastrulation in the name of yolk: Evolution of gastrulation in yolk-rich amniote eggs. *Mechanisms of Development*, *81*, 3–22.

Aszodi, A., Chan, D., Hunziker, E., Bateman, J. F., & Fassler, R. (1998). Collagen II is essential for the removal of the notochord and the formation of intervertebral discs. *The Journal of Cell Biology*, *143*, 1399–1412.

Aulehla, A., & Pourquie, O. (2010). Signaling gradients during paraxial mesoderm development. *Cold Spring Harbor Perspectives in Biology*, *2*, a000869.

Baffi, M. O., Moran, M. A., & Serra, R. (2006). Tgfbr2 regulates the maintenance of boundaries in the axial skeleton. *Developmental Biology*, *296*, 363–374.

Baffi, M. O., Slattery, E., Sohn, P., Moses, H. L., Chytil, A., & Serra, R. (2004). Conditional deletion of the TGF-beta type II receptor in Col2a expressing cells results in defects in the axial skeleton without alterations in chondrocyte differentiation or embryonic development of long bones. *Developmental Biology*, *276*, 124–142.

Bagnall, K. M., Higgins, S. J., & Sanders, E. J. (1988). The contribution made by a single somite to the vertebral column: Experimental evidence in support of resegmentation using the chick-quail chimaera model. *Development (Cambridge, England)*, *103*, 69–85.

Barrios, A., Poole, R. J., Durbin, L., Brennan, C., Holder, N., & Wilson, S. W. (2003). Eph/Ephrin signaling regulates the mesenchymal-to-epithelial transition of the paraxial mesoderm during somite morphogenesis. *Current Biology*, *13*, 1571–1582.

Beckett, M. C., Ralphs, J. R., Caterson, B., & Hayes, A. J. (2015). The transmembrane heparan sulphate proteoglycan syndecan-4 is involved in establishment of the lamellar structure of the annulus fibrosus of the intervertebral disc. *European Cells & Materials*, *30*, 69–88. discussion 88.

Beddington, R. S. (1994). Induction of a second neural axis by the mouse node. *Development*, *120*, 613–620.

Bedore, J., Sha, W., McCann, M. R., Liu, S., Leask, A., & Seguin, C. A. (2013). Impaired intervertebral disc development and premature disc degeneration in mice with notochord-specific deletion of CCN2. *Arthritis and Rheumatism*, *65*, 2634–2644.

Bellairs, R., Curtis, A. S., & Sanders, E. J. (1978). Cell adhesiveness and embryonic differentiation. *Journal of Embryology and Experimental Morphology*, *46*, 207–213.

Besnard, V., Wert, S. E., Hull, W. M., & Whitsett, J. A. (2004). Immunohistochemical localization of Foxa1 and Foxa2 in mouse embryos and adult tissues. *Gene Expression Patterns*, *5*, 193–208.

Blitz, E., Sharir, A., Akiyama, H., & Zelzer, E. (2013). Tendon-bone attachment unit is formed modularly by a distinct pool of Scx- and Sox9-positive progenitors. *Development (Cambridge, England)*, *140*, 2680–2690.
Bonavita, R., Vincent, K., Pinelli, R., & Dahia, C. L. (2018). Formation of the sacrum requires down-regulation of sonic hedgehog signaling in the sacral intervertebral discs. *Biology Open*, *7*.
Borycki, A. G., Mendham, L., & Emerson, C. P., Jr. (1998). Control of somite patterning by sonic hedgehog and its downstream signal response genes. *Development (Cambridge, England)*, *125*, 777–790.
Brand-Saberi, B., & Christ, B. (2000). Evolution and development of distinct cell lineages derived from somites. *Current Topics in Developmental Biology*, *48*, 1–42.
Brent, A. E., Braun, T., & Tabin, C. J. (2005). Genetic analysis of interactions between the somitic muscle, cartilage and tendon cell lineages during mouse development. *Development*, *132*, 515–528.
Brent, A. E., Schweitzer, R., & Tabin, C. J. (2003). A somitic compartment of tendon progenitors. *Cell*, *113*, 235–248.
Brigstock, D. R. (2003). The CCN family: A new stimulus package. *The Journal of Endocrinology*, *178*, 169–175.
Bruggeman, B. J., Maier, J. A., Mohiuddin, Y. S., Powers, R., Lo, Y., Guimaraes-Camboa, N., et al. (2012). Avian intervertebral disc arises from rostral sclerotome and lacks a nucleus pulposus: Implications for evolution of the vertebrate disc. *Developmental Dynamics: An Official Publication of the American Association of Anatomists*, *241*, 675–683.
Burgess, R., Rawls, A., Brown, D., Bradley, A., & Olson, E. N. (1996). Requirement of the paraxis gene for somite formation and musculoskeletal patterning. *Nature*, *384*, 570–573.
Candia, A. F., Watabe, T., Hawley, S. H., Onichtchouk, D., Zhang, Y., Derynck, R., et al. (1997). Cellular interpretation of multiple TGF-beta signals: Intracellular antagonism between activin/BVg1 and BMP-2/4 signaling mediated by Smads. *Development*, *124*, 4467–4480. [Cambridge, England].
Chen, J., Lee, E. J., Jing, L., Christoforou, N., Leong, K. W., & Setton, L. A. (2013). Differentiation of mouse induced pluripotent stem cells (iPSCs) into nucleus pulposus-like cells in vitro. *PLoS One*, *8*, e75548.
Chiang, C., Litingtung, Y., Lee, E., Young, K. E., Corden, J. L., Westphal, H., et al. (1996). Cyclopia and defective axial patterning in mice lacking sonic hedgehog gene function. *Nature*, *383*, 407–413.
Choi, Y., Chung, H., Jung, H., Couchman, J. R., & Oh, E. S. (2011). Syndecans as cell surface receptors: Unique structure equates with functional diversity. *Matrix Biology: Journal of the International Society for Matrix Biology*, *30*, 93–99.
Choi, K. S., Cohn, M. J., & Harfe, B. D. (2008). Identification of nucleus pulposus precursor cells and notochordal remnants in the mouse: Implications for disk degeneration and chordoma formation. *Developmental Dynamics*, *237*, 3953–3958.
Choi, K. S., & Harfe, B. D. (2011). Hedgehog signaling is required for formation of the notochord sheath and patterning of nuclei pulposi within the intervertebral discs. *Proceedings of the National Academy of Sciences of the United States of America*, *108*, 9484–9489.
Christ, B., Huang, R., & Scaal, M. (2004). Formation and differentiation of the avian sclerotome. *Anatomy and Embryology (Berlin)*, *208*, 333–350.
Christ, B., Huang, R., & Scaal, M. (2007). Amniote somite derivatives. *Developmental Dynamics*, *236*, 2382–2396.
Christ, B., Huang, R., & Wilting, J. (2000). The development of the avian vertebral column. *Anatomy and Embryology (Berlin)*, *202*, 179–194.
Christ, B., & Ordahl, C. P. (1995). Early stages of chick somite development. *Anatomy and Embryology*, *191*, 381–396.

Cooke, J., & Zeeman, E. C. (1976). A clock and wavefront model for control of the number of repeated structures during animal morphogenesis. *Journal of Theoretical Biology, 58*, 455–476.

Cox, M. K., & Serra, R. (2014). Development of the intervertebral disc. In I. M. Shapiro & M. V. Risbud (Eds.), *The intervertebral disc* (pp. 33–53). Heidelberg New York Dordrecht London: Springer-Verlag Wien.

Craft, A. M., Ahmed, N., Rockel, J. S., Baht, G. S., Alman, B. A., Kandel, R. A., et al. (2013). Specification of chondrocytes and cartilage tissues from embryonic stem cells. *Development (Cambridge, England), 140*, 2597–2610.

Crossin, K. L., Hoffman, S., Grumet, M., Thiery, J. P., & Edelman, G. M. (1986). Site-restricted expression of cytotactin during development of the chicken embryo. *The Journal of Cell Biology, 102*, 1917–1930.

Dahia, C. L., Mahoney, E. J., Durrani, A. A., & Wylie, C. (2009). Intercellular signaling pathways active during intervertebral disc growth, differentiation, and aging. *Spine (Phila Pa 1976), 34*, 456–462.

Dahia, C. L., Mahoney, E., & Wylie, C. (2012). Shh signaling from the nucleus pulposus is required for the postnatal growth and differentiation of the mouse intervertebral disc. *PLoS One*, 7, e35944.

DiPaola, C. P., Farmer, J. C., Manova, K., & Niswander, L. A. (2005). Molecular signaling in intervertebral disk development. *Journal of Orthopaedic Research, 23*, 1112–1119.

Dobrovolskaia-Zavadskaia, N. (1927). Sur la mortification spotanee de la chez la souris nouvau-nee et sur l'existence d'un caractere (facteur) herededitaire, non-viable. *Comptes Rendus des Seances de la Societe de Biologie, 97*, 114–116.

Dockter, J. L. (2000). Sclerotome induction and differentiation. *Current Topics in Developmental Biology, 48*, 77–127.

Dockter, J., & Ordahl, C. P. (2000). Dorsoventral axis determination in the somite: A re-examination. *Development (Cambridge, England), 127*, 2201–2206.

Duband, J. L., Dufour, S., Hatta, K., Takeichi, M., Edelman, G. M., & Thiery, J. P. (1987). Adhesion molecules during somitogenesis in the avian embryo. *The Journal of Cell Biology, 104*, 1361–1374.

Erwin, W. M., Ashman, K., O'Donnel, P., & Inman, R. D. (2006). Nucleus pulposus notochord cells secrete connective tissue growth factor and up-regulate proteoglycan expression by intervertebral disc chondrocytes. *Arthritis and Rheumatism, 54*, 3859–3867.

Erwin, W. M., & Inman, R. D. (2006). Notochord cells regulate intervertebral disc chondrocyte proteoglycan production and cell proliferation. *Spine (Phila Pa 1976), 31*, 1094–1099.

Fan, C. M., & Tessier-Lavigne, M. (1994). Patterning of mammalian somites by surface ectoderm and notochord: Evidence for sclerotome induction by a hedgehog homolog. *Cell, 79*, 1175–1186.

Ferrer-Vaquer, A., Viotti, M., & Hadjantonakis, A. K. (2010). Transitions between epithelial and mesenchymal states and the morphogenesis of the early mouse embryo. *Cell Adhesion & Migration, 4*, 447–457.

Furukawa, T., Ito, K., Nuka, S., Hashimoto, J., Takei, H., Takahara, M., et al. (2009). Absence of biglycan accelerates the degenerative process in mouse intervertebral disc. *Spine, 34*, E911–E917.

Gadjanski, I., Spiller, K., & Vunjak-Novakovic, G. (2012). Time-dependent processes in stem cell-based tissue engineering of articular cartilage. *Stem Cell Reviews, 8*, 863–881.

Ghazanfari, S., Werner, A., Ghazanfari, S., Weaver, J. C., & Smit, T. H. (2018). Morphogenesis of aligned collagen fibers in the annulus fibrosus: Mammals versus avians. *Biochemical and Biophysical Research Communications, 503*, 1168–1173.

Glazier, J. A., Zhang, Y., Swat, M., Zaitlen, B., & Schnell, S. (2008). Coordinated action of N-CAM, N-cadherin, EphA4, and ephrinB2 translates genetic prepatterns into structure during somitogenesis in chick. *Current Topics in Developmental Biology, 81*, 205–247.

Goldstein, R. S., & Kalcheim, C. (1992). Determination of epithelial half-somites in skeletal morphogenesis. *Development (Cambridge, England)*, *116*, 441–445.

Gotz, W., Osmers, R., & Herken, R. (1995). Localisation of extracellular matrix components in the embryonic human notochord and axial mesenchyme. *Journal of Anatomy*, *186*(Pt. 1), 111–121.

Han, Y., & Lefebvre, V. (2008). L-Sox5 and Sox6 drive expression of the aggrecan gene in cartilage by securing binding of Sox9 to a far-upstream enhancer. *Molecular and Cellular Biology*, *28*, 4999–5013.

Hayes, A. J., Benjamin, M., & Ralphs, J. R. (1999). Role of actin stress fibres in the development of the intervertebral disc: Cytoskeletal control of extracellular matrix assembly. *Developmental Dynamics: An Official Publication of the American Association of Anatomists*, *215*, 179–189.

Hayes, A. J., Hughes, C. E., Ralphs, J. R., & Caterson, B. (2011). Chondroitin sulphate sulphation motif expression in the ontogeny of the intervertebral disc. *European Cells & Materials*, *21*, 1–14.

Hayes, A. J., Isaacs, M. D., Hughes, C., Caterson, B., & Ralphs, J. R. (2011). Collagen fibrillogenesis in the development of the annulus fibrosus of the intervertebral disc. *European Cells & Materials*, *22*, 226–241.

Hayes, A. J., & Ralphs, J. R. (2011). The response of foetal annulus fibrosus cells to growth factors: Modulation of matrix synthesis by TGF-beta1 and IGF-1. *Histochemistry and Cell Biology*, *136*, 163–175.

Huang, A. H., Lu, H. H., & Schweitzer, R. (2015). Molecular regulation of tendon cell fate during development. *Journal of Orthopaedic Research*, *33*, 800–812.

Huang, R., Zhi, Q., Brand-Saberi, B., & Christ, B. (2000). New experimental evidence for somite resegmentation. *Anatomy and Embryology*, *202*, 195–200.

Kaestner, K. H., Hiemisch, H., Luckow, B., & Schutz, G. (1994). The HNF-3 gene family of transcription factors in mice: Gene structure, cDNA sequence, and mRNA distribution. *Genomics*, *20*, 377–385.

Kahn, A. J. (2015). FOXO3 and related transcription factors in development, aging, and exceptional longevity. *The Journals of Gerontology. Series A, Biological Sciences and Medical Sciences*, *70*, 421–425.

Kalamajski, S., Bihan, D., Bonna, A., Rubin, K., & Farndale, R. W. (2016). Fibromodulin interacts with collagen cross-linking sites and activates Lysyl oxidase. *The Journal of Biological Chemistry*, *291*, 7951–7960.

Kalcheim, C., & Ben-Yair, R. (2005). Cell rearrangements during development of the somite and its derivatives. *Current Opinion in Genetics & Development*, *15*, 371–380.

Karasugi, T., Semba, K., Hirose, Y., Kelempisioti, A., Nakajima, M., Miyake, A., et al. (2009). Association of the tag SNPs in the human SKT gene (KIAA1217) with lumbar disc herniation. *Journal of Bone and Mineral Research*, *24*, 1537–1543.

Kawamura, A., Koshida, S., & Takada, S. (2008). Activator-to-repressor conversion of T-box transcription factors by the Ripply family of Groucho/TLE-associated mediators. *Molecular and Cellular Biology*, *28*, 3236–3244.

Kispert, A., Herrmann, B. G., Leptin, M., & Reuter, R. (1994). Homologs of the mouse Brachyury gene are involved in the specification of posterior terminal structures in Drosophila, Tribolium, and Locusta. *Genes & Development*, *8*, 2137–2150.

Kraus, P., Sivakamasundari, V., Olsen, V., Villeneuve, V., Hinds, A., & Lufkin, T. (2018). Klhl14 antisense RNA is a target of key Skeletogenic transcription factors in the developing intervertebral disc. *Spine*, https://doi.org/10.1097/BRS.0000000000002827. [Epub ahead of print].

Kuta, A., Mao, Y., Martin, T., Ferreira de Sousa, C., Whiting, D., Zakaria, S., et al. (2016). Fat4-Dchs1 signalling controls cell proliferation in developing vertebrae. *Development (Cambridge, England)*, *143*, 2367–2375.

Lawson, L., & Harfe, B. D. (2015). Notochord to nucleus pulposus transition. *Current Osteoporosis Reports*, *13*, 336–341.
Lee, C. R., Sakai, D., Nakai, T., Toyama, K., Mochida, J., Alini, M., et al. (2007). A phenotypic comparison of intervertebral disc and articular cartilage cells in the rat. *European Spine Journal*, *16*, 2174–2185.
Lefebvre, V., Behringer, R. R., & de Crombrugghe, B. (2001). L-Sox5, Sox6 and Sox9 control essential steps of the chondrocyte differentiation pathway. *Osteoarthritis and Cartilage*, *9*(Suppl. A), S69–S75.
Leitges, M., Neidhardt, L., Haenig, B., Herrmann, B. G., & Kispert, A. (2000). The paired homeobox gene Uncx4.1 specifies pedicles, transverse processes and proximal ribs of the vertebral column. *Development (Cambridge, England)*, *127*, 2259–2267.
Lenas, P., Luyten, F. P., Doblare, M., Nicodemou-Lena, E., & Lanzara, A. E. (2011). Modularity in developmental biology and artificial organs: A missing concept in tissue engineering. *Artificial Organs*, *35*, 656–662.
Lenas, P., Moos, M., & Luyten, F. P. (2009a). Developmental engineering: A new paradigm for the design and manufacturing of cell-based products. Part I: From three-dimensional cell growth to biomimetics of in vivo development. *Tissue Engineering. Part B, Reviews*, *15*, 381–394.
Lenas, P., Moos, M., & Luyten, F. P. (2009b). Developmental engineering: A new paradigm for the design and manufacturing of cell-based products. Part II: From genes to networks: Tissue engineering from the viewpoint of systems biology and network science. *Tissue Engineering. Part B, Reviews*, *15*, 395–422.
Li, T. F., Darowish, M., Zuscik, M. J., Chen, D., Schwarz, E. M., Rosier, R. N., et al. (2006). Smad3-deficient chondrocytes have enhanced BMP signaling and accelerated differentiation. *Journal of Bone and Mineral Research: The Official Journal of the American Society for Bone and Mineral Research*, *21*, 4–16.
Liu, Y., Rahaman, M. N., & Bal, B. S. (2014). Modulating notochordal differentiation of human induced pluripotent stem cells using natural nucleus pulposus tissue matrix. *PLoS One*, *9*, e100885.
Loh, K. M., Chen, A., Koh, P. W., Deng, T. Z., Sinha, R., Tsai, J. M., et al. (2016). Mapping the pairwise choices leading from pluripotency to human bone, heart, and other mesoderm cell types. *Cell*, *166*, 451–467.
Maier, J. A., Lo, Y., & Harfe, B. D. (2013). Foxa1 and Foxa2 are required for formation of the intervertebral discs. *PLoS One*, *8*, e55528.
Mansouri, A., Pla, P., Larue, L., & Gruss, P. (2001). Pax3 acts cell autonomously in the neural tube and somites by controlling cell surface properties. *Development (Cambridge, England)*, *128*, 1995–2005.
Martins, R., Lithgow, G. J., & Link, W. (2016). Long live FOXO: Unraveling the role of FOXO proteins in aging and longevity. *Aging Cell*, *15*, 196–207.
McCann, M. R., Tamplin, O. J., Rossant, J., & Seguin, C. A. (2012). Tracing notochord-derived cells using a Noto-cre mouse: Implications for intervertebral disc development. *Disease Models & Mechanisms*, *5*, 73–82.
Merceron, C., Mangiavini, L., Robling, A., Wilson, T. L., Giaccia, A. J., Shapiro, I. M., et al. (2014). Loss of HIF-1alpha in the notochord results in cell death and complete disappearance of the nucleus pulposus. *PLoS One*, *9*, e110768.
Mikawa, T., Poh, A. M., Kelly, K. A., Ishii, Y., & Reese, D. E. (2004). Induction and patterning of the primitive streak, an organizing center of gastrulation in the amniote. *Developmental Dynamics: An Official Publication of the American Association of Anatomists*, *229*, 422–432.
Minogue, B. M., Richardson, S. M., Zeef, L. A., Freemont, A. J., & Hoyland, J. A. (2010). Characterization of the human nucleus pulposus cell phenotype and evaluation of novel marker gene expression to define adult stem cell differentiation. *Arthritis and Rheumatism*, *62*, 3695–3705.

Mittapalli, V. R., Huang, R., Patel, K., Christ, B., & Scaal, M. (2005). Arthrotome: A specific joint forming compartment in the avian somite. *Developmental Dynamics: An Official Publication of the American Association of Anatomists*, *234*, 48–53.

Mitter, D., Krakow, D., Farrington-Rock, C., & Meinecke, P. (2008). Expanded clinical spectrum of spondylocarpotarsal synostosis syndrome and possible manifestation in a heterozygous father. *American Journal of Medical Genetics. Part A*, *146a*, 779–783.

Monaghan, A. P., Kaestner, K. H., Grau, E., & Schutz, G. (1993). Postimplantation expression patterns indicate a role for the mouse forkhead/HNF-3 alpha, beta and gamma genes in determination of the definitive endoderm, chordamesoderm and neuroectoderm. *Development*, *119*, 567–578.

Monsoro-Burq, A. H. (2005). Sclerotome development and morphogenesis: When experimental embryology meets genetics. *The International Journal of Developmental Biology*, *49*, 301–308.

Morimoto, M., Sasaki, N., Oginuma, M., Kiso, M., Igarashi, K., Aizaki, K., et al. (2007). The negative regulation of Mesp2 by mouse Ripply2 is required to establish the rostro-caudal patterning within a somite. *Development (Cambridge, England)*, *134*, 1561–1569.

Moussad, E. E., & Brigstock, D. R. (2000). Connective tissue growth factor: What's in a name? *Molecular Genetics and Metabolism*, *71*, 276–292.

Murchison, N. D., Price, B. A., Conner, D. A., Keene, D. R., Olson, E. N., Tabin, C. J., et al. (2007). Regulation of tendon differentiation by scleraxis distinguishes force-transmitting tendons from muscle-anchoring tendons. *Development (Cambridge, England)*, *134*, 2697–2708.

Murtaugh, L. C., Chyung, J. H., & Lassar, A. B. (1999). Sonic hedgehog promotes somitic chondrogenesis by altering the cellular response to BMP signaling. *Genes & Development*, *13*, 225–237.

Murtaugh, L. C., Zeng, L., Chyung, J. H., & Lassar, A. B. (2001). The chick transcriptional repressor Nkx3.2 acts downstream of Shh to promote BMP-dependent axial chondrogenesis. *Developmental Cell*, *1*, 411–422.

Muthukumar, N. (2016). Proatlas segmentation anomalies: Surgical management of five cases and review of the literature. *Journal of Pediatric Neurosciences*, *11*, 14–19.

Nakamichi, R., Ito, Y., Inui, M., Onizuka, N., Kayama, T., Kataoka, K., et al. (2016). Mohawk promotes the maintenance and regeneration of the outer annulus fibrosus of intervertebral discs. *Nature Communications*, 7, 12503.

Neubuser, A., Koseki, H., & Balling, R. (1995). Characterization and developmental expression of Pax9, a paired-box-containing gene related to Pax1. *Developmental Biology*, *170*, 701–716.

Ohba, S., He, X., Hojo, H., & McMahon, A. P. (2015). Distinct transcriptional programs underlie Sox9 regulation of the mammalian chondrocyte. *Cell Reports*, *12*, 229–243.

Oldershaw, R. A., Baxter, M. A., Lowe, E. T., Bates, N., Grady, L. M., Soncin, F., et al. (2010). Directed differentiation of human embryonic stem cells toward chondrocytes. *Nature Biotechnology*, *28*, 1187–1194.

Paavola, L. G., Wilson, D. B., & Center, E. M. (1980). Histochemistry of the developing notochord, perichordal sheath and vertebrae in Danforth's short-tail (sd) and normal C57BL/6 mice. *Journal of Embryology and Experimental Morphology*, *55*, 227–245.

Peacock, A. (1951). Observations on the prenatal development of the intervertebral disc in man. *Journal of Anatomy*, *85*, 260–274.

Peck, S. H., McKee, K. K., Tobias, J. W., Malhotra, N. R., Harfe, B. D., & Smith, L. J. (2017). Whole transcriptome analysis of notochord-derived cells during embryonic formation of the nucleus pulposus. *Scientific Reports*, 7, 10504.

Perbal, B. (2004). CCN proteins: Multifunctional signalling regulators. *Lancet*, *363*, 62–64.

Peters, H., Wilm, B., Sakai, N., Imai, K., Maas, R., & Balling, R. (1999). Pax1 and Pax9 synergistically regulate vertebral column development. *Development (Cambridge, England)*, *126*, 5399–5408.

Pourquie, O. (2011). Vertebrate segmentation: From cyclic gene networks to scoliosis. *Cell, 145*, 650–663.
Pryce, B. A., Watson, S. S., Murchison, N. D., Staverosky, J. A., Dunker, N., & Schweitzer, R. (2009). Recruitment and maintenance of tendon progenitors by TGFbeta signaling are essential for tendon formation. *Development, 136*, 1351–1361.
Purmessur, D., Guterl, C. C., Cho, S. K., Cornejo, M. C., Lam, Y. W., Ballif, B. A., et al. (2013). Dynamic pressurization induces transition of notochordal cells to a mature phenotype while retaining production of important patterning ligands from development. *Arthritis Research & Therapy, 15*, R122.
Richardson, S. M., Ludwinski, F. E., Gnanalingham, K. K., Atkinson, R. A., Freemont, A. J., & Hoyland, J. A. (2017). Notochordal and nucleus pulposus marker expression is maintained by sub-populations of adult human nucleus pulposus cells through aging and degeneration. *Scientific Reports, 7*, 1501.
Rider, C. C., & Mulloy, B. (2010). Bone morphogenetic protein and growth differentiation factor cytokine families and their protein antagonists. *The Biochemical Journal, 429*, 1–12.
Rodrigues-Pinto, R., Ward, L., Humphreys, M., Zeef, L. A. H., Berry, A., Hanley, K. P., et al. (2018). Human notochordal cell transcriptome unveils potential regulators of cell function in the developing intervertebral disc. *Scientific Reports, 8*, 12866.
Rufai, A., Benjamin, M., & Ralphs, J. R. (1995). The development of fibrocartilage in the rat intervertebral disc. *Anatomy and Embryology (Berlin), 192*, 53–62.
Sakai, D., Nakai, T., Mochida, J., Alini, M., & Grad, S. (2009). Differential phenotype of intervertebral disc cells: Microarray and immunohistochemical analysis of canine nucleus pulposus and anulus fibrosus. *Spine (Phila Pa 1976), 34*, 1448–1456.
Salisbury, J. R. (1993). The pathology of the human notochord. *The Journal of Pathology, 171*, 253–255.
Sasaki, H., & Hogan, B. L. (1993). Differential expression of multiple fork head related genes during gastrulation and axial pattern formation in the mouse embryo. *Development, 118*, 47–59.
Schweitzer, R., Chyung, J. H., Murtaugh, L. C., Brent, A. E., Rosen, V., Olson, E. N., et al. (2001). Analysis of the tendon cell fate using Scleraxis, a specific marker for tendons and ligaments. *Development (Cambridge, England), 128*, 3855–3866.
Scotti, C., Tonnarelli, B., Papadimitropoulos, A., Scherberich, A., Schaeren, S., Schauerte, A., et al. (2010). Recapitulation of endochondral bone formation using human adult mesenchymal stem cells as a paradigm for developmental engineering. *Proceedings of the National Academy of Sciences of the United States of America, 107*, 7251–7256.
Sekiya, I., Larson, B. L., Vuoristo, J. T., Reger, R. L., & Prockop, D. J. (2005). Comparison of effect of BMP-2, -4, and -6 on in vitro cartilage formation of human adult stem cells from bone marrow stroma. *Cell and Tissue Research, 320*, 269–276.
Semba, K., Araki, K., Li, Z., Matsumoto, K., Suzuki, M., Nakagata, N., et al. (2006). A novel murine gene, sickle tail, linked to the Danforth's short tail locus, is required for normal development of the intervertebral disc. *Genetics, 172*, 445–456.
Senthinathan, B., Sousa, C., Tannahill, D., & Keynes, R. (2012). The generation of vertebral segmental patterning in the chick embryo. *Journal of Anatomy, 220*, 591–602.
Sivakamasundari, V., Kraus, P., Sun, W., Hu, X., Lim, S. L., Prabhakar, S., et al. (2017). A developmental transcriptomic analysis of Pax1 and Pax9 in embryonic intervertebral disc development. *Biology Open, 6*, 187–199.
Smits, P., & Lefebvre, V. (2003). Sox5 and Sox6 are required for notochord extracellular matrix sheath formation, notochord cell survival and development of the nucleus pulposus of intervertebral discs. *Development, 130*, 1135–1148.
Sohn, P., Cox, M., Chen, D., & Serra, R. (2010). Molecular profiling of the developing mouse axial skeleton: A role for Tgfbr2 in the development of the intervertebral disc. *BMC Developmental Biology, 10*, 29.

Spittank, H., Goehmann, U., Hage, H., & Sacher, R. (2016). Persistent proatlas with additional segmentation of the craniovertebral junction—The Tsuang-Goehmann-malformation. *Journal of Radiology Case Reports, 10*, 15–23.

Stafford, D. A., Brunet, L. J., Khokha, M. K., Economides, A. N., & Harland, R. M. (2011). Cooperative activity of noggin and gremlin 1 in axial skeleton development. *Development (Cambridge, England), 138*, 1005–1014.

Stafford, D. A., Monica, S. D., & Harland, R. M. (2014). Follistatin interacts with noggin in the development of the axial skeleton. *Mechanisms of Development, 131*, 78–85.

Stemple, D. L. (2005). Structure and function of the notochord: An essential organ for chordate development. *Development, 132*, 2503–2512.

Stockdale, F. E., Nikovits, W., Jr., & Christ, B. (2000). Molecular and cellular biology of avian somite development. *Developmental Dynamics: An Official Publication of the American Association of Anatomists, 219*, 304–321.

Stott, D., Kispert, A., & Herrmann, B. G. (1993). Rescue of the tail defect of Brachyury mice. *Genes & Development, 7*, 197–203.

Sugimoto, Y., Takimoto, A., Akiyama, H., Kist, R., Scherer, G., Nakamura, T., et al. (2013). Scx+/Sox9+ progenitors contribute to the establishment of the junction between cartilage and tendon/ligament. *Development (Cambridge, England), 140*, 2280–2288.

Sulik, K., Dehart, D. B., Iangaki, T., Carson, J. L., Vrablic, T., Gesteland, K., et al. (1994). Morphogenesis of the murine node and notochordal plate. *Developmental Dynamics, 201*, 260–278.

Swiderski, R. E., & Solursh, M. (1992). Localization of type II collagen, long form alpha 1 (IX) collagen, and short form alpha 1(IX) collagen transcripts in the developing chick notochord and axial skeleton. *Developmental Dynamics, 194*, 118–127.

Tam, P. P., & Behringer, R. R. (1997). Mouse gastrulation: The formation of a mammalian body plan. *Mechanisms of Development, 68*, 3–25.

Tamplin, O. J., Cox, B. J., & Rossant, J. (2011). Integrated microarray and ChIP analysis identifies multiple Foxa2 dependent target genes in the notochord. *Developmental Biology, 360*, 415–425.

Tanaka, T., & Uhthoff, H. K. (1981). Significance of resegmentation in the pathogenesis of vertebral body malformation. *Acta Orthopaedica Scandinavica, 52*, 331–338.

Tang, R., Jing, L., Willard, V. P., Wu, C. L., Guilak, F., Chen, J., et al. (2018). Differentiation of human induced pluripotent stem cells into nucleus pulposus-like cells. *Stem Cell Research & Therapy, 9*, 61.

Tassabehji, M., Fang, Z. M., Hilton, E. N., McGaughran, J., Zhao, Z., de Bock, C. E., et al. (2008). Mutations in GDF6 are associated with vertebral segmentation defects in Klippel-Feil syndrome. *Human Mutation, 29*, 1017–1027.

Theiler, K. (1988). Vertebral malformations. *Advances in Anatomy, Embryology, and Cell Biology, 112*, 1–99.

Tribioli, C., & Lufkin, T. (1999). The murine Bapx1 homeobox gene plays a critical role in embryonic development of the axial skeleton and spleen. *Development (Cambridge, England), 126*, 5699–5711.

Umegaki, M., Miyao, Y., Sasaki, M., Yoshimura, K., Iwatsuki, K., & Yoshimine, T. (2017). Report of a familial case of proatlas segmentation abnormality with late clinical onset. *Journal of Clinical Neuroscience: Official Journal of the Neurosurgical Society of Australasia, 39*, 79–81.

van den Akker, G. G. H., Koenders, M. I., van de Loo, F. A. J., van Lent, P., Blaney Davidson, E., & van der Kraan, P. M. (2017). Transcriptional profiling distinguishes inner and outer annulus fibrosus from nucleus pulposus in the bovine intervertebral disc. *European Spine Journal: Official Publication of the European Spine Society, the European Spinal Deformity Society, and the European Section of the Cervical Spine Research Society, 26*, 2053–2062.

van den Akker, G. G., Surtel, D. A., Cremers, A., Hoes, M. F., Caron, M. M., Richardson, S. M., et al. (2016). EGR1 controls divergent cellular responses of distinctive nucleus pulposus cell types. *BMC Musculoskeletal Disorders*, *17*, 124.
van den Akker, G. G., Surtel, D. A., Cremers, A., Rodrigues-Pinto, R., Richardson, S. M., Hoyland, J. A., et al. (2014). Novel immortal human cell lines reveal subpopulations in the nucleus pulposus. *Arthritis Research & Therapy*, *16*, R135.
Ward, L., Evans, S. E., & Stern, C. D. (2017). A resegmentation-shift model for vertebral patterning. *Journal of Anatomy*, *230*, 290–296.
Ward, L., Pang, A. S. W., Evans, S. E., & Stern, C. D. (2018). The role of the notochord in amniote vertebral column segmentation. *Developmental Biology*, *439*, 3–18.
Wiggan, O., Fadel, M. P., & Hamel, P. A. (2002). Pax3 induces cell aggregation and regulates phenotypic mesenchymal-epithelial interconversion. *Journal of Cell Science*, *115*, 517–529.
Yamaguchi, T. P., Takada, S., Yoshikawa, Y., Wu, N., & McMahon, A. P. (1999). T (Brachyury) is a direct target of Wnt3a during paraxial mesoderm specification. *Genes & Development*, *13*, 3185–3190.
Yoon, B. S., Ovchinnikov, D. A., Yoshii, I., Mishina, Y., Behringer, R. R., & Lyons, K. M. (2005). Bmpr1a and Bmpr1b have overlapping functions and are essential for chondrogenesis in vivo. *Proceedings of the National Academy of Sciences of the United States of America*, *102*, 5062–5067.
Yoshimoto, Y., Takimoto, A., Watanabe, H., Hiraki, Y., Kondoh, G., & Shukunami, C. (2017). *Scleraxis* is required for maturation of tissue domains for proper integration of the musculoskeletal system. *Scientific Reports*, 7. article number 45010.
Zakin, L., Chang, E. Y., Plouhinec, J. L., & De Robertis, E. M. (2010). Crossveinless-2 is required for the relocalization of chordin protein within the vertebral field in mouse embryos. *Developmental Biology*, *347*, 204–215.
Zakin, L., Metzinger, C. A., Chang, E. Y., Coffinier, C., & De Robertis, E. M. (2008). Development of the vertebral morphogenetic field in the mouse: Interactions between Crossveinless-2 and twisted gastrulation. *Developmental Biology*, *323*, 6–18.
Zeng, L., Kempf, H., Murtaugh, L. C., Sato, M. E., & Lassar, A. B. (2002). Shh establishes an Nkx3.2/Sox9 autoregulatory loop that is maintained by BMP signals to induce somitic chondrogenesis. *Genes & Development*, *16*, 1990–2005.
Zhao, J., Li, S., Trilok, S., Tanaka, M., Jokubaitis-Jameson, V., Wang, B., et al. (2014). Small molecule-directed specification of sclerotome-like chondroprogenitors and induction of a somitic chondrogenesis program from embryonic stem cells. *Development (Cambridge, England)*, *141*, 3848–3858.
Zieba, J., Forlenza, K. N., Khatra, J. S., Sarukhanov, A., Duran, I., Rigueur, D., et al. (2016). TGFbeta and BMP dependent cell fate changes due to loss of Filamin B produces disc degeneration and progressive vertebral fusions. *PLoS Genetics*, *12*, e1005936.

CHAPTER FOUR

Regulatory mechanisms of jaw bone and tooth development

Yuan Yuan*, Yang Chai*
Center for Craniofacial Molecular Biology, University of Southern California, Los Angeles, CA, United States
*Corresponding authors: e-mail address: yuan164@usc.edu; ychai@usc.edu

Contents

Abstract

Jaw bones and teeth originate from the first pharyngeal arch and develop in closely related ways. Reciprocal epithelial-mesenchymal interactions are required for the early patterning and morphogenesis of both tissues. Here we review the cellular contribution during the development of the jaw bones and teeth. We also highlight signaling networks as well as transcription factors mediating tissue-tissue interactions that are essential for jaw bone and tooth development. Finally, we discuss the potential for stem cell mediated regenerative therapies to mitigate disorders and injuries that affect these organs.

Current Topics in Developmental Biology, Volume 133
ISSN 0070-2153
https://doi.org/10.1016/bs.ctdb.2018.12.013

1. An overview of jaw bone and tooth development

The maxilla and mandible together form the lower part of the facial skeleton, which performs important functions in our daily life. The jaw bones serve as anchors for the teeth, which are critical for mastication and speech.

In vertebrates, the maxilla and mandible, like most of the other craniofacial bones, are derived from cranial neural crest cells (CNCCs). These cells are known for their multipotency and their extensive migration through the embryo (Chai et al., 2000; Le Douarin, Creuzet, Couly, & Dupin, 2004; Noden, 1975; Thiery, Duband, & Delouvee, 1982). During early development, CNCCs migrate out from the hindbrain (rhombomere segments r1–r7), traveling along the dorsal-ventral axis as loosely connected streams that ultimately come to populate the pharyngeal arches. Shortly after first pharyngeal arch (PA1) patterning, a group of mesenchymal cells condenses and develops into Meckel's cartilage (MC). The MC in each half of the mandible lengthens ventromedially and dorsolaterally, until the two eventually come together to fuse at the distal tip of the mandibular arch (Richany, Bast, & Anson, 1956). Meanwhile, lateral to the MC, mandibular bone starts to form. In the maxilla, the ossification process begins slightly later than in the mandible. At the cellular level, condensed mesenchymal cells undergo differentiation into osteoblasts with the guidance of a series of osteogenic transcriptional regulators, such as Dlx5, Runx2, and Osterix (Baek, Kim, de Crombrugghe, & Kim, 2013; Zhang, 2010). As mandibular ossification progresses, the bony tissue approaches and ultimately wraps around the MC, while the cartilaginous tissue of the MC becomes hypertrophic and degenerates in a process similar to endochondral ossification. Eventually, multinuclear phagocytotic cells called chondroclasts resorb the calcified cartilaginous matrix. In the most distal and proximal regions, the symphysis, condyle, and mandibular angle are formed through endochondral ossification. The rest of the posterior portion of MC may contribute to the formation of the sphenomandibular ligament (Moore, Persaud, & Samperio, 1999). Jaw bone development continues postnatally and ceases around 20 years of age (Love, Murray, & Mamandras, 1990).

Tooth development can be roughly divided into two major events: crown formation, which happens mainly at the embryonic stages, and root development, which begins around postnatal day 3 in the mouse. The first morphological sign of tooth initiation in the mouse is evident at around embryonic day (E)11.5 (Theiler stage 19), with the thickening of epithelial tissue called the dental placode. This tissue continues to proliferate and form the tooth bud. Meanwhile, the mesenchymal tissue around the tooth bud condenses and forms the tooth germ. With the proliferation and in-folding of the epithelium, the tooth bud progresses through the cap and bell stages. During these stages, stem cells residing in the dental mesenchyme and dental epithelium become committed and form odontoblasts and ameloblasts, respectively. Odontoblasts form dentin whereas ameloblasts contribute to enamel formation.

After the crown has formed, the dental epithelium elongates and grows apically to form a bilayered epithelial structure between the dental papilla and dental follicle called Hertwig's epithelial root sheath (HERS), which functions as a signaling center to guide root formation. In mammals, HERS is a transient structure. After its movement to the cervical loop of the enamel organ, it undergoes perforation and eventually apoptosis, leaving a mesh-like matrix on the root surface. CNC-derived dental mesenchyme is also critical for this developmental event. It gives rise to multiple tissue types including odontoblasts, dental pulp cells, cementoblasts and periodontal ligament (PDL) cells. Traditionally, researchers believed that mesenchymal cells receive signals from the HERS for tooth root elongation (Cate, 1996). Recently, using an inducible Cre line, researchers began to uncover important cell populations as well as signaling within mesenchymal tissue that also play essential roles during tooth root development (Feng et al., 2017; Li, Parada, & Chai, 2017).

In the following sections, we will take a closer look at the different cellular components and molecular networks that regulate different stages of jaw bone development, then turn to tooth development. We will also discuss the potential for stem cell mediated regenerative therapies to mitigate disorders and injuries that affect these organs.

2. Early development of the first pharyngeal arch

2.1 Cellular contributions to mandible and maxilla development

The neural crest is a fascinating and extensively studied cell population largely due to its unique properties. Neural crest cells (NCCs) originate at the ectodermal border of the neural plate. As the neural tube closes, the NCCs undergo epithelial to mesenchymal transition (EMT) and migrate into the mesodermal mesenchyme ventrolaterally; therefore, they are referred to as ectomesenchymal cells (Loring & Erickson, 1987; Teillet, Kalcheim, & Le Douarin, 1987). Based on their original location along the rostral-caudal axis, NCCs can be further divided into four populations: cranial, cardiac, vagal, and trunk (Gilbert, 2000). CNCCs contribute to most of the craniofacial bones, including the maxilla and mandible, as well as cartilage, nerves, and connective tissue in the face. They migrate out of the dorsal neural tube and soon divide into streams which will later enter the pharyngeal arches. This striking pattern of NCC migration is closely related to the rhombomeric organization of the hindbrain. Two gene families are particularly critical for establishing unique segmental identities for hindbrain: the Hox genes and Ephrin/Eph receptors. Extensive studies have shown that Hox genes account for the antero-posterior identity of rhombomeric segments (Barrow, Stadler, & Capecchi, 2000; Krumlauf, 1994; Lumsden & Krumlauf, 1996; McGinnis & Krumlauf, 1992; Studer et al., 1998). During the period of craniofacial morphogenesis in the mouse embryo, Hox gene expression is not detectable in NCCs derived from rhombomeres 1 and 2, which later form the entire facial skeleton. Targeted inactivation of *Hoxa2* results in homeotic transformation of skeletal elements derived from the second branchial arch into more anterior structures, leading to a duplication of MC adjacent to the otic capsule (Gendron-Maguire, Mallo, Zhang, & Gridley, 1993; Rijli et al., 1993). Interestingly, NCCs derived from both Hox-negative and Hox-positive regions can further differentiate into cartilage and bone; however, intramembranous ossification only takes place in the Hox-negative region. Higher osteogenic capacity and more robust *in vivo* bone regeneration have also been observed in progenitor cells derived from Hox-negative CNCCs compared to skeletal progenitor cells from the mesoderm (Chung et al., 2009; Leucht et al., 2008).

The Ephrins and Eph receptors are expressed in the different rhombomeres in a non-overlapping pattern that mediates cell sorting at the boundaries of odd- and even-numbered rhombomeres. This cell sorting process is essential for the segmental streams of migrating NCCs and for preventing the intermingling of cells between adjacent rhombomeres (Mellitzer, Xu, & Wilkinson, 1999; Xu, Mellitzer, Robinson, & Wilkinson, 1999). Inactivation of this signaling pathway using truncated Eph receptors disturbs the boundaries and leads to abnormal migration of third arch NCCs into the second and fourth arch territories (Smith, Robinson, Patel, & Wilkinson, 1997).

Other mechanisms that govern CNCC migration include contact inhibition of locomotion and cell repolarization, co-attraction, and chemotaxis controlled by multiple signaling molecules, such as Complement3a (C3a) and the C3a receptor, stromal cell-derived factor (SDF), VEGF, GDNF, and endothelin (Carmona-Fontaine et al., 2008; Olesnicky Killian, Birkholz, & Artinger, 2009). Simoes-Costa and Bronner identified a hierarchical gene regulatory network composed of a series of transcriptional factors that are specifically expressed in CNCCs in a spatially and temporally restricted manner during CNCC induction and early migration (Simoes-Costa & Bronner, 2016).

Cells from paraxial mesoderm give rise to the muscle component and some skeletal tissue in the posterior part of the head. In the mandibular arch, mesodermal tissue in the center is surrounded by CNCCs, with a clear cell-cell boundary between the two; together they form the mesenchymal core of the pharyngeal arch (Chai & Maxson, 2006). Even though mesodermal tissue does not directly contribute to jaw bone formation, it can still affect the development of the maxilla and mandible through tissue-tissue interactions. Ablation of *Tbx1*, which is exclusively expressed in the mesoderm, leads to defects in the formation of the proximal mandible (Aggarwal et al., 2010). Pharyngeal ectoderm and endoderm together cover the mesenchymal core. The ectoderm is essential for regulating the fate of CNCCs during mandibular morphogenesis, whereas the establishment of ectodermal identity is independent of CNCCs (Veitch, Begbie, Schilling, Smith, & Graham, 1999). The pharyngeal endoderm makes a limited contribution to craniofacial development. However, the pharyngeal pouch, which is formed by endodermal tissue, serves as a signaling center for tissue-tissue interaction. Using lineage tracing techniques, cells from different origins can be visualized in whole embryos and in cross-sections of the first pharyngeal arch. Post-migratory CNCCs, mesoderm- and ectoderm-derived cells are detectable in *Wnt1Cre;R26R*, *Myf5Cre;R26R*, and *K14Cre;R26R* embryos, respectively, which provide valuable information that can help us gain a better understanding of dynamic cell-cell interactions and identify the regulatory mechanisms that are active during craniofacial development (Chai & Maxson, 2006).

2.2 Molecular identity of the developing mandible and maxilla

Multiple heterotopic graft experiments in birds and amphibians have clearly suggested that jaw patterning information is passively carried by the NCCs and maintained throughout subsequent development (Noden, 1978a, 1978b, 1983). However, other experiments also demonstrated the importance of environmental cues from other tissues, including cephalic ectoderm, neuroectoderm, and pharyngeal endoderm. For example, tissue-specific loss-of-function of Fgf8 in the first arch ectoderm of murine embryos results in a severe mandible phenotype with loss of the majority of the bone structure (Trumpp, Depew, Rubenstein, Bishop, & Martin, 1999). The fact that numerous genes involved in jaw development are turned on only after CNCCs reach PA1 suggests that patterning of CNCCs within PA1 also relies on environmental cues.

Taking these findings into account, the next step is to address how CNCCs are patterned within PA1. Depew and Compagnucci proposed a very interesting predictive model, known as the "hinge and caps" model, to explain the jaw patterning process (Depew & Compagnucci, 2008). Genes such as *Satb2*, expressed in the "caps" located near the distal midline of the mandibular process of the first arch and the lambdoidal junctions where the frontonasal prominence meets the maxillary process, are important for the coordination and evolution of the jaws (Depew & Compagnucci, 2008). This model explains some of the similarities between the maxilla and mandible during development. However, how the distinct identities of the upper and lower jaws are established at this early patterning stage is a question still needing to be answered. Endothelin signaling-mediated expression of distal-less genes is critical for establishing the difference between the maxilla and mandible. During PA1 patterning, Endothelin 1 is expressed in the ectoderm at the distal end of the mandibular process whereas Endothelin receptor A (Ednra) is expressed exclusively in the mesenchyme with an intensity gradient from the distal to the proximal region, suggesting that endothelin signaling mediates mandible patterning through epithelial-mesenchymal interaction. Loss of Ednra results in a homeotic transformation of mandible to maxilla, supporting its important function in mandible identity establishment (Sato et al., 2008). Molecularly, Dlx5/6 expression is downregulated in the mandibular processes of *Ednra* mutant mice (Ruest, Xiang, Lim, Levi, & Clouthier, 2004). *Dlx5/6*$^{-/-}$ mice show a similar phenotype, suggesting that Dlx5/6 are downstream transcriptional regulators of endothelin signaling (Depew, Lufkin, & Rubenstein, 2002). *Six1,* which is expressed on the oral side of both the maxillary and mandibular processes, negatively regulates endothelin signaling to maintain maxillary identity. In *Six1*$^{-/-}$ mice, ectopic endothelin signaling is found in the proximal end of first pharyngeal arch, which leads to the formation of a cartilage-capped, rod-shaped bone at the zygomatic arch (Tavares, Cox, Maxson, Ford, & Clouthier, 2017). Recently, using a zebrafish model

Barske and colleagues found that Nr2f nuclear receptor plays a crucial role independent of endothelin signaling in the patterning of the upper jaw (Barske et al., 2018). This finding broadened the general consensus regarding patterning of the maxilla and mandible and strongly suggests that multiple signaling pathways contribute to inter-pharyngeal arch patterning.

Within the mandibular process of the first pharyngeal arch, the CNCC-derived mesenchyme is patterned along the proximal-distal and oral-aboral axes (Chai & Maxson, 2006). During this process, mesenchymal cells receive signals secreted from mandibular ectoderm, activating downstream patterning genes. Major signaling pathways such as Bmp, Fgf, Shh and Wnt form a complex regulatory network to control the establishment of different domains (Fig. 1). Fgf8 is expressed at the proximal end of the mandibular process epithelium, activating multiple patterning genes expressed in the oral mesenchyme, including two specific Lim-homeobox domain genes, *Lhx6* and *Lhx7*. Meanwhile, Bmp4 expressed in the distal epithelium

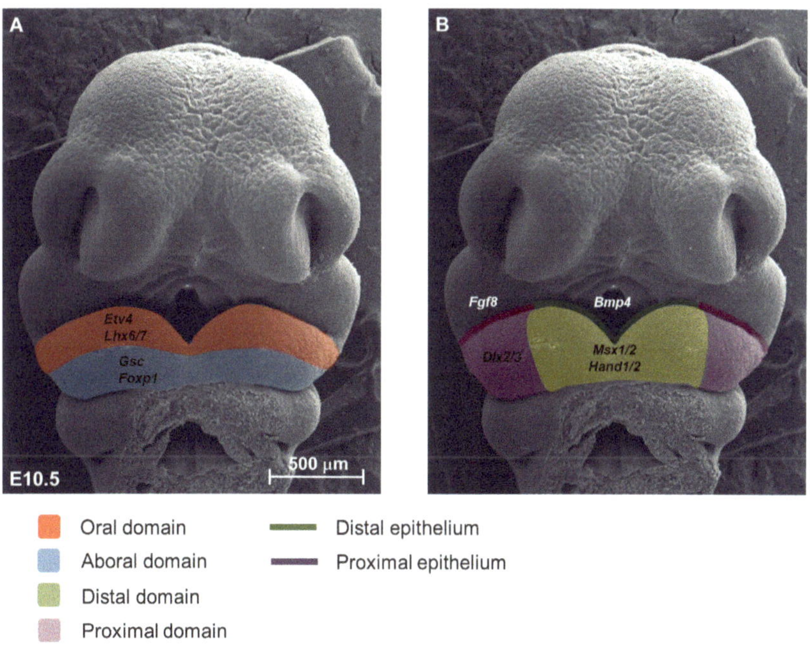

Fig. 1 Patterning of the first branchial arch. Frontal view of a scanning electron microscopic image of an embryonic day (E) 10.5 mouse embryo shows that the first branchial arch can be divided into oral/aboral domains (A) and proximal/distal domains (B). In the oral domain, Fgf signaling from the proximal oral epithelium regulates the expression of *Etv4* and *Lhx6/7*, which prevent the expansion of *Gsc* expression in the aboral domain. *Bmp4* is expressed in the distal oral epithelium, regulating the expression of *Msx1* and *Msx2*.

antagonizes Fgf signaling to establish the distal domain (Fig. 1B). The genes *Msx1* and *Alx*, which are expressed in the distal mesenchyme, have also been identified as downstream targets of Bmp signaling. Wnt signaling overlaps with Bmp signaling, and loss of Wnt signaling by knocking out R-spondin2 (Rspo2), a canonical Wnt signaling activator, results in downregulation of both Bmp and Fgf signaling, suggesting a tight signaling interaction between the proximal and distal domains (Jin, Turcotte, Crocker, Han, & Yoon, 2011). Shh ligand, which is expressed in the pharyngeal endoderm during patterning, is also a survival factor for CNC-derived mesenchymal cells. Proximal Shh signaling is downstream of Wnt signaling in the distal domain, whereas both Wnt and Shh signaling are mediated by epithelial Islet expression (Li, Fu, et al., 2017). Tucker and colleagues have demonstrated the biological significance of proximal-distal domain establishment by blocking Bmp signaling at the distal end of the mandibular process. They found a transformation of tooth identity from incisor to molar with ectopic expression of Barx1 in the distal mesenchyme (Tucker, Matthews, & Sharpe, 1998).

Unlike the proximal-distal axis, the regulatory mechanism of oral-aboral patterning remains elusive. Fgf signaling is certainly critical for multiple genes expressed in the oral domain of the CNC-derived mesenchyme. Specifically, Fgf8 is expressed in the proximal oral ectoderm of both the maxillary and mandibular processes, and its downstream read-out, Etv4, is expressed in the oral half of the mandibular mesenchyme. Lhx6/7, which is directly regulated by Fgf8, is also expressed in the oral mesenchyme (Fig. 1A). Interestingly, Fgf8 has also been reported to positively regulate goosecoid (Gsc) expression, which is present in the Lhx6/7-negative aboral region (Tucker, Yamada, Grigoriou, Pachnis, & Sharpe, 1999). Endothelin 1 (Et-1) is another signaling molecule that might control oral-aboral axis establishment. Et-1 is expressed in the aboral ectoderm of the mandibular process. Multiple studies have shown that endothelin signaling is important for restricting oral mesenchyme gene expansion into the aboral side. For example, endothelin can directly induce Gsc expression and negatively regulate Six1 expression (Tavares et al., 2017; Tucker et al., 1999). The biological function of oral-aboral axis patterning is still not very clear. Interestingly, both *Gsc* and *Lhx6/7* knockout mice have shortened mandibles; however, *Gsc* mutant mice show a defect at the proximal end whereas *Lhx6/7* mutant mice have a defect at the distal end (Denaxa, Sharpe, & Pachnis, 2009; Rivera-Perez, Wakamiya, & Behringer, 1999). This suggests that there might be a dynamic interaction between proximal-distal and oral-aboral axis patterning.

3. Jaw bone development

3.1 Meckel's cartilage

After PA1 is patterned and before any ossification center starts to form, a group of CNC-derived mesenchymal cells condense and differentiate into chondrocytes to form a pair of symmetric, rod-shaped cartilages named Meckel's cartilage (MC). The MCs elongate along the dorsal-ventral axis and fuse at their distal ends to form the mandibular symphysis. Most mammalian cartilages ossify and become bony structures, although some retain their cartilaginous character, such as the tracheal, nasal, and articular cartilages. However, the MC has a more complex fate. Both its distal and proximal extremities undergo ossification to form part of the anterior portion of the mandible bone as well as the incus and malleus bones of the middle ear. The intermediate part of MC undergoes dedifferentiation to become fibrous tissue.

Concerning the functional significance of MC, the consensus is that the presence of MC is indispensable for mandibular development, because it serves as a template for mandibular formation. Mutant mouse models that have abnormal MC formation often show later mandibular defects as well (Li, Fu, et al., 2017; Matsui & Klingensmith, 2014; Yahiro, Higashihori, & Moriyama, 2017). However, the molecular mechanisms and interactions that regulate MC are still largely unknown. Numerous genes and signaling pathways have been reported to relate to the formation and degradation of the MC. Like other cartilaginous tissues, MC formation is also mediated by Sox9, which is a well-known chondrogenic transcriptional regulator expressed in chondroblasts and mature chondrocytes. Loss of Sox9 in the CNCC lineage inevitably affects the formation of MC, causing complete absence of MC during the entire course of mandibular development (Mori-Akiyama, Akiyama, Rowitch, & de Crombrugghe, 2003). Interestingly, although the mandibular bone is severely defective in these mutant mice, early osteogenic markers are found ectopically expressed in the craniofacial region at E15.5. This study suggests that the presence of MC is necessary but not sufficient for mandibular bone development.

Shh signaling is also critical for MC induction. Targeted deletion of Shh expression in these tissues leads to an increase of cell death in the mandibular arch and complete absence of MC (Billmyre & Klingensmith, 2015). This phenotype is also associated with severe mandibular defects. The importance of Shh signaling for MC formation is also revealed by other experiments. Ablation of Islet expression in oral epithelium results in micrognathia with defective MC morphology. Hh signaling is down-regulated in these mutants and the mandible phenotype can be partially rescued using *Isl1*$^{Shh\text{-}Cre}$*;Tg-pmes-Ihh* compound mice, which overexpress transgenic Ihh (Li, Fu, et al., 2017).

During degradation of the MC, autophagy and chondrocyte apoptosis play crucial roles. Starting from E15, Beclin1, a central regulator of autophagy, can be detected in prehypertrophic and hypertrophic chondrocytes located in the central portion facing the proximal end of the incisor teeth. LC3 and Caspase 3 expression is detectable at a slightly later stage in the same location, which suggests that autophagy occurs prior to hypertrophic chondrocyte cell death; this is the final fate of the majority of MC (Yang, Zhang, Liu, Zhou, & Li, 2012). Bmp signaling is involved in the degeneration of MC. In *Noggin*$^{-/-}$ mutant mice, in which Bmp signaling is over-activated, MC is significantly thickened due to elevated cell proliferation and remains in an unossified state at the caudal end at E18.5. With sustained Bmp signaling, the middle portion of MC fails to degrade and undergoes endochondral ossification to form mandibular bone (Wang, Zheng, Chen, & Chen, 2013). Recently, epigenetic regulation, specifically histone methylation, has been revealed as another factor that affects MC degradation. In the absence of Setdb1, an enzyme that methylates the lysine 9 residue of the histone H3 protein (H3K9), the MC develops a similar phenotype to that of *Noggin*$^{-/-}$ mutants. Enlargement and persistence of MC are found with over-activated Bmp signaling (Yahiro et al., 2017). Because methylation of the lysine residues of histone H3 negatively regulates gene expression, this study indicates that the withdrawal of Bmp signaling from MC during degeneration is controlled by Setdb1-mediated histone methylation and is required for normal mandibular development.

3.2 Mandibular bone osteogenesis

Both intramembranous and endochondral ossification contribute to the formation of mandibular bone (Lee et al., 2001). The majority of the mandible (the intermediate portion) is ossified in an intramembranous fashion, which is characterized by mesenchymal stem cells initially proliferating and forming a small, dense cluster. These stem cells then undergo differentiation into osteoblasts with an associated morphological change from spindle-shaped to columnar. Meanwhile, the osteoblasts create an extracellular matrix called osteoid tissue, which contains Type-I collagen fibrils and is able to bind calcium salts (Amano et al., 2010). Finally, the osteoid tissue mineralizes to form rudimentary bone tissue with mature osteocytes in the middle and active osteoblasts at the osteogenic front. During this process, a series of well-defined osteogenic markers are expressed at each stage of differentiation. Dlx5 has been identified as one of the earliest markers to be expressed in committed osteoprogenitor cells. Dlx5 also induces Runx2, which is a master regulator for activating the program of osteoblastogenesis and is also expressed in early committed osteoprogenitor cells (Kawane et al., 2014). Runx2-deficient mice completely lack bone tissue due to the arrested differentiation of osteoblasts. Interestingly, cartilage formation in these mutants is only mildly affected, notably including MC, again suggesting that the presence of MC is not sufficient for mandibular bone development (Shibata et al., 2004). Alp and Osterix (Osx) are two other factors that are involved in the later stages of osteogenesis. They are expressed in the differentiated osteoblasts and are downstream of Runx2. Loss of Osx in CNCC derivatives leads to the absence of almost all craniofacial skeletal structures, suggesting that Osx is required for craniofacial bone formation by CNC-derived cells (Baek et al., 2013).

The distal-most region of the mandibular bone is ossified through endochondral ossification, the other essential bone-forming process that occurs during embryonic development. The onset of this process is similar to chondrogenesis, in which a group of mesenchymal cells condense and differentiate into Sox9-positive chondrocytes to secrete collagen types II, IX, and XI and aggrecan. Then these chondrocytes undergo maturation from a proliferative stage to a hypertrophic stage and eventually undergo apoptosis due to a drastic change in the micro-environment, leaving the cartilaginous remnants as the scaffold for the osteoblasts laying down bone matrix. Recently, lineage tracing studies have enabled us to observe the fate of these hypertrophic chondrocytes (HCs) directly, and there is some recent evidence showing that these HCs may undergo transdifferentiation and continue with a new role in the osteoblast lineage. *Col10a1-Cre* can specifically target HCs, including late HCs that express Mmp13 and Osx. By crossing these mice with a reporter line, it has been found that HC-derived cells become Col1a1-expressing osteoblasts and sclerostin (Sost)-expressing osteocytes during bone tissue development as well as in bone injury repair (Yang, Tsang, Tang, Chan, & Cheah, 2014). Many major signaling pathways are associated with endochondral ossification, including Bmp, Hh, Wnt, Notch, and retinoic acid signaling. In the distal mandible, *Ihh*-null mice show reduced chondroprogenitor cell proliferation that results in altered endochondral ossification and an abnormal mandibular symphysis.

This phenotype is partially rescued by ablation of Gli3 expression, which is a negative regulator of symphyseal development (Sugito et al., 2011). Golgi-associated N-sulfotransferase 1 (Ndst1) also regulates mandibular symphysis development. Ndst1 catalyzes sulfation of heparan sulfate proteoglycan (HS-PG) glycosaminoglycan chains, which are found on the cell surface as well as in the extracellular matrix and mediate numerous developmental processes. *Ndst1*-null mice have severe craniofacial skeletal defects including a fused mandibular symphysis due to ectopic osteogenesis at the newborn stage (Yasuda et al., 2010). Ihh signaling was also found to be expanded during the ossification of the mandible in these mice, which might contribute to the up-regulation of Osterix and collagen I expression seen in these mutants.

The proximal end of the mandible is composed of three eminences, namely the coronoid, condyle, and angular processes. At the osteogenic front, unlike in the primary cartilage, progenitor cells express both osteogenic and chondrogenic markers, such as Runx2, Osterix, and Sox9. Since these cells have the potential to differentiate into either osteoblasts or chondrocytes, they are called osteochondroprogenitor cells. During the fate determination of these progenitor cells, Tgf-β signaling is critical. Conditional inactivation of *Tgfbr2* in CNCCs leads to increased osteoprogenitor differentiation and disrupted chondrogenesis in the proximal region of the mandible. Enhanced Col I expression and weakened Sox9 expression are found in the same region, suggesting that osteochondroprogenitor cells lean toward the osteogenic rather than chondrogenic lineage. Moreover, by ablating Dlx5, which is an early osteogenic regulator, the mandibular phenotype of these mutants can be partially rescued (Oka et al., 2007).

Ihh signaling is also required in temporomandibular joint (TMJ) formation and condyle growth. In *Ihh*-null mice, TMJ development is severely compromised. Condylar cartilage growth, polymorphic cell proliferation, and PTHrP expression are all inhibited in these mice. This phenotype can be partially reversed by ablation of Gli3, a natural inhibitor of Hh signaling (Shibukawa et al., 2007).

3.3 Hemifacial microsomia

Hemifacial microsomia (HFM) is a common congenital defect that has an incidence ranging from 1:3500 to 1: 5600 (Hartsfield, 2007). It is primarily characterized by unilateral hypoplasia of the mandible and ear; other craniofacial malformations sometimes associated with HFM include facial palsy, cleft lip/palate and orbital defects. Three possible hypotheses have been raised to explain HFM: vascular abnormality and hemorrhage, disrupted development of MC, and abnormal development of CNCCs (Chen, Zhao, Shen, & Dai, 2018). Small molecule drugs such as thalidomide can cause local hemorrhage or vascular abnormalities, and accordingly, Poswillo established an animal model that mimics the phenotypes of HFM using triazene and thalidomide treatment (Poswillo, 1973). However, the exact mechanism that leads to the HFM phenotypes is still unclear. MC is closely associated with the formation of the mandible and middle ear. Disruption of the development of MC often leads to mandibular hypoplasia. A recent study showed that loss of VEGF expression in the CNCCs impairs blood vessel growth, leading to insufficient blood supply to MC and ultimately causing mandibular hypoplasia (Wiszniak et al., 2015). This study revealed the internal connections among these three possible pathogenic mechanism models and provides insight into potential prevention and treatment strategies for HFM.

3.4 Quantitative analysis using dynamic imaging and anatomical landmarks

Traditionally, researchers have relied on regular histology and whole-mount skeletal staining to document normal and abnormal craniofacial development. Recently, microCT imaging utilizing defined anatomical landmarks has made it possible to analyze defects quantitatively in the mandible and maxilla. Based on the landmarks that have been established with morphometrics, we are able to identify nuanced differences between normal and abnormal skeletal growth, which provides a basis for understanding the localized and overall influence of mutations associated with disease (Ho et al., 2015). Percival and colleagues systematically analyzed the craniofacial skeletal structures of control and $Fgfr2^{+/P253R}$ mice at multiple embryonic stages. They found that certain bones are significantly reduced in volume in the mutants, while others are not. Interestingly, they also found that the density of bone tissues formed through intramembranous and endochondral ossification differ during development (Percival, Huang, Jabs, Li, & Richtsmeier, 2014).

To uncover maxillary and mandibular phenotypes that are uniquely associated with specific mutant models, precise measurement and comparison between microCT data of different samples are required. FaceBase, which is a collaborative NIDCR-funded consortium, has developed a web-based platform (available at facebase.org) that allows users to rotate and view each facial bone in any position. One can select any facial bone and view its anatomical landmarks based on *Mouse Development* (Rossant & Tam, 2002), and also calculate the distance between any two anatomical landmarks. These measurements can serve as the basis for evaluating normal/abnormal facial bone development.

4. Tooth development

4.1 Early interaction between odontogenic ectoderm and ectomesenchyme

Similar to mandible patterning, odontogenic signaling for tooth initiation also relies on epithelial-mesenchymal interactions. As early as E10, Bmp4 expression is restricted to the oral ectoderm where the incisors will form. By the time of tooth initiation, Bmp4 expression switches from the epithelium to the underlying mesenchyme corresponding to the condensation beneath the epithelial thickening. In the later bud and cap stages, Bmp4 expression is restricted to the tooth germ. Msx1, one of the downstream targets of Bmp4, is also expressed in the CNC-derived ectomesenchyme (Vainio, Karavanova, Jowett, & Thesleff, 1993). Interestingly, Msx1 can also positively regulate Bmp4 expression, and this feedback loop is critical for tooth initiation and morphogenesis (Chen, Bei, Woo, Satokata, & Maas, 1996). *Msx1* null mice show an arrest of molar tooth development at the bud stage with reduced Bmp4 expression (Satokata & Maas, 1994). This data suggests that Msx1 is not only expressed in response to signals from the dental epithelium, but also regulates downstream target genes to provide feedback to the dental epithelium.

Pax9 is another transcription factor that is essential for tooth initiation. At E11, it is also expressed in the ectomesenchyme corresponding to the future dental mesenchyme condensation. Pax9-deficient mice experience arrest of tooth development at the bud stage with reduced mesenchymal cell condensation. Bmp4, Msx1 and Lef1 all are downregulated in these mutant mice (Peters, Neubuser, Kratochwil, & Balling, 1998). Pax9 expression can be induced by Fgf8 secreted from the oral epithelium, and Pax9 expression is downregulated in *Fgf8*$^{-/-}$ mice (Neubuser, Peters, Balling, & Martin, 1997). Interestingly, Fgf8 cannot induce Bmp4 expression or *vice versa*. Instead, they work antagonistically to establish incisor and molar formation domains (Chai & Maxson, 2006).

Shh expression in the tooth-forming regions of PA1 starts at E11.5. Shh expression is highly restricted to the epithelial thickening of the future tooth germ (Bitgood & McMahon, 1995). Shh signaling is responsible for the localized proliferation of the epithelial tissue, which in turns invaginates into the underlying mesenchyme to form a tooth bud. The localization of Shh expression is accomplished through antagonism of Wnt7b expression in the non-tooth-forming epithelium (Sarkar et al., 2000). Ectopic application of Shh protein to the non-dental oral ectoderm of E10.5 mandible explants leads to formation of multiple ectopic epithelial invaginations after 3 days in culture. Moreover, inhibition of Shh signaling in a mandible culture with blocking antibody results in failed tooth bud formation with reduced cell proliferation and increased apoptosis after 3 days of culture, again suggesting that Shh regulates epithelial cell proliferation and survival in the developing tooth germ (Cobourne, Hardcastle, & Sharpe, 2001). *In vivo* studies using transgenic mouse models have also demonstrated the importance of Shh for tooth development. Loss of *Shh* in the oral epithelium ($K14\text{-}Cre;Shh^{fl/fl}$) severely affects tooth development, including retardation of tooth growth, abnormal placement of teeth in the jaw, and disrupted tooth morphogenesis. Interestingly, Shh can also regulate the pattern of developing cusps; the lingual side of the tooth is more severely affected than the buccal side when it is disrupted (Dassule, Lewis, Bei, Maas, & McMahon, 2000).

4.2 Signaling regulating dentin and enamel formation

In the late bell stage, mineralized dentin and enamel tissues are formed by odontoblasts from the dental mesenchyme and ameloblasts from the dental epithelium. The formation of dentin and enamel takes place at the interface between the mesenchyme and epithelium and is regulated by multiple signaling pathways through tissue-tissue interactions. The enamel knot, which lies at the tip of the future cusp, serves as a signaling center that mediates the differentiation of odontoblasts and ameloblasts (Fig. 2). More than 10 signaling molecules belonging to the BMP, FGF, Hh, and Wnt families are expressed in the primary enamel knot to guide not only dentin and enamel formation, but also cusp patterning during tooth morphogenesis (Jussila & Thesleff, 2012).

Even though the search for master transcriptional regulators for odontogenesis is still ongoing, the importance of Bmp signaling has been well characterized. Multiple growth factors in the Bmp family have been shown to induce terminal differentiation of odontoblasts *in vitro* (Nakashima, 1994; Tasli, Aydin, Yalvac, & Sahin, 2014; Zhu et al., 2018). Loss of *Smad4*, a common mediator of Bmp and Tgf-β signaling, in the CNC-derived mesenchyme leads to defects in odontoblast differentiation and formation of ectopic bone-like structure in the dentin-forming region. More interestingly, despite the defect in dentin formation, enamel formation appears unaffected in these mutant mice, suggesting that functional odontoblast differentiation is not required for ameloblast differentiation (Li et al., 2011).

There is also evidence showing that Wnt signaling is associated with dentinogenesis. Wnt10a is specifically expressed in the primary and secondary enamel knots. In addition, Axin2, the canonical Wnt signaling transducer and read-out, is expressed in developing odontoblasts and dental pulp cells. Both loss- and gain-of-function of Wnt signaling in early odontoblasts have been studied using $OC\text{-}Cre;Wls^{CO/CO}$ and $OC\text{-}Cre;Catnb^{lox(ex3)/+}$ mutant mice, respectively. Loss of Wnt signaling through inactivation of *Wntless (Wls)* leads to compromised odontoblast maturation,

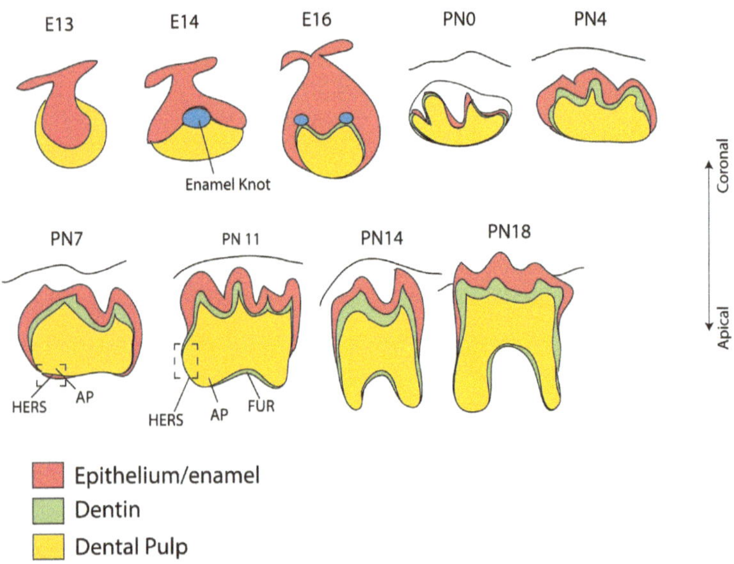

Fig. 2 Scheme of tooth development. Tooth development begins with the evagination of the epithelium into the underling mesenchyme to form the tooth bud. During crown formation, enamel knots serve as the signaling center to mediate the differentiation of odontoblasts and ameloblasts, and the patterning of the cusps. Starting from postnastal day 3.5, bilayered HERS grows apically and guides the tooth root development. HERS, Hertwig's epithelial root sheath; AP, apical papilla; FUR, furcation.

down-regulation of the terminal differentiated odontoblast marker Dspp, and reduced dentin thickness, whereas over-activation of Wnt signaling through overexpression of β-catenin results in excessive dentin and cementum formation, which may be due to prematurely differentiated odontoblasts (Bae et al., 2015; Kim et al., 2011). These studies collectively indicate that temporo-spatial regulation of Wnt/beta-catenin signaling is essential for normal odontoblast differentiation and dentin formation.

Bmp signaling, which is one of the few signaling pathways that transmits bidirectional signals between epithelial and mesenchymal tissues, also regulates ameloblast differentiation and enamel formation. Ablation of Bmp2 in odontoblasts and dental pulp cells using *Osx-Cre;Bmp2*$^{fl/fl}$ results in a severe phenotype of both incisors and molars with a thin, hypomineralized enamel layer (Feng, Yang, et al., 2011). Two factors downstream of Bmp signaling, Runx2 and Osterix, are also related to ameloblast differentiation. *In vitro* culture of mouse ameloblast lineage cells (mALCs) revealed that Runx2 physically interacts with Fam50a to increase its binding affinity to the ameloblastin (*Ambn*) promoter (Kim et al., 2018). Tooth morphogenesis initially progresses normally in *Osterix* null mice; however, markers of mature ameloblasts (Enam, Amelx, Mmp20, Amtn, Klk4) have limited expression in incisors and molar tissues of these mice, and they lack enamel matrix (Bae et al., 2018).

Proper ion exchange is critical for the demineralization and remineralization processes during dentin and enamel formation because it regulates and maintains the required calcium and pH homeostasis. Multiple human syndromes and diseases including Timothy syndrome, Olmsted syndrome and osteopetrosis are associated with defects in enamel and dentin formation with impaired biomineralization. Patients with these conditions have mutations in genes encoding various ion channel-related proteins, such as *CACNA1*, *TRPV3*, *CLCN7* and *AE2* (Duan, 2014). ClC-5 functions as a Cl^-/H^+ exchanger and plays an important role in pH regulation. *ClC-5* knockout mice have abnormal dentin, similar to the characteristics of dentinogenesis imperfecta in humans, possibly due to an overexpression of Tgf-β signaling (Duan et al., 2009). The cystic fibrosis transmembrane conductance regulator (CFTR), which is a transporter-class ion channel, also regulates Cl^- exchange. CFTR is expressed in ameloblasts during amelogenesis. *CFTR-ΔF508* pigs have hypomineralized and visibly disorganized enamel tissue, and a similar phenotype is observed in Cftr-deficient mice, with soft, chalky white incisor enamel that degenerates shortly after completion of the secretory phase of amelogenesis (Lacruz et al., 2012).

4.3 Tooth root development

Tooth root development is mainly driven by two cell populations: the dental epithelium-derived HERS and the CNC-derived dental pulp stem cells. The HERS contains a transient cell population that serves as a signaling center and provides different types of factors that trigger tooth root elongation (Fig. 2). Dental pulp stem cells can be identified using Gli1 in the apical portion of the dental papilla at postnatal day 3.5 (P3.5), and then later take up permanent residence in the apical region of the elongated tooth root (Feng et al., 2017; Liu et al., 2015). Lineage tracing studies showed that these cells proliferate and populate the entire tooth root mesenchyme during development. A large body of research has shown that disruption of the HERS or dental pulp stem cells through targeted genetic modification disturbs the development of the tooth root. An array of growth and transcription factors has been uncovered based on their expression patterns, and multiple mutant animal models with tooth root defects have been generated (Li, Parada, & Chai, 2017).

During HERS formation, Bmp signaling is activated in both the dental epithelium and mesenchyme. Recent studies have shown that a Bmp-Smad4-Shh-Gli1 signaling network regulates the fate of the transient dental epithelial stem cells, which are Sox2+, in the mouse molar (Li et al., 2015). Specifically, loss of Bmp signaling in the dental epithelium leads to an expansion of Shh signaling, which in turn causes the maintenance of the cervical loop structure and retention of Sox2+ dental epithelial stem cells postnatally. As a result, HERS formation is delayed and tooth root development is arrested. Interestingly, loss of Shh ligand expression in the dental epithelium rescues this tooth root phenotype, suggesting that the Bmp/Shh signaling cascade is critical for HERS formation and root development. Consistent with this, ablation of *Msx2*, which is a direct downstream target of Smad-mediated Bmp signaling, also leads to shortened molar roots (Aioub et al., 2007). As mentioned above, multiple Bmp ligands including Bmp2, 3, 4 and 7 as well as phosphorylated Smad 1/5/8, which indicates activation of Bmp signaling, are expressed in the dental mesenchyme, implying that mesenchymal Bmp signaling plays a role in tooth root development. Indeed, using $Gli1\text{-}Cre^{ERT};Bmpr1a^{fl/fl}$ mice to specifically knock out Bmp type I receptor in dental pulp stem cells results in impaired tooth root formation. An odontoblast differentiation defect is also observed in these mice, as indicated by loss of Klf4 expression in the pre-odontoblast region; Klf4 may serve as a switch for regulating odontoblast differentiation (Feng et al., 2017).

Evidence shows that an Nfic/Hh signaling cascade also regulates tooth root development. *Nfic*$^{-/-}$ mice show a tooth root defect and downregulation of Hh signaling attenuator Hhip. Therefore, these mice have an expansion of Hh signaling similar to that of *KRT14-rtTA;tetO-Cre; Smad4*$^{fl/fl}$ mutant mice. Treatment of *Nfic*$^{-/-}$ mice with Hh inhibitor partially rescues cell proliferation and root morphology (Liu et al., 2015). These data suggest that the proper regulation of Hh signaling and activation through the Bmp/Nfic/Hh network is critical for tooth root development.

4.4 Tooth and jaw bone interaction

Teeth and jaw bones have a common origin: they both arise from the first pharyngeal arch and they develop in closely related ways. Multiple mouse models show that mutations affecting early mandible development also have an impact on tooth formation (Denaxa et al., 2009; Peters et al., 1998; Satokata & Maas, 1994). *Msx1* is a critical patterning gene expressed in the distal half of the mandibular arch during early stages of tooth development. Later, *Msx1* is strongly expressed in the dental mesenchyme. *Msx1*$^{-/-}$ mice exhibit both mandibular abnormalities and tooth defects. The overall length of the mandible is slightly shorter and the alveolar ridge is absent in these *Msx1* mutant mice, and tooth development fails to progress past the bud stage (Satokata & Maas, 1994). This phenotype suggests that Msx1 is not only needed for the differentiation of dental follicle cells into alveolar bone osteoblasts, but also required as feedback to the epithelium for progression of the tooth bud to the cap and bell stages. Similarly, Pax9 is also expressed in the mandibular arch at the early patterning stage and in the dental mesenchyme. In Pax9-deficient mice, alveolar bones and coronoid processes are missing and tooth development stalls at the bud stage. Interestingly, Msx1 and Pax9 not only show closely overlapping expression patterns, but also physically interact with each other to guide tooth formation (Ogawa, Kapadia, Wang, & D'Souza, 2005).

The teeth and jaw bones also interact through mechanical force. After tooth extraction, a reduction of the alveolar ridge is commonly observed. One explanation for this bone resorption is that the forces on the bone are reduced after tooth loss so that less bone is needed (Hansson & Halldin, 2012). Clinically, a bone allograft in combination with a membrane is used to improve the ridge dimensions in these patients who will require dental implants to restore their dentition (Iasella et al., 2003).

4.5 Dental stem cells

Unlike mouse molars and all human teeth, mouse incisors grow continuously throughout the lifetime of the animal due to adult stem cells residing within the tissue. These stem cells undergo self-renewal and maintain the homeostasis of the dentin and enamel. Harada and colleagues first identified the stellate reticulum, which is located in the cervical loop area, as the putative site of dental epithelial stem cells (Harada et al., 1999). Starting from E14.5, Sox2 expression is found in the labial cervical loop of the mouse incisor, and from E16.5 to E18.5, Sox2 expression is more restricted to the proximal tip of the cervical loop, which corresponds to the location of the putative epithelial stem cell population. Lineage tracing showed that Sox2 positive cells contribute to all epithelial lineages of the tooth (Juuri et al., 2012). This differentiation process involves a group of multipotent progenitor cell progeny called transit-amplifying (TA) cells, which are adjacent to the stem cell population. Notch signaling has been reported to be critical for stem cell maintenance as well as fate determination. Notch1, Notch2, and lunatic fringe, a Notch homolog found in Drosophila, are all expressed by cervical loop epithelial cells (Harada et al., 1999). In explant culture, inhibition of Notch signaling using DAPT leads to a reduction of cell proliferation and increased apoptosis in the epithelial stem cell niche (Felszeghy, Suomalainen, & Thesleff, 2010). Signals from the mesenchyme play a role in the self-renewal and differentiation of the stem cell population in the incisor epithelium. Fgf3 is expressed in the mesenchyme underlying the cervical loop region. Fgf10-expressing cells partially overlap with the population that expresses Fgf3 and surround the whole cervical loop epithelium, while Fgfr1b is strongly expressed in the basal epithelial cells and stratum intermedium, suggesting that Fgf signaling might regulate the continuous growth of the incisor epithelium. Interestingly, Fgf3-deficient mice have relatively normal tooth development, which may be due to compensation by Fgf10. In *Fgf3*$^{-/-}$;*Fgf10*$^{+/-}$ compound mutant mice, the lower incisors are shorter and frequently broken. They also have hypoplastic morphology of the cervical loop and either very thin or missing enamel (Wang et al., 2007). In addition, even though Fgf3 is not required for early cervical loop morphogenesis, its asymmetric expression pattern may be important for the difference in size between the labial and lingual portions of the cervical loop. Fgf3 induces cell proliferation in the incisor epithelium. Consistent with this, *Follistatin*$^{-/-}$ mice exhibit an enlarged lingual cervical loop along with ectopic expression of Fgf3 in the lingual dental mesenchyme underlying the epithelium (Wang et al., 2004).

Recent studies have revealed the identity of mesenchymal stem cells (MSCs) in mouse incisors. When dental pulp-derived cells are cultured, a group of cells with MSC characteristics such as clonogenic, multi-lineage differentiation and expression of defining markers such as CD90, CD73, and CD105 can be rapidly isolated. Despite this, it took more than 10 years to identify the *in vivo* location of this heterogeneous cell population. Feng and colleagues first showed that NG2-labeled pericytes contribute to odontoblasts during growth as well as after damage to the dental pulp (Fig. 3) (Feng, Mantesso, De Bari, Nishiyama, & Sharpe, 2011). However, in both cases, pericyte-derived odontoblasts only account for 15% of the whole population, suggesting that another source of MSCs may contribute the majority of odontoblasts and dental pulp. In addition, NG2+ pericytes do not contribute to incisor homeostasis, suggesting that different stem cell populations are primarily responsible for tissue homeostasis and injury repair.

Since this initial finding, peripheral nerve-associated glia were identified as another source of dental MSCs (Fig. 3). Using two different *ERT2-Cre* drivers targeting glial cells (*Plp1* and *Sox10*) combined with reporters, lineage tracing studies showed that these cells can contribute to the odontoblast population during incisor growth, homeostasis, and injury repair.

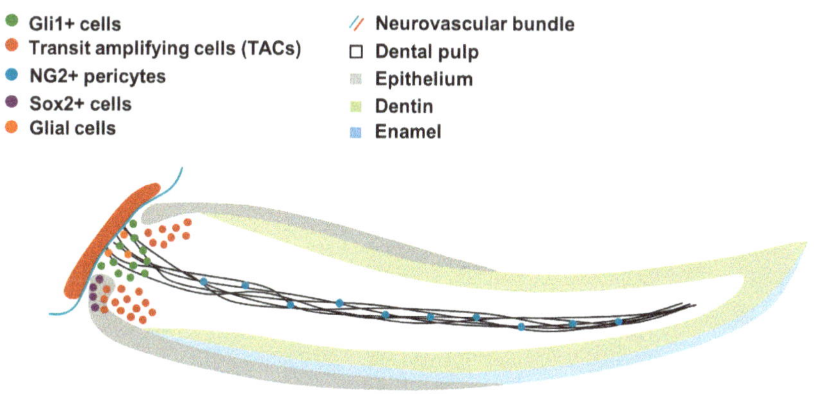

Fig. 3 Stem cell population in the mouse incisor. In the labial cervical loop, Sox2+ epithelial stem cells are responsible for replenishing the enamel and epithelial tissue. Gli1+ and glial cells residing in the mesenchyme between the labial and lingual cervical loop regions close to the neurovascular bundle represent slow-cycling stem cells. Transit amplifying cells located either in the mesenchyme close to the labial and lingual cervical loop region or in the labial cervical loop are derived from self-renewing stem cells. NG2+ pericytes located on the abluminal surface of endothelial cells contribute to mesenchymal tissue repair after injury.

However, depending on the dosage of tamoxifen injection given to these mice, the amount of Schwann-cell-derived progeny varies from 4% to 47%, again suggesting the existence of another source of MSCs (Kaukua et al., 2014).

General consensus holds that the most proximal end of the mouse incisor, namely the mesenchyme between the lingual and labial epithelial cervical loops, serves as a stem cell niche that supplies new cells for replacement of tissue loss due to occlusion and abrasion. This model was confirmed through identification of a small group of label-retaining cells detectable in this location after a 4-week chase period. These Gli1+ cells surround the neurovascular bundle and receive Shh signal from the sensory nerve (Fig. 3) (Zhao et al., 2014). After 4 weeks of lineage tracing, the typical turn over time for odontoblasts and ameloblasts in mouse incisors, almost 100% of odontoblasts and pulp cells are derived from these Gli1+ cells. Interestingly, these Gli1+ cells do not express surface markers that define MSCs *in vitro*, such as CD105, CD146, and Sca1, which are highly expressed in the NG2+ pericytes that are derived from these Gli1+ cells. This finding will have an important impact on the definition and identification of MSCs *in vivo* (Zhao et al., 2014).

5. Stem cells and regenerative therapies

5.1 Mandibular distraction osteogenesis, growth factors, and stem cell treatment

Mandibular hypoplasia is one of the most common congenital malformations, and can be either non-syndromic or associated with other anomalies, as in Pierre Robin sequence or Marfan syndrome. Surgical intervention is required in many cases due to the breathing and swallowing difficulties caused by posterior tongue displacement and the resulting airway compromise. Mandibular distraction osteogenesis (MDO) is the current standard treatment for micrognathia and proceeds in three stages: (i) the initial latency stage, which starts after an osteotomy is created, allowing the initial healing and callus formation; (ii) the activation stage, during which the ends of the bone are gradually moved apart, allowing new bone to form in the gap; and (iii) the consolidation stage, once the mandible reaches the optimal length, allowing for the final maturation of the newly formed bone.

Different growth factors, stem cells, and adjuvant procedures have been tested in an attempt to promote the healing process during distraction osteogenesis in several animal models. Multiple bone morphogenetic proteins including Bmp2, Bmp4, and Bmp7 enhance bone volume, remodeling, and mature bone formation in both rat and rabbit models (Mizumoto, Moseley, Drews, Cooper, & Reddi, 2003; Yonezawa, Harada, Ikebe, Shinohara, & Enomoto, 2006). Nerve growth factor induces bone formation around the regenerating axons and improves mechanical strength and histomorphometric outcomes in a rabbit model (Cao et al., 2012). Insulin-like growth factor appears to increase the mineral deposition rate, suggesting a positive anabolic effect (Stewart et al., 1999).

Bone marrow-derived mesenchymal stem cells (BMMSCs) undergo osteogenic differentiation following stimulation from certain biological signals and are relatively easy to harvest and amplify in culture. In several MDO models, BMMSC-treated groups showed significant higher radio-density and bone volume and thickness in the early consolidation stage (Aykan et al., 2013; Kim, Cho, Lee, & Hwang, 2013; Ma et al., 2013). In another study, stromal cell-derived factor-1 (SDF-1) was shown to facilitate migration of MSCs *in vitro* and *in vivo*. In a rat model of MDO, the recruitment of endogenous MSCs to the injury site was significantly enhanced by SDF-1 treatment, which provides a new insight into how manipulation of endogenous MSCs could be used to enhance the bone healing process (Cao et al., 2013).

5.2 Dentin repair and regeneration

Following an injury or lesion in a tooth, odontoblasts in and around the injury site are damaged and the tooth is at risk of infection. In this case, isolated pericytes in the dental pulp and glial cells close to the injury site undergo rapid proliferation and differentiate into odontoblast-like cells that generate reparative dentin to protect the exposed pulp (Pang Yvonne et al., 2015). At the same time, stem cells residing in the proximal incisor niche also contribute to replenishing the odontoblasts. The signaling mechanism that guides this reparative process remains unclear. Wnt signaling, which is activated in TA cells as well as in the region of newly differentiated odontoblasts, is critical for the transition from slow-cycling stem cells to rapidly proliferating progenitor cells. Moreover, mice with an activated Wnt pathway due to loss of Axin2, which represses Wnt signaling in a ligand-dependent manner, show a much stronger repair response than controls (Hunter Daniel et al., 2015). Tgf-β signaling has also been shown to play a role in dental stem cell differentiation. Non-ionizing, low-power laser treatment activates latent, endogenous

Tgf-β1 *via* a specific methionine residue, thereby promoting stem cell differentiation, which subsequently significantly increases the amount of dentin regeneration in rat injury model (Arany et al., 2014).

6. Conclusion and future directions

The development of the jaws and teeth are closely related, not only due to their common origins involving cranial neural crest cells, but also the regulatory mechanisms and factors that they share. Many transgenic mouse models exhibit malformations in both the teeth and jaws, suggesting a close relationship between them. Despite the significant progress we have made in understanding the regulatory mechanisms behind jaw and tooth development over the past few decades, there are still several unanswered questions, especially regarding the early stages of their development, including how patterning processes contribute to mesenchymal cell fate determination as well as bone and cartilage formation. Future studies, with the help of emerging techniques such as single-cell RNA sequencing, will enable us to identify and trace heterogeneous cell populations within the mandibular process, which could shed light on the regulatory mechanism of cell fate determination. Understanding the molecular regulation of each stage of the development of the jaws and teeth will not only improve our knowledge of the etiology of developmental defects, but also facilitate the treatment of related diseases using stem cells and tissue regeneration.

Acknowledgments

We would like to thank Julie Mayo and Bridget Samuels for critical reading of the manuscript and discussion, and funding support from the National Institute of Dental and Craniofacial Research, NIH (R37 DE012711 and U01 DE024421) to Yang Chai.

References

Aggarwal, V. S., Carpenter, C., Freyer, L., Liao, J., Petti, M., & Morrow, B. E. (2010). *Developmental Biology*, *344*, 669–681.

Aioub, M., Lezot, F., Molla, M., Castaneda, B., Robert, B., Goubin, G., et al. (2007). *Bone*, *41*, 851–859.

Amano, O., Doi, T., Yamada, T., Sasaki, A., Sakiyama, K., Kanegae, H., et al. (2010). *Journal of Oral Biosciences*, *52*, 125–135.

Arany, P. R., Cho, A., Hunt, T. D., Sidhu, G., Shin, K., Hahm, E., et al. (2014). *Science Translational Medicine*, *6*, 238ra69.

Aykan, A., Ozturk, S., Sahin, I., Gurses, S., Ural, A. U., Oren, N. C., et al. (2013). *The Journal of Craniofacial Surgery*, *24*, e169–e175.

Bae, J. M., Clarke, J. C., Rashid, H., Adhami, M. D., McCullough, K., Scott, J. S., et al. (2018). *Journal of Bone and Mineral Research, 33*, 1126–1140.
Bae, C. H., Kim, T. H., Ko, S. O., Lee, J. C., Yang, X., & Cho, E. S. (2015). *Journal of Dental Research, 94*, 439–445.
Baek, W. Y., Kim, Y. J., de Crombrugghe, B., & Kim, J. E. (2013). *Biochemical and Biophysical Research Communications, 432*, 188–192.
Barrow, J. R., Stadler, H. S., & Capecchi, M. R. (2000). *Development, 127*, 933–944.
Barske, L., Rataud, P., Behizad, K., Del Rio, L., Cox, S. G., & Crump, J. G. (2018). *Developmental Cell, 44*, 337–347. e5.
Billmyre, K. K., & Klingensmith, J. (2015). *Developmental Dynamics, 244*, 564–576.
Bitgood, M. J., & McMahon, A. P. (1995). *Developmental Biology, 172*, 126–138.
Cao, J., Wang, L., Du, Z. J., Liu, P., Zhang, Y. B., Sui, J. F., et al. (2013). *The British Journal of Oral & Maxillofacial Surgery, 51*, 937–941.
Cao, J., Wang, L., Lei, D. L., Liu, Y. P., Du, Z. J., & Cui, F. Z. (2012). *Oral Surgery, Oral Medicine, Oral Pathology, and Oral Radiology, 113*, 48–53.
Carmona-Fontaine, C., Matthews, H. K., Kuriyama, S., Moreno, M., Dunn, G. A., Parsons, M., et al. (2008). *Nature, 456*, 957.
Cate, A. T. (1996). *Oral Diseases, 2*, 55–62.
Chai, Y., Jiang, X., Ito, Y., Bringas, P., Jr., Han, J., Rowitch, D. H., et al. (2000). *Development, 127*, 1671–1679.
Chai, Y., & Maxson, R. E., Jr. (2006). *Developmental Dynamics, 235*, 2353–2375.
Chen, Y., Bei, M., Woo, I., Satokata, I., & Maas, R. (1996). *Development, 122*, 3035–3044.
Chen, Q., Zhao, Y., Shen, G., & Dai, J. (2018). *Journal of Dental Research, 97*, 1297–1305.
Chung, I. H., Yamaza, T., Zhao, H., Choung, P. H., Shi, S., & Chai, Y. (2009). *Stem Cells, 27*, 866–877.
Cobourne, M. T., Hardcastle, Z., & Sharpe, P. T. (2001). *Journal of Dental Research, 80*, 1974–1979.
Dassule, H. R., Lewis, P., Bei, M., Maas, R., & McMahon, A. P. (2000). *Development, 127*, 4775–4785.
Denaxa, M., Sharpe, P. T., & Pachnis, V. (2009). *Developmental Biology, 333*, 324–336.
Depew, M. J., & Compagnucci, C. (2008). *Journal of Experimental Zoology. Part B, Molecular and Developmental Evolution, 310*, 315–335.
Depew, M. J., Lufkin, T., & Rubenstein, J. L. (2002). *Science, 298*, 381–385.
Duan, X. (2014). *Journal of Dental Research, 93*, 117–125.
Duan, X., Mao, Y., Yang, T., Wen, X., Wang, H., Hou, J., et al. (2009). *Archives of Oral Biology, 54*, 1118–1124.
Felszeghy, S., Suomalainen, M., & Thesleff, I. (2010). *Differentiation, 80*, 241–248.
Feng, J., Jing, J., Li, J., Zhao, H., Punj, V., Zhang, T., et al. (2017). *Development, 144*, 2560–2569.
Feng, J., Mantesso, A., De Bari, C., Nishiyama, A., & Sharpe, P. T. (2011). *Proceedings of the National Academy of Sciences of the United States of America, 108*, 6503–6508.
Feng, J., Yang, G., Yuan, G., Gluhak-Heinrich, J., Yang, W., Wang, L., et al. (2011). *Cells, Tissues, Organs, 194*, 216–221.
Gendron-Maguire, M., Mallo, M., Zhang, M., & Gridley, T. (1993). *Cell, 75*, 1317–1331.
Gilbert, S. F. (2000). *Developmental biology* (6th ed.). Sunderland, MA: Sinauer Associates.
Hansson, S., & Halldin, A. (2012). *Journal of Dental Biomechanics, 3* 1758736012456543.
Harada, H., Kettunen, P., Jung, H.-S., Mustonen, T., Wang, Y. A., & Thesleff, I. (1999). *The Journal of Cell Biology, 147*, 105.
Hartsfield, J. K. (2007). *Orthodontics & Craniofacial Research, 10*, 121–128.
Ho, T. V., Iwata, J., Ho, H. A., Grimes, W. C., Park, S., Sanchez-Lara, P. A., et al. (2015). *Developmental Biology, 400*, 180–190.
Hunter Daniel, J., Bardet, C., Mouraret, S., Liu, B., Singh, G., Sadoine, J., et al. (2015). *Journal of Bone and Mineral Research, 30*, 1150–1159.

Iasella, J. M., Greenwell, H., Miller, R. L., Hill, M., Drisko, C., Bohra, A. A., et al. (2003). *Journal of Periodontology, 74*, 990–999.
Jin, Y. R., Turcotte, T. J., Crocker, A. L., Han, X. H., & Yoon, J. K. (2011). *Developmental Biology, 352*, 1–13.
Jussila, M., & Thesleff, I. (2012). *Cold Spring Harbor Perspectives in Biology, 4*, a008425.
Juuri, E., Saito, K., Ahtiainen, L., Seidel, K., Tummers, M., Hochedlinger, K., et al. (2012). *Developmental Cell, 23*, 317–328.
Kaukua, N., Shahidi, M. K., Konstantinidou, C., Dyachuk, V., Kaucka, M., Furlan, A., et al. (2014). *Nature, 513*, 551–554.
Kawane, T., Komori, H., Liu, W., Moriishi, T., Miyazaki, T., Mori, M., et al. (2014). *Journal of Bone and Mineral Research, 29*, 1960–1969.
Kim, I. S., Cho, T. H., Lee, Z. H., & Hwang, S. J. (2013). *Tissue Engineering. Part A, 19*, 66–78.
Kim, Y., Hur, S. W., Jeong, B. C., Oh, S. H., Hwang, Y. C., Kim, S. H., et al. (2018). *Journal of Cellular Physiology, 233*, 1512–1522.
Kim, T. H., Lee, J. Y., Baek, J. A., Lee, J. C., Yang, X., Taketo, M. M., et al. (2011). *Biochemical and Biophysical Research Communications, 412*, 549–555.
Krumlauf, R. (1994). *Cell, 78*, 191–201.
Lacruz, R. S., Smith, C. E., Moffatt, P., Chang, E. H., Bromage, T. G., Bringas, P., Jr., et al. (2012). *Journal of Cellular Physiology, 227*, 1776–1785.
Le Douarin, N. M., Creuzet, S., Couly, G., & Dupin, E. (2004). *Development, 131*, 4637–4650.
Lee, S. K., Kim, Y. S., Oh, H. S., Yang, K. H., Kim, E. C., & Chi, J. G. (2001). *The Anatomical Record, 263*, 314–325.
Leucht, P., Kim, J.-B., Amasha, R., James, A. W., Girod, S., & Helms, J. A. (2008). *Development, 135*, 2845–2854.
Li, J., Feng, J., Liu, Y., Ho, T. V., Grimes, W., Ho, H. A., et al. (2015). *Developmental Cell, 33*, 125–135.
Li, F., Fu, G., Liu, Y., Miao, X., Li, Y., Yang, X., et al. (2017). *Molecular and Cellular Biology, 37*, e00590-16.
Li, J., Huang, X., Xu, X., Mayo, J., Bringas, P., Jr., Jiang, R., et al. (2011). *Development, 138*, 1977–1989.
Li, J., Parada, C., & Chai, Y. (2017). *Development, 144*, 374–384.
Liu, Y., Feng, J., Li, J., Zhao, H., Ho, T. V., & Chai, Y. (2015). *Development, 142*, 3374–3382.
Loring, J. F., & Erickson, C. A. (1987). *Developmental Biology, 121*, 220–236.
Love, R. J., Murray, J. M., & Mamandras, A. H. (1990). *American Journal of Orthodontics and Dentofacial Orthopedics, 97*, 200–206.
Lumsden, A., & Krumlauf, R. (1996). *Science, 274*, 1109–1115.
Ma, D., Ren, L., Yao, H., Tian, W., Chen, F., Zhang, J., et al. (2013). *Journal of Orthopaedic Research, 31*, 1082–1088.
Matsui, M., & Klingensmith, J. (2014). *Developmental Biology, 392*, 168–181.
McGinnis, W., & Krumlauf, R. (1992). *Cell, 68*, 283–302.
Mellitzer, G., Xu, Q., & Wilkinson, D. G. (1999). *Nature, 400*, 77–81.
Mizumoto, Y., Moseley, T., Drews, M., Cooper, V. N., 3rd, & Reddi, A. H. (2003). *The Journal of Bone and Joint Surgery. American Volume, 85-A*(Suppl. 3), 124–130.
Moore, K. L., Persaud, T. V. N., & Samperio, J. O. (1999). *Embriología clínica*. McGraw-Hill Interamericana.
Mori-Akiyama, Y., Akiyama, H., Rowitch, D. H., & de Crombrugghe, B. (2003). *Proceedings of the National Academy of Sciences of the United States of America, 100*, 9360–9365.
Nakashima, M. (1994). *Journal of Dental Research, 73*, 1515–1522.
Neubuser, A., Peters, H., Balling, R., & Martin, G. R. (1997). *Cell, 90*, 247–255.

Noden, D. M. (1975). *Developmental Biology, 42*, 106–130.
Noden, D. M. (1978a). *Developmental Biology, 67*, 313–329.
Noden, D. M. (1978b). *Developmental Biology, 67*, 296–312.
Noden, D. M. (1983). *Developmental Biology, 96*, 144–165.
Ogawa, T., Kapadia, H., Wang, B., & D'Souza, R. N. (2005). *Archives of Oral Biology, 50*, 141–145.
Oka, K., Oka, S., Sasaki, T., Ito, Y., Bringas, P., Jr., Nonaka, K., et al. (2007). *Developmental Biology, 303*, 391–404.
Olesnicky Killian, E. C., Birkholz, D. A., & Artinger, K. B. (2009). *Developmental Biology, 333*, 161–172.
Pang Yvonne, W. Y., Feng, J., Daltoe, F., Fatscher, R., Gentleman, E., Gentleman Molly, M., et al. (2015). *Journal of Bone and Mineral Research, 31*, 514–523.
Percival, C. J., Huang, Y., Jabs, E. W., Li, R., & Richtsmeier, J. T. (2014). *Developmental Dynamics, 243*, 541–551.
Peters, H., Neubuser, A., Kratochwil, K., & Balling, R. (1998). *Genes & Development, 12*, 2735–2747.
Poswillo, D. (1973). *Oral Surgery, Oral Medicine, and Oral Pathology, 35*, 302–328.
Richany, S. F., Bast, T. H., & Anson, B. J. (1956). *Quarterly Bulletin of Northwestern University Medical School, 30*, 331–355.
Rijli, F. M., Mark, M., Lakkaraju, S., Dierich, A., Dolle, P., & Chambon, P. (1993). *Cell, 75*, 1333–1349.
Rivera-Perez, J. A., Wakamiya, M., & Behringer, R. R. (1999). *Development, 126*, 3811–3821.
Rossant, J., & Tam, P. T. (2002). *Mouse development: Patterning, morphogenesis, and organogenesis.* Elsevier Science.
Ruest, L.-B., Xiang, X., Lim, K.-C., Levi, G., & Clouthier, D. E. (2004). *Development (Cambridge, England), 131*, 4413.
Sarkar, L., Cobourne, M., Naylor, S., Smalley, M., Dale, T., & Sharpe, P. T. (2000). *Proceedings of the National Academy of Sciences of the United States of America, 97*, 4520–4524.
Sato, T., Kurihara, Y., Asai, R., Kawamura, Y., Tonami, K., Uchijima, Y., et al. (2008). *Proceedings of the National Academy of Sciences of the United States of America, 105*, 18806–18811.
Satokata, I., & Maas, R. (1994). *Nature Genetics, 6*, 348–356.
Shibata, S., Suda, N., Yoda, S., Fukuoka, H., Ohyama, K., Yamashita, Y., et al. (2004). *Anatomy and Embryology (Berlin), 208*, 273–280.
Shibukawa, Y., Young, B., Wu, C., Yamada, S., Long, F., Pacifici, M., et al. (2007). *Developmental Dynamics, 236*, 426–434.
Simoes-Costa, M., & Bronner, M. E. (2016). *Science, 352*, 1570–1573.
Smith, A., Robinson, V., Patel, K., & Wilkinson, D. G. (1997). *Current Biology, 7*, 561–570.
Stewart, K. J., Weyand, B., van't Hof, R. J., White, S. A., Lvoff, G. O., Maffulli, N., et al. (1999). *British Journal of Plastic Surgery, 52*, 343–350.
Studer, M., Gavalas, A., Marshall, H., Ariza-McNaughton, L., Rijli, F. M., Chambon, P., et al. (1998). *Development, 125*, 1025–1036.
Sugito, H., Shibukawa, Y., Kinumatsu, T., Yasuda, T., Nagayama, M., Yamada, S., et al. (2011). *Journal of Dental Research, 90*, 625–631.
Tasli, P. N., Aydin, S., Yalvac, M. E., & Sahin, F. (2014). *Applied Biochemistry and Biotechnology, 172*, 3016–3025.
Tavares, A. L. P., Cox, T. C., Maxson, R. M., Ford, H. L., & Clouthier, D. E. (2017). *Development, 144*, 2021–2031.
Teillet, M. A., Kalcheim, C., & Le Douarin, N. M. (1987). *Developmental Biology, 120*, 329–347.
Thiery, J. P., Duband, J. L., & Delouvee, A. (1982). *Developmental Biology, 93*, 324–343.

Trumpp, A., Depew, M. J., Rubenstein, J. L. R., Bishop, J. M., & Martin, G. R. (1999). *Genes & Development, 13*, 3136–3148.
Tucker, A. S., Matthews, K. L., & Sharpe, P. T. (1998). *Science, 282*, 1136–1138.
Tucker, A. S., Yamada, G., Grigoriou, M., Pachnis, V., & Sharpe, P. T. (1999). *Development, 126*, 51–61.
Vainio, S., Karavanova, I., Jowett, A., & Thesleff, I. (1993). *Cell, 75*, 45–58.
Veitch, E., Begbie, J., Schilling, T. F., Smith, M. M., & Graham, A. (1999). *Current Biology, 9*, 1481–1484.
Wang, X.-P., Suomalainen, M., Felszeghy, S., Zelarayan, L. C., Alonso, M. T., Plikus, M. V., et al. (2007). *PLoS Biology, 5,* e159.
Wang, X. P., Suomalainen, M., Jorgez, C. J., Matzuk, M. M., Werner, S., & Thesleff, I. (2004). *Developmental Cell, 7*, 719–730.
Wang, Y., Zheng, Y., Chen, D., & Chen, Y. (2013). *Developmental Biology, 381*, 301–311.
Wiszniak, S., Mackenzie, F. E., Anderson, P., Kabbara, S., Ruhrberg, C., & Schwarz, Q. (2015). *Proceedings of the National Academy of Sciences of the United States of America, 112*, 6086–6091.
Xu, Q., Mellitzer, G., Robinson, V., & Wilkinson, D. G. (1999). *Nature, 399*, 267–271.
Yahiro, K., Higashihori, N., & Moriyama, K. (2017). *Biochemical and Biophysical Research Communications, 482*, 883–888.
Yang, L., Tsang, K. Y., Tang, H. C., Chan, D., & Cheah, K. S. E. (2014). *Proceedings of the National Academy of Sciences of the United States of America, 111*, 12097–12102.
Yang, R. T., Zhang, C., Liu, Y., Zhou, H. H., & Li, Z. B. (2012). *Anatomical Record (Hoboken), 295*, 734–741.
Yasuda, T., Mundy, C., Kinumatsu, T., Shibukawa, Y., Shibutani, T., Grobe, K., et al. (2010). *Journal of Dental Research, 89*, 1111–1116.
Yonezawa, H., Harada, K., Ikebe, T., Shinohara, M., & Enomoto, S. (2006). *Journal of Cranio-Maxillofacial Surgery, 34*, 270–276.
Zhang, C. (2010). *Journal of Orthopaedic Surgery and Research, 5*, 37.
Zhao, H., Feng, J., Seidel, K., Shi, S., Klein, O., Sharpe, P., et al. (2014). *Cell Stem Cell, 14*, 160–173.
Zhu, L., Ma, J., Mu, R., Zhu, R., Chen, F., Wei, X., et al. (2018). *Life Sciences, 202*, 175–181.

CHAPTER FIVE

Joints in the appendicular skeleton: Developmental mechanisms and evolutionary influences

Danielle Rux[a,*], Rebekah S. Decker[b], Eiki Koyama[a], Maurizio Pacifici[a]
[a]Translational Research Program in Pediatric Orthopaedics, Division of Orthopaedic Surgery, The Children's Hospital of Philadelphia, Philadelphia, PA, United States
[b]Genomics Institute of the Novartis Research Foundation, San Diego, CA, United States
*Corresponding author: e-mail address: ruxd@email.chop.edu

Contents

Abstract

The joints are a diverse group of skeletal structures, and their genesis, morphogenesis, and acquisition of specialized tissues have intrigued biologists for decades. Here we review past and recent studies on important aspects of joint development, including the roles of the interzone and morphogenesis of articular cartilage. Studies have documented the requirement of interzone cells in limb joint initiation and formation of most, if not all, joint tissues. We highlight these studies and also report more detailed interzone dissection experiments in chick embryos. Articular cartilage has always received special attention owing to its complex architecture and phenotype and its importance in long-term joint function. We pay particular attention to mechanisms by which neonatal articular cartilage grows and thickens over time and eventually acquires its multizone structure and becomes mechanically fit in adults. These and other studies are placed in the context of evolutionary biology, specifically regarding the dramatic changes in limb joint organization during transition from aquatic to land life. We describe previous studies, and include new data, on the knee joints of aquatic axolotls that unlike those in higher vertebrates, are not cavitated, are filled with rigid fibrous tissues and resemble amphiarthroses. We show that when axolotls metamorph to life on

Current Topics in Developmental Biology, Volume 133
ISSN 0070-2153
https://doi.org/10.1016/bs.ctdb.2018.11.002

land, their intra-knee fibrous tissue becomes sparse and seemingly more flexible and the articular cartilage becomes distinct and acquires a tidemark. In sum, there have been considerable advances toward a better understanding of limb joint development, biological responsiveness, and evolutionary influences, though much remains unclear. Future progress in these fields should also lead to creation of new developmental biology-based tools to repair and regenerate joint tissues in acute and chronic conditions.

1. Introduction

The joints are a diverse and multi-faceted group of skeletal structures. They differ not only in anatomical location, architecture and size, but also in the type and degree of movement they allow and the nature and structure of their components. One current and useful classification of joint diversity is largely based on the degree of joint movement (Gray, 1988). Synarthroses allow minimum if any movement and consist of dense connective tissue separating the opposing skeletal elements, one example being the joints between cranial bones. Amphiarthroses permit some but delimited movement and display a fibrocartilaginous structure between the adjacent skeletal elements. Examples are the intervertebral joints and the pubic symphysis. Lastly, diarthroses permit free, reciprocal and nearly friction-less movement, and prominent members of this subgroup are the synovial joints in the appendicular skeleton. This classification emphasizes the strict relationship between the structural organization of the joints and their functional properties, each combination producing an organ exquisitely fitted to diverse anatomical locations and fulfilling specific biological and mechanical requirements. As it will be described later in this chapter, joint diversity also reflects evolutionary influences, processes and traits.

The synovial joints in the limbs have long attracted strong research attention not only for their importance in daily activities, overall skeletal function and quality of life but also for their susceptibility to acquired and congenital diseases, including osteoarthritis (OA), symphalangism, and developmental dysplasia of the hip (Archer, Caterson, Benjamin, & Ralphs, 1999; Goldring & Goldring, 2007; Hunziker, 2002; Seemann et al., 2005). These joints are composed of multiple tissues and structures. They all share a fibrous capsule that is continuous with tissues attached to the flanking skeletal elements including periosteum, insulating the joint from the internal body environment. The capsule's inner portion is covered by a synovial membrane—a distinct tissue composed of tightly-assembled and flat shaped synovial fibroblasts—which is rich in stem cells (Kurth et al., 2011). The joint cavity is filled with synovial fluid that contains phospholipids, hyaluronan, and glycoproteins such as *Prg4*/lubricin, each component contributing in its own manner to joint lubrication and friction-less motion (Jones & Flannery, 2007; Kosinska et al., 2012; Seror, Zhu, Goldberg, Day, & Klein, 2015; Temple-Wong et al., 2016). The epiphyseal ends of the opposing skeletal elements are covered by articular cartilage, a complex multi-zone tissue that is rich in collagen II, aggrecan, and other extracellular matrix molecules and provides resilience during movement (Bhosale & Richardson, 2008; Hunziker, Kapfinger, & Geiss, 2007). The limb joints also contain components required for certain joint-specific functions, including the anterior cruciate ligament and the patella in the knee and the teres in the hip, that are essential for regulation of motion directionality and joint stabilization and contribute also to proprioception (Ellison & Berg, 1985; Schutte, Dabezies, Zimny, & Happel, 1987). In sum, limb joint functioning requires the orchestrated contributions and efforts of multiple tissues and structures that potentially, last throughout life. As indicated above however, these physiologic traits and mechanisms often succumb to disease or injury since the innate repair capacity of limb joints—and articular cartilage in particular—is notoriously poor. This situation remains a major healthcare problem and challenge and has vexed scientists and clinicians for years. Enormous efforts have been—and are being—devoted to finding therapeutic means by which joint tissues could be repaired or regenerated by biological and bioengineering approaches, but this laudable task has yet to be fulfilled (Johnstone et al., 2013; Makris, Gomoll, Malizos, Hu, & Athanasiou, 2015). Thus, there has been much interest in recent years in deciphering the developmental biology of limb joints, with the hope that detailed information and understanding in this area may elicit the conception and creation of new repair means mimicking or reproducing developmental mechanisms of joint tissue formation (Caldwell & Wang, 2015; Longobardi et al., 2015). Our own group has contributed to studies on synovial joint development, growth, and morphogenesis (Decker et al., 2017; Koyama et al., 2008, 2010), and this chapter focuses mainly on early joint determination events, tissue morphogenetic mechanisms, and evolutionary influences. Other important aspects of limb joint formation have been reviewed elsewhere (Decker, 2017; Longobardi et al., 2015; Pacifici, Decker, & Koyama, 2018; Pitsillides & Ashhurst, 2008; Salva & Merrill, 2017).

2. Onset of limb synovial joint formation: The interzone

For over a century, the developing limb in mammalian and avian embryos has served as a popular system to study skeletogenesis, owing to its experimental accessibility and relative simplicity compared to the trunk and head. Embryologists realized long ago that the limb skeletal primordia are initially laid down largely as uninterrupted mesenchymal cell condensations with no obvious traits of where the joints would form (Haines, 1947; Hinchliffe & Johnson, 1980). The primordia include the Y-shaped preskeletal condensation that depicts the prospective stylopod element (humerus/femur) continuous with the two zeugopod elements (radius and ulna or tibia and fibula), and the digit rays that correspond to carpal/tarsal and phalangeal elements. It was found that the first morphological indication of joint formation became histologically evident with the emergence at each prospective joint site of an "interzone" (Haines, 1947), also called "zwischenmasse" or "articular disk" by earlier investigators (Whillis, 1940). The interzone consists of mesenchymal cells that initially are closely bound to each other and are apparently interconnected, in a manner distinct from that of surrounding mesenchymal cells and chondrocytic cells within flanking long bone anlagen. One study indicated that the initially close cell-cell contacts among interzone cells reflect the presence of tight junctions and expression of connexin 32 and 43 (Archer, Dowthwaite, & Francis-West, 2003). Soon after, the interzone acquires a typical tri-layer configuration with two compacted cell layers, each bound to the epiphyseal ends of flanking long bone anlagen and one central and more dispersed cell layer. This arrangement is quite obvious in avian limbs where the overall interzone is thick and highly cellular, but is more subtle in mammalian limbs where the interzone is rather thin and sparse (Mitrovic, 1977, 1978). These species-specific differences in thickness, structure, and cellularity have remained largely unexplained in terms of possible developmental significance, but we provide some new insights below.

Though important, the studies above left several important questions unanswered, in particular whether the interzone merely represents an otherwise passive signpost or landmark specifying the anatomical location of the future joint or whether its cells and/or their progenies would have active roles in joint formation. A first experimental attempt to address this key question was carried out in a study in chick embryos (Holder, 1977). Microsurgical procedures were used to remove the prospective interzone from the developing right elbow joint site in stage 24–26 chick embryos *in ovo* (about 4.5 days old). Effects were monitored over time by anatomical examination and compared to the phenotype of unoperated contralateral left limb in the same embryos. The elbow joint failed to form in operated limbs and the epiphyseal cartilaginous ends of humerus, radius, and ulna did not separate and became fused, with penetrance of the phenotype of about 100%. These novel observations showed for the first time that interzone cells are required for formation of synovial joints and their roles cannot be compensated for by surrounding cells nor cells within the flanking long bone cartilaginous anlagen. Despite its novelty and great importance, this study has not been extended and further verified. In addition, (i) the joints were analyzed anatomically but not histologically or by other means and (ii) it was not clear how selective the removal of the interzone had been effected and whether adjacent tissues were accidentally removed as well. Thus, we carried out similar microsurgical manipulations on developing stage 25–26 chick embryo elbow joints (Fig. 1), but varied our interventions as follows. In a first group of embryos, we mechanically disturbed the interzone with a dissecting needle, but left it in place. In a second group, we removed it, but carefully limited the resected tissue slice to about 100 μm in thickness, very close to the calculated thickness of native interzone at that stage. In a third set, we resected a larger segment of about 300–350 μm in thickness, thus removing not only the interzone but also an epiphyseal portion of flanking cartilaginous elements. After tissue resection, skeletal elements were stabilized by insertion of tungsten pins. Embryos were re-incubated and examined over time. In day 10 control unoperated elbows on the left side of the embryos, the joints displayed typical mature characteristics that included a well-developed articular cartilage, a large capsule, and synovial lining (Fig. 1A). Of note is the fact that a fairly conspicuous fibrous layer covered articular cartilage (Fig. 1B, yellow brackets), a trait typical of avian limb joints largely absent in mammalian joints. Mechanical disturbance of the interzone deranged joint formation (Fig. 1C and D). This was exemplified by delayed separation of the opposing skeletal elements and a poor definition of the epiphyseal ends of each element (Fig. 1D), indicating that the interzone is extremely sensitive to manipulation and cannot recover its function easily. Whole resection of the interzone completely prevented joint formation, and there was cartilaginous continuity between the humerus on one side and ulna and radius on the opposite side (Fig. 1E and F). The intervening ectopic cartilaginous tissue was composed of small-sized chondrocytes, resembling those present at the flanking epiphyses at that stage (Fig. 1F).

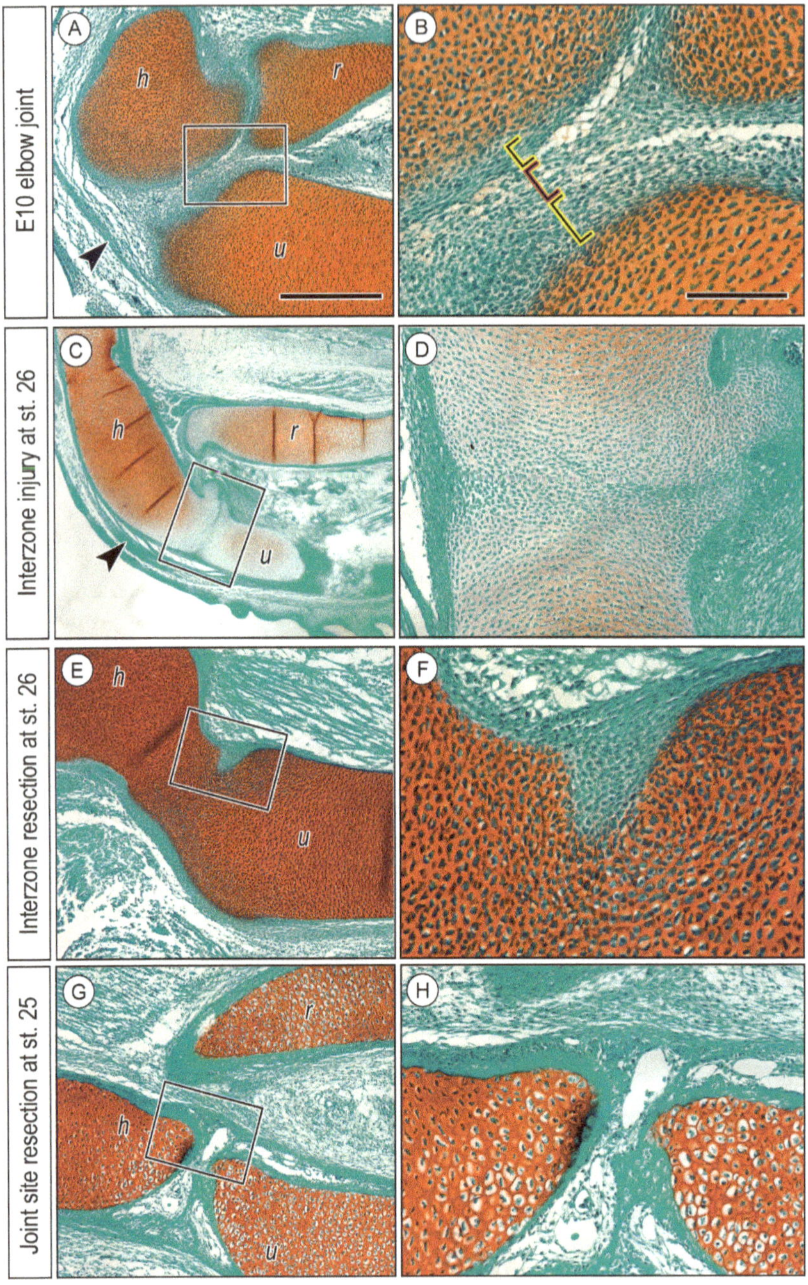

Fig. 1 The interzone is essential for limb joint formation. Stage 25–26 chick embryos *in ovo* were subjected to microsurgical intervention to alter or remove the elbow joint interzone on the right side. The left elbow was left untouched and served as internal

A similar cartilaginous continuity was observed in a related study after removal of a 150-μm thick tissue slice from the prospective elbow joint in stage 27 chick embryos, though details about surgical approach and tissue slice identity were scanty (Ozpolat et al., 2012). In comparison, elbow sites from which a large and thick tissue segment had been removed not only lacked joints but also the flanking long bone cartilaginous elements remained largely separated from each other and were composed of hypertrophic chondrocytes undergoing endochondral ossification (Fig. 1G and H). Removal of such large prospective joint tissue segment thus appeared to have caused a conversion of the epiphyseal ends into a neodiaphysis. Overall, the data agree with Holder's observations that the interzone per se exerts a required and non-redundant function in limb joint development and its removal alters the developmental program of flanking skeletal cells. The data are also in line with intriguing observations and concepts reported several decades ago. For example, Hampe and Wolff transplanted thick slices of stage 20–22 hind limb buds containing the prospective knee region onto the chorioallantoic membrane of host chick embryos and found that a seemingly normal joint formed over time (Hampe, 1956; Wolff, 1958). Perceptively, they concluded that the cells within the transplanted tissue had acted autonomously and possessed all the determination cues and morphogenetic

---

control. Embryos were re-incubated until day 10 (E10) at which point the elbows were dissected out and processed for histological analysis and Safranin O-fast green staining. (A) Images of E10 control elbow showing the prominent Safranin O-positive humerus (*h*), radius (*r*), and ulna (*u*) epiphyses flanking the developing synovial tissues and cavity. Note also the well-developed capsule (arrowhead). (B) High magnification image of boxed area in (A) showing the compact fibrous tissue (yellow brackets) covering the cartilaginous ends of the skeletal elements and an intervening and slightly less dense tissue (pink bracket). (C and D) Images of E10 elbow in which the interzone had been mechanically damaged but not removed. Area boxed in (C) is shown at higher magnification in (D). Note the considerable delay in joint development depicted by substandard separation of the opposing elements and by defective definition of the epiphyses compared to controls. Capsule appears to be unaffected (arrowhead). (E and F) Images showing absence of elbow joint and fusion of the opposing cartilaginous epiphyses after resection of the interzone. Area boxed in (E) is shown at higher magnification in (F). Note the small and relatively uniform size of chondrocytes in both intervening tissue and flanking epiphyses. (G and H) Images showing that removal of interzone and neighboring tissue led to absence of a joint and physical separation of neighboring skeletal elements. Note also the hypertrophic phenotype of chondrocytes in the opposing and truncated elements suggesting that the epiphyses were undergoing ectopic endochondral ossification typical of the diaphysis at this stage. Scale bar in (A) for C, E, and G, 250 μm; scale bar in (B) for D, F, and H, 100 μm.

information needed to create a joint, a concept subsequently reiterated by others (Cosden-Decker, Bickett, Lattermann, & MacLeod, 2012; Wolpert, 1969).

3. Interzone cell function and fate

Given that interzone cells are needed for joint formation, what do they actually do? Some answers to this lingering central question have been provided by genetic cell lineage trace-track studies carried out in the last decade. Kingsley and coworkers were the first to show that the transforming growth factor-β superfamily member *Gdf5* is selectively expressed by interzone cells at the very onset of limb joint development in mouse and chick embryos (Storm & Kingsley, 1996, 1999). The group subsequently created a transgenic BAC-based *Gdf5-Cre* mouse line for conditional gene ablation studies, but used it also to genetically label the *Gdf5*-expressing cells with a reporter and monitor them (Rountree et al., 2004). Their data showed that reporter-positive cells persisted at the joint sites at a later examined time point, were present in joint tissues including articular cartilage, and did not appear to contribute much to formation of long bone shafts or other surrounding tissues. In collaborative follow-up efforts with that group, we used *Gdf5-Cre;R26R-LacZ* double transgenic mice and systematically monitored the topographical distribution and fate of β-galactosidase-expressing cells and their progenies (collectively termed *Gdf5*+ cells) over embryogenesis and postnatal life up to about 2 months of age (Koyama et al., 2007, 2008). We found that the *Gdf5*+ cells generated articular cartilage, intra-joint ligaments, meniscus, synovial lining, and the inner half of the capsule in limb joints including hip, knee, elbow, carpal/tarsal elements, and digits. The cells were restricted to joint tissues prenatally and postnatally and made little if any contribution to formation of surrounding tissues and structures, though occasional *Gdf5*+ cells could be observed in underlying cartilaginous shafts. Similar trends were observed in prenatal and postnatal wrists where the *Gdf5*+ cells were confined to the articulating surfaces, though wrist skeletal development is quite distinct from that of long bones (Hogg, 1980). The data agreed well with concurrent genetic cell lineage tracing studies in developing mouse embryo knees carried out by others indicating that *Col2*+ articular chondrocytes arise as a population distinct from *Matrillin-1*+ chondrocytes constituting the bulk of the long bone elements (Hyde, Dover, Aszodi, Wallis, & Boot-Handford, 2007). Considered together with the interzone extirpation data above (Fig. 1) (Holder, 1977), the genetic cell lineage studies showed that interzone cells are not only required for joint formation but also survive and thrive over time to produce most if not all joint tissues, supporting the notion that they represent a specialized cohort of progenitor cells with innate joint formation capacity.

Because the interzones emerge within preskeletal mesenchymal condensations that are initially uninterrupted but differentiate to chondrocytes soon after, it has long been debated whether interzone cells represent mesenchymal cells present (and determined) at each prospective joint site or are generated via de-differentiation of chondrocytes formed at those sites. Spatiotemporal analyses in developing day 4.0–4.5 chick limbs using immunohistochemistry indicated that the cartilage markers collagen II and keratan sulfate were present over the entire prospective metatarsalphalangeal digital rays prior to interzone formation (Craig, Bentley, & Archer, 1987). We reported similar *Col2a1* gene expression patterns at those stages as revealed by *in situ* hybridization (Koyama, Leatherman, Shimazu, Nah, & Pacifici, 1995). Mouse studies elicited comparable observations. Hyde et al. made use of *Col2a1-Cre;R26R-LacZ* mouse embryos and compared the spatiotemporal distribution of β-galactosidase-positive cells (that includes progeny cells and collectively termed *Col2*+ cells here) with the distribution of cells actively expressing *Col2a1* in developing knee joints (Hyde, Boot-Handford, & Wallis, 2008). At embryonic day 12.5 (E12.5) when the interzone had not become evident yet, there was continuous and overlapping distribution of *Col2*+ cells with *Col2a1* transcripts throughout the Y-shaped preskeletal condensations. The interzone became appreciable by E13.5; at this stage, most interzone cells did not express *Col2a1*, but were still positive for β-galactosidase activity. These data agree with observations by Soeda and coworkers using knock-in *Sox9-LacZ* reporter mouse embryos (Soeda et al., 2010) in which the reporter is driven by the chondrogenic master regulator gene *Sox9* (Bi, Deng, Zhang, Behringer, & de Crombrugghe, 1999). They found that reporter positive, *Sox9*-expressing cells were present in incipient interzones in E13.5 mouse embryo knee joints. Considered collectively then, the above studies indicate that the cells founding and constituting the initial interzone at the very onset of joint formation have a chondrogenic history and express chondrogenic markers, giving support to the notion that they largely derive from local de-differentiated chondrocytes.

As the cell lineage tracing-tracking studies above indicate, interzone cells and their progeny give rise to diverse joint tissues over time, including inner capsule, meniscus, intra-joint ligaments, and articular cartilage

(Koyama et al., 2008; Rountree et al., 2004). However, it had remained unclear whether the cells are multipotent and produce different tissues within the developing joint or whether they are pre-specified into subgroups by local morphogenetic cues and endogenous determination mechanisms to produce different tissues at appropriate locations and time. Progress and insights into this complex but fundamental question have been reported recently. In a study from our group (Decker et al., 2017), we carried out additional analyses of joint progenitors with our new BAC-based *Gdf5-CreER*^{T2} transgenic mice mated with ROSA reporter mice, aiming to trace and track the cells starting at different stages of joint development. Pregnant *Gdf5-CreER;R26-zsGreen* mice received a single injection of tamoxifen at E13.5, E15.5, or E17.5, and the resulting *Gdf5-CreER*+ cells and possible progenies were monitored in their pups over postnatal time. Focusing on digit joints that are structurally simpler, we found that cells labeled at E13.5 were mostly present in the developing capsule and synovial lining by postnatal day 0 (P0) and those patterns did not change much by P28 or thereafter. Cells labeled at E15.5 or E17.5 was present in both articular layers and capsule by P0, and these patterns also did not change significantly with postnatal age. In addition, the overall number of *Gdf5-CreER*+ cells increased over time but not in a major way, indicating limited proliferative activity. To verify the latter and potentially important observation, we carried out experiments with (unbiased) *ROSA-CreER* mice mated with *R26-Confetti* reporter mice in which green (GFP), yellow (YFP), red (RFP), or cyan (CFP) fluorescent protein can be alternatively activated upon *Cre* action in different cells (Snippert et al., 2010). *ROSA-CreER/Confetti* mice were injected with tamoxifen once at E13.5, P0, P7, or P14, and we estimated proliferation by quantifying cells expressing the same reporter (representing daughter cells) over increasing postnatal time. For example, analysis at 2 months showed that *ROSA-CreER*+ cells initially labeled at E13.5 produced the largest progeny, with cell clusters containing as many as 18 daughter cells. Cell cluster size declined markedly thereafter, and averaged about 3–4 cells after P0 induction and only about 1–2 cells with increasing age. Taken together, the studies above indicate that *Gdf5*+joint progenitors become quickly determined and committed within the developing joint, do not migrate much from their initial birth location, produce relatively small progenies, and give rise to distinct joint tissues over time.

Data and conclusions agreed well with a recent and very important study by Zelzer and coworkers (Shwartz, Viukov, Krief, & Zelzer, 2016). The authors created knock-in *Gdf5-CreER*^{T2} mice and crossed them with reporter mice to further analyze joint cell progenitor origin and fate. Reporter induction by a single tamoxifen injection at E10.5 failed to elicit *Gdf5*+ (*tdTomato*-positive) cells over the entire initial interzone by E11.5; only most epiphyseal cells were labeled in knee, elbow, and carpalphalangeal joints by E18.5. When tamoxifen was given three consecutive times at E11.5, E13.5, and E15.5, *Gdf5*+ cells were present in most joint tissues by E18.5. Comparison of reporter expression after a single tamoxifen injection versus real-time endogenous *Gdf5* expression pointed to a highly dynamic behavior of joint-forming cells. Groups of cells turned *Gdf5* expression of while maintaining *tdTomato* expression and other groups turned it on while all contributing to formation of different joint tissues. Cell proliferation measured by BrdU incorporation indicated that the mitotic activity of *Gdf5*+ cells decreased significantly between E13.5 and E18.5. The data led the authors to conclude that while *Gdf5*+ cells constituting the initial interzone are direct descendants of de-differentiated *Sox9*+ chondrocytes, additional *Sox9*+ progenitors are recruited from the joint's immediate surroundings, and generate additional *Gdf5*+ cohorts contributing to distinct tissue joint development. The data fit well with observations in previous studies suggesting that local progenitors immediately surrounding the incipient joint site are recruited into the *Gdf5*+ lineage over time and take part in joint formation (Hyde et al., 2008; Koyama et al., 2007; Niedermaier et al., 2005). This mechanism may contribute not only to specification of given tissues within the developing joint but may also sustain the considerable growth and lateral expansion of the developing joints to accommodate and cover the fast expanding epiphyses of long bones during late prenatal as well as postnatal time.

We mentioned above that developing chick embryo limbs display thick interzones with a clear triple-layer tissue organization, while the interzone in mammalian embryonic limbs is rather thin, but the possible implications of this structural difference have remained unclear. Fig. 2 shows the histology and collagen gene expression patterns in chick and mouse embryo knees at day 17 of embryogenesis. Clearly, the proximal epiphysis in chick embryo tibia (Fig. 2A and B) displays a developing articular cartilage expressing *collagen II* (Fig. 2D, *ac*) covered by a thick fibrous layer strongly expressing *collagen I* and facing the emerging joint cavity (Fig. 2C, *fl*), a dual tissue structure that persists through adulthood in this species (Pacifici, 1995). In comparison, mouse embryo tibia (Fig. 2E and F) contains similar incipient *collagen II*-expressing articular chondrocytes (Fig. 2H, *ac*) that exhibit some residual *collagen I* expression (Fig. 2G, *ac*). The latter is rapidly lost after birth,

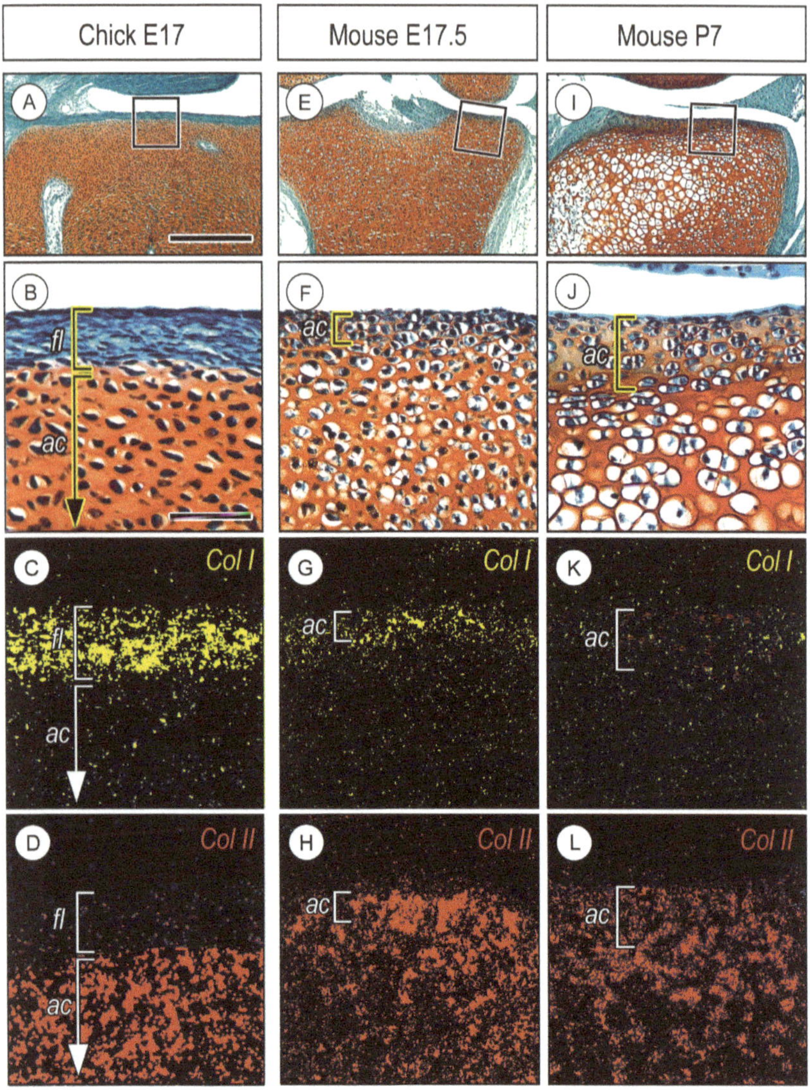

Fig. 2 Tissue structure and gene expression in developing tibia articular cartilage in chick and mouse. (A and B) Images show Safranin O/fast green-stained proximal tibia epiphysis in E17 chick embryo. Area boxed in (A) is shown at higher magnification in (B). Note the presence of a prominent, thick, and fast green-positive fibrous layer (*fl*) facing the synovial cavity and overlaying the Safranin O-positive articular cartilage (*ac*). (C and D) *In situ* hybridization images of serial sections showing strong expression of *collagen I* (C, *Col I*) and *collagen II* (D, *Col II*) in fibrous layer and cartilage, respectively. (E and H) Images show Safranin O/fast green-stained proximal tibia epiphysis in E17.5 mouse embryo. Area boxed in (E) is shown at higher magnification in (F). Note that incipient articular cartilage (*ac*, yellow bracket in F) is rather narrow at this stage and is identified

and the maturing articular cartilage expresses *collagen II* only already by P7 (Fig. 2I–L). Possibly then, the initial differences in organization, cellularity, and structure of interzones in chick and mouse may have developmental consequences, with the conspicuous interzone in chick leading to formation of a thick fibrous layer and thick cartilage and with a subtler mammalian interzone leading to absence of fibrous layer and thinner cartilage. We include data below suggesting that differences in interzone characteristics and cell fates may have evolutionary significance as well.

4. Articular cartilage postnatal growth and morphogenesis

The key function of articular cartilage is to provide resilience during cycles of compression and relaxation during daily activities and to contribute to friction-less joint motion via elaboration of anti-adhesive and lubricant macromolecules (Hunziker et al., 2007). These properties are directly assignable to the unique composition and structure of the tissue. In limb joints such as the knee in adult mammals including humans, the tissue displays a characteristic zonal organization that includes: (i) a thin superficial zone abutting the synovial space that is composed of flat, elongated, and tightly-bound cells oriented along the major direction of movement, contains scant, and isotropic matrix, and contributes to joint lubrication by producing molecules such as lubricin (Prg4); (ii) a thick intermediate/deep zone containing large round chondrocytes in a columnar organization perpendicular to the surface and surrounded by abundant and anisotropic matrix in which collagen II fibrils and aggrecan-hyaluronan complexes provide tensile strength and elasticity, respectively; and (iii) a mineralized zone below the tidemark with very large chondrocytes that is bound to subchondral bone (Hunziker et al., 2007). These structural and organizational features allow articular cartilage to exert its biomechanical function and to endure and maintain tissue homeostasis through life. Unfortunately, structural

based on *Gdf5*+ cell lineage tracing (Decker et al., 2017). Its cells strongly express *collagen II* (H, *Col II*) and exhibit some residual *collagen I* expression (G, *Col I*), possibly stemming from their former mesenchymal character. (I–L) Images show Safranin O/fast green-stained proximal tibia epiphysis in neonatal P7 mouse. Area boxed in (I) is shown at higher magnification in (J). Note that the developing articular cartilage has grown in thickness (*ac*, yellow bracket in J), still expresses *collagen II* strongly (L, *Col II*) but no longer expresses *collagen I* (K, *Col I*). Scale bar in (A) for E and I, 350 μm; scale bar in (B) for all remaining panels, 50 μm.

derangements of articular cartilage due to acute injury or chronic conditions can lead to irreversible changes and joint pathologies including OA (Mow, Ratcliffe, & Poole, 1992). There are also pediatric joint pathologies such as osteochondritis dissecans in which large segments of articular cartilage become detached from the subchondral bone, causing pain, and joint damage and often requiring surgery (Shea, Jacobs, Carey, Anderson, & Oxford, 2013). As indicated above, current remedies for both acute and chronic joint maladies are unable to restore native structure and function (Johnstone et al., 2013; Makris et al., 2015). A key problem in this field is that we know relatively little about how articular cartilage acquires its functional zonal organization by adult age. In addition, it is not understood how articular cartilage remains phenotypically stable and persists through life under normal circumstances, while the remaining and preponderant cartilaginous skeleton is transient and is all replaced by endochondral bone by end of puberty with closure of the growth plates.

As exemplified by mouse studies, incipient articular cartilage in late embryogenesis and neonatal stages is rather thin and composed of small, randomly distributed, and variously shaped chondrocytes and scant isotropic extracellular matrix (see Fig. 2F) (Li et al., 2016; Rhee et al., 2005). The tissue undergoes remarkable growth in thickness over the first 2–3 weeks of age. While small flat cells continue to populate its incipient surface zone, the bulk of underlying chondrocytes grow in average size and become progressively separated by ever increasing cartilage matrix. By 6–8 weeks of age, the tissue had acquired its characteristic and fully functional mature multi-zonal organization with a clear superficial zone, an intermediate/deep zone with chondrocyte columns and abundant anisotropic matrix, and a mineralized zone at the bottom (Decker et al., 2017). Similar postnatal maturation and structuring of articular cartilage characterize other and larger mammalian species (Hunziker et al., 2007). Thus, how does articular cartilage evolve from a thin and matrix-poor tissue at neonatal stages to a highly structured, thick, and zonal tissue in adults?

One explanation was originally suggested by the identification of cells with a distinct character in postnatal articular cartilage (Dowthwaite et al., 2004). The authors isolated cells from the superficial, intermediate, and deep zones from juvenile bovine articular cartilage by sequential enzymatic treatment and then characterized the cells by *in vitro* and *in vivo* assays. Superficial zone cells displayed high affinity binding to substrate-bound fibronectin and effective colony unit-forming ability *in vitro*, and expressed the progenitor cell marker gene *Notch1 in vitro* and *in vivo*. The cells displayed developmental plasticity in that they engrafted into, and became part of, several tissues once transplanted into chick embryos, including bone, tendon, and perimysium. In comparison, cells in the intermediate and deep zones exhibited the expected chondrocytic phenotypes. These and other data led the authors to propose that cells in the superficial zone have a progenitor/stem character. Over postnatal time, the cells would be responsible for articular cartilage growth and establishment of its zonal structure by a mechanism of appositional growth. The cells would proliferate and produce vertical columns of overlapping daughter chondrocytes spanning the entire tissue thickness from superficial to deep zone, a notion proposed also in previous studies (Hayes, MacPherson, Morrison, Dowthwaite, & Archer, 2001). Cells with a progenitor/stem character expressing markers such as CD90 and STRO-1 have been identified in human articular cartilage as well (Williams et al., 2010).

Appositional growth has also been invoked in a recent cell lineage tracing study to similarly account for postnatal articular cartilage growth and acquisition of its zonal organization in adults (Kozhemyakina et al., 2015). The study was based on previous data showing that *Prg4* is expressed in developing limb joints by late mouse embryogenesis (Rhee et al., 2005). The authors created knock-in *Prg4-CreER*^{T2} mice (*Prg4* heterozygous null) and mated them with *LacZ* reporter mice. After a single tamoxifen injection of pregnant mice at E17.5 of gestation, reporter-positive *Prg4*+ cells were found to be located exclusively in a single cell layer at the very surface of incipient tibial articular tissue in P0 pups. With increasing postnatal time, the *Prg4*+ cells became more numerous and were then found to be present in vertical chondrocyte columns spanning the whole tissue thickness in adult mice. When mice were injected with tamoxifen at postnatal times, the reporter-positive chondrocyte columns appeared to be partial and did not span the whole tissue thickness in adults. BrdU incorporation was used to estimate mitotic activity, and the data indicated that over 70% of superficial cells were proliferative even in 1-month-old mice. These and other data led to the conclusion that the *Prg4*+ cells located at the articular surface of neonatal tibia serve as stem cells and elicit postnatal growth of articular cartilage by proliferation and apposition, generating columns of daughter cells spanning the entire thickness of articular cartilage in adults. Data and conclusions were further examined in a follow-up study by Chagin and coworkers (Li et al., 2016). Using the same knock-in *Prg4-CreER*^{T2} mice but mated to *Rosa-Confetti* reporter mice, the authors found that *Prg4*+ cells present at the tibia articular tissue surface expressed stem cell traits and renewed themselves or generated chondrocytes by symmetric and asymmetric cell divisions at neonatal stages. With increasing postnatal time, the progeny of those cells formed daughter cell clusters and could be found within the growing and thickening articular cartilage. This was accompanied by increases in average cell volume, but at variance with the previous study (Kozhemyakina et al., 2015), no columns of daughter cells (expressing the same reporter) were found spanning the entire articular cartilage in adults. The authors concluded that articular superficial cells are self-renewing progenitors able not only to maintain themselves but also to produce progenies reconstituting articular cartilage over postnatal life.

Because of the broad relevance of this research area, we carried out analogous studies, but with the following considerations in mind. The exclusive presence of *Prg4*+ cells in a single and most superficial articular cell layer observed at P0 (Kozhemyakina et al., 2015; Li et al., 2016) does not actually match endogenous *Prg4* that is expressed throughout the 6–8 layers of incipient articular cartilage at that stage (Koyama et al., 2008; Rhee et al., 2005), raising some concern about the reliability of knock-in heterozygous null *Prg4-CreER*^{T2} mice. Appositional growth of postnatal articular cartilage would require high rates of cell proliferation, but proliferation in neonatal and juvenile cartilage is relatively modest (Decker et al., 2017; Li et al., 2016). It would also require considerable matrix turnover when in fact the articular matrix—and collagen in particular—is quite stable (Eyre, 2002; Poole et al., 2001). To carry out follow-up studies, we made use of our new BAC-based *Prg4-CreER*^{T2} transgenic mice (in which the endogenous *Prg4* alleles are intact) mated with single and *Rosa-Confetti* reporter mice. We found that when *Prg4-CreER;R26-tdTomato* mice were injected with tamoxifen once at E17.5, reporter-positive cells were present throughout the 6–8 cell layers in tibia articular cartilage in P0 pups, matching endogenous *Prg4* expression at that stage. Identical patterns were observed in *Prg4-CreER;R26-Confetti* mice injected at E17.5 and harvested at P0, though recombination efficiency and reporter expression were overall lower as expected (Snippert et al., 2010). *Prg4-CreER;R26-Confetti* mice injected at E17.5 were then harvested at successive postnatal time points up to 2 months of age. With increasing time, the *Prg4*+ cells produced small and locally-restricted groups of daughter cells (same reporter color) averaging 3–6 cells/cluster, suggesting that their mitotic activity was low as was their mobility. There was evidence of horizontal and vertical cluster expansion over time, but single color cells or clusters did not span the full thickness of adult articular cartilage nor did they form columns of same-color daughter cells. Indeed, stereological imaging and 3D reconstruction throughout articular cartilage in late juvenile and adult mice indicated that articular chondrocytes were not perfectly aligned and did not fully overlap each other, thus forming stacks rather than columns. The data were confirmed and verified using unbiased *ROSA-CreER;Confetti* mice injected once at E17.5 and *Gdf5Cre;R26-Confetti* mice, both harvested at successive time points as above.

Additional stereological and reconstruction analyses showed that the average cell volume of articular chondrocytes increased markedly over postnatal time and did so in a zone-specific manner. Computation of data actually indicated that such increases were a major driver of overall articular cartilage thickening and growth over postnatal life. Interestingly, this mechanism is reminiscent of chondrocyte hypertrophy in the growth plate that is by far the major driver of skeletal elongation and growth while cell proliferation plays a minor role (Breur, VanEnkevort, Farnum, & Wilsman, 1991). Importantly also, the maximum volume displayed by deep zone chondrocytes did not exceed about 60% of that of growth plate hypertrophic chondrocytes, suggesting that chondrocytes can set and maintain distinct volumes in different settings and that articular chondrocytes normally avoid full hypertrophy. Second harmonic generation with two-photon microscopy indicated that the collagen fibrils in the matrix were isotropic and scattered at neonatal stages and became anisotropic and aligned along the chondrocyte stacks in adult articular cartilage (Decker et al., 2017). These and other data led us to conclude that rather than by apposition, articular cartilage in large joints such as the knee grows and thickens mainly by formation of non-daughter cell stacks and increases in average cell volume, with important contribution by matrix deposition and accumulation but modest contribution by proliferation. Because the chondrocyte stacks are made of non-daughter cells, it is possible that they are produced by a process of realignment and reorientation of neighboring cells and clusters, possibly aided by mechanisms such as convergent extension (Tada & Heisenberg, 2012). Our data and conclusions differ in some respects from those reported in the studies above (Kozhemyakina et al., 2015; Li et al., 2016), and further studies are thus needed to sort out and explain these differences and move ahead in this research field. It will also be important to determine whether articular cartilage grows and matures by similar mechanisms in different joints in the limbs and other skeletal sites.

5. Evolutionary considerations

Joints have undergone significant and remarkable changes through the evolutionary transition of vertebrates from a cartilaginous to a bony skeleton, and also from aquatic to terrestrial habitats. Analyses of such evolutionary changes offer unique insights into mechanisms of joint determination, diversification, and organization as a function of changes in environments, body locomotion and physical demands. The evolution of joints in the appendicular skeleton followed the transition from water to land to air over hundreds of millions of years as it adapted to the unique and ever more sophisticated capabilities in higher species culminating in primates and humans. Others have analyzed and previously reviewed important themes regarding the conserved signaling pathways involved in joint development and the transcriptional control of joint diversification among vertebrate and non-vertebrate species (Salva & Merrill, 2017; Shubin, Tabin, & Carroll, 1997). Here we highlight salient aspects of joint organization, structure, architecture, and function in the appendicular skeleton during the progression from fins to limbs as vertebrates adjusted to terrestrial life and how these evolutionary considerations provide insight into the developmental mechanisms of synovial joints.

Fishes are among the oldest documented vertebrates. These animals have stood the test of time and have adapted to drastic climate and environmental changes over hundreds of millions of years. The colonization of marine habitats by vertebrate life began around 450 million years ago (MYA) and dominated the following Devonian era, commonly referred to as the "era of the fishes," lasting from about 415 to 360 MYA (Becker, Gradstein, & Hammer, 2012). Cartilaginous fishes were the first to evolve from primitive vertebrate animals that lacked appendages. The skeleton of those fishes was not ossified, contained mineralized cartilaginous structures that were flexible and therefore lacked the necessity for cavitated joints. Sharks are the living descendants of these primitive ancestors and first evolved around 425 MYA (Becker et al., 2012). Bony fishes first evolved around 420 MYA with ray-finned fish first followed by lobe-finned fish that evolved around 415 MYA (Fig. 3) (Becker et al., 2012; Botella, Blom, Dorka, Ahlberg, & Janvier, 2007; Clack, 2013). As their names suggest, these two classes of fishes differed in the structure of their fins. Ray-finned fish exhibited fins that extended from the body wall as a series of spines and were connected together by thin tissue to form the overall fin (Fig. 3). Living examples of

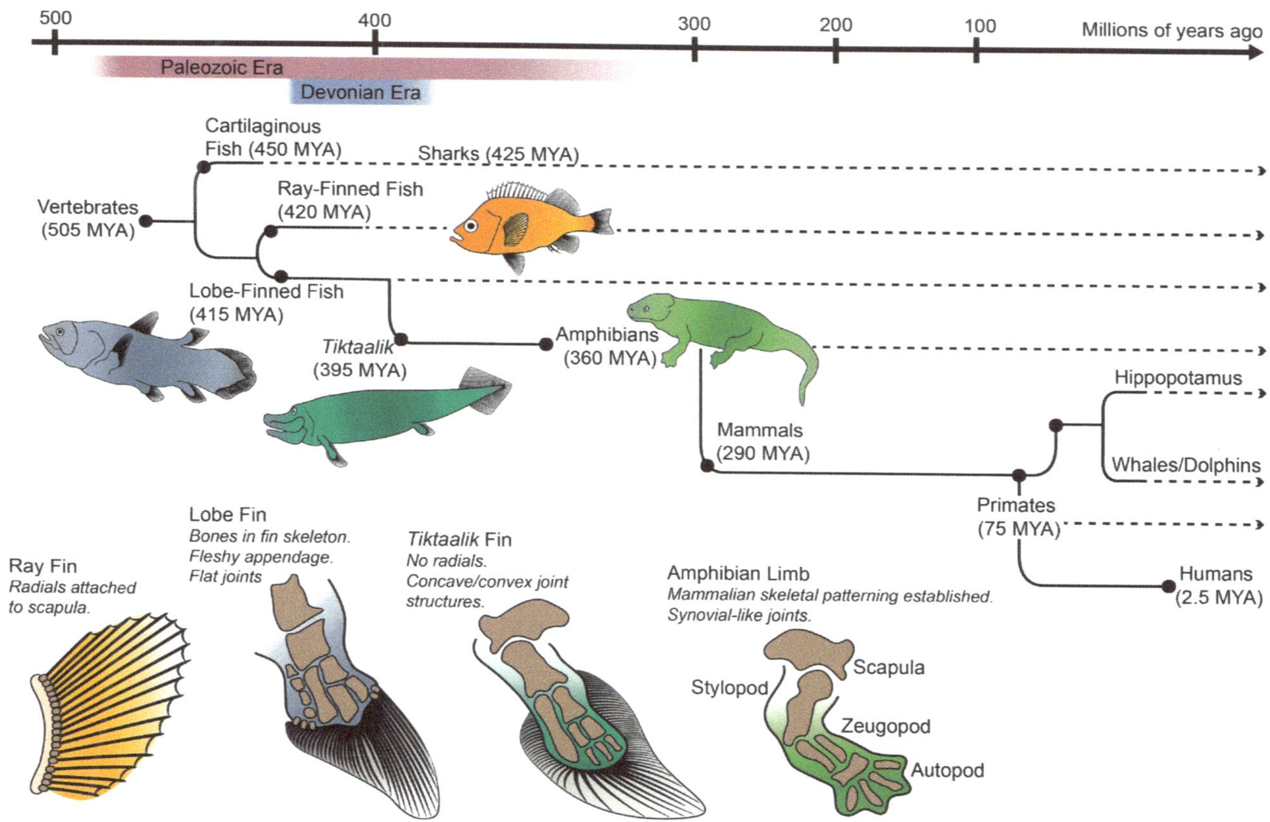

Fig. 3 Timeline of vertebrate evolution through the fin-to-limb transition. Solid and dashed lines represent evolutionary timelines and currently extant species. Pictured at the bottom are enlarged cartoons of the pectoral fins/limbs of the animals indicated in the timeline and depicting the evolutionary transition from ray fin to lobe fin to *Tiktaalik* fin to amphibian limb. Key aspects of skeletal transitions at each stage are highlighted within these cartoons.

these fish include commonly hunted species such as tuna and salmon, and also lab-maintained species such as spotted gar and zebrafish. By contrast, lobe-finned fish displayed much thicker, fleshy fins attached to the body wall by a single skeletal element (evolved to the humerus in higher species) that articulated with several smaller, and more distal, bones. The connection of the fin to the trunk by a single bone in lobe-finned fish is a key trait that has been maintained through evolution of lobe-finned fishes and is a defining feature of all tetrapods including humans (Clack, 2013; Romer, 1958; Shubin, Daeschler, & Coates, 2004; William & Raven, 1941). Over additional time, the articulated bony elements of the lobe fins transitioned to the stereotypic limb pattern maintained by all amphibians, birds, and mammals: the proximal stylopod (humerus/femur), the medial zeugopod (radius/ulna and tibia/fibula), and the distal autopod (with several bones composing the wrist/ankle and digits) (Fig. 3) (Hinchliffe, 1994; Hinchliffe & Johnson, 1980).

The transition of lobe-finned fish from water to land represents one of the critical evolutionary advances in vertebrate history that required not only major changes in the skeleton and joints but also adaptation and diversification of body functions including hearing and breathing. The now extinct genus of lobe-finned fish *Panderichthys* ultimately gave rise to terrestrial tetrapods that colonized land in the post-Devonian era (Boisvert, Mark-Kurik, & Ahlberg, 2008; Vorobyeva & Schultze, 1991); however, a clear evolutionary path from fin-to-limb remained elusive until the landmark discovery of a novel fossilized animal that was first reported in 2006. Two studies published that year described *Tiktaalik roseae*, a fish that exhibited unequivocal transitional features of both fish and terrestrial tetrapods (Daeschler, Shubin, & Jenkins Jr., 2006; Shubin, Daeschler, & Jenkins Jr., 2006). A series of fossils were uncovered in the Devonian sediment of Arctic Canada on Ellesmere Island placing the evolution of this animal in the correct timeline during aquatic to terrestrial transition. Based on the fossilized structures, it is hypothesized that *Tiktaalik* displayed gills, lungs, and a neck that swiveled, and its pectoral fins revealed a clear progression toward amphibian limbs: a single proximal bone attached to the trunk (ancient stylopod), two smaller and more distal bones (ancient zeugopod), and several even smaller bones that shaped the remaining fin/limb-like elements (Fig. 3). It was dubbed "the fish that crawled out of water."

The overall geometry and spatial orientation of the joint, curvatures of the opposing skeletal surfaces, and other anatomical features make it possible to generate hypotheses concerning the mobility and function of the skeletal elements in fossilized animals. The glenohumeral joint of *Tiktaalik* likely permitted a significant degree of motion freedom given the deep concave-convex contours of the opposing articulating surfaces. Such architecture suggests that this joint could have been fully cavitated and, perhaps, synovial in nature. The distal epiphysis of the humerus was part of an elbow-like joint that articulated with—and provided independent mobility to—two medial short bones (representing primitive radius and ulna). The two articulating surfaces on the humerus facing the medial bones were well separated from each other, a feature not seen in more primitive fishes (Fig. 3) (Andrews & Westoll, 1968; Shubin et al., 2004). However, they were not fully oriented along the longitudinal axis suggesting that those medial elements were oriented laterally and experienced some motion limitation. There were several small bones distal to the primitive ulna and radius—including ulnare and radials—that displayed intervening joints with rather shallow concavities and convexities in between, suggesting some but probably limited mobility and flexibility (Shubin et al., 2006). Subsequent analyses of the pelvic girdle in *Tikaalik* by the same groups further bridged the gap between fish and terrestrial tetrapods (Shubin, Daeschler, & Jenkins Jr., 2014). In addition, multiple discoveries of tetrapod trackways embedded in sediment from the Devonian era scattered throughout Europe, China, and Australia corroborate evidence of the fin-to-limb transition of the *Tiktaalik* fossils. Specifically, fossilized trackways in Poland allowed Paleontologists to predict that ancient tetrapods existing in that region likely ambulated with amphibian-like movements consistent with predicted mobility of *Tiktaalik* appendicular joints (Niedzwiedzki, Szrek, Narkiewicz, Narkiewicz, & Ahlberg, 2010). Discrepancies in the evolutionary timeline of the fin-to-limb transition do exist: *Tiktaalkik* dating to the late-Devonian era, the Polish trackways to the mid-Devonian era, and other trackways ranging from early- to late-Devonian (Clack, 1997; Gouramanis, Webb, & Warren, 2003; Lu et al., 2012; Stossel, 1995; Warren, Jupp, & Bolton, 1985; Warren & Wakefield, 1972; Williams, Sergeev, Stossel, & Ford, 1997). These require further analysis, however, the combined studies provide consistent evidence that requirements for ambulation on land necessitated the evolution of joints to provide novel functions that are not seen in entirely aquatic fishes.

The above discoveries and insights are indeed striking, but the lack of preservation of soft tissues in fossils make it quite difficult to evaluate more specific joint features, including cavitation, mobility range, lubrication, and biomechanical tissue traits. One way to address these limitations has long been the analyses of anatomical features in extant animals carrying ancestral characteristics. The axolotl salamander (*Ambystoma mexicanum*) is one such species and remains a popular model of skeletal development, growth, and regeneration, and has been recognized more recently as a model for joint development/evolution and repair (Cosden et al., 2011; Cosden-Decker et al., 2012; Thampi, Liu, Zeng, & MacLeod, 2018). Axolotls display neoteny (Shaffer & Voss, 1996) and retain larval salamander characteristics (i.e., maintenance of gills) and an aquatic habitat. In rare cases axolotls undergo metamorphosis to terrestrial life in response to environmental stress (Thampi et al., 2018) or by administration of thyroxine hormone (Page, Monaghan, Walker, & Voss, 2009). This real-time transition offers insight on the biologic changes that occur in the transition from an aquatic to a terrestrial environment. Previously, one of us and colleagues carried out anatomical and histological analyses of the limb joints in aquatic axolotls of different ages. In 1-year-old skeletally mature axolotls, the hip, shoulder, and elbow joints were found to be cavitated, thus resembling full-fledged synovial joints. In sharp contrast, the knee, carpal/tarsal, and interphalangeal joints were not cavitated and were filled with a dense fibrous and highly cellular tissue interconnecting and bridging the epiphyseal cartilaginous ends of the opposing skeletal elements, thus resembling amphiarthroses. The fibrous tissue was still present at 2 years of age (the last time point examined) and as revealed by immunohistochemistry, was rich in collagen I and GDF5, but poor in aggrecan (Cosden-Decker et al., 2012). Fig. 4A and B shows additional histological images of knees from 2- to 10-year-old aquatic axolotls. At 2 years of age, the epiphyseal and metaphyseal regions of femur, tibia, and fibula were cartilaginous and displayed a continuous uninterrupted histological organization, without an obvious distinction or boundary between articular cartilage area and underlying growth plate (Fig. 4A). By 10 years, the metaphysis was ossified, but the epiphyseal cartilage still displayed a seamless organization along with the remaining and shallower growth plate (Fig. 4B and C).

Major joint differences were observed in axolotls that had undergone spontaneous metamorphosis and adjustment to a terrestrial habitat. At 2 years of age, the articular cartilage area in their knees was compacted, narrower, and had slightly lower chondrocyte density, thus becoming more distinguishable from the underlying growth plate (Fig. 4D, yellow brackets). This trend had advanced and became quite obvious in 10-year-old animals. The articular cartilage area was not only further compacted and structured but also now separated by a clear tidemark from the underlying residual growth

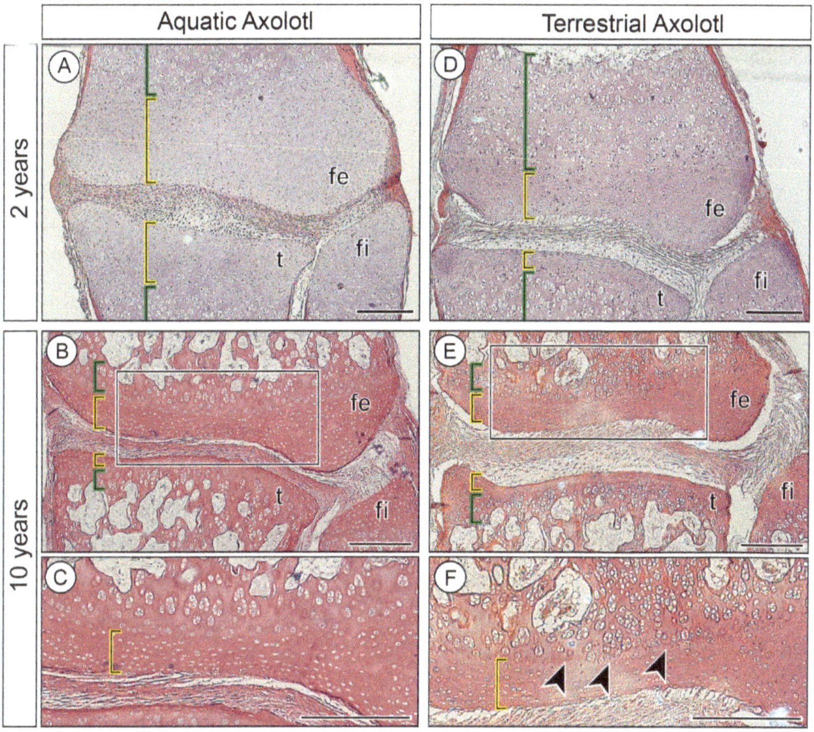

Fig. 4 Epiphyseal cartilage changes with adaptation to terrestrial life in spontaneously metamorphosed axolotl salamanders. Hematoxylin and eosin-stained sections of aquatic and terrestrial axolotl knee joints are depicted: femur (fe), tibia (t), and fibula (fi). (A) In 2-year-old aquatic animals, the presumptive epiphyseal cartilage (yellow bracket) is thick and continuous with the underlying growth plate (green bracket). (B and C) In 10-year-old aquatic animals, the epiphyseal cartilage is reduced in thickness (yellow brackets) but maintains morphological continuity with growth plate cartilage (green bracket). (D) In a 2-year-old metamorphosed terrestrial sibling, epiphyseal cartilage (green bracket) is much reduced in thickness compared to aquatic animal shown in (A). (E and F) By 10 years of age, a clear tidemark is visible and creates a clear histological separation from the underlying ossified growth plate (F, arrowheads). Scale bar for all panels, 100 μm.

plate cartilage (Fig. 4E and F, arrowheads). Additionally, the intra-knee fibrous tissue exhibited reduced cellularity and histological density at 2 years in terrestrial axolotls (Fig. 5C) and displayed a significantly expanded overall thickness at 10 years (Fig. 5D, green bracket) compared to the compacted, crowded, and highly structured fibrous tissue in aquatic counterparts (Fig. 5A and B, green bracket). These data correlate quite well with a recent morphometric and gene expression study of 1-year-old aquatic and terrestrial axolotls where authors report reduced cell density in the articular

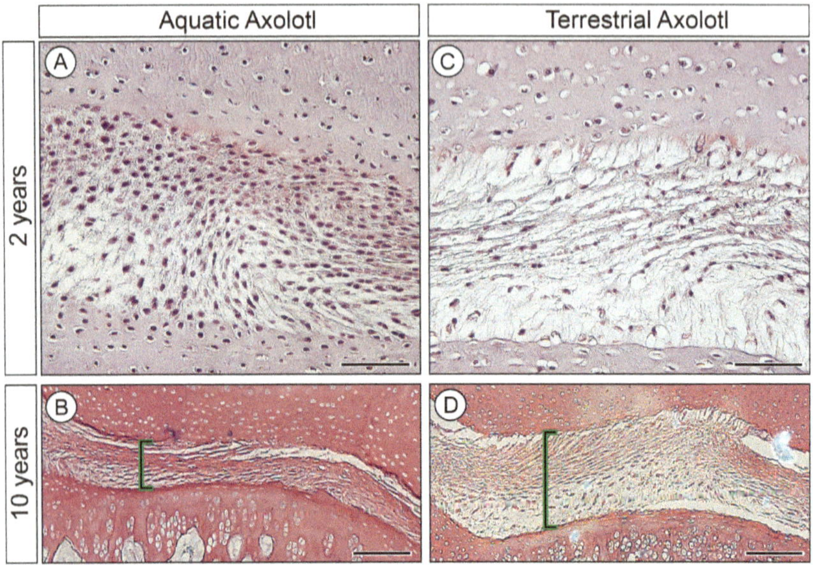

Fig. 5 Intra-joint tissue modifications during aquatic to terrestrial transition in axolotl salamanders. Images from specimens shown in Fig. 4. (A and B) The intra-joint tissue in aquatic axolotls is densely packed with fibrous cells (A) and thickness is reduced with age (B, green bracket). (C and D) By contrast, the intra-joint tissue in terrestrial axolotls displays reduced cell density, increased matrix deposition (C), and maintains thickness with age (D, green bracket). All tissue sections are stained with hematoxylin and eosin. Scale bar for panels A and C, 50 μm; scale bar for panels B and D, 150 μm.

cartilage and in the fibrous intra-joint tissue in the metamorphosed animals (Thampi et al., 2018). Interestingly, the authors also reported that fibrous tissue retained appreciable expression of *Gdf5* in these adult animals, expression that is rapidly lost by early postnatal stages in the limb joints of mammals (Koyama et al., 2008). Taken together, the combined data provide suggestive and intriguing evidence that as axolotls transition to the new physical and locomotive demands of life on land, knee joint tissue organization, and probably biomechanical function, is altered accordingly. Most impressive is the response of epiphyseal end cartilaginous tissue that is rather generic and undefined in aquatic animals, but becomes distinct and separated by a tidemark in terrestrial animals. The accompanying reduction in cell number and histological density in intra-joint fibrous tissue in the latter animals point to higher elasticity, joint flexibility, and range of motion and may even depict morphogenetic steps in the direction of cavitation. The maintenance of strong *Gdf5* expression in both aquatic and terrestrial axolotls highlights the likelihood that these cells are descendants of embryonic interzone cells, as they are in mammals. However, these axolotl joint cells may persist and multiply throughout postnatal life underscoring their distinct capacity for tissue regeneration (Cosden et al., 2011).

As we have highlighted above, it has long been held that full-fledged synovial joints first emerged in the appendicular skeleton of lobe-finned fish and were eventually passed on to amphibians and terrestrial tetrapods (Bemis, 1986). However, due to their superiority in providing mobility and possible evolutionary advantages, they could have evolved earlier. Indeed, synovial joint-like structures were proposed to occur in the jaw (temporomandibular joint) of ray-finned fishes such as longnose gar and sturgeon (Haines, 1942). While this early study fell short of describing the presence of defining synovial features including capsule, articular cartilage, and lubrication, a very recent study has provided strong evidence that the jaw and pectoral fin joints in ray-finned zebrafish, stickleback, and gar are synovial in nature and express *Prg4/lubricin* homologs (Askary et al., 2016). Genetic deletion of zebrafish *Prg4b* gene resulted in joint degeneration similar to that seen in *Prg4*-deficient mice and humans (Koyama et al., 2014; Marcelino et al., 1999; Rhee et al., 2005). Evolution of synovial joints thus appears to have preceded the ray-finned to lobe-finned fish divergence. In this context, it is relevant and thought-provoking to consider the fin joints in dolphins and whales, mammals that evolved from a terrestrial tetrapod around 40 MYA. While their scapular-humeral joints are synovial and freely movable (Klima, Oelschlager, & Wunsch, 1980; Rommel, 1990), the more distal joints are not and contain fibrous disks and resemble amphiarthroses (Cozzi, Huggenberger, & Oelschlager, 2016). In aquatic mammals then, distal limb joints appear to have reverted to more ancestral and less mobile forms, possibly caused by reduced weight bearing and range of motion requirements in the aquatic environment. On the other hand, when extant aquatic axolotls undergo metamorphosis into terrestrial animals, their joints quickly adapt and acquire characteristics of tetrapod joints, including a well-defined articular cartilage. In sum, the genetic instructions to create a synovial joint first evolved in early fishes and have been passed along through evolution while endowed with considerable adaptability to divergent physical and environmental demands.

6. Conclusions and implications

The above synopsis of past and most recent literature on key aspects of limb joint determination, growth, and morphogenesis shows that much has been learned in these important areas of research. It is clear now that mesenchymal interzone cells are endowed with the ability to generate joint tissues, and their microsurgical removal prevents joint formation and can alter the developmental behavior and fate of chondrocytes in adjacent long bone shafts. Genetic cell lineage tracing studies have provided evidence that interzone cells are heterogeneous in origin, with the centrally located ones derived directly from chondrocytes and the more lateral ones recruited from joint site-flanking mesenchyme. The data raise the tantalizing possibility that this diversity of origin reflects diversity of fate and function, with the laterally recruited cells largely engaged in formation of capsule and synovial lining and expansion of the joint, and with the centrally located cells mainly engaged in intra-joint ligament and articular cartilage formation. More work will of course be needed to verify and extend these predictions. If correct however, they would signify that interzone cells are not equivalent and quickly become committed to generating distinct local tissues, despite homogenous expression of genes including *Gdf5* and *Wnt9a* (Koyama et al., 2008). In this regard, it will be crucial to clarify how interzone cells in different joints are able to produce and/or elicit formation of joint-specific structures, such as the meniscus only in the knee. These studies could also contribute to understanding the developmental basis and mechanisms of joint diversity along the limbs, possibly clarifying whether the cells also influence overall joint morphogenesis, shape, orientation, and configuration.

Among the various joint tissues, articular cartilage has attracted the most attention over the years, a reflection of its indispensable biomechanical function and unfortunate susceptibility to malfunction with aging and poor repair capacity after injury (Trippel, Ghivizzani, & Nixon, 2004). Major efforts continue in order to address this medical need and to find biological and bioengineered tools to solve it. As addressed above, the current limited success in that field likely reflects our current poor understanding of basic developmental aspects of articular cartilage and most importantly, how articular chondrocytes become permanent and avoid the transient fate of chondrocytes that form the bulk of the growing skeleton. In addition, we lack a mechanistic understanding of how articular cartilage acquires its intricate and functional stratified architecture by maturity. Studies we describe above have provided a measure of progress regarding the basis of articular cartilage zonal structure, but we remain utterly ignorant on how articular chondrocytes acquire a permanent state. The latter is a major issue since hypertrophic degeneration of articular cartilage and its switch to a transient fate are often observed during osteoarthritis (OA), leading to the demise of tissue and cells (Buckwalter & Mankin, 1998). Indeed, unwanted chondrocyte hypertrophy often undermines bioengineered attempts to repair articular cartilage with mesenchymal stem cells isolated from marrow, fat, or other tissues (Benz, Breit, Lukoschek, Mau, & Richter, 2002; Schnabel et al., 2002).

It is important to cite two recent studies contributing important new information on basic aspects of articular cartilage developmental biology. In the first study (Gamer et al., 2018), the authors examined the roles of BMP2 in postnatal maturation and maintenance of murine knee joints and found that *Bmp2* is expressed in both embryonic and adult articular cartilage, unlike *Gdf5* expression that predominates only during embryonic stages. Joint-specific ablation of *Bmp2* caused derangement of articular cartilage structure in adults and progression to early onset OA, in addition to disrupting meniscus organization and long-term functioning. A second study focused on the important question of how articular cartilage regulates its overall thickness (Kobayashi & Kozlova, 2018), an intriguing issue given that cartilage thickness is joint specific and can even vary significantly on the proximal and distal sides of a given joint (Koyama et al., 2008). Based on their previous studies (Papaioannou, Inloes, Nakamura, Paltrinieri, & Kobayashi, 2013), the authors further explored the roles of Lin28a, an RNA-binding protein regulating growth and metabolism, and found that *Lin28a* over-expression, starting in mouse embryos, up-regulated Erk kinase signaling and caused an increase in knee articular cartilage thickness in adults. The same was seen after over-expression of constitutive active *Kras* that also up-regulated Erk signaling. These studies represent important progress toward a much needed in depth understanding of how articular cartilage acquires its distinct multi-zone organization and how articular chondrocytes are endowed with a permanent phenotype. Progress in these areas will not only be of fundamental basic value but is likely to also have major repercussions in joint translational medicine.

The study of joint evolution is instructive in several respects. As delineated above, it appears clear that the first joint to cavitate and acquire a synovial character was the temporomandibular joint (Askary et al., 2016), likely providing a major asset for species survival, nutrition, and propagation. The emergence of similar joints at most proximal sites in the fins is likely to have derived from similar evolutionary demands and advantages, and it is truly remarkable that extant axolotls still display synovial joints in shoulder, hip, and elbow only, but not in the knee. We can surmise that the limited movement and range of motion permitted by the thick fibrous tissue in their knee are fully compatible with survival and thriving in water, but the sudden changes we describe here following metamorphosis to land speak volumes about joint adaptability and responsiveness to new environments and physical forces. In this context, axolotls could represent an ideal model system to decipher the mechanisms by which the interzone cells in shoulder, hip, and elbow sustain and permit cavitation while those in the knee and more distal joints do not. Cavitation requires production of joint lubricants and the insulation of the synovial fluid from the body environment by a capsule. Thus, those comparative analyses could shed light on mechanisms regulating the genesis of these traits as well, providing once again important basic information but also key clues on how to rebuild faulty joint tissues in disease.

Acknowledgments

The original studies upon which this chapter is based were supported by NIH grant AR062908 (M.P.), NIAMS grant F32AR064071 (R.D.), and NIAMS grant F32AR074227 (D.R.). We thank colleagues and collaborators who participated in those studies and co-authored previous research papers in this area. We would also like to express our gratitude to Dr. James N. MacLeod at the University of Kentucky for generously allowing us to include axolotl data that had been previously gathered in his lab by one of us (R.S.D.). Due to the concise nature of this review, not all relevant and deserving literature and authors could be cited.

References

Andrews, S. M., & Westoll, T. S. (1968). The postcranial skeleton of Eustenopteron foordi Whiteaves. *Transactions of the Royal Society of Edinburgh*, *68*, 207–329.

Archer, C. W., Caterson, B., Benjamin, M., & Ralphs, J. R. (1999). *The biology of the synovial joint*. London: Harwood Academics.

Archer, C. W., Dowthwaite, G. P., & Francis-West, P. (2003). Development of synovial joints. *Birth Defects Research. Part C, Embryo Today*, *69*, 144–155.

Askary, A., Smeeton, J., Paul, S., Schindler, S., Braasch, I., Ellis, N. A., et al. (2016). Ancient origin of lubricated joints in bony vertebrates. *eLife*, *5*, e16415.

Becker, R. T., Gradstein, F. M., & Hammer, O. (2012). *A geologic time scale*. Cambridge University Press.

Bemis, W. E. (1986). Vertebrate evolution: Evolutionary biology of primitive fishes. *Science*, *233*, 114–115.

Benz, K., Breit, S., Lukoschek, M., Mau, H., & Richter, W. (2002). Molecular analysis of expansion, differentiation, and growth factor treatment of human chondrocytes identifies differentiation markers and growth-related genes. *Biochemical and Biophysical Research Communications*, *293*, 284–292.

Bhosale, A. M., & Richardson, J. B. (2008). Articular cartilage: Structure, injuries and review of management. *British Medical Bulletin*, *87*, 77–95.

Bi, W., Deng, J. M., Zhang, Z., Behringer, R. R., & de Crombrugghe, B. (1999). Sox9 is required for cartilage formation. *Nature Genetics*, *22*, 85–89.

Boisvert, C. A., Mark-Kurik, E., & Ahlberg, P. E. (2008). The pectoral fin of Panderichthys and the origin of digits. *Nature, 456*, 636–638.
Botella, H., Blom, H., Dorka, M., Ahlberg, P. E., & Janvier, P. (2007). Jaws and teeth of the earliest bony fishes. *Nature, 448*, 583–586.
Breur, G. J., VanEnkevort, B. A., Farnum, C. E., & Wilsman, N. J. (1991). Linear relationship between the volume of hypertrophic chondrocytes and the rate of longitudinal bone growth in growth plates. *Journal of Orthopaedic Research, 9*, 348–359.
Buckwalter, J. A., & Mankin, H. J. (1998). Articular cartilage: Tissue design and chondrocyte-matrix interactions. *Instructional Course Lectures, 47*, 477–486.
Caldwell, K. L., & Wang, J. (2015). Cell-based articular cartilage repair: The link between development and regeneration. *Osteoarthritis and Cartilage, 23*, 351–362.
Clack, J. A. (1997). Devonian tetrapod trackways and trackmakers: A review of the fossils and footprints. *Palaeogeography, Palaeoclimatology, Palaeoecology, 130*, 227–250.
Clack, J. A. (2013). Gaining ground: The origin and evolution of tetrapods. *Choice: Current Reviews for Academic Libraries: vol. 50.* (pp. 1082–1083). .
Cosden, R. S., Lattermann, C., Romine, S., Gao, J., Voss, S. R., & MacLeod, J. N. (2011). Intrinsic repair of full-thickness articular cartilage defects in the axolotl salamander. *Osteoarthritis and Cartilage, 19*, 200–205.
Cosden-Decker, R. S., Bickett, M. M., Lattermann, C., & MacLeod, J. N. (2012). Structural and functional analysis of intra-articular interzone tissue in axolotl salamanders. *Osteoarthritis and Cartilage, 20*, 1347–1356.
Cozzi, B., Huggenberger, S., & Oelschlager, H. (2016). *Anatomy of dolphins: Insights into body structure and function.* Academic Press.
Craig, F. M., Bentley, G., & Archer, C. W. (1987). The spatial and temporal pattern of collagens I and II and keratan sulphate in the developing chick metatarsophalangeal joint. *Development, 99*, 383–391.
Daeschler, E. B., Shubin, N. H., & Jenkins, F. A., Jr. (2006). A Devonian tetrapod-like fish and the evolution of the tetrapod body plan. *Nature, 440*, 757–763.
Decker, R. S. (2017). Articular cartilage and joint development from embryogenesis to adulthood. *Seminars in Cell & Developmental Biology, 62*, 50–56.
Decker, R. S., Um, H.-B., Dyment, N. A., Cottingham, N., Usami, Y., Enomoto-Iwamoto, M., et al. (2017). Cell origin, volume and arrangement are drivers of articular cartilage formation, morphogenesis and response to injury in mouse limbs. *Developmental Biology, 426*, 56–68.
Dowthwaite, G. P., Bishop, J. C., Redman, S. N., Khan, I. M., Rooney, P., Evans, D. J. R., et al. (2004). The surface of articular cartilage contains a progenitor cell population. *Journal of Cell Science, 117*, 889–897.
Ellison, A., & Berg, E. (1985). Embryology, anatomy and function of the anterior cruciate ligament. *The Orthopedic Clinics of North America, 16*, 3–14.
Eyre, D. (2002). Review: Collagen of articular cartilage. *Arthritis Research, 4*, 30–35.
Gamer, L. W., Pregizer, S., Gamer, J., Feigenson, M., Ionescu, A., Li, Q., et al. (2018). The role of Bmp2 in the maturation and maintenance of the murine knee joint. *Journal of Bone and Mineral Research, 33*, 1–10.
Goldring, M. B., & Goldring, S. R. (2007). Osteoarthritis. *Journal of Cellular Physiology, 213*, 626–634.
Gouramanis, C., Webb, J. A., & Warren, A. A. (2003). Fluviodeltaic sedimentology and ichnology of part of the Silurian Grampians group, western Victoria. *Australian Journal of Earth Sciences, 50*, 811–825.
Gray, H. (1988). In T. P. Pick & R. Howden (Eds.), *Gray's anatomy* (pp. 219–220). London: Galley Press.
Haines, R. W. (1942). Eudiarthrodial joints in fishes. *Journal of Anatomy, 77*, 12–19.
Haines, R. W. (1947). The development of joints. *Journal of Anatomy, 81*, 33–55.

Hampe, A. (1956). Sur la regulation de pieces excedentaires dans le bourgeon de membre de l'embryon de poulet. *Comptes Rendus des Seances de la Societe de Biologie*, *150*, 1726–1729.

Hayes, A. J., MacPherson, S., Morrison, H., Dowthwaite, G. P., & Archer, C. W. (2001). The development of articular cartilage: Evidence for an appositional growth mechanism. *Anatomy and Embryology*, *203*, 469–479.

Hinchliffe, J. R. (1994). Evolutionary developmental biology of the tetrapod limb. *Development. Supplement*, *1994*, 163–168.

Hinchliffe, J. R., & Johnson, D. R. (1980). *The development of the vertebrate limb* (pp. 72–83). New York: Oxford University Press.

Hogg, D. (1980). A re-investigation of the centres of ossification in the avian skeleton at and after hatching. *Journal of Anatomy*, *130*, 725–743.

Holder, N. (1977). An experimental investigation into the early development of the chick elbow joint. *Journal of Embryology and Experimental Morphology*, *39*, 115–127.

Hunziker, E. B. (2002). Articular cartilage repair: Basic science and clinical progress. A review of the current status and prospects. *Osteoarthritis and Cartilage*, *10*, 432–463.

Hunziker, E. B., Kapfinger, E., & Geiss, M. D. (2007). The structural architecture of adult mammalian articular cartilage evolves by a synchronized process of tissue resorption and neoformation during postnatal development. *Osteoarthritis and Cartilage*, *15*, 403–413.

Hyde, G., Boot-Handford, R. P., & Wallis, G. A. (2008). Col2a1 lineage tracing reveals that the meniscus of the knee joint has a complex cellular origin. *Journal of Anatomy*, *213*, 531–538.

Hyde, G., Dover, S., Aszodi, A., Wallis, G. A., & Boot-Handford, R. P. (2007). Lineage tracing using matrilin-1 gene expression reveals that articular chondrocytes exist as the joint interzone forms. *Developmental Biology*, *304*, 825–833.

Johnstone, B., Alini, M., Cucchiarini, M., Dodge, G. R., Eglin, D., Guilak, F., et al. (2013). Tissue engineering for articular cartilage repair. The state of the art. *European Cells & Materials*, *25*, 248–267.

Jones, A. R. C., & Flannery, C. R. (2007). Bioregulation of lubricin expression by growth factors and cytokines. *European Cells & Materials*, *13*, 40–45.

Klima, M., Oelschlager, H. A., & Wunsch, D. (1980). Morphology of the pectoral girdle in the amazon dolphin inia geoffrensis with special reference to the shoulder joint and movements of the flippers. *Zeitschrift fur Saugetierkunde*, *45*, 288–309.

Kobayashi, T., & Kozlova, A. (2018). Lin28a overexpression reveals the role of Erk signaling in articular cartilage development. *Development*, *145*, 1–6.

Kosinska, M. K., Liebisch, G., Lochnit, G., Wilhelm, J., Klein, H., Kaesser, U., et al. (2012). A lipidomic study of phospholipid classes and species in human synovial fluid. *Arthritis and Rheumatism*, *65*, 2323–2333.

Koyama, E., Leatherman, J. L., Shimazu, A., Nah, H.-D., & Pacifici, M. (1995). Syndecan-3, tenascin-C and the development of cartilaginous skeletal elements and joints in chick limbs. *Developmental Dynamics*, *203*, 152–162.

Koyama, E., Ochiai, T., Rountree, R. B., Kingsley, D. M., Enomoto-Iwamoto, M., Iwamoto, M., et al. (2007). Synovial joint formation during mouse limb skeletogenesis. Roles of Indian hedgehog signaling. *Annals of the New York Academy of Sciences*, *1116*, 100–112.

Koyama, E., Saunders, C., Salhab, I., Decker, R. S., Chen, I., Um, H., et al. (2014). Lubricin is required for the structural integrity and post-natal maintenance of TMJ. *Journal of Dental Research*, *93*, 663–670.

Koyama, E., Shibukawa, Y., Nagayama, M., Sugito, H., Young, B., Yuasa, T., et al. (2008). A distinct cohort of progenitor cells participates in synovial joint and articular cartilage formation during mouse limb skeletogenesis. *Developmental Biology*, *316*, 62–73.

Koyama, E., Yasuda, T., Minugh-Purvis, N., Kinumatsu, T., Yallowitz, A. R., Wellik, D. M., et al. (2010). Hox11 genes establish synovial joint organization and phylogenetic characteristics in developing mouse zeugopod skeletal elements. *Development, 137*, 3795–3800.

Kozhemyakina, E., Zhang, M. Q., Ionescu, A., Kobayashi, A., Kronenberg, H. M., Warman, M. L., et al. (2015). Identification of a Prg4-expressing articular cartilage progenitor cell population in mice. *Arthritis and Rheumatism, 67*, 1261–1273.

Kurth, T. B., Dell' Accio, F., Crouch, V., Augelio, A., Sharpe, P. T., & De Bari, C. (2011). Functional mesenchymal stem cell niches in adult mouse knee joint synovium in vivo. *Arthritis and Rheumatism, 63*, 1289–1300.

Li, L., Newton, P. T., Bouderlique, T., Sejnohova, M., Zikmund, T., Kozhemyakina, E., et al. (2016). Superficial cells are self-renewing chondrocyte progenitors, which form the articular cartilage in juvenile mice. *The FASEB Journal, 31*, 1067–1084.

Longobardi, L., Li, T., Tagliaferro, L., Temple, J. D., Willicockson, H. H., Ye, P., et al. (2015). Synovial joints: From development to homeostasis. *Current Osteoporosis Reports, 13*, 41–51.

Lu, J., Zhu, M., Long, J. A., Zhao, W., Senden, T. J., Jia, L., et al. (2012). The earliest known stem-tetrapod from the lower Devonian of China. *Nature Communications, 3*, 1160.

Makris, E. A., Gomoll, A. H., Malizos, K. N., Hu, J. C., & Athanasiou, K. A. (2015). Repair and tissue engineering techniques for articular cartilage. *Nature Reviews Rheumatology, 11*, 21–34.

Marcelino, J., Carpten, J. D., Suwairi, W. M., Gutierrez, O. M., Schwartz, S., Robbins, C., et al. (1999). CACP, encoding a secreted proteoglycan, is mutated in camptodactyly-arthropathy-coxa vara-pericarditis syndrome. *Nature Genetics, 23*, 319–322.

Mitrovic, D. R. (1977). Development of the metatarsalphalangeal joint in the chick embryo: Morphological, ultrastructural and histochemical studies. *The American Journal of Anatomy, 150*, 333–348.

Mitrovic, D. R. (1978). Development of the diathrodial joints in the rat embryo. *The American Journal of Anatomy, 151*, 475–485.

Mow, V. C., Ratcliffe, A., & Poole, A. R. (1992). Cartilage and diartrodial joints as paradigms for hierarchical materials and structures. *Biomaterials, 13*, 67–97.

Niedermaier, M., Schwabe, G. C., Fees, S., Helmrich, A., Brieske, N., Seeman, P., et al. (2005). An inversion involving the mouse Shh locus results in brachydactyly through dysregulation of Shh expression. *The Journal of Clinical Investigation, 115*, 900–909.

Niedzwiedzki, G., Szrek, P., Narkiewicz, K., Narkiewicz, M., & Ahlberg, P. E. (2010). Tetrapod trackways from the early middle Devonian period of Poland. *Nature, 463*, 43–48.

Ozpolat, B. D., Zapata, M., Fruge, J. D., Coote, J., Lee, J., Muneoka, K., et al. (2012). Regeneration of the elbow joint in the developing chick embryo recapitulates development. *Developmental Biology, 372*, 229–238.

Pacifici, M. (1995). Tenascin-C and the development of articular cartilage. *Matrix Biology, 14*, 689–698.

Pacifici, M., Decker, R. S., & Koyama, E. (2018). Limb synovial joint development from the hips down: Implications for articular cartilage repair and regeneration. In M. J. Stoddart, A. M. Craft, G. Pattappa, & O. F. W. Gardner (Eds.), *Developmental biology and musculoskeletal tissue engineering. Principles and applications* (pp. 67–101). London: Academic Press.

Page, R. B., Monaghan, J. R., Walker, J. A., & Voss, S. R. (2009). A model of transcriptional and morphological changes during thyroid hormone-induced metamorphosis of the axolotl. *General and Comparative Endocrinology, 162*, 219–232.

Papaioannou, G., Inloes, J. B., Nakamura, Y., Paltrinieri, E., & Kobayashi, T. (2013). Let-7 and miR-140 microRNAs coordinately regulate skeletal development. *Proceedings of the National Academy of Sciences of the United States of America, 110*, E3291–E3300.

Pitsillides, A. A., & Ashhurst, D. E. (2008). A critical evaluation of specific aspects of joint development. *Developmental Dynamics, 237*, 2284–2294.

Poole, A. R., Kojima, T., Yasuda, T., Mwale, F., Kobayashi, M., & Laverty, S. (2001). Composition and structure of articular cartilage: A template for tissue repair. *Clinical Orthopaedics and Related Research, 391*, 526–533.

Rhee, D. K., Marcelino, J., Baker, M., Gong, Y., Smits, P., Lefebvre, V., et al. (2005). The secreted glycoprotein lubricin protects cartilage surfaces and inhibits synovial cell outgrowth. *The Journal of Clinical Investigation, 115*, 622–631.

Romer, A. S. (1958). Tetrapod limbs and early tetrapod life. *Evolution, 12*, 365.

Rommel, S. (1990). *Osteology of the bottlenose dolphin*. San Diego, CA: Academic Press.

Rountree, R. B., Schoor, M., Chen, H., Marks, M. E., Harley, V., Mishina, Y., et al. (2004). BMP receptor signaling is required for postnatal maintenance of articular cartilage. *PLoS Biology, 2*, 1815–1827.

Salva, J. E., & Merrill, A. E. (2017). Signaling networks in joint development. *Developmental Dynamics, 246*, 262–274.

Schnabel, M., Marlovits, S., Eckhoff, G., Fichtel, I., Gotzen, L., Vecsei, V., et al. (2002). Dedifferentiation-associated changes in morphology and gene expression in primary human articular chondrocytes in cell culture. *Osteoarthritis and Cartilage, 10*, 62–70.

Schutte, M. J., Dabezies, E. J., Zimny, M. L., & Happel, L. T. (1987). Neural anatomy of the human anterior cruciate ligament. *The Journal of Bone and Joint Surgery. American Volume, 69*, 243–247.

Seemann, P., Schwappecher, R., Kjaer, K. W., Krakow, D., Lehmann, K., Dawson, K., et al. (2005). Activating and deactivating mutations in the receptor interaction site of GDF5 cause symphalangism or brachydactyly type A2. *The Journal of Clinical Investigation, 115*, 2373–2381.

Seror, J., Zhu, L., Goldberg, R., Day, A. J., & Klein, J. (2015). Supramolecular synergy in the boundary lubrication of synovial joints. *Nature Communications, 6*, 6497.

Shaffer, H. B., & Voss, S. R. (1996). Phylogenetic and mechanistic analysis of a developmentally integrated character complex: Alternate life history modes in ambystomatid salamanders. *American Zoologist, 36*, 24–35.

Shea, K. G., Jacobs, J. C., Carey, J. L., Anderson, A. F., & Oxford, J. T. (2013). Osteochondritis dissecans knee histology studies have variable findings and theories of etiology. *Clinical Orthopaedics and Related Research, 471*, 1127–1136.

Shubin, N. H., Daeschler, E. B., & Coates, M. I. (2004). The early evolution of the tetrapod humerus. *Science, 304*, 90–93.

Shubin, N. H., Daeschler, E. B., & Jenkins, F. A., Jr. (2006). The pectoral fin of Tiktaalik roseae and the origin of the tetrapod limb. *Nature, 440*, 764–771.

Shubin, N. H., Daeschler, E. B., & Jenkins, F. A., Jr. (2014). Pelvic girdle and fin of Tiktaalik roseae. *Proceedings of the National Academy of Sciences of the United States of America, 111*, 893–899.

Shubin, N., Tabin, C., & Carroll, S. (1997). Fossils, genes and the evolution of animal limbs. *Nature, 388*, 639–648.

Shwartz, Y., Viukov, S., Krief, S., & Zelzer, E. (2016). Joint development involves a continuous influx of Gdf5-positive cells. *Cell Reports, 15*, 1–11.

Snippert, H. J., van der Flier, L. G., Sato, T., van Es, J. H., van der Born, M., Kroon-Veeboer, C., et al. (2010). Intestinal crypt homeostasis results from neutral competition between symmetrically dividing Lgr5 stem cells. *Cell, 143*, 134–144.

Soeda, T., Deng, J. M., de Crombrugghe, B., Behringer, R. R., Nakamura, T., & Akiyama, H. (2010). Sox9-expressing precursors are the cellular origin of the cruciate ligament of the knee joint and the limb tendons. *Genesis, 48*, 635–644.

Storm, E. E., & Kingsley, D. M. (1996). Joint patterning defects caused by single and double mutations in members of the bone morphogenetic protein (BMP) family. *Development, 122*, 3969–3979.

Storm, E. E., & Kingsley, D. M. (1999). GDF5 coordinates bone and joint formation during digit development. *Developmental Biology, 209*, 11–27.
Stossel, I. (1995). The discovery of a new Devonian tetrapod trackway in SW Ireland. *Journal of Geological Society, 152*, 407–413.
Tada, M., & Heisenberg, C.-P. (2012). Convergent extension: Using collective cell migration and cell intercalation to shape embryos. *Development, 139*, 3897–3904.
Temple-Wong, M. M., Ren, S., Quach, P., Hansen, B. C., Chen, A. C., Hasegawa, A., et al. (2016). Hyaluronan concentration and size distribution in human knee synovial fluid: Variations with age and cartilage degeneration. *Arthritis Research & Therapy, 18*, 18.
Thampi, P., Liu, J., Zeng, Z., & MacLeod, J. N. (2018). Changes in the appendicular skeleton during metamorphosis in the axolotl salamander (Ambystoma mexicanum). *Journal of Anatomy, 233*(4), 468–477.
Trippel, S. B., Ghivizzani, S. C., & Nixon, A. J. (2004). Gene-based approaches for the repair of articular cartilage. *Gene Therapy, 11*, 351–359.
Vorobyeva, E. I., & Schultze, H. P. (1991). *Origins of the higher groups of tetrapods*. Cornell.
Warren, A., Jupp, R., & Bolton, B. (1985). Earliest tetrapod trackway. *Alcheringa: An Australasian Journal of Palaeontology, 10*, 183–186.
Warren, J. W., & Wakefield, N. A. (1972). Trackways of tetrapod vertebrates from the upper Devonian of Victoria, Australia. *Nature*, 469–470.
Whillis, J. (1940). The development of synovial joints. *Journal of Anatomy, 74*, 277–283.
William, G. R., & Raven, H. C. (1941). Studies on the origin and early evolution of paired fins and limbs. *The Annals of the New York Academy of Sciences, 42*, 273–291.
Williams, R., Khan, I. M., Richardson, K., Nelson, L., McCarthy, H. E., Analbelsi, T., et al. (2010). Identification and clonal characterization of a progenitor cell sub-population in normal human articular cartilage. *PLoS One, e13246*, 5.
Williams, E. A., Sergeev, S. A., Stossel, I., & Ford, M. (1997). An Eifelian U-Pb zircon date for the Enagh tuff bed from the old red sandstone of the Munster Basin in NW Iveragh, SW Ireland. *Journal of Geological Society, 154*.
Wolff, E. (1958). Le principe de competition. *Bulletin de la Société Zoologique de France, 83*, 13–15.
Wolpert, L. (1969). Positional information and the spatial pattern of cellular differentiation. *Journal of Theoretical Biology, 25*, 1–47.

CHAPTER SIX

BMPs, TGFβ, and border security at the interzone

Karen M. Lyons[a], Vicki Rosen[b,*]
[a]Department of Orthopaedic Surgery, Geffen School of Medicine, UCLA, Los Angeles, CA, United States
[b]Department of Developmental Biology, Harvard School of Dental Medicine, Boston, MA, United States
*Corresponding author: e-mail address: Vicki_rosen@hsdm.harvard.edu

Contents

Abstract

Synovial joints enable movement and protect the integrity of the articular cartilage. Joints form within skeletal condensations destined to undergo chondrogenesis. The suppression of this chondrogenic program in the interzone is the first morphological sign of joint formation. While we have a fairly good understanding of the essential roles of BMP and TGFβ family members in promoting chondrogenic differentiation in developing skeletal elements, we know very little about how BMP activity is suppressed specifically within the interzone, a crucial step in joint development. The function of the BMP ligand Gdf5 has been especially difficult to decipher. On the one hand, Gdf5 is required to promote chondrogenesis of articular elements. On the other hand, Gdf5 is highly expressed in the joint interzone where chondrogenesis must be suppressed for the formation of many joints. Here we review the evidence that BMP signaling must be suppressed within the joint interzone for joint morphogenesis to progress, and consider how Gdf5 exerts its divergent effects on chondrogenesis and joint formation. We also consider how TGFβ signaling impacts formation of the interzone. Finally, we propose a model whereby Gdf5 exerts distinct effects in the interzone *vs.* surrounding cartilage based on the repertoire of BMP receptors available in these tissues. Understanding how BMP antagonists and counteracting TGFβ signals intersect with Gdf5 to sculpt the joint interzone is essential for understanding the origin of osteoarthritis and other diseases of joint tissues.

Current Topics in Developmental Biology, Volume 133
ISSN 0070-2153
https://doi.org/10.1016/bs.ctdb.2019.02.001

1. Introduction

Skeletons with articulated joints emerged approximately 400 million years ago coincident with the first appearance of bone tissue in Cambrian fishes. Since that time joints have fulfilled the same function of allowing for enhanced movement. It is remarkable to think that despite the functional importance of synovial joints in daily life, key details of the joint formation program and the signals required for its initiation remain unresolved. As joint diseases are among the most common and most debilitating of musculoskeletal disorders, understanding the molecular and cellular events that come together to form a joint is central to developing new treatment options for joint diseases.

The earliest morphological sign of joint formation is the emergence of the interzone (IZ), a tripartite structure consisting of a mid-density inner layer (the central intermediate lamina) and two high density outer layers that eventually give rise to articular cartilages found on the skeletal elements located at either side of the joint (Koyama et al., 2008; Pacifici et al., 2006) (Fig. 1). Initially composed of prechondrogenic cells recruited from the cartilage condensations in the limb bud, cells in the IZ adopt either a chondrogenic fate, giving rise to articular cartilage, undergo apoptosis or some other clearance mechanism to give rise to the joint space, or differentiate into joint-associated structures, such as the joint capsule, synovium, tendon and ligament. A number of signaling molecules have been implicated in joint formation (Decker, Koyama, & Pacifici, 2014; Smeeton, Askary, & Crump, 2016). Wnt signaling is both necessary and sufficient to specify joint forming domains in the appendicular skeleton, while Ihh signaling is required for limb and jaw joint development (Guo et al., 2004; Hartmann & Tabin, 2001; Pacifici et al., 2006; Ray, Singh, Sohaskey, Harland, & Bandyopadhyay, 2015). FGF signaling has also been reported to influence chondrogenic differentiation of IZ cells (Tang et al., 2017). These signaling families appear to interact in a highly complex manner and at present it is not entirely clear how they coordinate to prefigure joint locations and control IZ cell fate. Adding to the complexity of joint morphogenesis is the fact that not all joints are identical. Some, such as interphalangeal joints, have a relatively simple organization, consisting essentially of a joint cavity, the joint capsule, and associated tendons. Others are much more complex in their size, shape of the articulations, and types of accessory structures. For example, the knee joint consists of multiple interacting tissues, including complex articular

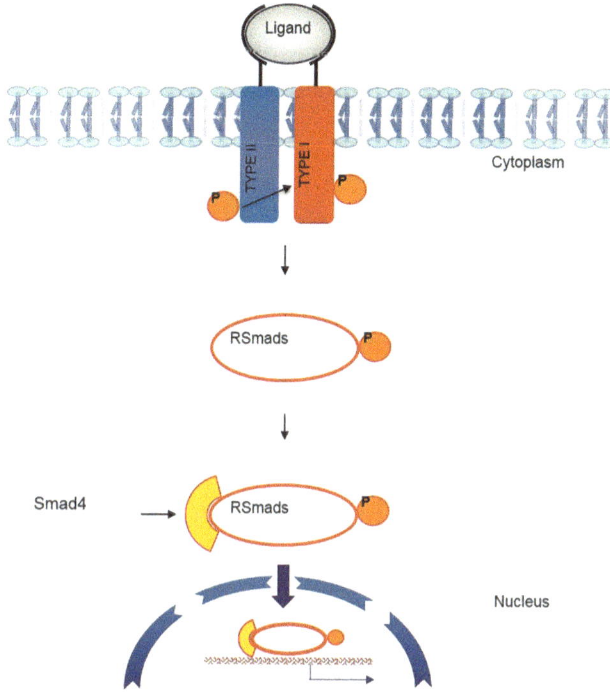

Fig. 1 Canonical BMP/TGFβ signaling. Canonical BMP/TGFβ signaling is initiated by ligand-induced oligomerization of receptor type I and type II serine/threonine kinases present on the surface of target cells. Type II BMP/TGFβ receptors are constitutively active and ligand binding brings the type II and type I receptors into close proximity, allowing type II receptors to phosphorylate type I receptors. Activation of the type I receptors leads to phosphorylation of either SMAD2/3 (TGFβ) or SMAD1/5 (BMP). These SMAD complexes interact with SMAD4 to activate or repress the transcription of a defined set of BMP or TGFβ target genes.

surfaces on the femur and tibia, a joint capsule, the meniscus, patella, and multiple ligaments. Clearly, the varying architectures of synovial joints lead to differences in the specific roles of individual genes in joint formation and in the timing of developmental milestones associated with each joint tissue. Nonetheless, a common feature of all joints is the formation of the IZ and much attention has been focused on understanding the mechanisms that specify the IZ as distinct from the surrounding cells that give rise to cartilage, ligaments, and other accessory structures.

Members of the TGFβ/BMP superfamily of ligands have received considerable attention as regulators of the joint formation program, owing to the fact that mutations in TGFβ/BMP ligands, receptors, and inhibitors are associated with defects in formation of joints in humans (reviewed in Salazar, Gamer, & Rosen, 2016). While data strongly support the necessity to suppress BMP signaling in the IZ (discussed below), BMPs are clearly required for the development of the adjacent articular cartilage and are abundantly expressed in the vicinity of the IZ. Furthermore, the BMP ligand Gdf5 is highly expressed within the IZ itself and in surrounding tissues (Storm, Huynh, Copeland, Kingsley, & Lee, 1994). Gdf5 is considered the most specific marker of IZ cell identity; it is surprising then that Gdf5 is actually a secreted protein and not present as a cell surface marker. Given that BMP signaling must be suppressed in the IZ, and Gdf5 is a BMP-like signaling molecule, the role of Gdf5 in the IZ has remained a mystery. In this review, we summarize the evidence that suppression of BMP signaling in IZ cells is essential for joint formation, and focus on how this inhibition may be achieved. We discuss potential mechanisms by which TGFβ pathways might inhibit BMP signaling in the IZ, and how Gdf5 might act as antagonist of other BMP ligands to repress BMP signaling in the IZ.

2. Overview of the BMP and TGFβ signaling pathways

In general terms, canonical BMP/TGFβ signaling is initiated by ligand-induced oligomerization of receptor type I and type II serine/threonine kinases, integral transmembrane proteins present on the surface of target cells (Fig. 2; reviewed in Lowery & Rosen, 2018; Salazar et al., 2016). Type II BMP/TGFβ receptors are constitutively active and ligand

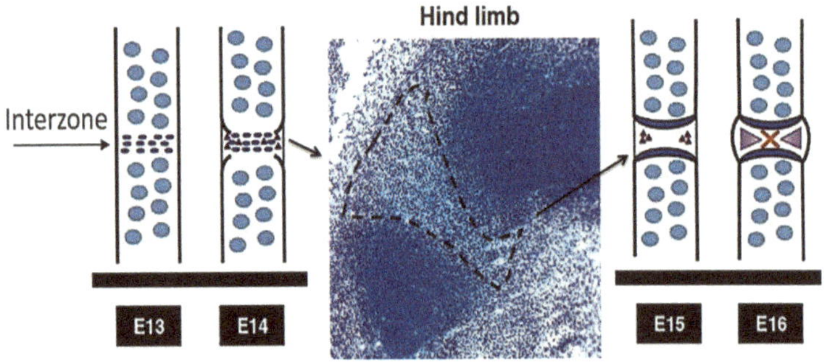

Fig. 2 Interzone marks the position of the presumptive knee joint. At this site, *Col2a1* expressing cells (*ovals*) cease production of ECM, and adopt either a chondrogenic fate, giving rise to articular cartilage (*solid lines*), are cleared to give rise to the joint space (*white*), or differentiate into joint-associated structures, shown here as meniscus (*triangles*), and cruciate ligaments (X in last panel).

binding brings the type II and type I receptors into close proximity, allowing type II receptors to phosphorylate type I receptors. Activation of the type I receptors leads to phosphorylation of the carboxy-terminal region of either SMAD2/3 (TGFβ) or SMAD1/5 (BMP). These SMAD complexes then interact with SMAD4 and activate or repress the transcription of a defined set of BMP or TGFβ target genes that appear to be distinct from each other. Non-canonical or SMAD4-independent BMP/TGFβ activities also require type I receptor activation, but in this case, ligand binding to the receptor initiates activation of other kinases such as MAPK, TAK1, RAS/RHO, and PI3K/AKT in a context dependent manner (Lowery & Rosen, 2018). The duration and amplitude of the BMP/TGFβ signal is controlled at the level of the ligands present, the receptors they occupy and intracellular signaling effectors. Signaling is further refined by the presence of transmembrane accessory receptors and antagonists that limit the interaction of specific ligands and receptors. Signaling is also regulated by the actions of inhibitory SMADs that antagonize SMAD-induced responses by targeting pathway components for ubiquitin-mediated destruction (Prunier, Baker, Ten Dijke, & Ritsma, 2019).

3. Inhibition of BMP signaling in IZ cells is a critical step in joint formation

BMP signaling, triggered by several distinct BMP proteins, directs multiple aspects of both endochondral and intramembranous ossification. Although necessary for chondrogenesis, BMP signaling must be precisely modulated for successful synovial joint formation. Any molecular manipulation leading to ectopic activation of BMP signaling such as overexpression of BMP ligands (Duprez et al., 1996; Tsumaki et al., 2002) or misexpression of constitutively active BMP receptors (Zhang et al., 2000) results in cartilage differentiation by presumptive IZ cells. Current evidence suggests that selective inhibition of BMP signaling may require several distinct mechanisms, the most prominent of which is the interaction of BMPs with noggin, a potent BMP antagonist that is produced by IZ cells (Brunet, McMahon, McMahon, & Harland, 1998). The domain of noggin expression within the IZ acts to insulate IZ cells from BMP ligands, consistent with the findings that mice lacking noggin fail to form synovial joints (Brunet et al., 1998). In fact, loss of noggin expression is the only instance in which all joint formation is altered, establishing modulation of BMP signaling as the most fundamental requirement for joint formation. In the absence of noggin, initial mesenchymal condensations are established but there is no evidence of IZ formation or Gdf5 expression at presumptive synovial joint locations. Instead, over-exuberant chondrogenesis occurs, most likely driven by unabated BMP signaling. Noggin expression within the IZ appears to follow a strictly defined spatial and temporal pattern. Mice with conditional deletions of noggin in unspecified limb mesenchyme phenocopy all limb features of noggin null mice, including a complete absence of fully formed synovial joints. However, selective ablation of noggin from Gdf5+ cells has only a modest effect on joint formation, with anomalies occurring primarily in distal joints in the wrists, ankles and digits (Salazar and Maridas, unpublished). These findings demonstrate that in most joints, BMP inhibition must occur prior to the formation of an overt IZ, but also show that there is a continuous requirement for BMP inhibition in the formation of distal joint structures.

While these studies provide strong evidence that BMP signaling must be inhibited in the IZ for joint formation to proceed, the identities of the ligands targeted by noggin in vivo are unknown. Most BMPs are not expressed in the IZ itself, but are present in adjacent areas during development (Gamer et al., 2018; Macias et al., 1997). Knockouts of specific BMP ligands in *Gdf5*-lineage cells will be needed to identify the BMPs that are targeted by noggin during IZ formation. Bmp2 is one of the few ligands expressed in the region of the IZ (Macias et al., 1997). Loss of Bmp2 in joint-forming cells has little or no impact on joint formation, but Bmp2 is not expressed until after the IZ has already undergone cavitation (Gamer et al., 2018). Gdf5, discussed below, is highly expressed in the region that will give rise to the IZ and remains at high levels in surrounding tissues following joint cavitation (Dyment et al., 2015; Koyama et al., 2008). Gdf5 is therefore a potential target of noggin. However, its role in joint formation is complex, as discussed below.

4. Genetic evidence that GDF5 has a role in joint formation

Gdf5, Gdf6 and Gdf7 form a distinct subgroup within the larger BMP family, and each is expressed in the IZ, with Gdf5 localized to all IZ (Storm et al., 1994; Wolfman et al., 1997). Lineage tracing studies in mice confirm that *Gdf5*-expressing cells give rise to mature joint components, including the articular cartilage, meniscus, ligaments, joint capsule, and enthesis (Shwartz, Viukov, Krief, & Zelzer, 2016). Loss of Gdf 5, 6 or 7 leads to fusions of specific joints, most frequently those in the lower extremities (Settle et al., 2003). The finding that loss of a BMP ligand leads to joint fusion appears paradoxical given that inhibition of BMP activity by noggin is needed to promote joint formation. Furthermore, ectopic Gdf 5, 6 or 7 promote chondrogenic differentiation rather than joint formation, and transgenic overexpression of *Gdf5* in mice leads to joint fusions throughout the skeleton (Tsumaki et al., 2002). These studies reveal potentially dual functions for Gdf5. Gdf5 can promote chondrogenesis in cells already committed to this fate, but can inhibit chondrogenesis in IZ cells. This duality of Gdf5 function is also evident in the spectrum of mutations found in humans and mice. Loss-of-function mutations in Gdf 5, 6 or 7 that lead to fusions in autopodal joints are accompanied by considerable shortening of skeletal elements, showing that these proteins promote both joint formation and chondrogenesis in vivo (Storm et al., 1994; Thomas et al., 1997, 1996). Gain-of-function mutations in humans that give Gdf5 and Gdf6 resistance to noggin inhibition cause joint fusions without affecting limb length (Dawson et al., 2006; Schwaerzer et al., 2012; Seemann et al., 2009; Wang, Yu, et al., 2016). These latter mutations suggest that Gdf5 is a target of noggin in autopodal joint IZs in vivo.

Regulation of Gdf5 expression during joint morphogenesis appears to be remarkably complex. Chen et al. (2016) have identified separate regions of the Gdf5 locus that control expression in axial tissues, in proximal vs distal joints in the limbs and in specific sub-sets of composite joints. These Gdf5 enhancers are distributed over a 100 kb of DNA, including regions both upstream and downstream of Gdf5 coding exons, in a manner consistent with the spatial specificity of joint patterning in vertebrates (Capellini et al., 2017). It is also clear that subtle changes in the location and level of Gdf5 expression may alter aspects of joint shape and thus be responsible for predisposing individuals to hip and knee joint instability, dislocation and adult-onset osteoarthritis (Kiapour, Cao, Young, & Capellini, 2018; Pregizer et al., 2018). The complexity of Gdf5 involvement in joint formation is further augmented by the likelihood that joints develop through a continuous influx of Gdf5+ cells into the IZ (Decker et al., 2017; Shwartz et al., 2016), although the source of these cells and how they become destined to the IZ remains to be uncovered.

5. How does Gdf5 signaling direct joint formation?

The above studies show that Gdf5 signaling promotes chondrogenesis in cells committed to that fate, yet is required to suppress chondrogenesis in the IZ in a subset of joints. The pathways, both canonical and non-canonical, through which Gdf5 mediates its divergent effects to mediate joint formation or chondrogenesis, are unknown. However, it is tempting to speculate that some of the effects of Gdf5 may be mediated through TAK1, based on the finding that mice in which TAK1 was deleted in limb mesenchyme exhibit multiple joint fusions (Gunnell et al., 2010). A subset of joint abnormalities was also seen in $Tak1^{Col2}$ mutants (Gunnell et al., 2010; Shim et al., 2009). The fact that TAK1 acts downstream of both BMP and TGFβ pathways (discussed below), may explain why joint defects are more widespread in *Tak1* mutants compared to those seen with loss of Gdf5 or with the TGFβ pathway components analyzed to date (discussed below). More work on TAK1 function in the IZ is warranted.

In addition to functions for Gdf5 via activation of signaling pathways in the IZ, recent studies suggest that Gdf5 may mediate its effects by blocking signal transduction. It has been proposed that the high levels of Gdf5 in the IZ act to sequester noggin in this region, preventing the spread of noggin across the IZ to more distal cells, which must be able to undergo chondrogenesis on the other side of the developing joint (Huang et al., 2016). A second potential function for Gdf5 in the IZ may be to act as a direct competitor to osteogenic BMPs. The preferred type 1 receptor for Gdf5 signaling is Bmpr1b (Nishitoh et al., 1996). The utilization of Bmpr1b by Gdf5 in vivo is corroborated by the similar phenotypes of mice lacking *Gdf5* or *Bmpr1b* (Baur, Mai, & Dymecki, 2000; Storm et al., 1994; Yi, Daluiski, Pederson, Rosen, & Lyons, 2000). These studies identified a regulatory role for Gdf5 signaling through Bmpr1b during joint formation. In contrast, Rountree et al. (2004) removed Bmpr1a, a receptor not efficiently utilized by Gdf5, from developing joints ($Bmpr1a^{Gdf5\Delta/\Delta}$) and reported that while initial joint specification and IZ formation were normal, as $Bmpr1a^{Gdf5\Delta/\Delta}$ mice aged, they developed severe OA due to the loss of expression of key extracellular matrix genes by articular chondrocytes, suggesting that after joint formation is complete, signaling through Bmpr1a is a universal mechanism for maintaining joint integrity.

The finding that Gdf5 is a less efficient activator of BMP signaling through Bmpr1a than Bmp2 raises the possibility that Gdf5 inhibits BMP signaling by competing with BMPs for occupancy of Bmpr1a. There is some evidence to support this possibility. In C2C12 cells, Gdf5 can antagonize the ability of Bmp2 to induce osteogenic differentiation, an effect mediated by Bmpr1a (Klammert et al., 2015). An antagonistic effect of Gdf5 on Bmp2-mediated ameloblast differentiation has also been noted (Liu et al., 2016). A mutation in a residue in Gdf5 located at an interaction surface with the type I BMP receptor (Seemann et al., 2009) reversed the antagonistic effect of Gdf5 on Bmp2 activity (Liu et al., 2016). The notion that Gdf5 may be a weak activator of BMP signaling through BMPR1a, thereby antagonizing the activities of more potent BMPs, is corroborated by a recent study by Antebi et al. (2017). This group investigated how various cell types respond to combinations of BMP ligands over a large concentration range. Some pairs acted additively, enhancing BMP signaling. Other pairs acted antagonistically or displayed more complex interactions. Of note, Gdf5 did not activate BMP signaling over a concentration range spanning four orders of magnitude, but it antagonized the activity of Bmp4 over this range. Mathematical modeling demonstrated that this type of antagonism occurs when a weakly activating ligand outcompetes a more highly activating ligand for the same receptor. As a direct test of the importance of receptor expression profiles, the authors showed that increasing levels of BMPr1b eliminated the antagonistic effect of Gdf5 on Bmp4 activity. This finding indicates that in cells expressing high levels of Bmpr1b, Gdf5 is an efficient activator of BMP signaling. Under these conditions, Gdf5 and Bmp4 can have additive effects because each ligand can bind to its preferred receptor. However, in cells with low levels of Bmpr1b expression, Gdf5 competes with Bmp4 for Bmpr1a.

How might this model account for the ability to Gdf5 to promote chondrogenesis in committed chondrocytes on the one hand, and to suppress chondrogenesis in the IZ on the other? (Fig. 3) IZ cells produce high levels of Gdf5 (Koyama et al., 2008). Bmpr1b is strongly expressed in cells destined to form the articular and growth plate chondrocytes (Baur et al., 2000; Wu et al., 2013). Expression of Bmpr1b in precursors of the epiphyseal surface would enable Gdf5 to act additively with other BMPs to stimulate chondrogenesis, without interfering with the activities of BMP ligands that bind preferentially to Bmpr1a. In contrast, IZ cells express Bmpr1a, but not Bmpr1b (Baur et al., 2000). Thus, Gdf5 would compete with other BMPs for binding to Bmpr1a in the IZ. Given that Gdf5 does not activate signaling effectively through Bmpr1a (Klammert et al., 2015; Seemann et al., 2009), the result would be decreased BMP signaling in the IZ relative to the adjacent cartilage. Consistent with this model, overexpression of constitutively

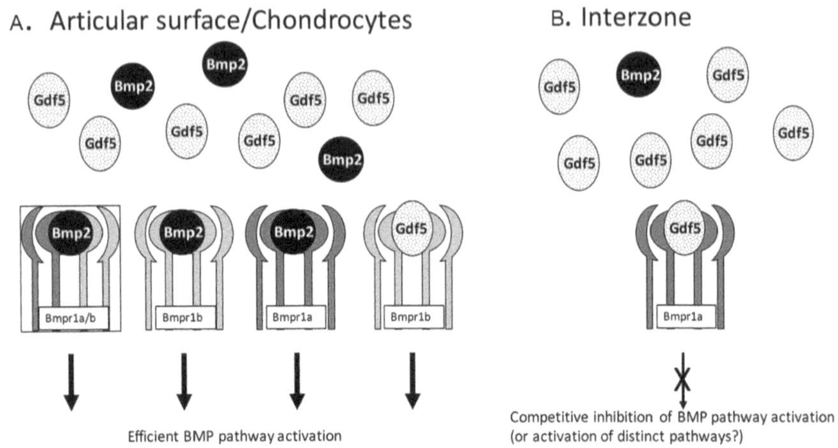

Fig. 3 Model for the divergent effects of Gdf5 based on receptor occupancy. (A) Nascent articular cartilage expresses both Bmpr1a and Bmpr1b. These form heteromeric complexes with type II receptors. Bmp2 and Gdf5 are expressed in the region of the joint interzone, but Gdf5 is expressed at higher levels. Bmp2 can bind with high affinity to either Bmpr1a or Bmpr1b-containing receptor complexes to activate BMP signaling efficiently. Mixed receptor complexes containing Bmpr1a and Bmpr1b can also be activated by Bmp2 (Gilboa et al., 2000), but whether Gdf5 can signal through these mixed complexes has not been reported. Gdf5 can bind to its preferred receptor, Bmpr1b in nascent articular cartilage. Because both ligands can bind to their preferred receptors, there is efficient activation of BMP signaling in articular chondrocytes. (B) Bmpr1a is expressed in the IZ but Bmpr1b is absent. The affinity of Gdf5 for Bmpr1a is much lower than is the affinity of Bmp2 for this receptor. However, because Gdf5 is present in much higher concentrations than other Bmps, Gdf5 occupies the Bmpr1a receptor complexes. This leads to inefficient BMP pathway activation. This model is purely speculative, but is consistent with multiple studies showing that Gdf5 acts as a competitive inhibitor of BMP signaling through Bmpr1a (Antebi et al., 2017; Klammert et al., 2015; Liu et al., 2016).

active Bmpr1b in the presumptive IZ led to chondrogenic differentiation of IZ cells and joint fusion (Ray et al., 2015). While this model is consistent with loss-of-function phenotypes for Gdf5, it does not explain why not all joints are affected by loss of Gdf5. Major defects that correlate with Gdf5 loss of function are confined to autopodal joints, in spite of the fact that Gdf5 is broadly expressed in all appendicular and craniofacial joints (Koyama et al., 2008). A logical possibility is that additional mechanisms must contribute to inhibition of BMP signaling in the IZ in other skeletal locations. Many studies have shown that TGFβ pathways antagonize BMP signaling in chondrocytes (Chen, Carrington, Hammonds, & Reddi, 1991; Hanada et al., 2001; Kawamura et al., 2012; Li et al., 2006), raising the possibility that this antagonism can also operate in the IZ.

6. TGFβ has a complex role in skeletal development

Most in vivo studies of TGFβ pathway function in skeletal development have focused on mice lacking *Tgfbr2* because TGFβRII is the only receptor that binds all TGFβ isoforms to elicit functional TGFβ signaling. Evidence for the functional importance of TGFβ pathways in limiting chondrogenic differentiation came from analysis of *Tgfbr2; Prrx1Cre* mice (Seo & Serra, 2007; Spagnoli et al., 2007). In these studies, chondrogenic differentiation was enhanced in micromass cultures from *Tgfbr2* mutant limb mesenchyme. Furthermore, chondrogenic condensations formed normally in *Tgfbr2; Prrx1Cre* mice, but accelerated differentiation was observed in the growth plate, indicative of enhanced BMP activity (Seo & Serra, 2007). Direct support for competition between TGFβ and BMP pathways in regulating the pace of chondrogenesis came from analysis of *Smad3−/−* chondrocytes, which have elevated BMP pathway activity in vitro (Li et al., 2006). Blockade of BMP activity could restore the pace of chondrocyte differentiation. Growth plates from *Smad3−/−* mice, *Smad2; Col2a1Cre* mice, and *Smad2/3* double mutants all exhibit postnatal dwarfism associated with accelerated chondrocyte differentiation (Wang, Song, et al., 2016; Yang et al., 2001), a phenotype attributed in part to repression of *Ihh* expression in the growth plate, which is mediated in part by direct binding of Smads 2 and 3 to distinct sites in the *Ihh* promoter (Wang, Song, et al., 2016). However, *Ihh* is directly activated by canonical BMP signaling (Minina et al., 2001; Retting, Song, Yoon, & Lyons, 2009), consistent with TGFβ and BMPs exerting opposing effects by impacting the expression of genes that regulate the pace of chondrocyte differentiation.

TGFβ signaling also regulates joint formation. A reporter transgene revealed high levels of *Tgfbr2* expression in interphalangeal joints at the time of cavitation in the mouse, indicating that TGFβs exert direct effects in these cells (Abe & Fujimori, 2013).

Mutations in TGFβRI, TGFβRII, Smad2, Smad3, TGFβ2, and TGFβ3 result in Loeys-Dietz Syndrome-like features, one of which is joint laxity. In these studies, joint laxity is associated with elevated TGFβ signaling. The converse phenotype is seen in loss-of-function studies in the mouse. Formation of interphalangeal IZs is defective in *Tgfbr2; Prrx1Cre* mice (Seo & Serra, 2007; Spagnoli et al., 2007). Marker analysis showed that IZ formation is initiated, but expression of the joint markers *Wnt9a* and *Gdf5* is not maintained and IZ cells differentiate into chondrocytes

(Seo & Serra, 2007; Spagnoli et al., 2007), consistent with the idea that TGFβRII maintains IZ fate by opposing chondrogenesis. Suppression of expression of the chemokine MCP-5 in IZ cells was identified as a key event in the suppression of chondrogenesis and joint cell fate, and this suppression was mediated by both TGFβRII-dependent and independent mechanisms (Longobardi et al., 2012). Joint phenotypes were not observed in *Tgfbr2; Col2a1* mice (Baffi et al., 2004) despite the expression of *Col2a1* in cells destined to undergo cavitation. Whether the joint formation defects in *Tgfbr2; Prrx1Cre* mice reflect a role for TGFβRII prior to the onset of overt chondrogenesis, or in a subset of cells that never express *Col2a1Cre* requires clarification. *Tgfbr2* expression is strong in shoulder, elbow, hip and knee joints (Spagnoli et al., 2007). However, aside from the absence of the meniscus and several tendons/ligaments, all of these joints form in *Tgfbr2; Prrx1Cre* (Pryce et al., 2009). As is the case for mice and humans bearing mutations in *Gdf5*, it is unclear why only a subset of joints that express TGFβRII are affected. One possibility is that studies focused on TGFβRII may underestimate the potential involvement of TGFβ pathways in joint formation.

Another important consideration is that TGFβ ligands can activate non-canonical TGFβ signaling in the absence of TGFβRII through TGFβRI-mediated signaling pathways (Iwata et al., 2012); Gdf8 utilizes TGFβRI/ActRIIB complexes and Gdf11 interacts with TGFβRI/ActRIIA or TGFβRI/ActRIIB to induce phosphorylation of Smad2/Smad3 (Aykul & Martinez-Hackert, 2016; Rebbapragada, Benchabane, Wrana, Celeste, & Attisano, 2003). Evidence for a broader role for TGFβRI than TGFβRII in joint formation is provided by comparisons of the phenotypes of *Tgfbr1; Dermo1Cre* and *Tgfbr2; Prrx1Cre* mice (Matsunobu et al., 2009; Seo & Serra, 2007; Spagnoli et al., 2007). Both *Prrx1* and *Dermo1* are expressed in precartilaginous mesenchyme prior to the onset of joint formation (Logan et al., 2002; Yu et al., 2003). *Tgfbr1; Dermo1Cre* mice had considerably more severe skeletal defects than *Tgfbr2; Prrx1Cre* mice. Ectopic cartilage protrusions were seen in the *Tgfbr1* mutants, associated with a very thin perichondrium (Matsunobu et al., 2009). These phenotypes are absent from *Tgfbr2; Prrx1Cre* mice. Although joints were not analyzed in detail, *Tgfbr1; Dermo1Cre* mice exhibited a partial fusion of the knee that was not seen in *Tgfbr2; Prrx1Cre* mice (Matsunobu et al., 2009). These findings suggest that TGFβRI may have roles independent of TGFβRII in growth plate and joint formation; direct comparisons of *Tgfbr1* and *Tgfbr2* phenotypes, using both *Prrx1Cre* and *Dermo1Cre,* will be needed to test this.

Furthermore, because these studies demonstrated enhanced chondrogenic differentiation, a hallmark of BMP activity, a direct analysis of BMP signaling activity is warranted.

7. How might TGFβ signaling interfere with BMP signaling in the IZ?

A myriad of interactions between TGFβ and BMP signaling have been documented in other systems (reviewed in Hudnall, Arthur, & Lowery, 2016). These include the induction of extracellular BMP antagonists by TGFβ (Knickmeyer et al., 2018), the ability of TGFβ pathway ligands to compete for type II receptors shared with BMP ligands (ActRIIA and ActRIIB) (Aykul & Martinez-Hackert, 2016; Hatsell et al., 2015), competition for co-Smad4 (Candia et al., 1997; Wakefield & Hill, 2013), TGFβ-mediated induction of genes that restrict BMP signaling (Oshimori & Fuchs, 2012), formation of mixed Smad2/3-Smad1/5 complexes (Gronroos et al., 2012; Hill, 2016), TGFβ-mediated induction of inhibitory Smads (iSmads) that target Smads1/5 (Afrakhte et al., 1998) and TGFβ-induced expression of SnoN and Ski repressors that can target BMP pathways (Ehnert et al., 2012; Kawamura et al., 2012). Additional, yet to be uncovered mechanisms are also possible. All remain to be examined before a clear picture emerges of the importance of each of these potential pathway interactions during IZ formation, and it may be that individual joints employ specific combinations of BMP/TGFβ interactions in order to form structures specific for that joint.

8. Conclusions and future directions

1. The IZ is best characterized as both a signaling center required for joint morphogenesis and as a primary source for the cells that give rise to the individual tissues found within newly formed synovial joints.
2. Joint formation requires both activation and repression of BMP signaling. BMP signaling is required for chondrogenesis and formation of articular cartilage; BMP inhibition is required for formation of an IZ. Intermediate levels of signaling likely drive formation of associated joint structures, such as tendons, ligaments, and joint capsules.
3. The fact that multiple BMP ligands and receptors are present in the IZ and that individual BMP ligands have specific receptor preferences,

creates the likelihood of formation of distinct BMP signaling complexes during joint morphogenesis and the possibility that these complexes will provide heterogeneity of signaling outputs.

4. When GDF5 is present with the other BMPs in the IZ, it serves to reduce BMP signal strength.
5. The presence of TGFβ signaling pathway components in the IZ allows for the possibility of novel interactions between BMPs, GDF5 and TGFβ in joint formation.
6. Much remains to be discovered about synovial joint morphogenesis and this information is vital to the development of strategies aimed at joint tissue repair and regeneration and the prevention of osteoarthritis.

References

Abe, T., & Fujimori, T. (2013). Reporter mouse lines for fluorescence imaging. *Development, Growth & Differentiation*, *55*, 390–405.

Afrakhte, M., Moren, A., Jossan, S., Itoh, S., Sampath, K., Westermark, B., et al. (1998). Induction of inhibitory Smad6 and Smad7 mRNA by TGF-beta family members. *Biochemical and Biophysical Research Communications*, *249*, 505–511.

Antebi, Y. E., Linton, J. M., Klumpe, H., Bintu, B., Gong, M., Su, C., et al. (2017). Combinatorial signal perception in the BMP pathway. *Cell*, *170*(1184–96), e24.

Aykul, S., & Martinez-Hackert, E. (2016). Transforming growth factor-beta family ligands can function as antagonists by competing for type II receptor binding. *The Journal of Biological Chemistry*, *291*, 10792–10804.

Baffi, M. O., Slattery, E., Sohn, P., Moses, H. L., Chytil, A., & Serra, R. (2004). Conditional deletion of the TGF-beta type II receptor in Col2a expressing cells results in defects in the axial skeleton without alterations in chondrocyte differentiation or embryonic development of long bones. *Developmental Biology*, *276*, 124–142.

Baur, S. T., Mai, J. J., & Dymecki, S. M. (2000). Combinatorial signaling through BMP receptor IB and GDF5: Shaping of the distal mouse limb and the genetics of distal limb diversity. *Development*, *127*, 605–619.

Brunet, L. J., McMahon, J. A., McMahon, A. P., & Harland, R. M. (1998). Noggin, cartilage morphogenesis, and joint formation in the mammalian skeleton. *Science*, *280*, 1455–1457.

Candia, A. F., Watabe, T., Hawley, S. H. B., Onichtchouk, D., Zhang, Y., Derynck, R., et al. (1997). Cellular interpretation of multiple TGF-ß signals: Intracellular antagonism between activin/BVg1 and BMP-2/4 signaling mediated by Smads. *Development*, *124*, 4467–4480.

Capellini, T. D., Chen, H., Cao, J., Doxey, A. C., Kiapour, A. M., Schoor, M., et al. (2017). Ancient selection for derived alleles at a GDF5 enhancer influencing human growth and osteoarthritis risk. *Nature Genetics*, *49*, 1202–1210.

Chen, H., Capellini, T. D., Schoor, M., Mortlock, D. P., Reddi, A. H., & Kingsley, D. M. (2016). Heads, shoulders, elbows, knees, and toes: Modular Gdf5 enhancers control different joints in the vertebrate skeleton. *PLoS Genetics*, *12*, e1006454.

Chen, P., Carrington, J. L., Hammonds, R. G., & Reddi, A. H. (1991). Stimulation of chondrogenesis in limb bud mesoderm cells by recombinant human bone morphogenetic protein 2B (BMP-2B) and modulation by transforming growth factor beta 1 and beta 2′. *Experimental Cell Research*, *195*, 509–515.

Dawson, K., Seeman, P., Sebald, E., King, L., Edwards, M., Williams, J., 3rd, et al. (2006). GDF5 is a second locus for multiple-synostosis syndrome. *American Journal of Human Genetics*, *78*, 708–712.

Decker, R. S., Koyama, E., & Pacifici, M. (2014). Genesis and morphogenesis of limb synovial joints and articular cartilage. *Matrix Biology*, *39*, 5–10.

Decker, R. S., Um, H. B., Dyment, N. A., Cottingham, N., Usami, Y., Enomoto-Iwamoto, M., et al. (2017). Cell origin, volume and arrangement are drivers of articular cartilage formation, morphogenesis and response to injury in mouse limbs. *Developmental Biology*, *426*, 56–68.

Duprez, D., Bell, E. J., Richardson, M. K., Archer, C. W., Wolpert, L., Brickell, P. M., et al. (1996). Overexpression of BMP-2 and BMP-4 alters the size and shape of developing skeletal elements in the chick limb. *Mechanisms of Development*, *57*, 145–157.

Dyment, N. A., Breidenbach, A. P., Schwartz, A. G., Russell, R. P., Aschbacher-Smith, L., Liu, H., et al. (2015). Gdf5 progenitors give rise to fibrocartilage cells that mineralize via hedgehog signaling to form the zonal enthesis. *Developmental Biology*, *405*, 96–107.

Ehnert, S., Zhao, J., Pscherer, S., Freude, T., Dooley, S., Kolk, A., et al. (2012). Transforming growth factor beta1 inhibits bone morphogenic protein (BMP)-2 and BMP-7 signaling via upregulation of ski-related novel protein N (SnoN): Possible mechanism for the failure of BMP therapy? *BMC Medicine*, *10*, 101.

Gamer, L. W., Pregizer, S., Gamer, J., Feigenson, M., Ionescu, A., Li, Q., et al. (2018). The role of Bmp2 in the maturation and maintenance of the murine knee joint. *Journal of Bone and Mineral Research*, *33*, 1708–1717.

Gilboa, L., Nohe, A., Geissendorfer, T., Sebald, W., Henis, Y. I., & Knaus, P. (2000). Bone morphogenetic protein receptor complexes on the surface of live cells: A new oligomerization mode for serine/threonine kinase receptors. *Molecular Biology of the Cell*, *11*, 1023–1035.

Gronroos, E., Kingston, I. J., Ramachandran, A., Randall, R. A., Vizan, P., & Hill, C. S. (2012). Transforming growth factor beta inhibits bone morphogenetic protein-induced transcription through novel phosphorylated Smad1/5-Smad3 complexes. *Molecular and Cellular Biology*, *32*, 2904–2916.

Gunnell, L. M., Jonason, J. H., Loiselle, A. E., Kohn, A., Schwarz, E. M., Hilton, M. J., et al. (2010). TAK1 regulates cartilage and joint development via the MAPK and BMP signaling pathways. *Journal of Bone and Mineral Research*, *25*, 1784–1797.

Guo, X., Day, T. F., Jiang, X., Garrett-Beal, L., Topol, L., & Yang, Y. (2004). Wnt/beta-catenin signaling is sufficient and necessary for synovial joint formation. *Genes & Development*, *18*, 2404–2417.

Hanada, K., Solchaga, L. A., Caplan, A. I., Hering, T. M., Goldberg, V. M., Yoo, J. U., et al. (2001). BMP-2 induction and TGF-beta 1 modulation of rat periosteal cell chondrogenesis. *Journal of Cellular Biochemistry*, *81*, 284–294.

Hartman, C., & Tabin, C. J. (2001). Wnt-14 plays a pivotal role in inducing synovial joint formation in the developing appendicular skeleton. *Cell*, *10*, 341–351.

Hatsell, S. J., Idone, V., Wolken, D. M., Huang, L., Kim, H. J., Wang, L., et al. (2015). ACVR1R206H receptor mutation causes fibrodysplasia ossificans progressiva by imparting responsiveness to activin A. *Science Translational Medicine*, 7, 303ra137.

Hill, C. S. (2016). Transcriptional control by the SMADs. *Cold Spring Harbor Perspectives in Biology*, *8*, a022079.

Huang, B. L., Trofka, A., Furusawa, A., Norrie, J. L., Rabinowitz, A. H., Vokes, S. A., et al. (2016). An interdigit signalling centre instructs coordinate phalanx-joint formation governed by 5'Hoxd-Gli3 antagonism. *Nature Communications*, 7, 12903.

Hudnall, A. M., Arthur, J. W., & Lowery, J. W. (2016). Clinical relevance and mechanisms of antagonism between the BMP and activin/TGF-beta signaling pathways. *The Journal of the American Osteopathic Association*, *116*, 452–461.

Iwata, J., Hacia, J. G., Suzuki, A., Sanchez-Lara, P. A., Urata, M., & Chai, Y. (2012). Modulation of noncanonical TGF-beta signaling prevents cleft palate in Tgfbr2 mutant mice. *The Journal of Clinical Investigation*, *122*, 873–885.

Kawamura, I., Maeda, S., Imamura, K., Setoguchi, T., Yokouchi, M., Ishidou, Y., et al. (2012). SnoN suppresses maturation of chondrocytes by mediating signal cross-talk between transforming growth factor-beta and bone morphogenetic protein pathways. *The Journal of Biological Chemistry*, *287*, 29101–29113.

Kiapour, A. M., Cao, J., Young, M., & Capellini, T. D. (2018). The role of Gdf5 regulatory regions in development of hip morphology. *PLoS One*, *13*, e0202785.

Klammert, U., Mueller, T. D., Hellmann, T. V., Wuerzler, K. K., Kotzsch, A., Schliermann, A., et al. (2015). GDF-5 can act as a context-dependent BMP-2 antagonist. *BMC Biology*, *13*, 77.

Knickmeyer, M. D., Mateo, J. L., Eckert, P., Roussa, E., Rahhal, B., Zuniga, A., et al. (2018). TGFbeta-facilitated optic fissure fusion and the role of bone morphogenetic protein antagonism. *Open Biology*, *8*, 170134.

Koyama, E., Shibukawa, Y., Nagayama, M., Sugito, H., Young, B., Yuasa, T., et al. (2008). A distinct cohort of progenitor cells participates in synovial joint and articular cartilage formation during mouse limb skeletogenesis. *Developmental Biology*, *316*, 62–73.

Li, T. F., Darowish, M., Zuscik, M. J., Chen, D., Schwarz, E. M., Rosier, R. N., et al. (2006). Smad3-deficient chondrocytes have enhanced BMP signaling and accelerated differentiation. *Journal of Bone and Mineral Research*, *21*, 4–16.

Liu, J., Saito, K., Maruya, Y., Nakamura, T., Yamada, A., Fukumoto, E., et al. (2016). Mutant GDF5 enhances ameloblast differentiation via accelerated BMP2-induced Smad1/5/8 phosphorylation. *Scientific Reports*, *6*, 23670.

Logan, M., Martin, J. F., Nagy, A., Lobe, C., Olson, E. N., & Tabin, C. J. (2002). Expression of Cre recombinase in the developing mouse limb bud driven by a Prxl enhancer. *Genesis*, *33*, 77–80.

Longobardi, L., Li, T., Myers, T. J., O'Rear, L., Ozkan, H., Li, Y., et al. (2012). TGF-beta type II receptor/MCP-5 axis: At the crossroad between joint and growth plate development. *Developmental Cell*, *23*, 71–81.

Lowery, J. W., & Rosen, V. (2018). The BMP pathway and its inhibitors in the skeleton. *Physiological Reviews*, *98*, 2431–2452.

Macias, D., Gañon, Y., Sampath, T. K., Piedra, M. E., Ros, M. A., & Hurle, J. M. (1997). Role of BMP-2 and OP-1 (BMP-7) in programmed cell death and skeletogenesis during chick limb development. *Development*, *124*, 1109–1117.

Matsunobu, T., Torigoe, K., Ishikawa, M., de Vega, S., Kulkarni, A. B., Iwamoto, Y., et al. (2009). Critical roles of the TGF-beta type I receptor ALK5 in perichondrial formation and function, cartilage integrity, and osteoblast differentiation during growth plate development. *Developmental Biology*, *332*, 325–338.

Minina, E., Wenzel, H. M., Kreschel, C., Karp, S., Gaffield, W., McMahon, A. P., et al. (2001). BMP and Ihh/PTHrP signaling interact to coordinate chondrocyte proliferation and differentiation. *Development*, *128*, 4523–4534.

Nishitoh, H., Ichijo, H., Kimura, M., Matsumoto, T., Makishima, F., Yamaguchi, A., et al. (1996). Identification of type I and type II serine/threonine kinase receptor for growth/differentiation factor-5. *Journal of Biological Chemistry*, *271*, 21345–21352.

Oshimori, N., & Fuchs, E. (2012). Paracrine TGF-beta signaling counterbalances BMP-mediated repression in hair follicle stem cell activation. *Cell Stem Cell*, *10*, 63–75.

Pacifici, M., Koyama, E., Shibukawa, Y., Wu, C., Tamamura, Y., & Enomoto-Iwamoto, M. (2006). Cellular and molecular mechanisms of synovial joint and articular cartilage formation. *Annals of the New York Academy of Sciences*, *1068*, 74–86.

Pryce, B. A., Watson, S. S., Murchison, N. D., Staverosky, J. A., Dunker, N., & Schweitzer, R. (2009). Recruitment and maintenance of tendon progenitors by TGFbeta signaling are essential for tendon formation. *Development*, *136*, 1351–1361.

Pregizer, S. K., Kiapour, A. M., Young, M., Chen, H., Schoor, M., Liu, Z., et al. (2018). Impact of broad regulatory regions on Gdf5 expression and function in knee development and susceptibility to osteoarthritis. *Annals of Rheumatic Disease, 77*, 212475.

Prunier, C., Baker, D., Ten Dijke, P., & Ritsma, L. (2019). TGF-β family signaling pathways in cellular dormancy. *Trends in Cancer, 5*, 66–78.

Ray, A., Singh, P. N., Sohaskey, M. L., Harland, R. M., & Bandyopadhyay, A. (2015). Precise spatial restriction of BMP signaling is essential for articular cartilage differentiation. *Development, 142*, 1169–1179.

Rebbapragada, A., Benchabane, H., Wrana, J. L., Celeste, A. J., & Attisano, L. (2003). Myostatin signals through a transforming growth factor beta-like signaling pathway to block adipogenesis. *Molecular and Cellular Biology, 23*, 7230–7242.

Retting, K. N., Song, B., Yoon, B. S., & Lyons, K. M. (2009). BMP canonical Smad signaling through Smad1 and Smad5 is required for endochondral bone formation. *Development, 136*, 1093–1104.

Rountree, R. B., Schoor, M., Chen, H., Marks, M. E., Harley, V., Mishina, Y., et al. (2004). BMP receptor signaling is required for postnatal maintenance of articular cartilage. *PLoS Biology, 2*, e355.

Salazar, V. S., Gamer, L. W., & Rosen, V. (2016). BMP signalling in skeletal development, disease and repair. *Nature Reviews. Endocrinology, 12*, 203–221.

Schwaerzer, G. K., Hiepen, C., Schrewe, H., Nickel, J., Ploeger, F., Sebald, W., et al. (2012). New insights into the molecular mechanism of multiple synostoses syndrome (SYNS): Mutation within the GDF5 knuckle epitope causes noggin-resistance. *Journal of Bone and Mineral Research, 27*, 429–442.

Seemann, P., Brehm, A., Konig, J., Reissner, C., Stricker, S., Kuss, P., et al. (2009). Mutations in GDF5 reveal a key residue mediating BMP inhibition by NOGGIN. *PLoS Genetics, 5*, e1000747.

Seo, H. S., & Serra, R. (2007). Deletion of Tgfbr2 in Prx1-cre expressing mesenchyme results in defects in development of the long bones and joints. *Developmental Biology, 310*, 304–316.

Settle, S. H., Rountree, R. B., Sinha, A., Thacker, A., Higgins, K., & Kingsley, D. M. (2003). Multiple joint and skeletal patterning defects caused by single and double mutations in the mouse Gdf6 and Gdf6 genes. *Developmental Biology, 254*, 116–130.

Shim, J. H., Greenblatt, M. B., Xie, M., Schneider, M. D., Zou, W., Zhai, B., et al. (2009). TAK1 is an essential regulator of BMP signalling in cartilage. *The EMBO Journal, 28*, 2028–2041.

Shwartz, Y., Viukov, S., Krief, S., & Zelzer, E. (2016). Joint development involves a continuous influx of Gdf5-positive cells. *Cell Reports, 15*, 2577–2587.

Smeeton, J., Askary, A., & Crump, J. G. (2016). Building and maintaining joints by exquisite local control of cell fate. *Wiley Interdisciplinary Reviews: Developmental Biology, 6*, e245.

Spagnoli, A., O'Rear, L., Chandler, R. L., Granero-Molto, F., Mortlock, D. P., Gorska, A. E., et al. (2007). TGF-beta signaling is essential for joint morphogenesis. *The Journal of Cell Biology, 177*, 1105–1117.

Storm, E. E., Huynh, T. V., Copeland, N. A., Kingsley, D. M., & Lee, S. J. (1994). Limb alterations in brachypodism mice due to mutations in a new member of the TGF beta-superfamily. *Nature, 368*, 639–643.

Tang, L., Wu, X., Zhang, H., Lu, S., Wu, M., Shen, C., et al. (2017). A point mutation in Fgf9 impedes joint interzone formation leading to multiple synostosis syndrome. *Human Molecular Genetics, 26*, 1280–1293.

Thomas, J. T., Kilpatrick, M. W., Lin, K., Erlacher, L., Lembessis, P., Costa, T., et al. (1997). Disruption of human limb morphogeneisis by a dominant negative mutation in CDMP1. *Nature Genetics, 17*, 58–64.

Thomas, J. T., Lin, K., Nandedkar, M., Camargo, M., Cervenka, J., & Luyten, F. P. (1996). A human chondrodysplasia due to a mutation in a TGF-beta superfamily member. *Nature Genetics*, *12*, 315–317.

Tsumaki, N., Nakase, T., Miyaji, T., Kakiuchi, M., Kimura, T., Ochi, T., et al. (2002). Bone morphogenetic protein signals are required for cartilage formation and differently regulate joint development during skeletogenesis. *Journal of Bone and Mineral Research*, *17*, 898–906.

Wakefield, L. M., & Hill, C. S. (2013). Beyond TGFbeta: Roles of other TGFbeta superfamily members in cancer. *Nature Reviews. Cancer*, *13*, 328–341.

Wang, W., Song, B., Anbarchian, T., Shirazyan, A., Sadik, J. E., & Lyons, K. M. (2016). Smad2 and Smad3 regulate chondrocyte proliferation and differentiation in the growth plate. *PLoS Genetics*, *12*, e1006352.

Wang, J., Yu, T., Wang, Z., Ohte, S., Yao, R. E., Zheng, Z., et al. (2016). A new subtype of multiple synostoses syndrome is caused by a mutation in GDF6 that decreases its sensitivity to noggin and enhances its potency as a BMP signal. *Journal of Bone and Mineral Research*, *31*, 882–889.

Wolfman, N. M., Hattersley, G., Cox, K., Celeste, A. J., Nelson, R., Yamaji, N., et al. (1997). Ectopic induction of tendon and ligament in rats by growth and differentiation factors 5, 6, and 7, members of the TGF-beta gene family. *Journal of Clinical Investigation*, *100*, 321–330.

Wu, L., Bluguermann, C., Kyupelyan, L., Latour, B., Gonzalez, S., Shah, S., et al. (2013). Human developmental chondrogenesis as a basis for engineering chondrocytes from pluripotent stem cells. *Stem Cell Reports*, *1*, 575–589.

Yang, X., Chen, L., Xu, X., Li, C., Huang, C., & Deng, C.-X. (2001). TGF-b/Smad3 signals repress chondrocyte hypertrophic differentiation and are required for maintaining articular cartilage. *The Journal of Cell Biology*, *153*, 35–46.

Yi, S. E., Daluiski, A., Pederson, R., Rosen, V., & Lyons, K. M. (2000). The type I BMP receptor BMPRIB is required for chondrogenesis in the mouse limb. *Development*, *127*, 621–630.

Yu, K., Xu, J., Liu, Z., Sosic, D., Shao, J., Olson, E. N., et al. (2003). Conditional inactivation of FGF receptor 2 reveals an essential role for FGF signaling in the regulation of osteoblast function and bone growth. *Development*, *130*, 3063–3074.

Zhang, Z., Yu, X., Zhang, Y., Geronimo, B., Lovlie, A., Fromm, S. H., et al. (2000). Targeted misexpression of constitutively active BMP receptor-1B causes bifurcation, duplication, and posterior transformation of digit in mouse limb. *Developmental Biology*, *220*, 154–167.

CHAPTER SEVEN

Roles and regulation of SOX transcription factors in skeletogenesis

Véronique Lefebvre*
The Children's Hospital of Philadelphia, Philadelphia, PA, United States
*Corresponding author: e-mail address: lefebvrev1@email.chop.edu

Contents

Abstract

SOX transcription factors participate in the specification, differentiation and activities of many cell types in development and beyond. The 20 mammalian family members are distributed into eight groups based on sequence identity, and while co-expressed same-group proteins often have redundant functions, different-group proteins typically have distinct functions. More than a handful of SOX proteins have pivotal roles in skeletogenesis. Heterozygous mutations in their genes cause human diseases, in which skeletal dysmorphism is a major feature, such as campomelic dysplasia (*SOX9*), or a minor feature, such as LAMSHF syndrome (*SOX5*) and Coffin-Siris-like syndromes (*SOX4* and *SOX11*). Loss- and gain-of-function experiments in animal models have revealed that SOX4 and SOX11 (SOXC group) promote skeletal progenitor survival and control skeleton patterning and growth; SOX8 (SOXE group) delays the differentiation of osteoblast progenitors; SOX9 (SOXE group) is essential for chondrocyte fate maintenance and differentiation, and works in cooperation with SOX5 and SOX6 (SOXD group) and other types of transcription factors. These and other SOX proteins have also been proposed, mainly through *in vitro* experiments, to have key roles in other aspects of skeletogenesis, such as SOX2 in osteoblast stem cell self-renewal. We here review

Current Topics in Developmental Biology, Volume 133
ISSN 0070-2153
https://doi.org/10.1016/bs.ctdb.2019.01.007

current knowledge of well-established and proposed skeletogenic roles of SOX proteins, their transcriptional and non-transcriptional actions, and their modes of regulation at the gene, RNA and protein levels. We also discuss gaps in knowledge and directions for future research to further decipher mechanisms underlying skeletogenesis in health and diseases and identify treatment options for skeletal malformation and degeneration diseases.

1. Introduction

The vertebrate skeleton is an edifice of many structures varying in composition, size, shape and anatomical position. Its development involves the specification and coordinated actions of highly specialized progenitor and differentiated cells (Berendsen & Olsen, 2015). Progenitors arise from the cranial neural crest, paraxial mesoderm and lateral plate mesoderm. Upon migrating to their destined locations, they form skeletogenic mesenchymal condensations. They then engage in multi-step differentiation programs to become chondrocytes, osteoblasts, synovial fibroblasts or tenocytes, which build the skeleton and ensure its growth and maturation. Subsets of progenitors persist throughout development within and around skeletal structures to produce new waves of differentiating cells and participate in intense patterning and differentiation cross talk with them. All cells' phenotypes rely on the implementation of specific genetic programs, and thus on proper expression and utilization of unique sets of transcription factors. The discovery three decades ago that forced expression of the transcription factor MYOD was sufficient to convert mesenchymal cells into myoblasts led to the proposition that each cell type would be governed by a single master transcription factor. Since then, it has been well proven that transcription factors work in sets rather than solo and that many families of transcription factors participate in cell type-specific functions. Each family is characterized by a unique DNA-binding domain, which typically recognizes a precise DNA sequence. Most transcription factors also feature domains that confer specific transcriptional activities. Pioneer transcription factors physically interact with naive chromatin and recruit chromatin-modifying enzymes to displace nucleosomes and poise gene loci for transcriptional activation (Iwafuchi-Doi & Zaret, 2016). Transactivators bind specific DNA sequences at open enhancers or promoters and recruit co-activators that contact the basal transcription machinery to effect transcription. Transrepressors, in contrast, recruit co-repressors to inhibit the basal transcriptional machinery.

Architectural factors promote the assembly of enhanceosomes (protein complexes bound to enhancers) by binding DNA near other factors and facilitating their physical and functional interactions.

The family of SOX proteins are key members of cell type-specific transcription factor sets in many lineages (Kamachi & Kondoh, 2013). We here review current knowledge and gaps in knowledge regarding skeletogenic SOX proteins. We introduce them in the context of their family, describe human diseases due to mutations in their genes, and review their roles and modes of regulation. We conclude with suggestions for future research on the SOX family to deepen understanding of skeletogenesis and related diseases.

2. Shared and distinctive features of SOX proteins

SOX proteins belong to the super-family of HMG (high-mobility-group) domain-containing proteins, as do the TCF/LEF WNT signaling targets and mediators (Kamachi & Kondoh, 2013). The HMG domain comprises three α-helices that bind DNA in the minor groove and force a 30–100° DNA bent (Fig. 1A). The latter property allows LEF1 to promote the assembly of enhanceosomes and might therefore be a property of SOX proteins too (Giese, Amsterdam, & Grosschedl, 1991).

SRY was the first SOX protein to be discovered. Encoded by the Sex-determining Region on the Y chromosome in mammals, it initiates a cascade of genetic events that lead to testis development and thus to male differentiation. Subsequently, all genes found to encode a protein with at least 50% identity with SRY in the HMG domain were called SOX, for SRY-related HMG box. SOX proteins share only 20% identity in this domain with other super-family members, but have conserved the residues necessary for characteristic DNA binding and bending. SOX genes exist in animals only, with a handful of them in invertebrates and 20 of them in humans and most mammals (Phochanukul & Russell, 2010; Schepers, Teasdale, & Koopman, 2002).

SOX proteins are distributed into eight groups (Fig. 1B). Those involved in skeletogenesis include the SOXB-group SOX2; the SOXC-group SOX4, SOX11, and SOX12; the SOXD-group SOX5 and SOX6, and the SOXE-group SOX8 and SOX9. Members of the same group share almost 100% identity in the HMG domain, but only about 50% with other-group proteins. Most SOX proteins exhibit at least one additional functional domain, conserved among group members only (Fig. 1C). SOX2, SOXC and SOXE

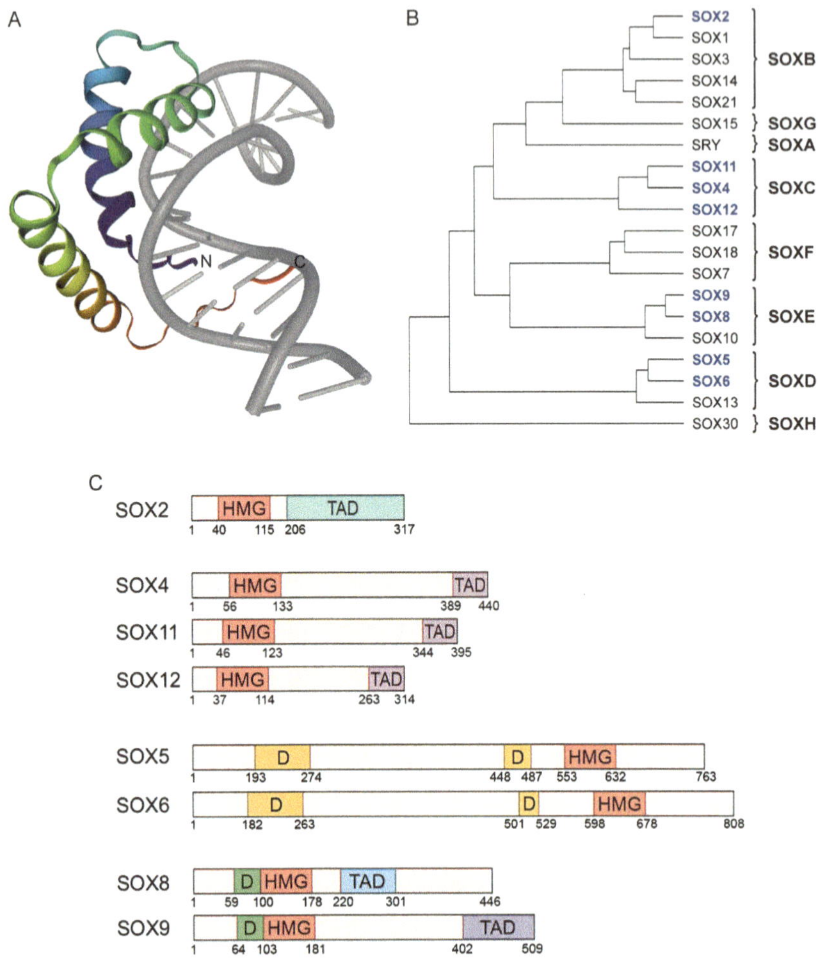

Fig. 1 General and specific properties of SOX transcription factors. (A) Rendering of the SOX4 HMG domain (rainbow-colored) bound to DNA (gray). The HMG domain has three α-helices that fold into an L-shape. It contacts DNA in the minor groove and induces a strong bend. Its N- and C-termini are indicated. This cartoon was generated by SWISS-MODEL using published data (Jauch, Ng, Narasimhan, & Kolatkar, 2012). (B) Phylogenic tree of the human SOX family members established based on sequence conservation in the HMG domain. It was generated using the UPGMA method in MacVector software (version 16.0.8). Skeletogenic SOX proteins are highlighted in blue. (C) Schematics of the domain structure of skeletogenic SOX proteins. HMG, DNA-binding domain; TAD, transactivation domain; D, dimerization domain. The same colors are used for the dimerization and transactivation domains of same-group proteins because of high conservation. Different colors are used for these domains for distinct-group proteins to reflect the lack of homology.

proteins feature group-specific transactivation domains and SOXD and SOXE proteins possess group-specific homodimerization domains (Barrionuevo & Scherer, 2010; Dy et al., 2008; Hoser et al., 2008; Lefebvre, Li, & de Crombrugghe, 1998; Sarkar & Hochedlinger, 2013).

All SOX proteins bind DNA motifs matching or resembling the $C^{A}/_{T}TTG^{A}/^{A}_{T}/_{T}$ sequence (Kamachi & Kondoh, 2013). They thus have the potential to select and compete for the same target genes. This mechanism was proposed to explain competition between SOXD and SOXE proteins in non-skeletal cell types, such as glial cells (Reiprich & Wegner, 2015), and sequential actions of SOX2 and SOXC proteins on the same targets in the neuronal lineage (Bergsland et al., 2011). SOX proteins, however, often select other targets than their family members do and distinct targets in different cell types or cell differentiation stages because of cooperativity with separate partners. This has been best documented for SOX2 (Sarkar & Hochedlinger, 2013) and is starting to be uncovered for other SOX proteins too.

Evolution has diversified the coding as well as regulatory sequences of SOX genes, such that each gene is expressed in a discrete set of cell types. These cell types can be far removed from each other. SOX9, for instance, is expressed in chondrocytes, Sertoli cells, neuronal cells, and many progenitor cell types. Same-group members often overlap in expression and thus functions, and different-group members can overlap in expression, but generally have distinct functions.

3. Skeletal dysmorphism due to SOX mutations

Mutations in 10 SOX genes, including skeletogenic ones, are known to date to cause a human developmental syndrome. The diseases are rare and most often due to *de novo* heterozygous mutations that result in gene or protein inactivation and thus reflect gene haploinsufficiency.

Mutations affecting *SOX9* cause Campomelic Dysplasia (CD), a severe skeletal malformation syndrome associated with XY sex reversal (Unger, Scherer, & Superti-Furga, 2008). Features include limb (melic) bending (campo), low-set ears, a flat nasal bridge, small jaw, cleft palate, and narrow chest. Most cases die neonatally from respiratory distress due to skeletal malformations. Survivors have short stature, flat face, micrognathia, kyphoscoliosis, hypoplastic nails, and hypotonia (Corbani et al., 2011). Most also have global developmental delay, mild mental retardation, hearing impairment, cardiac defects and hydronephrosis. CD-causing mutations abolish SOX9 protein production or function, or affect *SOX9* gene expression. Chromosomal translocations occurring up to 700 kb upstream of *SOX9* generally cause CD; up to 1 Mb cause acampomelic dysplasia, a mild form of CD; and up to 1.5 Mb cause Pierre Robin Sequence, featuring micrognathia and cleft palate (Gordon et al., 2009).

Heterozygous mutations in *SOX5* cause LAMSHF syndrome, a global developmental delay disorder manifested by marked speech and locomotor retardation, hypotonia, strabismus and skeletal malformations (Lamb et al., 2012; Nesbitt et al., 2015). The latter include short stature, frontal bossing, microretrognathia, clinodactyly, butterfly vertebrae, and scoliosis.

A balanced translocation (t(9;11)(q33;p15) disrupting *SOX6* (11p15) was found in a child with a complex craniofacial dysostosis including craniosynostosis (Tagariello et al., 2006). A child with a 9q32-q34 deletion had a similar phenotype, but without craniosynostosis, and a child with a missense mutation in *SOX6* only had craniosynostosis. These cases suggest that *SOX6* mutations could cause craniosynostosis, but additional cases are needed to definitively link *SOX6* mutations to this disease.

SOX11 heterozygous mutations were described to cause a mild Coffin-Siris-like syndrome (Hempel et al., 2016; Tsurusaki et al., 2014). Affected children had developmental delay, intellectual disability, short stature, microcephaly, fifth-finger clinodactyly, 2–3 toe syndactyly, nail hypoplasia, scoliosis, a wide mouth and prominent lips. One child also had a cleft palate (Khan, Study, Baker, & Clayton-Smith, 2018). Very recently, *SOX4* heterozygous mutations were found to cause a similar syndrome (Zawerton et al., 2019).

These autosomal-dominant diseases reveal the importance of SOX gene dosage in skeletogenesis and other processes, but do not reveal the full spectrum of SOX gene activities. The higher severity of campomelic dysplasia compared to the LAMSHF and Coffin-Siris-like syndromes does not mean that only SOX9 has key skeletogenic functions, but rather reflects the fact that SOX9, unlike SOX5/SOX6 and SOX4/SOX11, has no redundant partner in skeletogenesis. Reaching deep understanding of the SOX gene functions is necessary to better understand developmental and adult diseases directly or indirectly due to SOX deficiencies.

4. SOX genes and the control of skeletal progenitors

Embryogenesis involves a cellular hierarchy, whereby embryonic pluripotent stem (ES) cells give rise to progenitor cells with progressively more restricted lineage potential. ES cell programming, self-renewal, and activity are governed by a quartet made of the transcription factors SOX2, OCT3/4 (POU-domain protein), KLF4 (zinc-finger protein), and c-Myc (a basic-helix-loop-helix protein) (Sarkar & Hochedlinger, 2013). The quartet has both pioneer and transactivation functions. The proteins bind their targets on adjacent recognition sites in enhancers and promoters. They remain expressed and essential in several types of multipotent progenitors. Relevant to skeletogenesis, SOX2 controls the delamination of neural crest cells from the neural tube (Mandalos & Remboutsika, 2017). It is expressed *in vitro* in a subset of newborn mouse calvarium cells (Basu-Roy et al., 2010). These presumptive osteoblast stem cells rely on SOX2 for self-renewal. Other SOX genes are also expressed and have important roles in skeletal progenitors.

The SOXE genes are expressed in the cranial neural crest and redundantly specify these cells (Haldin & LaBonne, 2010). Only *SOX10* remains expressed during cell migration to definitive anatomic sites. It is then turned off in skeletal progenitors, but stays on and directs cell differentiation in the neuronal, glial and melanocyte lineages (Sommer, 2011).

The SOXC genes, primarily *Sox4* and *Sox11*, are expressed in many progenitor types, including skeletal progenitors (Dy et al., 2008). They are needed for cell survival at the time of mesenchyme formation and act at least in part by activating the gene for TEAD2, a HIPPO pathway mediator (Bhattaram et al., 2014, 2010). They may also act in skeletal progenitors as in cancers by promoting cell migration and epithelial-to-mesenchymal transition by upregulating genes for essential AKT, p53, WNT, and NOTCH signaling components (Kuo et al., 2015; Lourenco & Coffer, 2017).

Sox8 and *Sox9* are co-expressed with the SOXC genes in skeletogenic mesenchyme (Akiyama et al., 2005). Their single inactivation has no obvious effect before osteoblasts and chondrocytes are due to differentiate, respectively (Akiyama, Chaboissier, Martin, Schedl, & de Crombrugghe, 2002; Mori-Akiyama, Akiyama, Rowitch, & de Crombrugghe, 2003; Schmidt et al., 2005). Whole-transcriptome and whole-epigenome profiling in mouse limb buds before pre-cartilaginous condensation have revealed limited differences between wild-type and *Sox9*-null embryos (Liu, Angelozzi, Haseeb, & Lefebvre, 2018). While these findings suggest limited transcriptional activity of SOX9 in progenitor cells, other studies have suggested important non-transcriptional actions. For instance, competition between SOX9 and β-catenin, an anti-chondrogenic protein, contributes to specifying the chondrocytic versus non-chondrocytic cell fate (Akiyama et al., 2004; Topol, Chen, Song, Day, & Yang, 2009). Complexes formed between the two proteins are degraded through the proteasomal pathway, resulting in elimination of the less abundant protein (Akiyama et al., 2004; Topol et al., 2009). SOX9 can also physically interact with RUNX2 and may thereby delay the master osteogenic actions of this RUNT-domain transcription factor (Zhou et al., 2006).

Once skeletal tissues overtly develop, progenitor/stem cells are maintained in specific locations and involve SOX proteins in their regulation and activities. The SOXC genes are highly expressed in perichondrium and presumptive joint cells. They set the boundaries of individual skeletal elements by keeping these cells non-chondrogenic, and they also establish cross talk with chondrocytes to initiate growth plate formation (Bhattaram et al., 2014). They may favor articular cartilage development by upregulating *Gdf5* (growth and differentiation factor-5) in presumptive joint cells (Kan et al., 2013), and may induce growth plate formation and organization by upregulating *Wnt5a* (non-canonical WNT ligand) in perichondrial cells (Kato, Bhattaram, Penzo-Mendez, Gadi, & Lefebvre, 2015). They also efficiently bind to and stabilize β-catenin, aborting thereby chondrogenesis in perichondrium and presumptive joints (Bhattaram et al., 2014).

Pthlh (parathyroid hormone-related signaling factor)-positive cells have been identified as chondrocyte stem cells in mouse epiphyseal growth plates (Mizuhashi et al., 2018), and multipotent skeletal stem cells in mouse and human growth plates based on cell surface markers (Chan et al., 2018, 2015; Mizuhashi et al., 2018). The expression and roles of SOX genes in these cell types are likely, but remain undocumented.

Based on all data described above and for neurogenesis (Reiprich & Wegner, 2015), a SOX hierarchy might exist in skeletal progenitors. SOX2 would govern stem cell specification and self-renewal, SOXC proteins would ensure cell survival and cross talk with the environment, and SOXE proteins would be involved in controlling downstream lineage commitment (Fig. 2A).

5. Roles of SOX genes in chondrogenesis

SOX9 and SOX5/SOX6 have long been known to be essential for chondrogenesis (Hata, Takahata, Murakami, & Nishimura, 2017; Kozhemyakina, Lassar, & Zelzer, 2015; Lefebvre & Dvir-Ginzberg, 2017). Yet, several outstanding questions on their specific actions were answered only recently and others remain unanswered. Their genes are active in the chondrocyte lineage from the progenitor mesenchymal stage until the prehypertrophic stage in growth plates or throughout adulthood in permanent

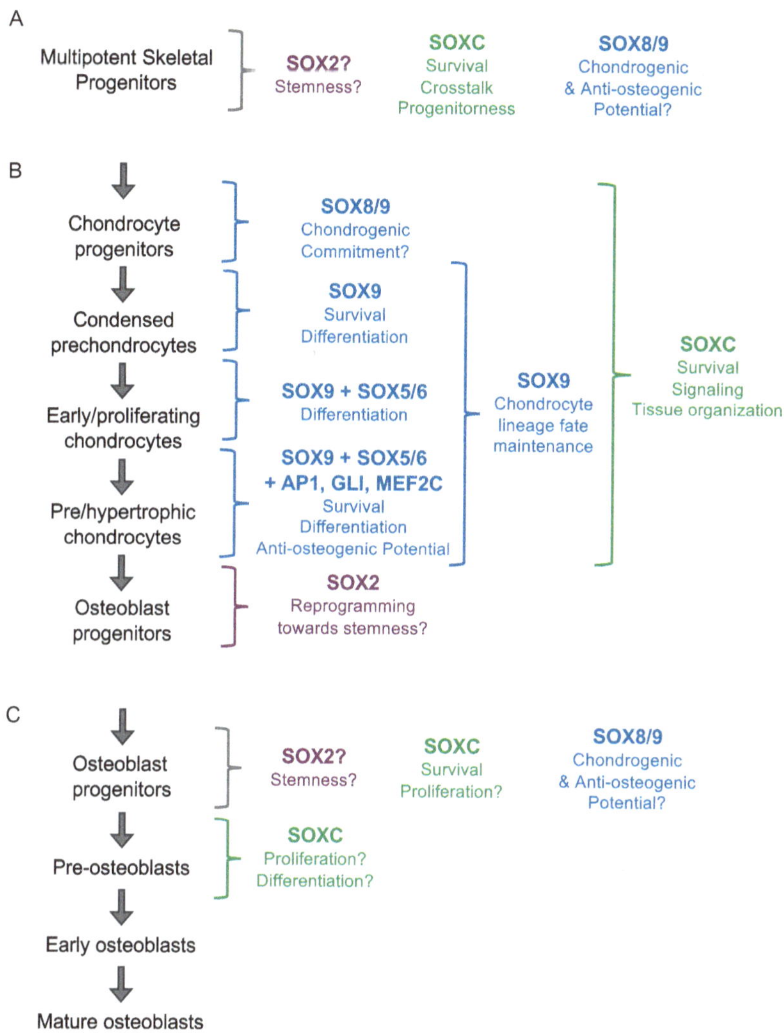

Fig. 2 Schematics summarizing current knowledge of the roles of SOX proteins in cell fate determination and differentiation during skeletogenesis. See the main text for detailed information.

cartilages (Dy et al., 2008; Henry, Liang, Akdemir, & de Crombrugghe, 2012; Lefebvre et al., 1998; Ng et al., 1997) (Fig. 2B). None is necessary for progenitor specification and colonization of skeletogenic sites (Akiyama et al., 2002; Bi, Deng, Zhang, Behringer, & de Crombrugghe, 1999; Smits et al., 2001), and although SOX9 was postulated to specify the chondrogenic fate of progenitors, it was recently found dispensable for this event (Liu et al., 2018). The pioneer factors that poise progenitors for chondrogenesis remain unknown, but it is possible that SOX9 is one of them and acts in redundancy with its co-expressed relative, SOX8 (Sock, Schmidt, Hermanns-Borgmeyer, Bosl, & Wegner, 2001). In its first or next role, SOX9 ensures prechondrocyte condensation and survival (Akiyama et al., 2002; Bi et al., 1999), likely by upregulating genes for cytoskeleton assembly, homotypic cell-cell adhesion and heterotypic cell-cell repulsion, such as *Sema3c* and *Sema3d* (Liu et al., 2018). SOX5 and SOX6 are dispensable at that stage (Smits et al., 2001).

SOX9 and SOX5/SOX6 are called the chondrogenic trio because they are essential for early chondrocyte differentiation. When *Sox9* is inactivated at the onset of chondrogenesis in mice, prechondrocytes fail to produce a cartilaginous extracellular matrix because they are unable to robustly express chondrocyte-specific genes, such as *Col2a1* (collagen type 2) and *Acan* (aggrecan) (Akiyama et al., 2002). Mice lacking either *Sox5* or *Sox6* have modest skeletal defects, whereas mice lacking both genes have severely underdeveloped cartilage primordia, indicating a large degree of gene redundancy (Smits et al., 2001). *Sox5/Sox6*-null chondrocytes weakly express cartilage-specific genes, despite expressing *Sox9* normally. These loss-of-function studies have thus demonstrated that the SOX trio is necessary for early chondrocyte differentiation. Complementing them, gain-of-function studies have demonstrated that the trio is also sufficient to convert progenitor cells into chondrocytes (Ikeda et al., 2004). The SOX trio has similar roles in the notochord. This embryonic structure secretes key morphogens for organogenesis and features a cartilage-like sheath that provides axial support. Without *Sox9* or *Sox5/Sox6*, notochord cells fail to produce this sheath and die before converting into intervertebral disc nuclei pulposi (Barrionuevo, Taketo, Scherer, & Kispert, 2006; Smits & Lefebvre, 2003).

The SOX trio is also indispensable for cartilage growth plate formation (Akiyama et al., 2004; Dy et al., 2012; Ikegami et al., 2011; Smits, Dy, Mitra, & Lefebvre, 2004). It continues to ensure extracellular matrix production as chondrocytes line up in longitudinal columns. Its expression increases in these columns, likely delaying cell proliferation arrest and prehypertrophic differentiation. It is abruptly turned off in hypertrophic chondrocytes, but the SOX9 protein survives its RNA (SOX5 and SOX6 have not been tested) and the trio is required for chondrocyte enlargement and expression of specific markers, such as *Col10a1* (Dy et al., 2012; He, Ohba, Hojo, & McMahon, 2016; Smits et al., 2004). *Sox9*-null prehypertrophic chondrocytes die or prematurely convert into osteoblasts. Conversely, chondrocytes overexpressing SOX9 in the *Col10a1* domain remain hypertrophic longer than normally

(Hattori et al., 2010). SOX9 thus maintains the chondrocyte fate throughout hypertrophy. It remains unknown whether SOX5/SOX6 are needed for this activity.

The approach of chromatin immunoprecipitation followed by high-throughput sequencing (ChIP-seq) has greatly helped provide detailed information on the transcriptional actions of the SOX trio. In non-hypertrophic chondrocytes, the trio binds enhancers and super-enhancers (enhancer clusters) associated with hundreds of cartilage-specific genes (Liu & Lefebvre, 2015; Ohba, He, Hojo, & McMahon, 2015). These genes encode extracellular matrix macromolecules (e.g., the collagen types II, IX and XI, aggrecan, and link protein), regulators of this matrix (e.g., chondroitin 4-sulfotransferase-11), and key signaling pathway components (e.g., fibroblast growth factor receptor-3). The trio also positively feedbacks its own genes. SOX9 binds most enhancers at pairs of SOX motifs oriented head-to-head and separated by 3–4 nucleotides. It may also contact enhancers through protein-protein interactions. This indirect mechanism was also proposed for its binding to promoters of broadly expressed genes. SOX5/SOX6 preferentially bind tandem pairs of SOX-like motifs close to SOX9. They thereby strengthen SOX9 binding to DNA and gene transactivation. Recognition motifs for forkhead, RUNT-domain, NFAT, zinc-finger and AP1 family members are enriched in SOX trio-bound enhancers. Accordingly, there is *in vivo* and *in vitro* evidence that GLI factors, which are zinc-finger proteins mediating Hedgehog signaling, functionally interact with SOX9 in proliferating and prehypertrophic chondrocytes, that the JUN and FOSL2 AP1 factors functionally interact with SOX9 in the transition to hypertrophy, and that competition between SOX9 and the forkhead FOXA2 factor may be determining in regulating hypertrophic markers, including *Col10a1* (He et al., 2016; Tan et al., 2018).

SOX2 was recently shown to be expressed, along with its stem cell partners OCT4 and NANOG, in bone fracture callus where hypertrophic chondrocytes transition into osteoblasts and to be involved in this event (Hu et al., 2017). This finding, along with evidence of *Sox2* expression in cartilage growth plates, suggests that SOX2 could have the same role in developmental endochondral ossification and that chondrocytes do not transdifferentiate into osteoblasts but rather revert to a progenitor state before undergoing osteogenesis.

As described earlier, SOXC expression is needed in progenitor cells around cartilage primordia to establish tissue boundaries and induce growth plate formation (Bhattaram et al., 2014). In addition, the SOXC trio also has cell-autonomous roles in growth plate chondrocytes. Its inactivation in chondrocytes results in short growth plates and mild dwarfism (Kato et al., 2015). Chondrocytes form partially disorganized columns and die at a low, but significant rate. The SOXC trio does not appear to control chondrocyte differentiation genes, but rather controls growth plate dynamics by promoting expression of signaling pathway mediators, including non-canonical WNT pathway components.

In conclusion, several SOX proteins are essential at multiple steps of chondrogenesis. SOX9 maintains the chondrocyte fate from progenitors to hypertrophic chondrocytes. It cooperates with SOX5/SOX6 to effect early chondrocyte differentiation and interacts with different transcription factors during growth plate chondrocyte maturation. The SOXC trio ensures proper patterning, growth and maturation of cartilage structures, and SOX2 may ensure growth plate chondrocyte transition into osteoblasts.

6. Roles of SOX genes in osteogenesis

Osteoblasts form bone and osteoclasts resorb bone. Their coordinated actions help ensure proper bone development and adult homeostasis. While no SOX gene is known to be expressed and critical in osteoclasts, several SOX genes are expressed in the osteoblast lineage and may functionally interact with master regulators, namely, RUNX2 (RUNT-domain protein) and OSX/SP7 (zinc-finger protein) (Liu & Lee, 2013).

As described earlier, SOX2 maintains a population of osteoblasts with stem cell properties *in vitro* (Basu-Roy et al., 2010) (Fig. 2C). Mice with inactivated *Sox2* in *Col1a1[2.3kb]Cre*-positive cells, which include osteoblasts, are small and osteopenic, suggesting important roles for SOX2 in osteoblasts too. However, since these mice also have non-skeletal phenotypes that might affect bones, further studies are warranted to definitively ascertain the roles of SOX2 in osteoblasts.

Sox8 and *Sox9* are co-expressed in osteochondroprogenitors (Akiyama et al., 2005; Schmidt et al., 2005). *Sox8*-null mice are osteopenic, likely because of reduced proliferation of progenitor cells and upregulation of *Runx2* leading to precocious differentiation of osteoblasts (Schmidt et al., 2005). Forced expression of SOX8 in differentiating osteoblasts in *Col1a1-SOX8* transgenic mice leads to drastic downregulation of *Runx2* and deficient bone formation. Similarly, *Col1a1-SOX9* transgenic mice are osteopenic and weakly express osteoblast markers (Zhou et al., 2006). *In vitro* experiments led to the proposition that SOX9 inhibits RUNX2 activity through physical interaction. Like SOX8, SOX9 may delay osteoblast differentiation of osteochondroprogenitors, but it remains unknown whether *in vivo* deletion of *Sox9* in these cells causes osteopenia and whether combined deletion of *Sox8* and *Sox9* has more severe consequences.

SOXC genes likely have important roles in osteoblastogenesis too. *Sox11*$^{-/-}$ mice die at birth with underdeveloped intramembranous and endochondral bones (Sock et al., 2004), *Sox4*$^{+/-}$ mice are osteopenic (Nissen-Meyer et al., 2007), and combined inactivation of all SOXC genes in mesodermal progenitors severely impairs skull formation (Bhattaram et al., 2014). *In vitro* studies have shown that *Sox11* and *Sox4* are highly expressed in osteoblast progenitors and that *Sox11*, but not *Sox4,* is downregulated during osteoblastogenesis (Gadi et al., 2013; Nissen-Meyer et al., 2007). *Sox4* knockdown in primary osteoblasts reduces progenitor cell proliferation and delays osteoblast differentiation without affecting *Runx2* expression (Nissen-Meyer et al., 2007). Similarly, *Sox11* knockdown in MC3T3-E1 cells reduces cell numbers and delays osteoblastogenesis (Gadi et al., 2013). As for other SOX genes, additional investigations are needed to pinpoint the specific activities of SOXC proteins in osteoblastogenesis.

7. Regulation of SOX genes and RNAs in skeletal cells

Each SOX gene is expressed in a discrete number of cell types. This pattern is specific to each one, and likely involves complex regulatory mechanisms.

Sox2 expression is upregulated downstream of fibroblast growth factor signaling in calvarium osteoblast progenitors *in vitro* (Basu-Roy et al., 2010). This result is consistent with the importance of FGF signaling in the development of skull and other bones (Ornitz & Marie, 2015), but it remains to be validated *in vivo*.

Disease-causing genomic alterations occurring in the 2-Mb region around *SOX9*, the analysis of topologically associated domains in this region, and transgenic mouse reporter assays with yeast artificial chromosomes have compellingly suggested that *SOX9* transcription is regulated by multiple, widely spread regulatory modules (Franke et al., 2016; Gordon et al., 2009; Wunderle, Critcher, Hastie, Goodfellow, & Schedl, 1998). Accordingly, a dozen enhancers active in chondrocytes at discrete differentiation stages and in other cells have been identified (Bagheri-Fam et al., 2006; Benko et al., 2009; Gonen et al., 2018; Mead et al., 2013; Mochizuki et al., 2018; Yao et al., 2015). SOX9 and SOX5/SOX6 regulate several of these enhancers and STAT3, a transcription factor activated downstream of many pathways, may control a rib cartilage-specific enhancer. The *SOX9* promoter region is insufficient to drive reporter expression in chondrocytes (Bagheri-Fam et al., 2006; Mead et al., 2013), but may nevertheless be involved in *SOX9* upregulation or downregulation in response to signaling pathways. This is suggested by evidence of HIF-1α (hypoxia-inducible factor-1 α) (Amarilio et al., 2007), CREB (cyclic-AMP response element-binding protein) (Piera-Velazquez et al., 2007) and STAT3 (Hall, Murray, Valdez, & Perantoni, 2017) binding to specific recognition motifs. In addition, two long non-coding RNAs may upregulate *SOX9* in chondrocytes. DA125942 (encoded by *CISTR-ACT*) interacts with *PTHLH* in *cis* and with *SOX9* in *trans* (Maass et al., 2012). *RORC* (regulator of chondrogenesis RNA), located 94 kb upstream of *SOX9*, is critical for *SOX9* expression in mesenchymal stem cells *in vitro* and successful differentiation of the cells into chondrocytes (Barter et al., 2017). Finally, the SOX9 RNA level may also be directly regulated by several microRNAs, whose roles remain untested *in vivo* (Lefebvre & Dvir-Ginzberg, 2017).

SOX5 and *SOX6* do not require SOX9 for activation at the onset of chondrogenesis (Liu et al., 2018). Multiple enhancers are active and bound by the SOX trio within and around them in chondrocytes, suggesting positive feedback regulation (Liu & Lefebvre, 2015). Other transcription factors involved in their expression remain unknown. Among signaling pathways, bone morphogenetic protein signaling is required for their expression as well as *Sox9* expression at the onset of chondrogenesis *in vivo* and *in vitro*, but no direct link has been established yet (Uusitalo et al., 2001; Yoon et al., 2005). *Mir-194* and *miR-146b* downregulate the *SOX5* RNA level and affect chondrogenic differentiation of human adipose stem cells and human bone marrow-derived skeletal stem cells *in vitro*, respectively, but their roles are unknown *in vivo* (Budd, de Andres, Sanchez-Elsner, & Oreffo, 2017; Xu, Kang, Liao, & Yu, 2012).

The *cis*-acting elements controlling *SOX4* and *SOX11* expression in skeletogenic cells are uncharacterized. Whereas signaling pathways regulating *SOX11* remain unknown too, *SOX4* expression was shown to be stimulated by parathyroid hormone in osteoblastic cells *in vitro* (Reppe et al., 2000). *SOX4* is upregulated by TGFβ signaling in cancer (David et al., 2016; Lourenco & Coffer, 2017; Vervoort et al., 2018) and its expression in skeletogenic cells that are dependent upon TGFβ signaling suggests that it could be a target of this pathway in skeletogenesis too. Several LncRNAs and miRNAs have been linked to *SOX4* in cancers and the findings suggest that they could also control *SOX4* expression in skeletogenesis. For instance, *LncSOX4* was shown to promote liver tumor-initiating cell self-renewal by binding to the *SOX4* promoter and recruiting STAT3 to transactivation (Chen et al., 2016). This LncRNA was proposed to promote cell proliferation and migration in osteosarcomas through stabilizing β-catenin, a property also attributed to SOX4 (Tian et al., 2017). *miR-188* and *SOX4* expressions inversely correlate in osteosarcomas (Pan, Meng, Liang, & Cao, 2018). The miRNA directly targets SOX4, and its inhibition of cell survival, proliferation and migration is restored by SOX4 overexpression. Similarly, *MiR-129-5p* targets the *SOX4* RNA in chondrosarcomas and may thereby inhibit canonical WNT signaling, cell survival, proliferation and migration (Zhang, Li, Song, & Wang, 2017).

Further studies are clearly needed to fully uncover the mechanisms controlling *SOX* genes in skeletogenesis. These mechanisms are likely more sophisticated than currently appreciated. Their deciphering and the analysis of genetic variants in *cis*-regulatory elements may help uncover the basis of skeletal malformation diseases as well as the complexity of the skeleton, including size and shape differences, among vertebrate species and among human individuals.

8. Post-translational regulation of SOX proteins in skeletal cells

Various types of post-translational modifications have been shown to affect SOX protein stability, intracellular localization, or activity, but few, reviewed below, have been validated in skeletogenesis *in vivo* to this date.

PKA (cAMP-dependent protein kinase A) increases SOX9 activity *in vitro* by phosphorylating the protein at Ser64 (upstream of the dimerization domain) and Ser181 (C-terminal to the HMG domain) (Huang, Zhou, Lefebvre, & de Crombrugghe, 2000). The latter event occurs in growth plate chondrocytes downstream of PTHrP signaling, where it could help delay chondrocyte hypertrophy (Huang, Chung, Kronenberg, & de Crombrugghe, 2001). These phosphorylation events, however, remain untested functionally *in vivo* and could be driven by several kinases downstream of many pathways (Lefebvre & Dvir-Ginzberg, 2017). Interestingly, SOX9 phosphorylation at Ser181 was found necessary for SUMOylation at Lys398 (N-terminal to the transactivation domain) downstream of BMP and canonical WNT signaling (Liu et al., 2013). SUMOylation enhances the ability of overexpressed SOX9 to promote neural crest delamination in chicken embryos. *Ptpn1* (protein-tyrosine phosphatase SHP2) inactivation in limb bud skeletal progenitors was shown to perturb cartilage patterning, growth and endochondral ossification and to increase SOX9 protein, but not RNA level (Zuo et al., 2018). SHP2 was proposed to limit SOX9 protein level and activity by preventing Ser181 phosphorylation and hence Lys398 SUMOylation. In other studies, SOX9 ubiquitination was proposed as an important mechanism to limit SOX9 protein level by inducing proteasomal degradation downstream of canonical WNT signaling (Akiyama et al., 2002). DDRGK1 (DDRGK domain-containing protein 1) was shown to physically interact with SOX9 and thereby to inhibit SOX9 ubiquitination and proteasomal degradation at the onset of chondrogenesis (Egunsola et al., 2017). This mechanism may explain why humans with a homozygous loss-of-function mutation in *DDRGK1* have Shohat-type spondyloepimetaphyseal dysplasia (SEMD).

Many types of post-translational modifications have been described for SOX2 (Ramakrishna, Kim, & Baek, 2014; Sarkar & Hochedlinger, 2013), but their relevance to skeletogenesis remains unknown. In contrast, few modifications have been reported for SOX5, SOX6, SOX8, and the SOXC proteins, and none is directly relevant to skeletogenesis. There is little doubt, however, that these proteins are subjected to post-translational regulation. Supporting this concept, inhibition of SOXC proteasomal degradation was proposed as a major mechanism whereby inflammatory cytokines mediate synovial fibroblast transformation in arthritis (Bhattaram, Muschler, Wixler, & Lefebvre, 2018). A post-translational mechanism was postulated but remains to be identified. The involvement of a similar mechanism for developmental pathways remains to be tested.

These studies and others carried out *in vitro* infer that adequate post-translational modifications of SOX proteins by various mechanisms must occur to ensure proper skeletogenesis. Their identification could help decipher mechanisms underlying skeletal malformation diseases and design treatments for diseases dependent directly or indirectly upon changes in SOX activities.

9. Conclusions and perspectives

The efforts of many research teams over the last three decades have uncovered or postulated central roles, distinct and complementary, for several SOX family members in pivotal cell fate determination and differentiation events in skeletogenesis. The SOXC proteins—SOX4 and

SOX11—control progenitor cell survival, undifferentiated status, and ability to cross talk with other cells to properly pattern, grow and mature skeletal structures. SOX8 maintains osteoblast progenitors at the proliferating, undifferentiated stage. SOX9 is mandatory to keep the chondrocyte fate from early to late differentiation stages and cooperates with SOX5 and SOX6 in driving robust expression of most early chondrocyte differentiation markers. It also cooperates with transcription factors from several families in the activation of stage-specific markers during growth plate chondrocyte maturation. Other roles have been proposed, but remain to be documented *in vivo*. They include roles for SOX2 in the maintenance of skeletal progenitor/stem cells and for SOXC proteins in osteoblast differentiation. This review has focused on skeletal progenitors, chondrocytes, and osteoblasts, but additional cells contribute to skeletogenesis, namely, osteoclasts, synovial fibroblasts, tenocytes, and intervertebral disc cells. Further studies are necessary to determine whether and how SOX proteins govern these cells. The deciphering of the molecular actions of SOX9 and SOX5/SOX6 in chondrogenesis has significantly progressed in recent years largely thanks to high-throughput sequencing approaches. Further studies are needed to complete the knowledge of the actions of SOX proteins in skeletogenesis, to better understand functional interactions with other transcription factors, to determine whether they have pioneer roles to poise genomes for cell fate changes, and to define non-transcriptional activities. To date, our understanding of modes of regulation of SOX genes and proteins remains in its infancy. Several pieces of a certainly large and complex puzzle have been uncovered and assembled, but more systematic and wide-ranging proteomic and genetic approaches *in vitro* and *in vivo* are warranted to identify all pieces and completely assemble the puzzle. Current knowledge and constantly evolving experimental approaches allow us to predict that ongoing and future research efforts on SOX genes and proteins will help identify key nodes in the normal process of skeletogenesis, the basis of normal and pathological skeletal variations among individuals, and targets for the development of efficient treatments for skeletal malformation disorders as well as adult-onset degeneration diseases and cancers involving similar mechanisms.

Acknowledgments

We thank B. Olsen for advice on the manuscript. Work in the Lefebvre lab was supported by the NIH/NIAMS Grants AR68308 and AR72649 (to V.L.).

References

Akiyama, H., Chaboissier, M. C., Martin, J. F., Schedl, A., & de Crombrugghe, B. (2002). The transcription factor Sox9 has essential roles in successive steps of the chondrocyte differentiation pathway and is required for expression of Sox5 and Sox6. *Genes & Development, 16*, 2813–2828.

Akiyama, H., Kim, J. E., Nakashima, K., Balmes, G., Iwai, N., Deng, J. M., et al. (2005). Osteo-chondroprogenitor cells are derived from Sox9 expressing precursors. *Proceedings of the National Academy of Sciences of the United States of America, 102*, 14665–14670.

Akiyama, H., Lyons, J. P., Mori-Akiyama, Y., Yang, X., Zhang, R., Zhang, Z., et al. (2004). Interactions between Sox9 and beta-catenin control chondrocyte differentiation. *Genes & Development, 18*, 1072–1087.

Amarilio, R., Viukov, S. V., Sharir, A., Eshkar-Oren, I., Johnson, R. S., & Zelzer, E. (2007). HIF1alpha regulation of Sox9 is necessary to maintain differentiation of hypoxic prechondrogenic cells during early skeletogenesis. *Development, 134*, 3917–3928.

Bagheri-Fam, S., Barrionuevo, F., Dohrmann, U., Gunther, T., Schule, R., Kemler, R., et al. (2006). Long-range upstream and downstream enhancers control distinct subsets of the complex spatiotemporal Sox9 expression pattern. *Developmental Biology, 291*, 382–397.

Barrionuevo, F., & Scherer, G. (2010). SOX E genes: SOX9 and SOX8 in mammalian testis development. *The International Journal of Biochemistry & Cell Biology, 42*, 433–436.

Barrionuevo, F., Taketo, M. M., Scherer, G., & Kispert, A. (2006). Sox9 is required for notochord maintenance in mice. *Developmental Biology, 295*, 128–140.

Barter, M. J., Gomez, R., Hyatt, S., Cheung, K., Skelton, A. J., Xu, Y., et al. (2017). The long non-coding RNA ROCR contributes to SOX9 expression and chondrogenic differentiation of human mesenchymal stem cells. *Development, 144*, 4510–4521.

Basu-Roy, U., Ambrosetti, D., Favaro, R., Nicolis, S. K., Mansukhani, A., & Basilico, C. (2010). The transcription factor Sox2 is required for osteoblast self-renewal. *Cell Death and Differentiation, 17*, 1345–1353.

Benko, S., Fantes, J. A., Amiel, J., Kleinjan, D. J., Thomas, S., Ramsay, J., et al. (2009). Highly conserved non-coding elements on either side of SOX9 associated with pierre robin sequence. *Nature Genetics, 41*, 359–364.

Berendsen, A. D., & Olsen, B. R. (2015). Bone development. *Bone, 80*, 14–18.

Bergsland, M., Ramskold, D., Zaouter, C., Klum, S., Sandberg, R., & Muhr, J. (2011). Sequentially acting Sox transcription factors in neural lineage development. *Genes & Development, 25*, 2453–2464.

Bhattaram, P., Muschler, G., Wixler, V., & Lefebvre, V. (2018). Inflammatory cytokines stabilize SOXC transcription factors to mediate the transformation of fibroblast-like synoviocytes in arthritic disease. *Arthritis & Rhematology, 70*, 371–382.

Bhattaram, P., Penzo-Mendez, A., Kato, K., Bandyopadhyay, K., Gadi, A., Taketo, M. M., et al. (2014). SOXC proteins amplify canonical WNT signaling to secure nonchondrocytic fates in skeletogenesis. *The Journal of Cell Biology, 207*, 657–671.

Bhattaram, P., Penzo-Mendez, A., Sock, E., Colmenares, C., Kaneko, K. J., Vassilev, A., et al. (2010). Organogenesis relies on SoxC transcription factors for the survival of neural and mesenchymal progenitors. *Nature Communications, 1*, 9.

Bi, W., Deng, J. M., Zhang, Z., Behringer, R. R., & de Crombrugghe, B. (1999). Sox9 is required for cartilage formation. *Nature Genetics, 22*, 85–89.

Budd, E., de Andres, M. C., Sanchez-Elsner, T., & Oreffo, R. O. C. (2017). MiR-146b is down-regulated during the chondrogenic differentiation of human bone marrow derived skeletal stem cells and up-regulated in osteoarthritis. *Scientific Reports*, 7, 46704.

Chan, C. K. F., Gulati, G. S., Sinha, R., Tompkins, J. V., Lopez, M., Carter, A. C., et al. (2018). Identification of the human skeletal stem cell. *Cell, 175*, 43–56.e21.

Chan, C. K., Seo, E. Y., Chen, J. Y., Lo, D., McArdle, A., Sinha, R., et al. (2015). Identification and specification of the mouse skeletal stem cell. *Cell, 160*, 285–298.

Chen, Z. Z., Huang, L., Wu, Y. H., Zhai, W. J., Zhu, P. P., & Gao, Y. F. (2016). LncSox4 promotes the self-renewal of liver tumour-initiating cells through Stat3-mediated Sox4 expression. *Nature Communications*, *7*, 12598.

Corbani, S., Chouery, E., Eid, B., Jalkh, N., Ghoch, J. A., & Megarbane, A. (2011). Mild campomelic dysplasia: Report on a case and review. *Molecular Syndromology*, *1*, 163–168.

David, C. J., Huang, Y. H., Chen, M., Su, J., Zou, Y., Bardeesy, N., et al. (2016). TGF-beta tumor suppression through a lethal EMT. *Cell*, *164*, 1015–1030.

Dy, P., Penzo-Mendez, A., Wang, H., Pedraza, C. E., Macklin, W. B., & Lefebvre, V. (2008). The three SoxC proteins—Sox4, Sox11 and Sox12—exhibit overlapping expression patterns and molecular properties. *Nucleic Acids Research*, *36*, 3101–3117.

Dy, P., Wang, W., Bhattaram, P., Wang, Q., Wang, L., Ballock, R. T., et al. (2012). Sox9 directs hypertrophic maturation and blocks osteoblast differentiation of growth plate chondrocytes. *Developmental Cell*, *22*, 597–609.

Egunsola, A. T., Bae, Y., Jiang, M. M., Liu, D. S., Chen-Evenson, Y., Bertin, T., et al. (2017). Loss of DDRGK1 modulates SOX9 ubiquitination in spondyloepimetaphyseal dysplasia. *The Journal of Clinical Investigation*, *127*, 1475–1484.

Franke, M., Ibrahim, D. M., Andrey, G., Schwarzer, W., Heinrich, V., Schopflin, R., et al. (2016). Formation of new chromatin domains determines pathogenicity of genomic duplications. *Nature*, *538*, 265–269.

Gadi, J., Jung, S. H., Lee, M. J., Jami, A., Ruthala, K., Kim, K. M., et al. (2013). The transcription factor protein Sox11 enhances early osteoblast differentiation by facilitating proliferation and the survival of mesenchymal and osteoblast progenitors. *The Journal of Biological Chemistry*, *288*, 25400–25413.

Giese, K., Amsterdam, A., & Grosschedl, R. (1991). DNA-binding properties of the HMG domain of the lymphoid-specific transcriptional regulator LEF-1. *Genes & Development*, *5*, 2567–2578.

Gonen, N., Futtner, C. R., Wood, S., Garcia-Moreno, S. A., Salamone, I. M., Samson, S. C., et al. (2018). Sex reversal following deletion of a single distal enhancer of Sox9. *Science*, *360*, 1469–1473.

Gordon, C. T., Tan, T. Y., Benko, S., Fitzpatrick, D., Lyonnet, S., & Farlie, P. G. (2009). Long-range regulation at the SOX9 locus in development and disease. *Journal of Medical Genetics*, *46*, 649–656.

Haldin, C. E., & LaBonne, C. (2010). SoxE factors as multifunctional neural crest regulatory factors. *The International Journal of Biochemistry & Cell Biology*, *42*, 441–444.

Hall, M. D., Murray, C. A., Valdez, M. J., & Perantoni, A. O. (2017). Mesoderm-specific Stat3 deletion affects expression of Sox9 yielding Sox9-dependent phenotypes. *PLoS Genetics*, *13*, e1006610.

Hata, K., Takahata, Y., Murakami, T., & Nishimura, R. (2017). Transcriptional network controlling endochondral ossification. *Journal of Bone Metabolism*, *24*, 75–82.

Hattori, T., Muller, C., Gebhard, S., Bauer, E., Pausch, F., Schlund, B., et al. (2010). SOX9 is a major negative regulator of cartilage vascularization, bone marrow formation and endochondral ossification. *Development*, *137*, 901–911.

He, X., Ohba, S., Hojo, H., & McMahon, A. P. (2016). AP-1 family members act with Sox9 to promote chondrocyte hypertrophy. *Development*, *143*, 3012–3023.

Hempel, A., Pagnamenta, A. T., Blyth, M., Mansour, S., McConnell, V., Kou, I., et al. (2016). Deletions and de novo mutations of SOX11 are associated with a neurodevelopmental disorder with features of coffin-siris syndrome. *Journal of Medical Genetics*, *53*, 152–162.

Henry, S. P., Liang, S., Akdemir, K. C., & de Crombrugghe, B. (2012). The postnatal role of Sox9 in cartilage. *Journal of Bone and Mineral Research*, *27*, 2511–2525.

Hoser, M., Potzner, M. R., Koch, J. M., Bosl, M. R., Wegner, M., & Sock, E. (2008). Sox12 deletion in the mouse reveals nonreciprocal redundancy with the related Sox4 and Sox11 transcription factors. *Molecular and Cellular Biology*, *28*, 4675–4687.

Hu, D. P., Ferro, F., Yang, F., Taylor, A. J., Chang, W., Miclau, T., et al. (2017). Cartilage to bone transformation during fracture healing is coordinated by the invading vasculature and induction of the core pluripotency genes. *Development*, *144*, 221–234.

Huang, W., Chung, U. I., Kronenberg, H. M., & de Crombrugghe, B. (2001). The chondrogenic transcription factor Sox9 is a target of signaling by the parathyroid hormone-related peptide in the growth plate of endochondral bones. *Proceedings of the National Academy of Sciences of the United States of America*, *98*, 160–165.

Huang, W., Zhou, X., Lefebvre, V., & de Crombrugghe, B. (2000). Phosphorylation of SOX9 by cyclic AMP-dependent protein kinase A enhances SOX9's ability to transactivate a Col2a1 chondrocyte-specific enhancer. *Molecular and Cellular Biology*, *20*, 4149–4158.

Ikeda, T., Kamekura, S., Mabuchi, A., Kou, I., Seki, S., Takato, T., et al. (2004). The combination of SOX5, SOX6, and SOX9 (the SOX trio) provides signals sufficient for induction of permanent cartilage. *Arthritis and Rheumatism*, *50*, 3561–3573.

Ikegami, D., Akiyama, H., Suzuki, A., Nakamura, T., Nakano, T., Yoshikawa, H., et al. (2011). Sox9 sustains chondrocyte survival and hypertrophy in part through Pik3ca-Akt pathways. *Development*, *138*, 1507–1519.

Iwafuchi-Doi, M., & Zaret, K. S. (2016). Cell fate control by pioneer transcription factors. *Development*, *143*, 1833–1837.

Jauch, R., Ng, C. K., Narasimhan, K., & Kolatkar, P. R. (2012). The crystal structure of the Sox4 HMG domain-DNA complex suggests a mechanism for positional interdependence in DNA recognition. *The Biochemical Journal*, *443*, 39–47.

Kamachi, Y., & Kondoh, H. (2013). Sox proteins: Regulators of cell fate specification and differentiation. *Development*, *140*, 4129–4144.

Kan, A., Ikeda, T., Fukai, A., Nakagawa, T., Nakamura, K., Chung, U. I., et al. (2013). SOX11 contributes to the regulation of GDF5 in joint maintenance. *BMC Developmental Biology*, *13*, 4.

Kato, K., Bhattaram, P., Penzo-Mendez, A., Gadi, A., & Lefebvre, V. (2015). SOXC transcription factors induce cartilage growth plate formation in mouse embryos by promoting noncanonical WNT signaling. *Journal of Bone and Mineral Research*, *30*, 1560–1571.

Khan, U., Study, D., Baker, E., & Clayton-Smith, J. (2018). Observation of cleft palate in an individual with SOX11 mutation: Indication of a role for SOX11 in human palatogenesis. *The Cleft Palate-Craniofacial Journal*, *55*, 456–461.

Kozhemyakina, E., Lassar, A. B., & Zelzer, E. (2015). A pathway to bone: Signaling molecules and transcription factors involved in chondrocyte development and maturation. *Development*, *142*, 817–831.

Kuo, P. Y., Leshchenko, V. V., Fazzari, M. J., Perumal, D., Gellen, T., He, T., et al. (2015). High-resolution chromatin immunoprecipitation (ChIP) sequencing reveals novel binding targets and prognostic role for SOX11 in mantle cell lymphoma. *Oncogene*, *34*, 1231–1240.

Lamb, A. N., Rosenfeld, J. A., Neill, N. J., Talkowski, M. E., Blumenthal, I., Girirajan, S., et al. (2012). Haploinsufficiency of SOX5 at 12p12.1 is associated with developmental delays with prominent language delay, behavior problems, and mild dysmorphic features. *Human Mutation*, *33*, 728–740.

Lefebvre, V., & Dvir-Ginzberg, M. (2017). SOX9 and the many facets of its regulation in the chondrocyte lineage. *Connective Tissue Research*, *58*, 2–14.

Lefebvre, V., Li, P., & de Crombrugghe, B. (1998). A new long form of Sox5 (L-Sox5), Sox6 and Sox9 are coexpressed in chondrogenesis and cooperatively activate the type II collagen gene. *The EMBO Journal*, *17*, 5718–5733.

Liu, C. F., Angelozzi, M., Haseeb, A., & Lefebvre, V. (2018). SOX9 is dispensable for the initiation of epigenetic remodeling and the activation of marker genes at the onset of chondrogenesis. *Development, 145*, dev164459.

Liu, T. M., & Lee, E. H. (2013). Transcriptional regulatory cascades in Runx2-dependent bone development. *Tissue Engineering. Part B, Reviews, 19*, 254–263.

Liu, C. F., & Lefebvre, V. (2015). The transcription factors SOX9 and SOX5/SOX6 cooperate genome-wide through super-enhancers to drive chondrogenesis. *Nucleic Acids Research, 43*, 8183–8203.

Liu, J. A., Wu, M. H., Yan, C. H., Chau, B. K., So, H., Ng, A., et al. (2013). Phosphorylation of Sox9 is required for neural crest delamination and is regulated downstream of BMP and canonical Wnt signaling. *Proceedings of the National Academy of Sciences of the United States of America, 110*, 2882–2887.

Lourenco, A. R., & Coffer, P. J. (2017). SOX4: Joining the master regulators of epithelial-to-mesenchymal transition? *Trends in Cancer, 3*, 571–582.

Maass, P. G., Rump, A., Schulz, H., Stricker, S., Schulze, L., Platzer, K., et al. (2012). A misplaced lncRNA causes brachydactyly in humans. *The Journal of Clinical Investigation, 122*, 3990–4002.

Mandalos, N. P., & Remboutsika, E. (2017). Sox2: To crest or not to crest? *Seminars in Cell & Developmental Biology, 63*, 43–49.

Mead, T. J., Wang, Q., Bhattaram, P., Dy, P., Afelik, S., Jensen, J., et al. (2013). A far-upstream (−70 kb) enhancer mediates Sox9 auto-regulation in somatic tissues during development and adult regeneration. *Nucleic Acids Research, 41*, 4459–4469.

Mizuhashi, K., Ono, W., Matsushita, Y., Sakagami, N., Takahashi, A., Saunders, T. L., et al. (2018). Resting zone of the growth plate houses a unique class of skeletal stem cells. *Nature, 563*, 254–258.

Mochizuki, Y., Chiba, T., Kataoka, K., Yamashita, S., Sato, T., Kato, T., et al. (2018). Combinatorial CRISPR/Cas9 approach to elucidate a far-upstream enhancer complex for tissue-specific Sox9 expression. *Developmental Cell, 46*, 794–806 e796.

Mori-Akiyama, Y., Akiyama, H., Rowitch, D. H., & de Crombrugghe, B. (2003). Sox9 is required for determination of the chondrogenic cell lineage in the cranial neural crest. *Proceedings of the National Academy of Sciences of the United States of America, 100*, 9360–9365.

Nesbitt, A., Bhoj, E. J., McDonald Gibson, K., Yu, Z., Denenberg, E., Sarmady, M., et al. (2015). Exome sequencing expands the mechanism of SOX5-associated intellectual disability: A case presentation with review of sox-related disorders. *American Journal of Medical Genetics. Part A, 167A*, 2548–2554.

Ng, L. J., Wheatley, S., Muscat, G. E., Conway-Campbell, J., Bowles, J., Wright, E., et al. (1997). SOX9 binds DNA, activates transcription, and coexpresses with type II collagen during chondrogenesis in the mouse. *Developmental Biology, 183*, 108–121.

Nissen-Meyer, L. S., Jemtland, R., Gautvik, V. T., Pedersen, M. E., Paro, R., Fortunati, D., et al. (2007). Osteopenia, decreased bone formation and impaired osteoblast development in Sox4 heterozygous mice. *Journal of Cell Science, 120*, 2785–2795.

Ohba, S., He, X., Hojo, H., & McMahon, A. P. (2015). Distinct transcriptional programs underlie Sox9 regulation of the mammalian chondrocyte. *Cell Reports, 12*, 229–243.

Ornitz, D. M., & Marie, P. J. (2015). Fibroblast growth factor signaling in skeletal development and disease. *Genes & Development, 29*, 1463–1486.

Pan, L., Meng, L., Liang, F., & Cao, L. (2018). miR188 suppresses tumor progression by targeting SOX4 in pediatric osteosarcoma. *Molecular Medicine Reports, 18*, 441–446.

Phochanukul, N., & Russell, S. (2010). No backbone but lots of Sox: Invertebrate sox genes. *The International Journal of Biochemistry & Cell Biology, 42*, 453–464.

Piera-Velazquez, S., Hawkins, D. F., Whitecavage, M. K., Colter, D. C., Stokes, D. G., & Jimenez, S. A. (2007). Regulation of the human SOX9 promoter by Sp1 and CREB. *Experimental Cell Research, 313*, 1069–1079.

Ramakrishna, S., Kim, K. S., & Baek, K. H. (2014). Posttranslational modifications of defined embryonic reprogramming transcription factors. *Cellular Reprogramming, 16*, 108–120.

Reiprich, S., & Wegner, M. (2015). From CNS stem cells to neurons and glia: Sox for everyone. *Cell and Tissue Research, 359*, 111–124.

Reppe, S., Rian, E., Jemtland, R., Olstad, O. K., Gautvik, V. T., & Gautvik, K. M. (2000). Sox-4 messenger RNA is expressed in the embryonic growth plate and regulated via the parathyroid hormone/parathyroid hormone-related protein receptor in osteoblast-like cells. *Journal of Bone and Mineral Research, 15*, 2402–2412.

Sarkar, A., & Hochedlinger, K. (2013). The sox family of transcription factors: Versatile regulators of stem and progenitor cell fate. *Cell Stem Cell, 12*, 15–30.

Schepers, G. E., Teasdale, R. D., & Koopman, P. (2002). Twenty pairs of sox: Extent, homology, and nomenclature of the mouse and human sox transcription factor gene families. *Developmental Cell, 3*, 167–170.

Schmidt, K., Schinke, T., Haberland, M., Priemel, M., Schilling, A. F., Mueldner, C., et al. (2005). The high mobility group transcription factor Sox8 is a negative regulator of osteoblast differentiation. *The Journal of Cell Biology, 168*, 899–910.

Smits, P., Dy, P., Mitra, S., & Lefebvre, V. (2004). Sox5 and Sox6 are needed to develop and maintain source, columnar, and hypertrophic chondrocytes in the cartilage growth plate. *The Journal of Cell Biology, 164*, 747–758.

Smits, P., & Lefebvre, V. (2003). Sox5 and Sox6 are required for notochord extracellular matrix sheath formation, notochord cell survival and development of the nucleus pulposus of intervertebral discs. *Development, 130*, 1135–1148.

Smits, P., Li, P., Mandel, J., Zhang, Z., Deng, J. M., Behringer, R. R., et al. (2001). The transcription factors L-Sox5 and Sox6 are essential for cartilage formation. *Developmental Cell, 1*, 277–290.

Sock, E., Rettig, S. D., Enderich, J., Bosl, M. R., Tamm, E. R., & Wegner, M. (2004). Gene targeting reveals a widespread role for the high-mobility-group transcription factor Sox11 in tissue remodeling. *Molecular and Cellular Biology, 24*, 6635–6644.

Sock, E., Schmidt, K., Hermanns-Borgmeyer, I., Bosl, M. R., & Wegner, M. (2001). Idiopathic weight reduction in mice deficient in the high-mobility-group transcription factor Sox8. *Molecular and Cellular Biology, 21*, 6951–6959.

Sommer, L. (2011). Generation of melanocytes from neural crest cells. *Pigment Cell & Melanoma Research, 24*, 411–421.

Tagariello, A., Heller, R., Greven, A., Kalscheuer, V. M., Molter, T., Rauch, A., et al. (2006). Balanced translocation in a patient with craniosynostosis disrupts the SOX6 gene and an evolutionarily conserved non-transcribed region. *Journal of Medical Genetics, 43*, 534–540.

Tan, Z., Niu, B., Tsang, K. Y., Melhado, I. G., Ohba, S., He, X., et al. (2018). Synergistic co-regulation and competition by a SOX9-GLI-FOXA phasic transcriptional network coordinate chondrocyte differentiation transitions. *PLoS Genetics, 14*, e1007346.

Tian, Z., Yang, G., Jiang, P., Zhang, L., Wang, J., & Sun, S. (2017). Long non-coding RNA Sox4 promotes proliferation and migration by activating Wnt/beta-catenin signaling pathway in osteosarcoma. *Pharmazie, 72*, 537–542.

Topol, L., Chen, W., Song, H., Day, T. F., & Yang, Y. (2009). Sox9 inhibits Wnt signaling by promoting beta-catenin phosphorylation in the nucleus. *The Journal of Biological Chemistry, 284*, 3323–3333.

Tsurusaki, Y., Koshimizu, E., Ohashi, H., Phadke, S., Kou, I., Shiina, M., et al. (2014). De novo SOX11 mutations cause Coffin-Siris syndrome. *Nature Communications, 5*, 4011.

Unger, S., Scherer, G., & Superti-Furga, A. (2008). Campomelic dysplasia. In *GeneReviews* (pp. 1–17). Seattle: University of Washington. [updated: 2013].
Uusitalo, H., Hiltunen, A., Ahonen, M., Gao, T. J., Lefebvre, V., Harley, V., et al. (2001). Accelerated up-regulation of L-Sox5, Sox6, and Sox9 by BMP-2 gene transfer during murine fracture healing. *Journal of Bone and Mineral Research, 16*, 1837–1845.
Vervoort, S. J., Lourenco, A. R., Tufegdzic Vidakovic, A., Mocholi, E., Sandoval, J. L., Rueda, O. M., et al. (2018). SOX4 can redirect TGF-beta-mediated SMAD3-transcriptional output in a context-dependent manner to promote tumorigenesis. *Nucleic Acids Research, 46*, 9578–9590.
Wunderle, V. M., Critcher, R., Hastie, N., Goodfellow, P. N., & Schedl, A. (1998). Deletion of long-range regulatory elements upstream of SOX9 causes campomelic dysplasia. *Proceedings of the National Academy of Sciences of the United States of America, 95*, 10649–10654.
Xu, J., Kang, Y., Liao, W. M., & Yu, L. (2012). MiR-194 regulates chondrogenic differentiation of human adipose-derived stem cells by targeting Sox5. *PLoS One, 7*, e31861.
Yao, B., Wang, Q., Liu, C. F., Bhattaram, P., Li, W., Mead, T. J., et al. (2015). The SOX9 upstream region prone to chromosomal aberrations causing campomelic dysplasia contains multiple cartilage enhancers. *Nucleic Acids Research, 43*, 5394–5408.
Yoon, B. S., Ovchinnikov, D. A., Yoshii, I., Mishina, Y., Behringer, R. R., & Lyons, K. M. (2005). Bmpr1a and Bmpr1b have overlapping functions and are essential for chondrogenesis in vivo. *Proceedings of the National Academy of Sciences of the United States of America, 102*, 5062–5067.
Zawerton, A., Yao, B., Yeager, J. P., Pippucci, T., Haseeb, A., Smith, J. D., et al. (2019). De novo SOX4 variants cause a neurodevelopmental diseases with mild dysmorphism. *American Journal of Human Genetics, 104*, 246–259.
Zhang, P., Li, J., Song, Y., & Wang, X. (2017). MiR-129-5p inhibits proliferation and invasion of chondrosarcoma cells by regulating SOX4/Wnt/beta-catenin signaling pathway. *Cellular Physiology and Biochemistry, 42*, 242–253.
Zhou, G., Zheng, Q., Engin, F., Munivez, E., Chen, Y., Sebald, E., et al. (2006). Dominance of SOX9 function over RUNX2 during skeletogenesis. *Proceedings of the National Academy of Sciences of the United States of America, 103*, 19004–19009.
Zuo, C., Wang, L., Kamalesh, R. M., Bowen, M. E., Moore, D. C., Dooner, M. S., et al. (2018). SHP2 regulates skeletal cell fate by modifying SOX9 expression and transcriptional activity. *Bone Research, 6*, 12.

CHAPTER EIGHT

Fibroblast growth factors in skeletal development

David M. Ornitz[a,*], Pierre J. Marie[b]
[a]Department of Developmental Biology, Washington University School of Medicine, St. Louis, MO, United States
[b]UMR-1132 Inserm (Institut national de la Santé et de la Recherche Médicale) and University Paris Diderot, Sorbonne Paris Cité, Hôpital Lariboisière, Paris, France
*Corresponding author: e-mail address: dornitz@wustl.edu

Contents

Abstract

Fibroblast growth factors (FGFs) and their receptors (FGFRs) are expressed throughout all stages of skeletal development. In the limb bud and in cranial mesenchyme, FGF signaling is important for formation of mesenchymal condensations that give rise to bone. Once skeletal elements are initiated and patterned, FGFs regulate both endochondral and intramembranous ossification programs. In this chapter, we review functions of the FGF signaling pathway during these critical stages of skeletogenesis, and explore skeletal malformations in humans that are caused by mutations in FGF signaling molecules.

Current Topics in Developmental Biology, Volume 133
ISSN 0070-2153
https://doi.org/10.1016/bs.ctdb.2018.11.020

1. Fibroblast growth factor signaling pathways

Fibroblast growth factor signaling has potent effects on many cell types, and are thus regulated at multiple levels (Brewer, Mazot, & Soriano, 2016; Ornitz & Itoh, 2015; Turner & Grose, 2010). At the transcriptional level, FGF ligands and FGF receptors (FGFRs) have precise tissue- and cell-specific expression patterns. Additionally, the level of expression and timing of expression during development and in response to physiological or pathological stress is tightly regulated. Once FGF ligands are made and released from cells, their access to responding cells is regulated through binding to extracellular matrix heparan sulfate proteoglycans (HSPGs) and FGF binding proteins, which control diffusion through the extracellular matrix and serve as cofactors for receptor binding and activation (Belov & Mohammadi, 2013; Ornitz, 2000; Ornitz & Itoh, 2015; Shimokawa et al., 2011; Wu et al., 2003; Xu & Esko, 2014).

FGFRs are a subclass of receptor tyrosine kinases that have an extracellular ligand binding domain, a single transmembrane domain, and an intracellular tyrosine kinase domain (Belov & Mohammadi, 2013; Lemmon & Schlessinger, 2010). Activation of an FGFR is induced by high affinity ligand binding which results in dimerization of the tyrosine kinase domains and trans-phosphorylation of tyrosine residues, which creates binding sites for adaptor proteins and activates tyrosine kinase activity. FGFR substrate 2α (FRS2α) is a docked adaptor protein that is phosphorylated by the activated FGFR kinase domain. Other adaptor proteins that are recruited to the FGFR kinase domain include phospholipase Cγ (PLCγ) and signal transducer and activator of transcription (STAT) 1, 3, and 5 (Belov & Mohammadi, 2013; Brewer et al., 2016; Furdui, Lew, Schlessinger, & Anderson, 2006; Goetz & Mohammadi, 2013; Ornitz & Itoh, 2015). The phosphorylation of FRS2α leads to the recruitment of the adaptor protein, GRB2, which activates the MAP kinase and PI3K-AKT signaling pathways (Kouhara et al., 1997; Lamothe et al., 2004). FGFR signaling can activate several MAP kinases including ERK1/2, JNK, and p38 (House, Branch, Newman, Doetschman, & Schultz Jel, 2005; Kanazawa et al., 2010; Liao et al., 2007; Tan et al., 1996; Tsang & Dawid, 2004). FGFR signaling is also inhibited by GRB2, which blocks PLCγ binding to FGFRs, and Sprouty (Spry), which interacts with GRB2 to block MAPK and PI3K signaling (Hanafusa, Torii, Yasunaga, & Nishida, 2002; Timsah et al., 2014). GRB14 is an adaptor protein that binds to FGFR phospho-tyrosine 766 and interferes with phosphorylation and activation of PLCγ (Browaeys-Poly et al., 2010). Further downstream of the FGFR, SEF (Similar Expression to FGF) antagonizes the MAPK pathway by interacting with MEK (Torii, Kusakabe, Yamamoto, Maekawa, & Nishida, 2004) and DUSP6 (Dual-specificity phosphatase 6), which limits MAPK signaling through dephosphorylation of ERK1/2 (Camps et al., 1998).

2. FGF/FGFR expression

2.1 Expression of FGF and FGF receptors in the developing appendicular skeleton

In the developing appendicular skeleton, *Fgf* ligands and receptors are expressed and function at all stages, from formation of the limb bud through growth, remodeling, homeostasis, and repair of mature bone. In the distal limb bud, *Fgfr1* and *Fgfr2* are present in mesenchymal cells, before any morphological or molecular indication of a mesenchymal condensation (also called a chondrogenic condensation) (Orr-Urtreger, Givol, Yayon, Yarden, & Lonai, 1991; Sheeba, Andrade, Duprez, & Palmeirim, 2010). At this precondensation stage, expression of *Fgfr3* and *Fgfr4* is not detected (Sheeba et al., 2010).

The apical ectodermal ridge (AER) is a signaling center at the distal edge of the limb bud. The AER expresses several FGFs (primarily FGF4 and FGF8 but also FGF2, FGF9 and FGF17). One model of AER FGF function suggests that AER FGFs signal to limb mesenchymal FGFRs and function to delay cell differentiation, increasing the time cells have to proliferate, and thus promoting limb bud outgrowth (Martin, 1998; Sun et al., 2000; Tabin & Wolpert, 2007). In this model, mesenchymal cells begin to differentiate when they are too far away from the AER to receive an AER derived FGF signal (Benazet & Zeller, 2009; Tabin & Wolpert, 2007). Differentiation of the limb mesenchyme that is out of range of AER FGFs results in the formation of a mesenchymal (chondrogenic) condensation, the primary event that initiates appendicular skeletogenesis.

The formation of a chondrogenic condensation is marked by expression of Sox9 and increased expression of *Fgfr2* (compared to the surrounding mesenchyme) (Delezoide et al., 1998; Eswarakumar et al., 2002; Orr-Urtreger et al., 1991; Peters, Werner, Chen, & Williams, 1992; Sheeba et al., 2010; Szebenyi, Savage, Olwin, & Fallon, 1995; Yu & Ornitz, 2008). *Fgfr1* remains more uniformly expressed throughout limb bud mesenchyme. *Fgfr3* and *Fgfr4* are excluded from distal limb bud mesenchyme; however, these *Fgfrs* are expressed in more proximal locations in the growing limb in developing muscle tissue (Delezoide et al., 1998; Orr-Urtreger et al., 1991; Peters et al., 1992; Sheeba et al., 2010; Szebenyi et al., 1995). The perichondrium and periosteum are derived from cells in the periphery of the condensation. These cells express both *Fgfr1* and *Fgfr2* (Delezoide et al., 1998; Eswarakumar et al., 2002). Centrally, cells commit to a chondrogenic fate and begin to express *Fgfr3* along with Sox9 and type II collagen (Peters et al., 1992; Peters, Ornitz, Werner, & Williams, 1993; Purcell et al., 2009). As chondrocytes begin to hypertrophy at the center of the developing skeletal segment along the proximal-distal axis, *Fgfr3* expression is decreased and *Fgfr1* expression is increased (Deng, Wynshaw-Boris, Zhou, Kuo, & Leder, 1996; Jacob, Smith, Partanen, & Ornitz, 2006; Karolak, Yang, & Elefteriou, 2015; Naski, Colvin, Coffin, & Ornitz, 1998; Peters et al., 1993, 1992).

Although the initiation of the mesenchymal chondrogenic condensation requires escape from AER FGFs, the subsequent development of the condensation is at least partially dependent on FGFR signaling (Kumar & Lassar, 2014; Mariani, Ahn, & Martin, 2008; Murakami, Kan, McKeehan, & de Crombrugghe, 2000; Yu & Ornitz, 2008). In support of this idea, FGF signaling increases Sox9 expression in primary chondrocytes and in undifferentiated mesenchymal cell lines (Murakami et al., 2000; Shung, Ota, Zhou, Keene, & Hurlin, 2012). Additionally, ERK1/2 activation maintains competence of limb bud mesenchyme to differentiate into chondrocytes by blocking Wnt-induced methylation and silencing of the *Sox9* promoter (Kumar & Lassar, 2014; Ten Berge, Brugmann, Helms, & Nusse, 2008). *Fgfr3* expression in proliferating chondrocytes is maintained through Sox9 binding sites in the *Fgfr3* gene (Oh et al., 2014). The mechanisms that regulate the transition from AER derived FGF signaling to local FGF signaling in condensing mesenchyme and FGF signaling in the skeletal primordium are not known. However, factors that could regulate this transition include proximity of ligand sources and responding tissue, ligand binding specificity of different FGFs and FGFRs, and regulation of heparan sulfate sulfation patterns which could regulate FGF diffusion through the extracellular matrix and the binding affinity to FGFRs (Nogami et al., 2004; Ornitz, 2000).

The perichondrium and periosteum that will give rise to the bone collar and cortical bone express *Fgfr1* in mesenchymal progenitors and FGFR2 in differentiating osteoblasts (Britto, Evans, Hayward, & Jones, 2001; Coutu, Francois, & Galipeau, 2011; Jacob et al., 2006; Molteni,

Modrowski, Hott, & Marie, 1999b; Ohbayashi et al., 2002). *Fgfr3* is expressed more intensely in chondroprogenitor cells located in the groove of Ranvier and ring of LaCroix (Robinson et al., 1999) and FGFR1 and FGFR3 are expressed in mouse and human articular chondrocytes (Fig. 1A) (Weng et al., 2012; Yan et al., 2011).

The growth plate is established when chondrocytes in the center of the mesenchymal condensation begin to hypertrophy and when vascular invasion of the hypertrophic zone chondrocytes forms a primary ossification center. These immature chondroprogenitors express FGFR3. In the established growth plate, *Fgfr3* expression remains high in proliferating

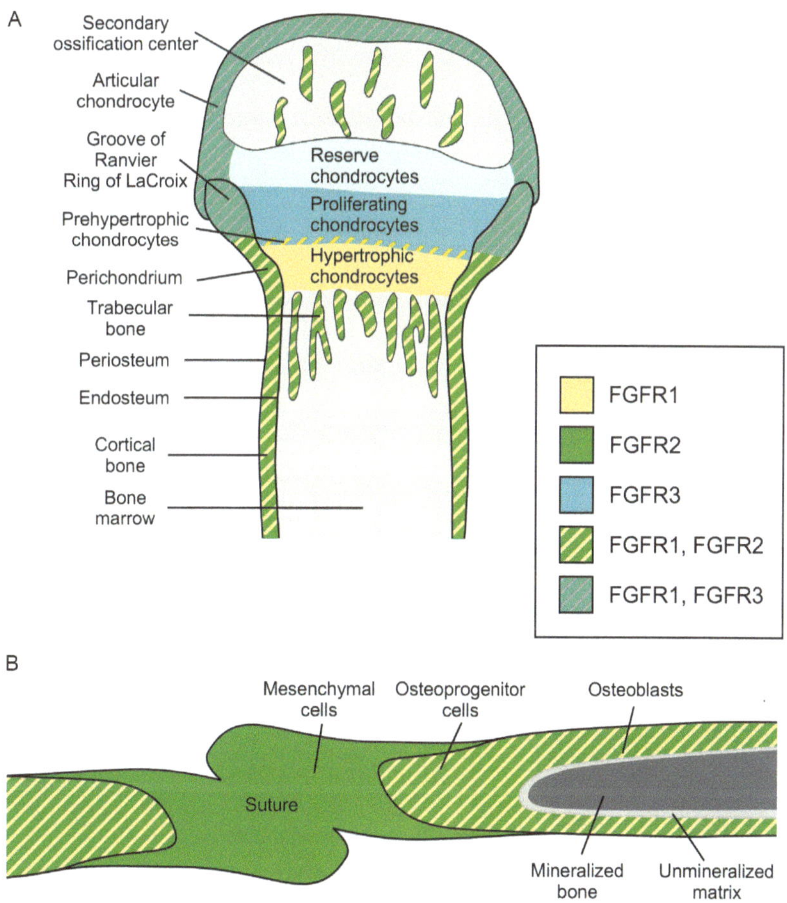

Fig. 1 Expression patterns of FGF receptors in endochondral bone (A) and membranous bone (B) during development. Diagram shows a schematic of a growth plate with color coded expression patterns of FGF receptors.

and prehypertrophic zone chondrocytes. As chondrocytes begin to hypertrophy, *Fgfr3* expression is shut down and *Fgfr1* expression is increased (Fig. 1) (Delezoide et al., 1998; Eswarakumar et al., 2002; Hamada, Suda, & Kuroda, 1999; Jacob et al., 2006; Karolak et al., 2015; Karuppaiah et al., 2016; Lazarus, Hegde, Andrade, Nilsson, & Baron, 2007; Ornitz & Marie, 2002; Peters et al., 1993; Yu et al., 2003).

2.2 Expression of FGF and FGF receptors in the developing axial skeleton

Several FGF ligands (*Fgf1, Fgf2, Fgf6, Fgf7, Fgf18,* and *Fgf23)* are differently expressed during development of the axial skeleton (Ornitz & Marie, 2015). *Fgfs 2, 4, 9,* and *18* are expressed during early stages of intramembranous bone formation (Britto et al., 2001; Kim, Rice, Kettunen, & Thesleff, 1998; Liu, Xu, Colvin, & Ornitz, 2002; Ohbayashi et al., 2002; Quarto, Behr, Li, & Longaker, 2009; Rice et al., 2000). *Fgf8* is expressed in developing calvarial osteoblasts (Xu, Lawshe, MacArthur, & Ornitz, 1999). *Fgf9* is expressed during mid to late stages of development in calvarial mesenchyme (Kim et al., 1998) and later in the perichondrium/periosteum and in the primary spongiosa (Garofalo et al., 1999; Hung, Yu, Lavine, & Ornitz, 2007). During cranial bone development, *Fgf18* is expressed in mesenchymal cells and differentiating osteoblasts; (Lazarus et al., 2007; Liu, Lavine, Hung, & Ornitz, 2007) whereas *Fgf23* is mainly produced by differentiated osteoblasts and osteocytes (Liu et al., 2003, 2006).

During axial skeletal development, *Fgfrs* are expressed in a space-dependent manner (Fig. 1B) (Ornitz & Marie, 2015). *Fgfrs 1, 2,* and *3* are co-expressed with *Fgfs 2, 4, 9,* and *18* during early stages of intramembranous bone formation (Britto et al., 2001; Kim et al., 1998; Liu et al., 2002; Ohbayashi et al., 2002; Quarto et al., 2009; Rice et al., 2000). *Fgfr1* and *Fgfr2* are expressed in pre-osteoblasts and osteoblasts in developing membranous bone (Rice, Rice, & Thesleff, 2003) and in the perichondrium and periosteum (Delezoide et al., 1998; Eswarakumar et al., 2002). At later stages, FGFR1 remains expressed in mesenchymal progenitors and FGFR2 in differentiating osteoblasts (Britto et al., 2001; Coutu et al., 2011; Jacob et al., 2006; Molteni et al., 1999b; Ohbayashi et al., 2002). Immature cultured osteoblasts express relatively higher levels of *Fgfr1*, whereas mature osteoblasts express relatively higher levels of *Fgfr2* (Rice et al., 2003). The developmental expression of these *Fgfrs* in specific cell types may mediate a differential response to FGF ligands (Cowan, Quarto, Warren, Salim, & Longaker, 2003).

3. FGF signaling in growth plate chondrocytes

FGF receptor 3 (FGFR3) is a key regulatory molecule functioning at all stages of chondrogenesis. Immature chondrocytes proliferate in response to FGFR3 signaling (Havens et al., 2008; Iwata et al., 2000; Iwata, Li, Deng, & Francomano, 2001). This response is likely activated by FGF9 and FGF18, which are expressed in adjacent mesenchyme (Hung et al., 2007; Liu et al., 2002, 2007; Ohbayashi et al., 2002). However, after formation of a growth plate with demarcated zones of proliferating and hypertrophic chondrocytes, FGFR3 signaling inhibits chondrogenesis by suppressing chondrocyte proliferation and differentiation into prehypertrophic and hypertrophic chondrocytes (Colvin, Bohne, Harding, McEwen, & Ornitz, 1996; Deng et al., 1996; Naski et al., 1998; Pannier et al., 2009). This inhibitory activity of FGFR3 on growth plate chondrocytes explains key aspects of the pathogenesis of chondrodysplasia syndromes (see below) in which gain-of-function mutations in FGFR3 cause decreased proliferation and differentiation of growth plate chondrocytes during pre-pubertal skeletal growth, resulting in skeletal dwarfism (Ornitz & Legeai-Mallet, 2017).

In growth plate chondrocytes, FGFR3 regulates several downstream signaling pathways that directly affect chondrocyte proliferation and differentiation (Ornitz & Legeai-Mallet, 2017; Ornitz & Marie, 2015). FGFR3 signaling activates STAT1, ERK1/2, and p38, decreases AKT phosphorylation, increases protein phosphatase 2a (PP2a), dephosphorylates (activates) p107 and p130, and increases expression of the cell cycle inhibitor, $p21^{Waf1/Cip1}$ (Fig. 2) (Aikawa, Segre, & Lee, 2001; Chapman et al., 2017; Cobrinik et al., 1996; Dailey, Laplantine, Priore, & Basilico, 2003; de Frutos et al., 2007; Kolupaeva, Daempfling, & Basilico, 2013; Kolupaeva, Laplantine, & Basilico, 2008; Kurimchak et al., 2013; Laplantine, Rossi, Sahni, Basilico, & Cobrinik, 2002; Legeai-Mallet, Benoist-Lasselin, Munnich, & Bonaventure, 2004; Priore, Dailey, & Basilico, 2006; Raucci, Laplantine, Mansukhani, & Basilico, 2004; Su et al., 1997). However, there is some debate as to which downstream molecules are primarily responsible for mediating FGFR3-regulated growth arrest (Krejci, Salazar, Goodridge, et al., 2008; Legeai-Mallet et al., 2004; Murakami et al., 2004).

Snail1 is a transcriptional repressor that is expressed in prehypertrophic chondrocytes and that is induced in Thanatophoric dysplasia bone tissue and in mouse models that activate FGFR3 (de Frutos et al., 2007;

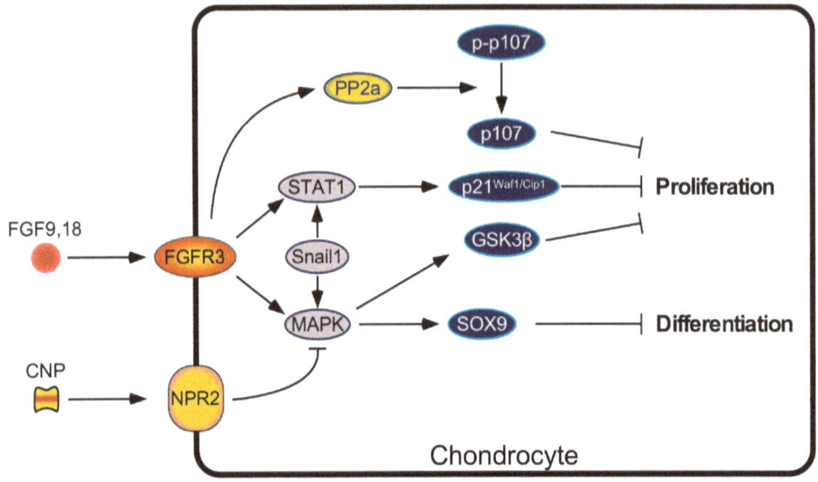

Fig. 2 FGF/FGFR signaling in chondrogenic cells. Diagram shows key signaling molecules downstream of FGFR3 in proliferating chondrocytes that regulate proliferation and differentiation. CNP-NPR2 is an interacting inhibitory pathway that is being exploited to treat Achondroplasia.

Karuppaiah et al., 2016; Seki et al., 2003). Ectopic activation of SNAIL1 in chondrocytes decreased chondrocyte proliferation and longitudinal bone growth (de Frutos et al., 2007). This phenotype required activation of STAT1 (which increases $p21^{Waf1/Cip1}$ expression) and MAPK, both of which act downstream of FGFR3 (de Frutos et al., 2007). Further supporting this model, conditional inactivation of both *Snail1* and *Snail2* in limb bud mesenchyme also increased $p21^{Waf1/Cip1}$ and decreased chondrocyte proliferation (Chen & Gridley, 2013). In cancer cells, activated ERK2 directly phosphorylates SNAIL1, leading to its nuclear accumulation and increased protein stability (Zhang et al., 2013) and SNAIL1 promotes the nuclear localization of p-ERK (Smith et al., 2014). However, it is not known if these activities occur in chondrocytes.

The requirement of STAT1 in the regulation of chondrocyte proliferation is however controversial. In support of a role for STAT1, inactivation of Stat1 *in vivo* partially rescues chondrocyte proliferation in mice with activating mutations in FGFR3. However, it does not rescue chondrocyte differentiation, which is under the control of MAPK signaling (Murakami et al., 2004). In further support of a role for STAT1 in regulating chondrocyte proliferation, and MAPK regulating chondrocyte differentiation, the c-natriuretic peptide (CNP) signaling pathway promotes bone growth through inhibition of MAPK signaling which increases extracellular matrix production and hypertrophic chondrocyte differentiation (Yasoda et al., 2004). The CNP signaling pathway has no effect on STAT1 activity or chondrocyte proliferation (Yasoda et al., 2004). One study, however, calls into question the role of STAT1. In rat chondrosarcoma cells, FGF does not activate STAT1 and FGF-induced growth arrest instead requires MAP kinase pathway activation (Krejci, Salazar, Goodridge, et al., 2008). However, this could be due to differences between a tumor cell line and normal growth plate chondrocytes.

FGFR3 also regulates chondrogenesis through interactions with other signaling pathways, including BMP, Wnt, IHH and PTHLH/PTH1R. In cultured chondrocytes, FGF signaling activates Wnt/β-catenin signaling through MAP kinase-mediated phosphorylation of LRP6 (Buchtova et al., 2015; Krejci et al., 2012). Phosphoproteomic analysis of rat chondrosarcoma cells identified Glycogen Synthase Kinase 3 beta (GSK3β) as a molecule that is required for the growth arrest response to FGF signaling (Chapman et al., 2017). These studies link FGF and Wnt/β-catenin signaling pathways to regulation of chondrogenesis, and may provide a link between MAPK signaling and inhibition of proliferation. However, differences between chondrosarcoma cells and growth plate chondrocytes still cloud the distinction between mitogenic and differentiation pathways. Future studies will be required to determine the precise relationship between FGFR3, Wnt/β-catenin, MAP kinase and STAT1 *in vivo* in growth plate chondrocytes.

FGFR3 signaling suppresses the expression of Bmp4, BMPR1a, Ihh, and Pth1r in the postnatal growth plate (Chen, Li, Qiao, Xu, & Deng, 2001; Naski et al., 1998; Qi et al., 2014). Additionally, in a chondrocyte cell line, overexpression of FGFR3 suppresses expression of both Pthlh and its receptor, Pth1r (Li et al., 2010). In addition to regulation of classical signaling pathways, FGFR3 has recently been shown to affect cilia length and intraflagellar transport in chondrocytes. This in turn may modulate chondrocyte response to hedgehog signaling (Kunova Bosakova et al., 2018).

4. FGF signaling in cortical, trabecular, and intramembranous bone

4.1 FGFR signaling in osteoblasts

FGF/FGFR signaling in osteogenic cells depends on the stage of cell maturation, the type of FGFR expressed, as well as the microenvironment (HSPGs and interacting proteins) that may enhance or attenuate FGF/FGFR

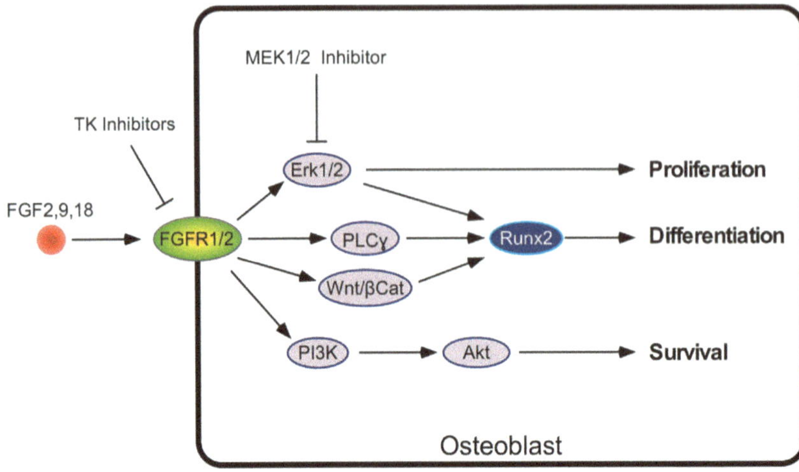

Fig. 3 FGF/FGFR signaling in osteogenic cells. FGF2, FGF9 or FGF18 interact with FGFR1 and FGFR2 to activate downstream signaling pathways that regulate proliferation, differentiation, and cell survival. Pharmacological inhibitors of tyrosine kinase (TK) or MEK1/2 are potential means to treat craniosynostosis.

signaling. Depending on the stage of cell differentiation, FGFR activation by FGF ligands leads to the activation of ERK1/2, PLC/PKC, and PI3K/Akt signaling, resulting in the control of cell proliferation, differentiation, and survival (Fig. 3) (Dailey, Ambrosetti, Mansukhani, & Basilico, 2005; Marie, Coffin, & Hurley, 2005). In cultured osteoblast precursor cells, activation of FGFR1/FGFR2 by FGF2 increases ERK1/2, MAPK, and cell proliferation (Choi et al., 2008; Miraoui et al., 2009). In pre-osteoblasts, PI3K/AKT activation by FGF2 (Debiais, Hott, Graulet, & Marie, 1998) or FGFR1-mediated increased Bcl2/Bax ratio (Agas et al., 2008) leads to increase cell survival, which contributes to enhance the pool of osteoprogenitors. In more mature osteoblasts, FGF/FGFR-mediated ERK1/2 activation leads to acetylation and stabilization of RUNX2, a key transcription factor involved in osteoblastogenesis (Park, Kim, Woo, Baek, & Ryoo, 2010; Xiao, Jiang, Gopalakrishnan, & Franceschi, 2002; Yoon et al., 2014). FGF2/FGFR2 activation in osteoblasts also increases RUNX2 expression through PKC activation (Kim et al., 2003; Niger et al., 2013), which promotes osteoblast differentiation. Consistently, mice lacking *Fgf2* show decreased bone formation due to altered osteoblast differentiation (Montero et al., 2000; Xiao, Sobue, et al., 2010). In cells of the perichondrium, PKC activation by FGF/FGFR leads to RUNX2-mediated expression of *Fgfr2*, *Fgf18*, and proteoglycans, which in turn enhances FGF signaling (Hinoi et al., 2006; Reinhold & Naski, 2007;

Teplyuk et al., 2009). *In vitro*, FGFR1/FGFR2 activation by FGF18 increases ERK1/2-mediated *Runx2* expression and osteoblast differentiation (Hamidouche et al., 2010). *In vivo*, mice lacking both *Fgf9* and *Fgf18* have severely deficient cranial bone formation (Hung, Schoenwolf, Lewandoski, & Ornitz, 2016) and *Fgf18*-deficient mice display reduced osteogenic mesenchymal cell proliferation, decreased osteoblast differentiation and delayed ossification (Liu et al., 2002; Ohbayashi et al., 2002), indicating that FGF18 is a positive regulator of osteogenic differentiation (Jeon et al., 2012). FGF treatment, or overexpression of FGF2 in transgenic mice, can also induce osteoblast apoptosis (Ignelzi, Wang, & Young, 2003; Mansukhani, Bellosta, Sahni, & Basilico, 2000), which limits the more mature osteoblast population. Finally, FGF2 promotes osteocyte differentiation *in vitro* by increasing the expression of E11/podoplanin, which is involved in the osteoblast-to-osteocyte transition (Ikpegbu et al., 2018). In an osteocytic cell line, FGF2-mediated ERK1/2 activation was found to increase the expression of *Dmp1*, a marker of osteocytes (Kyono, Avishai, Ouyang, Landreth, & Murakami, 2012), suggesting that FGF signaling controls osteocyte differentiation.

FGF signaling is highly dependent on the expression of FGFRs. *In vitro*, constitutive *Fgfr2* activation enhances ERK1/2 and PKCα signaling and *Runx2* expression, leading to increased osteoblast differentiation (Miraoui et al., 2009; Miraoui, Severe, Vaudin, Pages, & Marie, 2010). *In vivo*, mice conditionally lacking *Fgfr2*, or harboring a mutation in *Fgfr2c*, show decreased *Runx2* expression and bone mineral density due to reduced osteoprogenitor cell proliferation and altered function of more mature osteoblasts, indicating that *Fgfr2* positively regulates osteoblast differentiation (Eswarakumar et al., 2002; Yu et al., 2003). Conditional inactivation of *Fgfr1* in osteoprogenitor cells leads to increased cell proliferation and delayed cell differentiation and matrix mineralization, whereas *Fgfr1* deficiency in more differentiated osteoblasts causes increased mineralization, possibly through increased expression of *Fgfr3* (Jacob et al., 2006). Young adult *Fgfr3*-null mice show decreased trabecular bone mineralization and osteoblast mineralizing function *in vitro*, suggesting that FGFR3 may also play a role in postnatal bone formation (Valverde-Franco et al., 2004).

Some isoforms of FGF2 have distinct effects on osteoblasts and osteocytes. Overexpression of the low molecular weight isoform of FGF2 (LMWFGF2) in osteoblasts causes increased trabecular and cortical bone mass resulting from increased bone formation, whereas LMWFGF2 inactivation causes opposite effects (Xiao et al., 2009). In contrast, overexpression of nuclear localized high molecular weight FGF2 isoforms (HMWFGF2) in osteoblasts causes a phenotype similar to X-linked hypophosphatemia (XLH), a disease characterized by increased expression of FGF23, a phosphate wasting agent, by osteocytes. However, HMWFGF2 overexpression in osteoblast precursor cells inhibits osteoblast differentiation and matrix mineralization through FGF23/FGFR/MAPK signaling independently of phosphate wasting (Xiao, Esliger, & Hurley, 2013). Conversely, HMWFGF2 inactivation increases *FGF23* expression, osteoblast differentiation and bone mass (Homer-Bouthiette, Doetschman, Xiao, & Hurley, 2014). The expression of FGF23 in osteoblasts and osteocytes is controlled by FGFR1 signaling (Martin et al., 2011). HMWFGF2 isoforms interact with intranuclear FGFR1 to activate integrative nuclear FGFR1 signaling (INFS) (Clinkenbeard & White, 2016), leading to increased FGF23 production and hypophosphatemic rickets in mice (Xiao, Naganawa, et al., 2010), suggesting that HMWFGF2 isoforms affects bone homeostasis through FGF23. Mechanistically, HMWFGF2 promotes FGF23 promoter activity through cAMP-dependent binding of FGFR1 and CREB to a conserved cAMP response element in the FGF23 promoter, whereas LMWFGF2 stimulates FGF23 promoter activity through PLCγ/NFAT and MAPK signaling (Han, Xiao, & Quarles, 2015). Inhibition of FGFR signaling can partially rescue hypophosphatemic rickets in HMWFGF2 transgenic male mice (Xiao, Du, Homer-Bouthiette, & Hurley, 2017). Conditional deletion of *Fgfr1* in osteocytes in a mouse model of XLH (Hyp mice) leads to reduced FGF23 levels in osteocytes and partially rescues the abnormal bone phenotype (Xiao et al., 2014). Pharmacological activation of FGFR1 in osteoblasts enhances FGF23 secretion and hypophosphatemia in adult mice (Wu et al., 2013), whereas FGFR inhibition ameliorates FGF23-mediated hypophosphatemic rickets (Wohrle, Henninger, et al., 2013). Collectively, these experiments support a role for FGFR1/FGF23 signaling in the control of bone mineralization *in vivo*.

FGF/FGFR signaling in osteoblasts is controlled by HSPGs which act as co-receptors for FGFs (Eswarakumar, Lax, & Schlessinger, 2005; Ornitz & Itoh, 2015). In cultured osteogenic cells, cell surface and secreted HSPGs amplify FGF2 signaling and osteoblast differentiation (Jackson et al., 2007). The HSPG syndecans interact with FGFs to control osteoblastogenesis (Mansouri, Hay, Marie, & Modrowski, 2015). During development, syndecan-2 is expressed in the perichondrium and periosteum at the onset of osteogenesis (Molteni, Modrowski, Hott, & Marie, 1999a), whereas syndecan-3 is expressed in mesenchymal cells during limb bud formation (Gould, Upholt, & Kosher, 1992) and syndecan-4 is expressed in calvaria osteoblasts (Molteni et al., 1999b) and is upregulated by FGF2 (Song, Cool, & Nurcombe, 2007). Rat calvaria osteoblasts co-express syndecan 1, 2, and 4 with FGFR1 and FGFR2 (Jackson et al., 2007; Molteni et al., 1999a, 1999b). Functionally, FGFRs and syndecans are positively controlled by RUNX2 in cultured osteoblasts (Molteni et al., 1999a), and syndecan 2 enhances the response to FGF2 during *in vitro* osteogenesis (Fromigue, Modrowski, & Marie, 2004; Teplyuk et al., 2009). Another HSPG, glypican-3, is essential for *Runx2* expression and osteoblast differentiation *in vitro* (Haupt et al., 2009), further indicating that these HSPGs positively interact with FGFRs to promote FGF actions during osteogenesis.

Several intracellular mechanisms control FGF/FGFR signaling and osteogenesis. Sprouty2 (Spry2) acts as a negative regulator of FGFR signaling, and a dominant-negative mutant Spry2 p.Y55A promotes FGF2-induced ERK activation in murine osteoblasts, resulting in increased *Runx2* expression and osteoblast differentiation (Sanui et al., 2015). FGFR interaction with the docking protein FRS2α and the ubiquitin ligase c-CBL results in ubiquitination of FGFR and its subsequent degradation by the proteasome (Schlessinger, 2003). In cultured osteoblasts, CBL interaction with activated FGFR signaling tightly controls FGF/FGFR-mediated osteoblastogenesis (Severe, Dieudonne, & Marie, 2013). FGFR2 ubiquitination by c-CBL and subsequent FGFR2 downregulation results in decreased osteogenic differentiation (Kaabeche, Lemonnier, Le Mee, Caverzasio, & Marie, 2004), whereas inhibition of c-CBL-FGFR2 interaction promotes osteoblast differentiation and survival through increased ERK1/2 and PI3K signaling *in vitro* (Dufour et al., 2008; Severe et al., 2013).

4.2 FGF interactions with other pathways

FGF signaling interacts with several other pathways to control osteoblastogenesis. *In vitro*, FGF1 and FGF2 positively interact with TGFβ to control osteoblast cell proliferation (Globus, Patterson-Buckendahl, & Gospodarowicz, 1988), and FGF2 regulates TGFβ2 synthesis distinctly in immature and mature cultured osteoblasts (Debiais et al., 1998). Important interactions were found between FGF and BMP signaling. FGF2 stimulates *Bmp2* expression in osteoblasts *in vitro* and during cranial development (Choi et al., 2005; Fakhry et al., 2005) and correlatively, BMP2 expression is reduced in bones of *Fgf2* null mice (Naganawa et al., 2008). Additionally, BMP2 and FGF signaling interact positively to control osteogenesis *in vitro* and *in vivo* (Kuhn et al., 2013; Nakamura et al., 2005). BMP2 modulates FGF2 isoforms in mouse osteoblasts (Sabbieti et al., 2013). Mechanistically, FGF2 enhances BMP2 signaling in osteoblasts by promoting phospho-SMAD1/5/8 and RUNX2 transactivation (Agas, Sabbieti, Marchetti, Xiao, & Hurley, 2013), whereas FGF2, FGF9, and FGF18 act indirectly by inhibiting the expression of the BMP antagonist *Noggin* (Fakhry et al., 2005; Reinhold, Abe, Kapadia, Liao, & Naski, 2004; Warren, Brunet, Harland, Economides, & Longaker, 2003).

FGF signaling interacts negatively or positively with Wnt/β-catenin signaling to control osteogenic stem cell fate and differentiation, depending on the stage of differentiation (Dailey et al., 2005; Miraoui & Marie, 2010). In osteoprogenitor cells, FGF2 antagonizes Wnt/β-catenin signaling (Ambrosetti, Holmes, Mansukhani, & Basilico, 2008; Mansukhani, Ambrosetti, Holmes, Cornivelli, & Basilico, 2005). However, *Fgf2*$^{-/-}$ mice display decreased Wnt gene expression in osteoblasts and reduced osteoblast differentiation (Fei, Xiao, Doetschman, Coffin, & Hurley, 2011). Exogenous FGF2 promotes canonical Wnt/β-catenin signaling and rescues osteoblast differentiation in *Fgf2*$^{-/-}$ mice, indicating that FGF signaling activates Wnt/β-catenin signaling in osteoblasts to promote osteogenesis (Fei et al., 2011). FGF18 is a target of Wnt signaling in osteogenic cells and Wnt/β-catenin signaling induces *Fgf18* expression through a complex with RUNX2 and TCF/LEF transcription factors on the *Fgf18* promoter (Reinhold & Naski, 2007).

FGF signaling positively interacts with parathyroid hormone (PTH) signaling in osteogenic cells to promote osteogenesis. *In vitro*, PTH enhances *Fgf2, Fgfr1,* and *Fgfr2* expression in osteoblasts (Hurley et al., 1999) and endogenous FGF2 is required for the positive PTH effects on osteoblast proliferation, differentiation, and survival (Sabbieti et al., 2009). *In vivo*, the anabolic effect of intermittent PTH is impaired in *Fgf2* null or heterozygous mice (Hurley et al., 1999), suggesting that PTH promotes osteogenesis in part through FGF2 expression (Sabbieti et al., 2009).

5. Mutations in FGFRs in human skeletal disease

5.1 Chondrodysplasia syndromes

Achondroplasia is the most common form of skeletal dwarfism in humans (Baujat, Legeai-Mallet, Finidori, Cormier-Daire, & Le Merrer, 2008; Horton, Hall, & Hecht, 2007; Murdoch et al., 1970; Ornitz &

Legeai-Mallet, 2017). A characteristic feature of Achondroplasia is disproportional limb shortening, called rhizomelia, where the proximal limb segment is more affected than the distal limb segment. Almost all cases of Achondroplasia can be accounted for by an autosomal dominant missense mutation (G380R) in *FGFR3* (Rousseau et al., 1994; Shiang et al., 1994; Vajo, Francomano, & Wilkin, 2000; Wilkin et al., 1998). Other autosomal dominant chondrodysplasia syndromes with similar phenotypes to Achondroplasia are also caused by mutations in FGFR3. Thanatophoric dysplasia types I and II is a severe and usually lethal disease caused by K650E and R248C mutations, respectively, in *FGFR3*. Hypochondroplasia is a milder form of dwarfism, which is caused by an N540K or K650N mutation in *FGFR3* (Bellus et al., 1995, 2000; Bonaventure et al., 1996; Tavormina et al., 1995). SADDAN syndrome (Severe Achondroplasia with Developmental Delay and Acanthosis Nigricans) (Bellus et al., 1999; Montone et al., 2018; Tavormina et al., 1999) and Platyspondylic lethal skeletal dysplasia, San Diego type (Brodie et al., 1999) are both caused by a K650M mutation in FGFR3. Interestingly, a family with proportional short stature was found to have a M528I mutation in *FGFR3* (Kant et al., 2015). All of the mutations in FGFR3 associated with dwarfism or short stature activate the receptor to varying degrees that correlate with disease severity (Kant et al., 2015; Krejci, Salazar, Kashiwada, et al., 2008; Naski, Wang, Xu, & Ornitz, 1996; Ornitz & Legeai-Mallet, 2017). Rare cases of homozygous or biallelic mutations in FGFR3 have more severe phenotypes and can be lethal (Chang et al., 2018; De Rosa, Fano, Araoz, Chertkoff, & Obregon, 2014; Patel & Filly, 1995; Pauli et al., 1983). Although the M528I mutation in *FGFR3* is activating *in vitro*, it is not known why this family has proportionate limb shortening while other mutations in FGFR3 result in disproportional (rhizomelic) limb shortening.

5.2 Mouse models with mutations in Fgfr3

Several mouse models have been developed for Achondroplasia and related diseases (Ornitz & Legeai-Mallet, 2017; Ornitz & Marie, 2015; Su, Jin, & Chen, 2014). The first model made was a transgenic mouse in which the *Col2a1* promoter drives expression of a mutant *Fgfr3* cDNA (p.Gly380Arg) in chondrocytes (Naski et al., 1998; Shazeeb et al., 2018). These mice have been referred to as *Fgfr3*$^{Ach/+}$ mice. These mice have been widely used to test potential therapies for Achondroplasia (Garcia et al., 2013; Wendt et al., 2015; Yasoda et al., 2009). A similar model has been made using mouse *Fgfr3* promoter elements to drive expression of a human mutant cDNA (Segev et al., 2000), or a knockin of the human mutant cDNA into the mouse *Fgfr3* locus (Lee, Song, Pai, Chen, & Chen, 2017). A true knockin mouse model has also been made with the common G380R mutation (Wang et al., 1999). This mouse exhibited an Achondroplasia-like phenotype. Achondroplasia can also be caused by a rare (G375C) mutation. Mice harboring this mutation showed skeletal dysplasia that was similar to human Achondroplasia (Chen et al., 1999). A knockin mouse model harboring the FGFR3 p. Y367C mutation found in Thanatophoric dysplasia showed dwarfism and a fully penetrant deafness phenotype (Pannier et al., 2009). Femur explants from this mouse have been used to test pharmacological therapies (Jonquoy et al., 2012) and, *in vivo*, the function of a CNP agonist (Lorget et al., 2012). Additional Thanatophoric dysplasia knockin mice have been made with the FGFR3 K644E mutation (Li et al., 1999) and a model for SADDAN syndrome has been made with the FGFR3 K644M mutation (Iwata et al., 2001).

5.3 FGFR signaling pathway-based therapeutic strategies

The gain-of-function nature of mutations in FGFR3 that result in skeletal dwarfism immediately suggest potential therapeutic strategies. As activating mutations in FGFR3 are also involved in human cancer (e.g., bladder tumors, multiple myeloma) (Cappellen et al., 1999; Chesi et al., 1997; Patani et al., 2016; Turner & Grose, 2010), similar therapeutic strategies can be considered. However, several important differences between cancer and chondrodysplasia syndromes must also be considered: (1) Pharmacological access to the growth plate, which is an avascular tissue, may be more restricted compared to vascularized tumors; (2) The phenotype in Achondroplasia patients with activating mutations in FGFR3 develops over a long period of time, beginning in late fetal development and continuing until puberty. Thus, therapies may need to be administered over many years, with the potential for more off target effects in a growing child. In contrast, cancers may require more potent, but shorter-term therapies; (3) Inherited or acquired mutations in *FGFR3* causing Achondroplasia affect all skeletal elements, the inner ear, and other tissues. Whereas in cancer, acquired mutations are generally restricted to specific organs and the cancer cells that metastasize from these organs, thus allowing consideration of local or cell-targeted therapies.

Small molecule tyrosine kinase inhibitors (TKI) that can directly reduce the tyrosine kinase activity resulting from activating mutations in *FGFR3* are viable candidates, and can be borrowed from the oncology fields. For example, CHIR-258, a tyrosine kinase inhibitor with activity toward FGFR3, showed efficacy in a xenograft mouse model of FGFR3-induced multiple myeloma (MM) (Trudel et al., 2005) and A31 was effective in increasing the growth of femur explants from FGFR3 (p.Tyr367Cys) mutant mice (Jonquoy et al., 2012). PD173074 and SU5402 inhibit all FGFRs and have been shown to inhibit the growth and induce apoptosis of multiple myeloma cells (Dimitroff et al., 1999; Mohammadi et al., 1997). ARQ-087 is a pan-FGFR inhibitor with efficacy in tibial explant culture models for Achondroplasia, as well as *in vitro* models for FGFR1 and FGFR2 mediated craniosynostosis syndromes (Balek et al., 2017). NVP-BGJ398 is a TKI with some selectivity for FGFR3 over other FGFRs (Gudernova et al., 2016). In mouse models for several FGFR-related cancers such as malignant rhabdoid tumors (Wohrle, Weiss, et al., 2013), hepatocellular carcinoma (Scheller et al., 2015), and skeletal diseases including FGF23-mediated hypophosphatemic rickets (Wohrle, Henninger, et al., 2013) and Achondroplasia (Komla-Ebri et al., 2016), NVP-BGJ398 showed efficacy. In Achondroplasia mice, NVP-BGJ398 was used *in vivo* to reduce FGFR3 activity and improve the skeletal phenotype (Komla-Ebri et al., 2016).

Another approach to inhibit FGFR3 is to directly block receptor activity with monoclonal antibodies. Antibodies that bind to the extracellular domain of FGFR3 can directly block ligand binding, interfere with co-factor-receptor interactions, or inhibit receptor dimerization. FGFR3-specific monoclonal antibodies were effective in slowing the growth of bladder cancer cell lines, FGFR3-dependent tumors in mice, and FGFR3-expressing tumor xenografts (Gorbenko et al., 2009; Gust et al., 2013; Hadari & Schlessinger, 2009; Martinez-Torrecuadrada et al., 2005; Qing et al., 2009; Rauchenberger et al., 2003; Trudel et al., 2006; Yin et al., 2016). FGFR3-specific monoclonal antibodies have not yet been evaluated *in vivo* in mouse models for achondroplasia.

A third approach is to use soluble FGFR3 extracellular domain decoy receptors (sFGFR3) with the objective of binding and sequestering available FGF ligands. This approach has the advantage that ligands can be sequestered before they enter the growth plate, thus adverting the issues of drug access to avascular growth plate chondrocytes. Several ligands have been identified that functionally regulate chondrogenesis (Garcia et al., 2013; Hung et al., 2007; Liu et al., 2002, 2007; Ohbayashi et al., 2002). Mutational modifications of a soluble FGFR3 could be used to enhance ligand binding specificity for FGFs that specifically regulate chondrogenesis. In a proof of concept study, subcutaneous injections of recombinant sFGFR3 into a transgenic mouse model for Achondroplasia (Naski et al., 1998), decreased mortality and improved skeletal growth (Garcia et al., 2013).

5.4 CATSHL syndrome (loss of function of Fgfr3)

Skeletal overgrowth is a striking feature of mice lacking FGFR3 (Colvin et al., 1996; Deng et al., 1996; Eswarakumar & Schlessinger, 2007). Although FGFR3 is mitogenic for immature chondrocytes early in embryonic development, the inhibitory functions of FGFR3 on growth plate chondrocytes account for the dominant regulation of skeletal growth. In mice lacking *Fgfr3* the proliferating and hypertrophic chondrocyte zones are expanded and growth plate activity persists into adulthood. A consequence of deregulated growth plate activity is overgrowth of the appendicular skeleton. FGFR3 signaling also affects vertebral bone growth and mice lacking FGFR3 also develop elongation of vertebral bodies and scoliosis. Because mice lacking FGFR3 are viable, it was anticipated that rare homozygous loss of function mutations in *FGFR3* or heterozygous dominant negative mutations in *FGFR3* might also occur in humans. This prediction was validated by the discovery of a four-generation family with dominantly inherited camptodactyly, tall stature, and hearing loss (termed CATSHL syndrome). Affected members of this family were heterozygous for a missense mutation (R621H) in the catalytic loop of the FGFR3 tyrosine kinase domain (Escobar, Tucker, & Bamshad, 2016; Toydemir et al., 2006). This mutation inactivates the kinase domain and likely functions as a dominant negative mutation. Consistent with the original hypothesis, two siblings with CATSHL syndrome were found to be homozygous for a T546K mutation in *FGFR3* (Makrythanasis et al., 2014). A similar recessive mutation in FGFR3 (V700E) has been found in sheep with spider lamb syndrome, a disease characterized by long limbs, kyphoscoliosis, malformed ribs and sternebrae, Roman noses, lack of body fat, and muscular atrophy (Beever et al., 2006). Heterozygous sheep are normal but show a mild increase in skeletal growth (Smith, Dally, Sainz, Rodrigue, & Oberbauer, 2006).

5.5 Craniosynostosis syndromes

Several mutations in FGFRs 1–3 cause craniosynostosis, a skeletal phenotype characterized by premature fusion of one or more cranial sutures (Johnson & Wilkie, 2011; Ornitz & Marie, 2002). Multiple molecular mechanisms underlie the phenotype induced by mutations in *FGFR1 and FGFR2* in Apert, Crouzon, and other syndromes. These include ligand-independent or ligand-dependent FGFR activation, loss of function, or altered FGFR trafficking (Ibrahimi et al., 2001; Neben & Merrill, 2015; Yu, Herr, Waksman, & Ornitz, 2000; Yu & Ornitz, 2001). Mutations in *FGFR2* may also induce Bent Bone Dysplasia, a dysmorphology of the appendicular skeleton (Merrill et al., 2012). In this syndrome, *FGFR2* mutations (p.M391R or p.Y381D) reduce ligand-dependent receptor activation, which leads to increased nucleolar activity of the mutant *FGFR2*, causing increased proliferation and decreased differentiation of osteoprogenitor cells (Merrill et al., 2012; Neben et al., 2014; Neben, Tuzon, Mao, Lay, & Merrill, 2017).

Activating mutations in *FGFR1 and FGFR2* induce various abnormalities in cranial osteogenesis in mouse models and patients (Hajihosseini, 2008; Marie, Kaabeche, & Guenou, 2008; Senarath-Yapa et al., 2012; Su et al., 2014). The primary abnormality is the fusion of one or more of the sutures of the skull vault (Wilkie, 2005). The mechanisms underlying the premature suture fusion induced by activating *FGFR2* mutations have been widely investigated in humans and mice (Marie, Debiais, & Hay, 2002; Su et al., 2014). Conditional expression of the Apert *FGFR2* p.*S252W* mutation in cells derived from mesoderm is sufficient to cause coronal craniosynostosis (Heuze et al., 2014). In mice, Apert *FGFR2* mutations induce abnormal mesodermal progenitor cell proliferation, differentiation, and cell fate in cranial sutures, although the cellular abnormalities depend on the suture and the state of cell differentiation (Heuze et al., 2014; Holmes et al., 2009; Motch Perrine et al., 2014; Wang et al., 2005). At an early stage of development of cranial sutures in mice, osteoprogenitor cell proliferation may be unchanged (Chen, Li, Li, Engel, & Deng, 2003) or transiently increased by Apert and Crouzon *FGFR2* mutations (Mansukhani et al., 2000, 2005). At later stages, osteoblast maturation is increased (Eswarakumar, Horowitz, Locklin, Morriss-Kay, & Lonai, 2004; Holmes et al., 2009; Liu, Kwon, Nam, & Hatch, 2013; Liu, Nam, Wang, & Hatch, 2013; Morita et al., 2014; Yang et al., 2008; Yeh et al., 2012; Yin et al., 2008). Mechanistically, activating mutations in *FGFR1, FGFR2* lead to increased *Runx2* expression associated with premature osteoblast differentiation in cranial sutures in mice (Baroni et al., 2005; Eswarakumar et al., 2004; Zhou et al., 2000). In humans, activating mutations in *FGFR2* induce premature osteoblast differentiation (Marie, 2012). Human cranial suture osteoblasts harboring Apert and Crouzon *FGFR2* mutations display increased *Runx2* expression and enhanced osteoblast maturation and function *in vitro* and *in vivo* (Baroni et al., 2005; Lemonnier, Hay, Delannoy, Lomri, et al., 2001; Lomri et al., 1998; Marie, 2015; Tanimoto et al., 2004). Conversely, disruption of FGFR2c, the mesenchymal splice variant of FGFR2, decreases *Runx2* transcription and retards ossification in mice (Eswarakumar et al., 2002). *FGFR2* activating mutations also increase *Sox2* expression in murine osteoblasts and reduce the expression of some Wnt target genes, although the implication of these genes in craniosynostosis remains unclear (Mansukhani et al., 2005). In mice and humans, fused sutures show apoptosis in more mature osteoblasts (Chen et al., 2014; Kaabeche et al., 2005; Lemonnier, Hay, Delannoy, Fromigue, et al., 2001; Lomri, Lemonnier, Delannoy, & Marie, 2001; Mansukhani et al., 2005) presumably as a consequence of the premature osteoblast maturation induced by activating *FGFR2* mutations (Holmes et al., 2009).

Although *FGFR3* mutations in Achondroplasia were found to be associated with partial premature fusion of the coronal sutures (Di Rocco et al., 2014; Twigg et al., 2009), this effect results from indirect effects on other signaling pathways. Increased expression of activated *Fgfr3* in chondrocytes induces osteoblast differentiation and premature suture closure due to activation of MAPK and BMP signaling (Matsushita et al., 2009). Activating *FGFR3* mutations in chondrocytes decreases trabecular bone mass in long bones as a result of increased osteoclast recruitment (Mugniery et al., 2012; Su et al., 2010). Conversely, conditional knock-out of *Fgfr3* in chondrocytes causes increased bone mass due to reduced osteoclast number and increased osteoblast number and bone formation mediated by increased expression of *Ihh, Bmp2, Bmp4, Bmp7, Wnt4*, and *Tgf-β1* (Wen et al., 2016).

5.6 FGFR signaling and potential therapeutic strategies in craniosynostosis

Several signaling pathways are activated by mutations in *FGFR2* (Marie, 2012; Su et al., 2014). In several mouse models, *FGFR2* mutations activate ERK1/2, p38 MAPK, AKT, β-catenin, and PLCγ in osteogenic cells (Chen et al., 2003; Kim et al., 2003; Pfaff et al., 2016; Shukla,

Coumoul, Wang, Kim, & Deng, 2007; Suzuki et al., 2012; Wang et al., 2010; Yin et al., 2008). ERF is an inhibitory ETS transcription factor directly bound by ERK1/2. A role of ERF and ERK signaling is supported by the finding that reduced expression of ERF is associated with craniosynostosis in humans and mice (Twigg et al., 2013). The *FGFR2* p. *S252W* mutation also activates PLCγ and PKCα in osteoblasts, resulting in increased *Runx2* expression and premature osteoblast differentiation in mice and humans (Baroni et al., 2005; Eswarakumar et al., 2002; Guenou, Kaabeche, Mee, & Marie, 2005; Miraoui et al., 2009; Tanimoto et al., 2004; Zhou et al., 2000).

Gain-of-function mutations in FGFR2 activate feedback pathways that contribute to craniosynostosis (Miraoui & Marie, 2010). *FGFR2* mutations that cause Apert syndrome lead to increased c-CBL-mediated ubiquitination of the *Src* family members *LYN* and *FYN* (Kaabeche et al., 2004), which contributes to increased osteoblast differentiation in human calvarial osteoblasts derived from Apert patients. Increased c-CBL recruitment in Apert osteoblasts also enhances ubiquitination and degradation of α5β1 integrin (Kaabeche et al., 2005) and PI3K (Dufour et al., 2008), which cause apoptosis in mature osteoblasts in the fused sutures found in Apert syndrome. In both mice and humans, the *FGFR2* p.S252W mutation increases c-CBL-mediated ubiquitination and degradation of FGFR2 (Holmes et al., 2009; Kaabeche et al., 2004). This is associated with increased c-CBL-Spy2 interaction and c-CBL sequestration, resulting in increased EGFR levels and signaling and increased osteoblast gene transcription in human cranial osteoblasts (Miraoui, Ringe, Haupl, & Marie, 2010). The *FGFR2* p.S252W mutation also increases platelet-derived growth factor receptor *α* (PDGFRα) expression and signaling as a result of enhanced FGFR2-PKC*α*-mediated AP-1 transcription (Miraoui, Ringe, et al., 2010). Both of these signaling mechanisms functionally contribute to craniosynostosis in Apert patients (Miraoui & Marie, 2010). In mice, conditional expression of activated *PDGFRα* in neural crest cells induces premature cranial fusion at early postnatal stages (Moenning et al., 2009). These experiments indicate that FGFR2 and PDGFRα signaling interact to promote suture fusion in mice and humans and suggest that therapeutic strategies targeting PDGFR signaling may attenuate the abnormal signals induced by activating *FGFR* mutations.

Several therapeutic strategies were proposed to counteract the aberrant cellular signaling induced by activating mutations in *FGFR2*

(Melville, Wang, Taub, & Jabs, 2010; Wilkie, 2007). Initial experiments indicate that some glycosaminoglycans may attenuate activated *FGFR2 S252W* signaling *in vitro* (McDowell et al., 2006). Strategies using a small hairpin RNA targeting the Apert *FGFR2* p.*S252W* mutation (Shukla et al., 2007), or a soluble dominant negative mutant form of FGFR2 (Tanimoto et al., 2004) can also reduce activated FGFR2 p.*S252W* signaling. *In vivo*, selective uncoupling of FRS2α and the activated FGFR2c with a Crouzon-like mutation can attenuate FGFR2 signaling and prevent premature suture fusion in mice (Eswarakumar et al., 2006). The pharmacological use of FGFR inhibitors (Balek et al., 2017; Eswarakumar et al., 2006), TK inhibitors (Perlyn, Morriss-Kay, Darvann, Tenenbaum, & Ornitz, 2006) or a MEK1/2 inhibitor (Shukla et al., 2007) was also found to reduce FGFR2 signaling and prevent craniosynostosis. Recent studies indicate that treatment with BMN111, a C-type natriuretic peptide analog that inhibits FGFR signaling at the level of RAF1 kinase, can improve skull morphology in a heterozygous *Fgfr2c* p.*C342Y* Crouzon syndrome mouse model (Holmes et al., 2018). The Prolyl isomerase, PIN1, plays a role in mediating FGFR signaling, and downregulation of *Pin1* gene dosage, or the use of a PIN1 enzyme inhibitor, was recently found to attenuate RUNX2 expression and premature cranial suture closure in heterozygous *Fgfr2 p.S252W* mutant mice, suggesting that PIN1 modulation can be used to prevent craniosynostosis (Shin et al., 2018). Collectively, these experiments indicate that attenuation of FGFR signaling by various means may be applied for the prevention or treatment of craniosynostosis.

6. Conclusions and perspectives

FGF/FGFR signaling exerts complex time- and space-dependent roles during skeletal development. The cellular and molecular mechanisms underlying the control of endochondral and intramembranous ossification by FGF/FGFR signaling are now better understood. Genetic and functional analyses in mouse models and in humans have improved our understanding of the pathogenesis of the skeletal phenotype and the underlying aberrant signaling mechanisms induced by mutations in *FGFRs 1–3* that cause chondrodysplasia and craniosynostosis syndromes in humans. However, the specific role of FGF/FGFR signaling in cells of the chondrocyte and osteoblast lineage at various stages of development, and the interactions of FGF/FGFR signaling with other signaling pathways during normal and abnormal skeletal development need to be further investigated. A better knowledge of the mechanisms underlying FGFR signaling may translate into novel therapeutic approaches with the goal of attenuating the skeletal phenotype induced by aberrant FGFR signaling in human skeletal dysplasias.

Acknowledgments

The authors thank all team members who contributed to the work reviewed in this chapter. The authors apologize to the investigators whose work could not be cited due to space limitations. This work was supported by NIH grant HD049808 (D.M.O.), by the Department of Developmental Biology at Washington University, and by the Institut National de la Recherche Médicale (Inserm), University Paris Diderot Sorbonne Paris Cité, the DIM Stem Pôle Ile de France, the French Minister of Research and the Asssociation Prévention et Traitement des Décalcifications (Paris, France) (P.J.M.).

References

Agas, D., Marchetti, L., Menghi, G., Materazzi, S., Materazzi, G., Capacchietti, M., et al. (2008). Anti-apoptotic Bcl-2 enhancing requires FGF-2/FGF receptor 1 binding in mouse osteoblasts. *Journal of Cellular Physiology*, *214*, 145–152.

Agas, D., Sabbieti, M. G., Marchetti, L., Xiao, L., & Hurley, M. M. (2013). FGF-2 enhances Runx-2/Smads nuclear localization in BMP-2 canonical signaling in osteoblasts. *Journal of Cellular Physiology*, *228*, 2149–2158.

Aikawa, T., Segre, G. V., & Lee, K. (2001). Fibroblast growth factor inhibits chondrocytic growth through induction of p21 and subsequent inactivation of cyclin E-Cdk2. *The Journal of Biological Chemistry*, *276*, 29347–29352.

Ambrosetti, D., Holmes, G., Mansukhani, A., & Basilico, C. (2008). Fibroblast growth factor signaling uses multiple mechanisms to inhibit Wnt-induced transcription in osteoblasts. *Molecular and Cellular Biology*, *28*, 4759–4771.

Balek, L., Gudernova, I., Vesela, I., Hampl, M., Oralova, V., Kunova Bosakova, M., et al. (2017). ARQ 087 inhibits FGFR signaling and rescues aberrant cell proliferation and differentiation in experimental models of craniosynostoses and chondrodysplasias caused by activating mutations in FGFR1, FGFR2 and FGFR3. *Bone*, *105*, 57–66.

Baroni, T., Carinci, P., Lilli, C., Bellucci, C., Aisa, M. C., Scapoli, L., et al. (2005). P253R fibroblast growth factor receptor-2 mutation induces RUNX2 transcript variants and calvarial osteoblast differentiation. *Journal of Cellular Physiology*, *202*, 524–535.

Baujat, G., Legeai-Mallet, L., Finidori, G., Cormier-Daire, V., & Le Merrer, M. (2008). Achondroplasia. *Best Practice & Research. Clinical Rheumatology*, *22*, 3–18.

Beever, J. E., Smit, M. A., Meyers, S. N., Hadfield, T. S., Bottema, C., Albretsen, J., et al. (2006). A single-base change in the tyrosine kinase II domain of ovine FGFR3 causes hereditary chondrodysplasia in sheep. *Animal Genetics*, *37*, 66–71.

Bellus, G. A., Bamshad, M. J., Przylepa, K. A., Dorst, J., Lee, R. R., Hurko, O., et al. (1999). Severe achondroplasia with developmental delay and acanthosis nigricans (SADDAN): Phenotypic analysis of a new skeletal dysplasia caused by a Lys650Met mutation in fibroblast growth factor receptor 3. *American Journal of Medical Genetics*, *85*, 53–65.

Bellus, G. A., McIntosh, I., Smith, E. A., Aylsworth, A. S., Kaitila, I., Horton, W. A., et al. (1995). A recurrent mutation in the tyrosine kinase domain of fibroblast growth factor receptor 3 causes hypochondroplasia. *Nature Genetics*, *10*, 357–359.

Bellus, G. A., Spector, E. B., Speiser, P. W., Weaver, C. A., Garber, A. T., Bryke, C. R., et al. (2000). Distinct missense mutations of the FGFR3 lys650 codon modulate receptor kinase activation and the severity of the skeletal dysplasia phenotype. *American Journal of Human Genetics*, *67*, 1411–1421.
Belov, A. A., & Mohammadi, M. (2013). Molecular mechanisms of fibroblast growth factor signaling in physiology and pathology. *Cold Spring Harbor Perspectives in Biology*, *5*, 1–24.
Benazet, J. D., & Zeller, R. (2009). Vertebrate limb development: Moving from classical morphogen gradients to an integrated 4-dimensional patterning system. *Cold Spring Harbor Perspectives in Biology*, *1*, a001339.
Bonaventure, J., Rousseau, F., Legeai-Mallet, L., Le Merrer, M., Munnich, A., & Maroteaux, P. (1996). Common mutations in the fibroblast growth factor receptor 3 (FGFR 3) gene account for achondroplasia, hypochondroplasia, and thanatophoric dwarfism. *American Journal of Medical Genetics*, *63*, 148–154.
Brewer, J. R., Mazot, P., & Soriano, P. (2016). Genetic insights into the mechanisms of Fgf signaling. *Genes & Development*, *30*, 751–771.
Britto, J. A., Evans, R. D., Hayward, R. D., & Jones, B. M. (2001). From genotype to phenotype: The differential expression of FGF, FGFR, and TGFbeta genes characterizes human cranioskeletal development and reflects clinical presentation in FGFR syndromes. *Plastic and Reconstructive Surgery*, *108*, 2026–2039. discussion 2040–2026.
Brodie, S. G., Kitoh, H., Lachman, R. S., Nolasco, L. M., Mekikian, P. B., & Wilcox, W. R. (1999). Platyspondylic lethal skeletal dysplasia, San Diego type, is caused by FGFR3 mutations. *American Journal of Medical Genetics*, *84*, 476–480.
Browaeys-Poly, E., Blanquart, C., Perdereau, D., Antoine, A. F., Goenaga, D., Luzy, J. P., et al. (2010). Grb14 inhibits FGF receptor signaling through the regulation of PLCgamma recruitment and activation. *FEBS Letters*, *584*, 4383–4388.
Buchtova, M., Oralova, V., Aklian, A., Masek, J., Vesela, I., Ouyang, Z., et al. (2015). Fibroblast growth factor and canonical WNT/beta-catenin signaling cooperate in suppression of chondrocyte differentiation in experimental models of FGFR signaling in cartilage. *Biochimica et Biophysica Acta*, *1852*, 839–850.
Camps, M., Nichols, A., Gillieron, C., Antonsson, B., Muda, M., Chabert, C., et al. (1998). Catalytic activation of the phosphatase MKP-3 by ERK2 mitogen-activated protein kinase. *Science*, *280*, 1262–1265.
Cappellen, D., De Oliveira, C., Ricol, D., de Medina, S., Bourdin, J., Sastre-Garau, X., et al. (1999). Frequent activating mutations of FGFR3 in human bladder and cervix carcinomas. *Nature Genetics*, *23*, 18–20.
Chang, I. J., Sun, A., Bouchard, M. L., Kamps, S. E., Hale, S., Done, S., et al. (2018). Novel phenotype of achondroplasia due to biallelic FGFR3 pathogenic variants. *American Journal of Medical Genetics. Part A*, *176*, 1675–1679.
Chapman, J. R., Katsara, O., Ruoff, R., Morgenstern, D., Nayak, S., Basilico, C., et al. (2017). Phosphoproteomics of fibroblast growth factor 1 (FGF1) signaling in chondrocytes: Identifying the signature of inhibitory response. *Molecular & Cellular Proteomics*, *16*, 1126–1137.
Chen, L., Adar, R., Yang, X., Monsonego, E. O., Li, C., Hauschka, P. V., et al. (1999). Gly369Cys mutation in mouse FGFR3 causes achondroplasia by affecting both chondrogenesis and osteogenesis. *The Journal of Clinical Investigation*, *104*, 1517–1525.
Chen, Y., & Gridley, T. (2013). Compensatory regulation of the Snai1 and Snai2 genes during chondrogenesis. *Journal of Bone and Mineral Research*, *28*, 1412–1421.
Chen, L., Li, D., Li, C., Engel, A., & Deng, C. X. (2003). A Ser252Trp [corrected] substitution in mouse fibroblast growth factor receptor 2 (Fgfr2) results in craniosynostosis. *Bone*, *33*, 169–178.

Chen, L., Li, C., Qiao, W., Xu, X., & Deng, C. (2001). A Ser(365)→Cys mutation of fibroblast growth factor receptor 3 in mouse downregulates Ihh/PTHrP signals and causes severe achondroplasia. *Human Molecular Genetics*, *10*, 457–465.

Chen, P., Zhang, L., Weng, T., Zhang, S., Sun, S., Chang, M., et al. (2014). A Ser252Trp mutation in fibroblast growth factor receptor 2 (FGFR2) mimicking human Apert syndrome reveals an essential role for FGF signaling in the regulation of endochondral bone formation. *PLoS One*, *9*, e87311.

Chesi, M., Nardini, E., Brents, L. A., Schrock, E., Ried, T., Kuehl, W. M., et al. (1997). Frequent translocation t(4,14)(p16.3;q32.3) in multiple myeloma is associated with increased expression and activating mutations of fibroblast growth factor receptor 3. *Nature Genetics*, *16*, 260–264.

Choi, S. C., Kim, S. J., Choi, J. H., Park, C. Y., Shim, W. J., & Lim, D. S. (2008). Fibroblast growth factor-2 and -4 promote the proliferation of bone marrow mesenchymal stem cells by the activation of the PI3K-Akt and ERK1/2 signaling pathways. *Stem Cells and Development*, *17*, 725–736.

Choi, K. Y., Kim, H. J., Lee, M. H., Kwon, T. G., Nah, H. D., Furuichi, T., et al. (2005). Runx2 regulates FGF2-induced Bmp2 expression during cranial bone development. *Developmental Dynamics*, *233*, 115–121.

Clinkenbeard, E. L., & White, K. E. (2016). Systemic control of bone homeostasis by FGF23 signaling. *Current Molecular Biology Reports*, *2*, 62–71.

Cobrinik, D., Lee, M. H., Hannon, G., Mulligan, G., Bronson, R. T., Dyson, N., et al. (1996). Shared role of the pRB-related p130 and p107 proteins in limb development. *Genes & Development*, *10*, 1633–1644.

Colvin, J. S., Bohne, B. A., Harding, G. W., McEwen, D. G., & Ornitz, D. M. (1996). Skeletal overgrowth and deafness in mice lacking fibroblast growth factor receptor 3. *Nature Genetics*, *12*, 390–397.

Coutu, D. L., Francois, M., & Galipeau, J. (2011). Inhibition of cellular senescence by developmentally regulated FGF receptors in mesenchymal stem cells. *Blood*, *117*, 6801–6812.

Cowan, C. M., Quarto, N., Warren, S. M., Salim, A., & Longaker, M. T. (2003). Age-related changes in the biomolecular mechanisms of calvarial osteoblast biology affect fibroblast growth factor-2 signaling and osteogenesis. *The Journal of Biological Chemistry*, *278*, 32005–32013.

Dailey, L., Ambrosetti, D., Mansukhani, A., & Basilico, C. (2005). Mechanisms underlying differential responses to FGF signaling. *Cytokine & Growth Factor Reviews*, *16*, 233–247.

Dailey, L., Laplantine, E., Priore, R., & Basilico, C. (2003). A network of transcriptional and signaling events is activated by FGF to induce chondrocyte growth arrest and differentiation. *The Journal of Cell Biology*, *161*, 1053–1066.

de Frutos, C. A., Vega, S., Manzanares, M., Flores, J. M., Huertas, H., Martinez-Frias, M. L., et al. (2007). Snail1 is a transcriptional effector of FGFR3 signaling during chondrogenesis and achondroplasias. *Developmental Cell*, *13*, 872–883.

De Rosa, M. L., Fano, V., Araoz, H. V., Chertkoff, L., & Obregon, M. G. (2014). Homozygous N540K hypochondroplasia—First report: Radiological and clinical features. *American Journal of Medical Genetics. Part A*, *164a*, 1784–1788.

Debiais, F., Hott, M., Graulet, A. M., & Marie, P. J. (1998). The effects of fibroblast growth factor-2 on human neonatal calvaria osteoblastic cells are differentiation stage specific. *Journal of Bone and Mineral Research*, *13*, 645–654.

Delezoide, A. L., Benoist-Lasselin, C., Legeai-Mallet, L., Le Merrer, M., Munnich, A., Vekemans, M., et al. (1998). Spatio-temporal expression of FGFR 1, 2 and 3 genes during human embryo-fetal ossification. *Mechanisms of Development*, *77*, 19–30.

Deng, C., Wynshaw-Boris, A., Zhou, F., Kuo, A., & Leder, P. (1996). Fibroblast growth factor receptor 3 is a negative regulator of bone growth. *Cell, 84*, 911–921.

Di Rocco, F., Biosse Duplan, M., Heuze, Y., Kaci, N., Komla-Ebri, D., Munnich, A., et al. (2014). FGFR3 mutation causes abnormal membranous ossification in achondroplasia. *Human Molecular Genetics, 23*, 2914–2925.

Dimitroff, C. J., Klohs, W., Sharma, A., Pera, P., Driscoll, D., Veith, J., et al. (1999). Anti-angiogenic activity of selected receptor tyrosine kinase inhibitors, PD166285 and PD173074: Implications for combination treatment with photodynamic therapy. *Investigational New Drugs, 17*, 121–135.

Dufour, C., Guenou, H., Kaabeche, K., Bouvard, D., Sanjay, A., & Marie, P. J. (2008). FGFR2-Cbl interaction in lipid rafts triggers attenuation of PI3K/Akt signaling and osteoblast survival. *Bone, 42*, 1032–1039.

Escobar, L. F., Tucker, M., & Bamshad, M. (2016). A second family with CATSHL syndrome: Confirmatory report of another unique FGFR3 syndrome. *American Journal of Medical Genetics. Part A, 170*, 1908–1911.

Eswarakumar, V. P., Horowitz, M. C., Locklin, R., Morriss-Kay, G. M., & Lonai, P. (2004). A gain-of-function mutation of Fgfr2c demonstrates the roles of this receptor variant in osteogenesis. *Proceedings of the National Academy of Sciences of the United States of America, 101*, 12555–12560.

Eswarakumar, V. P., Lax, I., & Schlessinger, J. (2005). Cellular signaling by fibroblast growth factor receptors. *Cytokine & Growth Factor Reviews, 16*, 139–149.

Eswarakumar, V. P., Monsonego-Ornan, E., Pines, M., Antonopoulou, I., Morriss-Kay, G. M., & Lonai, P. (2002). The IIIc alternative of Fgfr2 is a positive regulator of bone formation. *Development, 129*, 3783–3793.

Eswarakumar, V. P., Ozcan, F., Lew, E. D., Bae, J. H., Tome, F., Booth, C. J., et al. (2006). Attenuation of signaling pathways stimulated by pathologically activated FGF-receptor 2 mutants prevents craniosynostosis. *Proceedings of the National Academy of Sciences of the United States of America, 103*, 18603–18608.

Eswarakumar, V. P., & Schlessinger, J. (2007). Skeletal overgrowth is mediated by deficiency in a specific isoform of fibroblast growth factor receptor 3. *Proceedings of the National Academy of Sciences of the United States of America, 104*, 3937–3942.

Fakhry, A., Ratisoontorn, C., Vedhachalam, C., Salhab, I., Koyama, E., Leboy, P., et al. (2005). Effects of FGF-2/-9 in calvarial bone cell cultures: Differentiation stage-dependent mitogenic effect, inverse regulation of BMP-2 and noggin, and enhancement of osteogenic potential. *Bone, 36*, 254–266.

Fei, Y., Xiao, L., Doetschman, T., Coffin, D. J., & Hurley, M. M. (2011). Fibroblast growth factor 2 stimulation of osteoblast differentiation and bone formation is mediated by modulation of the Wnt signaling pathway. *The Journal of Biological Chemistry, 286*, 40575–40583.

Fromigue, O., Modrowski, D., & Marie, P. J. (2004). Growth factors and bone formation in osteoporosis: Roles for fibroblast growth factor and transforming growth factor beta. *Current Pharmaceutical Design, 10*, 2593–2603.

Furdui, C. M., Lew, E. D., Schlessinger, J., & Anderson, K. S. (2006). Autophosphorylation of FGFR1 kinase is mediated by a sequential and precisely ordered reaction. *Molecular Cell, 21*, 711–717.

Garcia, S., Dirat, B., Tognacci, T., Rochet, N., Mouska, X., Bonnafous, S., et al. (2013). Postnatal soluble FGFR3 therapy rescues achondroplasia symptoms and restores bone growth in mice. *Science Translational Medicine, 5*, 203ra124.

Garofalo, S., Kliger-Spatz, M., Cooke, J. L., Wolstin, O., Lunstrum, G. P., Moshkovitz, S. M., et al. (1999). Skeletal dysplasia and defective chondrocyte differentiation by targeted overexpression of fibroblast growth factor 9 in transgenic mice. *Journal of Bone and Mineral Research, 14*, 1909–1915.

Globus, R. K., Patterson-Buckendahl, P., & Gospodarowicz, D. (1988). Regulation of bovine bone cell proliferation by fibroblast growth factor and transforming growth factor beta. *Endocrinology*, *123*, 98–105.

Goetz, R., & Mohammadi, M. (2013). Exploring mechanisms of FGF signalling through the lens of structural biology. *Nature Reviews. Molecular Cell Biology*, *14*, 166–180.

Gorbenko, O., Ovcharenko, G., Klymenko, T., Zhyvoloup, O., Gaman, N., Volkova, D., et al. (2009). Generation of monoclonal antibody targeting fibroblast growth factor receptor 3. *Hybridoma (Larchmt)*, *28*, 295–300.

Gould, S. E., Upholt, W. B., & Kosher, R. A. (1992). Syndecan 3: A member of the syndecan family of membrane-intercalated proteoglycans that is expressed in high amounts at the onset of chicken limb cartilage differentiation. *Proceedings of the National Academy of Sciences of the United States of America*, *89*, 3271–3275.

Gudernova, I., Vesela, I., Balek, L., Buchtova, M., Dosedelova, H., Kunova, M., et al. (2016). Multikinase activity of fibroblast growth factor receptor (FGFR) inhibitors SU5402, PD173074, AZD1480, AZD4547 and BGJ398 compromises the use of small chemicals targeting FGFR catalytic activity for therapy of short-stature syndromes. *Human Molecular Genetics*, *25*, 9–23.

Guenou, H., Kaabeche, K., Mee, S. L., & Marie, P. J. (2005). A role for fibroblast growth factor receptor-2 in the altered osteoblast phenotype induced by Twist haploinsufficiency in the Saethre-Chotzen syndrome. *Human Molecular Genetics*, *14*, 1429–1439.

Gust, K. M., McConkey, D. J., Awrey, S., Hegarty, P. K., Qing, J., Bondaruk, J., et al. (2013). Fibroblast growth factor receptor 3 is a rational therapeutic target in bladder cancer. *Molecular Cancer Therapeutics*, *12*, 1245–1254.

Hadari, Y., & Schlessinger, J. (2009). FGFR3-targeted mAb therapy for bladder cancer and multiple myeloma. *The Journal of Clinical Investigation*, *119*, 1077–1079.

Hajihosseini, M. K. (2008). Fibroblast growth factor signaling in cranial suture development and pathogenesis. *Frontiers of Oral Biology*, *12*, 160–177.

Hamada, T., Suda, N., & Kuroda, T. (1999). Immunohistochemical localization of fibroblast growth factor receptors in the rat mandibular condylar cartilage and tibial cartilage. *Journal of Bone and Mineral Metabolism*, *17*, 274–282.

Hamidouche, Z., Fromigue, O., Nuber, U., Vaudin, P., Pages, J. C., Ebert, R., et al. (2010). Autocrine fibroblast growth factor 18 mediates dexamethasone-induced osteogenic differentiation of murine mesenchymal stem cells. *Journal of Cellular Physiology*, *224*, 509–515.

Han, X., Xiao, Z., & Quarles, L. D. (2015). Membrane and integrative nuclear fibroblastic growth factor receptor (FGFR) regulation of FGF-23. *The Journal of Biological Chemistry*, *290*, 10447–10459.

Hanafusa, H., Torii, S., Yasunaga, T., & Nishida, E. (2002). Sprouty1 and Sprouty2 provide a control mechanism for the Ras/MAPK signalling pathway. *Nature Cell Biology*, *4*, 850–858.

Haupt, L. M., Murali, S., Mun, F. K., Teplyuk, N., Mei, L. F., Stein, G. S., et al. (2009). The heparan sulfate proteoglycan (HSPG) glypican-3 mediates commitment of MC3T3-E1 cells toward osteogenesis. *Journal of Cellular Physiology*, *220*, 780–791.

Havens, B. A., Velonis, D., Kronenberg, M. S., Lichtler, A. C., Oliver, B., & Mina, M. (2008). Roles of FGFR3 during morphogenesis of Meckel's cartilage and mandibular bones. *Developmental Biology*, *316*, 336–349.

Heuze, Y., Singh, N., Basilico, C., Jabs, E. W., Holmes, G., & Richtsmeier, J. T. (2014). Morphological comparison of the craniofacial phenotypes of mouse models expressing the Apert FGFR2 S252W mutation in neural crest- or mesoderm-derived tissues. *Bone*, *63*, 101–109.

Hinoi, E., Bialek, P., Chen, Y. T., Rached, M. T., Groner, Y., Behringer, R. R., et al. (2006). Runx2 inhibits chondrocyte proliferation and hypertrophy through its expression in the perichondrium. *Genes & Development*, *20*, 2937–2942.

Holmes, G., Rothschild, G., Roy, U. B., Deng, C. X., Mansukhani, A., & Basilico, C. (2009). Early onset of craniosynostosis in an Apert mouse model reveals critical features of this pathology. *Developmental Biology*, *328*, 273–284.
Holmes, G., Zhang, L., Rivera, J., Murphy, R., Assouline, C., Sullivan, L., et al. (2018). C-type natriuretic peptide analog treatment of craniosynostosis in a Crouzon syndrome mouse model. *PLoS One*, *13*, e0201492.
Homer-Bouthiette, C., Doetschman, T., Xiao, L., & Hurley, M. M. (2014). Knockout of nuclear high molecular weight FGF2 isoforms in mice modulates bone and phosphate homeostasis. *The Journal of Biological Chemistry*, *289*, 36303–36314.
Horton, W. A., Hall, J. G., & Hecht, J. T. (2007). Achondroplasia. *Lancet*, *370*, 162–172.
House, S. L., Branch, K., Newman, G., Doetschman, T., & Schultz Jel, J. (2005). Cardioprotection induced by cardiac-specific overexpression of fibroblast growth factor-2 is mediated by the MAPK cascade. *American Journal of Physiology. Heart and Circulatory Physiology*, *289*, H2167–H2175.
Hung, I. H., Schoenwolf, G. C., Lewandoski, M., & Ornitz, D. M. (2016). A combined series of Fgf9 and Fgf18 mutant alleles identifies unique and redundant roles in skeletal development. *Developmental Biology*, *411*, 72–84.
Hung, I. H., Yu, K., Lavine, K. J., & Ornitz, D. M. (2007). FGF9 regulates early hypertrophic chondrocyte differentiation and skeletal vascularization in the developing stylopod. *Developmental Biology*, *307*, 300–313.
Hurley, M. M., Tetradis, S., Huang, Y. F., Hock, J., Kream, B. E., Raisz, L. G., et al. (1999). Parathyroid hormone regulates the expression of fibroblast growth factor-2 mRNA and fibroblast growth factor receptor mRNA in osteoblastic cells. *Journal of Bone and Mineral Research*, *14*, 776–783.
Ibrahimi, O. A., Eliseenkova, A. V., Plotnikov, A. N., Yu, K., Ornitz, D. M., & Mohammadi, M. (2001). Structural basis for fibroblast growth factor receptor 2 activation in Apert syndrome. *Proceedings of the National Academy of Sciences of the United States of America*, *5*, 5.
Ignelzi, M. A., Jr., Wang, W., & Young, A. T. (2003). Fibroblast growth factors lead to increased Msx2 expression and fusion in calvarial sutures. *Journal of Bone and Mineral Research*, *18*, 751–759.
Ikpegbu, E., Basta, L., Clements, D. N., Fleming, R., Vincent, T. L., Buttle, D. J., et al. (2018). FGF-2 promotes osteocyte differentiation through increased E11/podoplanin expression. *Journal of Cellular Physiology*, *233*, 5334–5347.
Iwata, T., Chen, L., Li, C., Ovchinnikov, D. A., Behringer, R. R., Francomano, C. A., et al. (2000). A neonatal lethal mutation in FGFR3 uncouples proliferation and differentiation of growth plate chondrocytes in embryos. *Human Molecular Genetics*, *9*, 1603–1613.
Iwata, T., Li, C. L., Deng, C. X., & Francomano, C. A. (2001). Highly activated Fgfr3 with the K644M mutation causes prolonged survival in severe dwarf mice. *Human Molecular Genetics*, *10*, 1255–1264.
Jackson, R. A., Murali, S., van Wijnen, A. J., Stein, G. S., Nurcombe, V., & Cool, S. M. (2007). Heparan sulfate regulates the anabolic activity of MC3T3-E1 preosteoblast cells by induction of Runx2. *Journal of Cellular Physiology*, *210*, 38–50.
Jacob, A. L., Smith, C., Partanen, J., & Ornitz, D. M. (2006). Fibroblast growth factor receptor 1 signaling in the osteo-chondrogenic cell lineage regulates sequential steps of osteoblast maturation. *Developmental Biology*, *296*, 315–328.
Jeon, E., Yun, Y. R., Kang, W., Lee, S., Koh, Y. H., Kim, H. W., et al. (2012). Investigating the role of FGF18 in the cultivation and osteogenic differentiation of mesenchymal stem cells. *PLoS One*, 7, e43982.
Johnson, D., & Wilkie, A. O. (2011). Craniosynostosis. *European Journal of Human Genetics*, *19*, 369–376.

Jonquoy, A., Mugniery, E., Benoist-Lasselin, C., Kaci, N., Le Corre, L., Barbault, F., et al. (2012). A novel tyrosine kinase inhibitor restores chondrocyte differentiation and promotes bone growth in a gain-of-function Fgfr3 mouse model. *Human Molecular Genetics*, *21*, 841–851.

Kaabeche, K., Guenou, H., Bouvard, D., Didelot, N., Listrat, A., & Marie, P. J. (2005). Cbl-mediated ubiquitination of alpha5 integrin subunit mediates fibronectin-dependent osteoblast detachment and apoptosis induced by FGFR2 activation. *Journal of Cell Science*, *118*, 1223–1232.

Kaabeche, K., Lemonnier, J., Le Mee, S., Caverzasio, J., & Marie, P. J. (2004). Cbl-mediated degradation of Lyn and Fyn induced by constitutive fibroblast growth factor receptor-2 activation supports osteoblast differentiation. *The Journal of Biological Chemistry*, *279*, 36259–36267.

Kanazawa, S., Fujiwara, T., Matsuzaki, S., Shingaki, K., Taniguchi, M., Miyata, S., et al. (2010). bFGF regulates PI3-kinase-Rac1-JNK pathway and promotes fibroblast migration in wound healing. *PLoS One*, *5*, e12228.

Kant, S. G., Cervenkova, I., Balek, L., Trantirek, L., Santen, G. W., de Vries, M. C., et al. (2015). A novel variant of FGFR3 causes proportionate short stature. *European Journal of Endocrinology*, *172*, 763–770.

Karolak, M. R., Yang, X., & Elefteriou, F. (2015). FGFR1 signaling in hypertrophic chondrocytes is attenuated by the Ras-GAP neurofibromin during endochondral bone formation. *Human Molecular Genetics*, *24*, 2552–2564.

Karuppaiah, K., Yu, K., Lim, J., Chen, J., Smith, C., Long, F., et al. (2016). FGF signaling in the osteoprogenitor lineage non-autonomously regulates postnatal chondrocyte proliferation and skeletal growth. *Development*, *143*, 1811–1822.

Kim, H. J., Kim, J. H., Bae, S. C., Choi, J. Y., Kim, H. J., & Ryoo, H. M. (2003). The protein kinase C pathway plays a central role in the fibroblast growth factor-stimulated expression and transactivation activity of Runx2. *The Journal of Biological Chemistry*, *278*, 319–326.

Kim, H. J., Rice, D. P., Kettunen, P. J., & Thesleff, I. (1998). FGF-, BMP- and Shh-mediated signalling pathways in the regulation of cranial suture morphogenesis and calvarial bone development. *Development*, *125*, 1241–1251.

Kolupaeva, V., Daempfling, L., & Basilico, C. (2013). The B55alpha regulatory subunit of protein phosphatase 2A mediates fibroblast growth factor-induced p107 dephosphorylation and growth arrest in chondrocytes. *Molecular and Cellular Biology*, *33*, 2865–2878.

Kolupaeva, V., Laplantine, E., & Basilico, C. (2008). PP2A-mediated dephosphorylation of p107 plays a critical role in chondrocyte cell cycle arrest by FGF. *PLoS One*, *3*, e3447.

Komla-Ebri, D., Dambroise, E., Kramer, I., Benoist-Lasselin, C., Kaci, N., Le Gall, C., et al. (2016). Tyrosine kinase inhibitor NVP-BGJ398 functionally improves FGFR3-related dwarfism in mouse model. *The Journal of Clinical Investigation*, *126*, 1871–1884.

Kouhara, H., Hadari, Y. R., Spivak-Kroizman, T., Schilling, J., Bar-Sagi, D., Lax, I., et al. (1997). A lipid-anchored Grb2-binding protein that links FGF-receptor activation to the Ras/MAPK signaling pathway. *Cell*, *89*, 693–702.

Krejci, P., Aklian, A., Kaucka, M., Sevcikova, E., Prochazkova, J., Masek, J. K., et al. (2012). Receptor tyrosine kinases activate canonical WNT/beta-catenin signaling via MAP kinase/LRP6 pathway and direct beta-catenin phosphorylation. *PLoS One*, *7*, e35826.

Krejci, P., Salazar, L., Goodridge, H. S., Kashiwada, T. A., Schibler, M. J., Jelinkova, P., et al. (2008). STAT1 and STAT3 do not participate in FGF-mediated growth arrest in chondrocytes. *Journal of Cell Science*, *121*, 272–281.

Krejci, P., Salazar, L., Kashiwada, T. A., Chlebova, K., Salasova, A., Thompson, L. M., et al. (2008). Analysis of STAT1 activation by six FGFR3 mutants associated with skeletal

dysplasia undermines dominant role of STAT1 in FGFR3 signaling in cartilage. *PLoS One*, *3*, e3961.
Kuhn, L. T., Ou, G., Charles, L., Hurley, M. M., Rodner, C. M., & Gronowicz, G. (2013). Fibroblast growth factor-2 and bone morphogenetic protein-2 have a synergistic stimulatory effect on bone formation in cell cultures from elderly mouse and human bone. *The Journals of Gerontology. Series A, Biological Sciences and Medical Sciences*, *68*, 1170–1180.
Kumar, D., & Lassar, A. B. (2014). Fibroblast growth factor maintains chondrogenic potential of limb bud mesenchymal cells by modulating DNMT3A recruitment. *Cell Reports*, *8*, 1419–1431.
Kunova Bosakova, M., Varecha, M., Hampl, M., Duran, I., Nita, A., Buchtova, M., et al. (2018). Regulation of ciliary function by fibroblast growth factor signaling identifies FGFR3-related disorders achondroplasia and thanatophoric dysplasia as ciliopathies. *Human Molecular Genetics*, 27, 1093–1105.
Kurimchak, A., Haines, D. S., Garriga, J., Wu, S., De Luca, F., Sweredoski, M. J., et al. (2013). Activation of p107 by fibroblast growth factor, which is essential for chondrocyte cell cycle exit, is mediated by the protein phosphatase 2A/B55alpha holoenzyme. *Molecular and Cellular Biology*, *33*, 3330–3342.
Kyono, A., Avishai, N., Ouyang, Z., Landreth, G. E., & Murakami, S. (2012). FGF and ERK signaling coordinately regulate mineralization-related genes and play essential roles in osteocyte differentiation. *Journal of Bone and Mineral Metabolism*, *30*, 19–30.
Lamothe, B., Yamada, M., Schaeper, U., Birchmeier, W., Lax, I., & Schlessinger, J. (2004). The docking protein Gab1 is an essential component of an indirect mechanism for fibroblast growth factor stimulation of the phosphatidylinositol 3-kinase/Akt anti-apoptotic pathway. *Molecular and Cellular Biology*, *24*, 5657–5666.
Laplantine, E., Rossi, F., Sahni, M., Basilico, C., & Cobrinik, D. (2002). FGF signaling targets the pRb-related p107 and p130 proteins to induce chondrocyte growth arrest. *The Journal of Cell Biology*, *158*, 741–750.
Lazarus, J. E., Hegde, A., Andrade, A. C., Nilsson, O., & Baron, J. (2007). Fibroblast growth factor expression in the postnatal growth plate. *Bone*, *40*, 577–586.
Lee, Y. C., Song, I. W., Pai, Y. J., Chen, S. D., & Chen, Y. T. (2017). Knock-in human FGFR3 achondroplasia mutation as a mouse model for human skeletal dysplasia. *Scientific Reports*, 7, 43220.
Legeai-Mallet, L., Benoist-Lasselin, C., Munnich, A., & Bonaventure, J. (2004). Overexpression of FGFR3, Stat1, Stat5 and p21Cip1 correlates with phenotypic severity and defective chondrocyte differentiation in FGFR3-related chondrodysplasias. *Bone*, *34*, 26–36.
Lemmon, M. A., & Schlessinger, J. (2010). Cell signaling by receptor tyrosine kinases. *Cell*, *141*, 1117–1134.
Lemonnier, J., Hay, E., Delannoy, P., Fromigue, O., Lomri, A., Modrowski, D., et al. (2001). Increased osteoblast apoptosis in apert craniosynostosis: Role of protein kinase C and interleukin-1. *The American Journal of Pathology*, *158*, 1833–1842.
Lemonnier, J., Hay, E., Delannoy, P., Lomri, A., Modrowski, D., Caverzasio, J., et al. (2001). Role of N-cadherin and protein kinase C in osteoblast gene activation induced by the S252W fibroblast growth factor receptor 2 mutation in Apert craniosynostosis. *Journal of Bone and Mineral Research*, *16*, 832–845.
Li, C., Chen, L., Iwata, T., Kitagawa, M., Fu, X. Y., & Deng, C. X. (1999). A Lys644Glu substitution in fibroblast growth factor receptor 3 (FGFR3) causes dwarfism in mice by activation of STATs and ink4 cell cycle inhibitors. *Human Molecular Genetics*, *8*, 35–44.
Li, M., Seki, Y., Freitas, P. H., Nagata, M., Kojima, T., Sultana, S., et al. (2010). FGFR3 down-regulates PTH/PTHrP receptor gene expression by mediating JAK/STAT signaling in chondrocytic cell line. *Journal of Electron Microscopy*, *59*, 227–236.

Liao, S., Porter, D., Scott, A., Newman, G., Doetschman, T., & Schultz Jel, J. (2007). The cardioprotective effect of the low molecular weight isoform of fibroblast growth factor-2: The role of JNK signaling. *Journal of Molecular and Cellular Cardiology*, *42*, 106–120.
Liu, S., Guo, R., Simpson, L. G., Xiao, Z. S., Burnham, C. E., & Quarles, L. D. (2003). Regulation of fibroblastic growth factor 23 expression but not degradation by PHEX. *The Journal of Biological Chemistry*, *278*, 37419–37426.
Liu, J., Kwon, T. G., Nam, H. K., & Hatch, N. E. (2013). Craniosynostosis-associated Fgfr2 (C342Y) mutant bone marrow stromal cells exhibit cell autonomous abnormalities in osteoblast differentiation and bone formation. *BioMed Research International*, *2013*, 292506.
Liu, Z., Lavine, K. J., Hung, I. H., & Ornitz, D. M. (2007). FGF18 is required for early chondrocyte proliferation, hypertrophy and vascular invasion of the growth plate. *Developmental Biology*, *302*, 80–91.
Liu, J., Nam, H. K., Wang, E., & Hatch, N. E. (2013). Further analysis of the Crouzon mouse: Effects of the FGFR2(C342Y) mutation are cranial bone-dependent. *Calcified Tissue International*, *92*, 451–466.
Liu, Z., Xu, J., Colvin, J. S., & Ornitz, D. M. (2002). Coordination of chondrogenesis and osteogenesis by fibroblast growth factor 18. *Genes & Development*, *16*, 859–869.
Liu, S., Zhou, J., Tang, W., Jiang, X., Rowe, D. W., & Quarles, L. D. (2006). Pathogenic role of Fgf23 in Hyp mice. *American Journal of Physiology. Endocrinology and Metabolism*, *291*, E38–E49.
Lomri, A., Lemonnier, J., Delannoy, P., & Marie, P. J. (2001). Increased expression of protein kinase C alpha, interleukin-1 alpha, and RhoA guanosine 5 '-triphosphatase in osteoblasts expressing the Ser252Trp fibroblast growth factor 2 Apert mutation: Identification by analysis of complementary DNA microarray. *Journal of Bone and Mineral Research*, *16*, 705–712.
Lomri, A., Lemonnier, J., Hott, M., de Parseval, N., Lajeunie, E., Munnich, A., et al. (1998). Increased calvaria cell differentiation and bone matrix formation induced by fibroblast growth factor receptor 2 mutations in Apert syndrome. *The Journal of Clinical Investigation*, *101*, 1310–1317.
Lorget, F., Kaci, N., Peng, J., Benoist-Lasselin, C., Mugniery, E., Oppeneer, T., et al. (2012). Evaluation of the therapeutic potential of a CNP analog in a Fgfr3 mouse model recapitulating Achondroplasia. *American Journal of Human Genetics*, *91*, 1108–1114.
Makrythanasis, P., Temtamy, S., Aglan, M. S., Otaify, G. A., Hamamy, H., & Antonarakis, S. E. (2014). A novel homozygous mutation in FGFR3 causes tall stature, severe lateral tibial deviation, scoliosis, hearing impairment, camptodactyly, and arachnodactyly. *Human Mutation*, *35*, 959–963.
Mansouri, R., Hay, E., Marie, P. J., & Modrowski, D. (2015). Role of syndecan-2 in osteoblast biology and pathology. *BoneKEy Reports*, *4*, 666.
Mansukhani, A., Ambrosetti, D., Holmes, G., Cornivelli, L., & Basilico, C. (2005). Sox2 induction by FGF and FGFR2 activating mutations inhibits Wnt signaling and osteoblast differentiation. *The Journal of Cell Biology*, *168*, 1065–1076.
Mansukhani, A., Bellosta, P., Sahni, M., & Basilico, C. (2000). Signaling by fibroblast growth factors (FGF) and fibroblast growth factor receptor 2 (FGFR2)-activating mutations blocks mineralization and induces apoptosis in osteoblasts. *The Journal of Cell Biology*, *149*, 1297–1308.
Mariani, F. V., Ahn, C. P., & Martin, G. R. (2008). Genetic evidence that FGFs have an instructive role in limb proximal-distal patterning. *Nature*, *453*, 401–405.
Marie, P. J. (2012). Fibroblast growth factor signaling controlling bone formation: An update. *Gene*, *498*, 1–4.
Marie, P. J. (2015). Osteoblast dysfunctions in bone diseases: From cellular and molecular mechanisms to therapeutic strategies. *Cellular and Molecular Life Sciences*, *72*, 1347–1361.

Marie, P. J., Coffin, J. D., & Hurley, M. M. (2005). FGF and FGFR signaling in chondrodysplasias and craniosynostosis. *Journal of Cellular Biochemistry*, *96*, 888–896.

Marie, P. J., Debiais, F., & Hay, E. (2002). Regulation of human cranial osteoblast phenotype by FGF-2, FGFR-2 and BMP-2 signaling. *Histology and Histopathology*, *17*, 877–885.

Marie, P. J., Kaabeche, K., & Guenou, H. (2008). Roles of FGFR2 and twist in human craniosynostosis: Insights from genetic mutations in cranial osteoblasts. *Frontiers of Oral Biology*, *12*, 144–159.

Martin, G. R. (1998). The roles of FGFs in the early development of vertebrate limbs. *Genes & Development*, *12*, 1571–1586.

Martin, A., Liu, S., David, V., Li, H., Karydis, A., Feng, J. Q., et al. (2011). Bone proteins PHEX and DMP1 regulate fibroblastic growth factor Fgf23 expression in osteocytes through a common pathway involving FGF receptor (FGFR) signaling. *The FASEB Journal*, *25*, 2551–2562.

Martinez-Torrecuadrada, J., Cifuentes, G., Lopez-Serra, P., Saenz, P., Martinez, A., & Casal, J. I. (2005). Targeting the extracellular domain of fibroblast growth factor receptor 3 with human single-chain Fv antibodies inhibits bladder carcinoma cell line proliferation. *Clinical Cancer Research*, *11*, 6280–6290.

Matsushita, T., Wilcox, W. R., Chan, Y. Y., Kawanami, A., Bukulmez, H., Balmes, G., et al. (2009). FGFR3 promotes synchondrosis closure and fusion of ossification centers through the MAPK pathway. *Human Molecular Genetics*, *18*, 227–240.

McDowell, L. M., Frazier, B. A., Studelska, D. R., Giljum, K., Chen, J., Liu, J., et al. (2006). Inhibition or activation of Apert syndrome FGFR2 (S252W) signaling by specific glycosaminoglycans. *The Journal of Biological Chemistry*, *281*, 6924–6930.

Melville, H., Wang, Y., Taub, P. J., & Jabs, E. W. (2010). Genetic basis of potential therapeutic strategies for craniosynostosis. *American Journal of Medical Genetics. Part A*, *152A*, 3007–3015.

Merrill, A. E., Sarukhanov, A., Krejci, P., Idoni, B., Camacho, N., Estrada, K. D., et al. (2012). Bent bone dysplasia-FGFR2 type, a distinct skeletal disorder, has deficient canonical FGF signaling. *American Journal of Human Genetics*, *90*, 550–557.

Miraoui, H., & Marie, P. J. (2010). Fibroblast growth factor receptor signaling crosstalk in skeletogenesis. *Science Signaling*, *3*, re9.

Miraoui, H., Oudina, K., Petite, H., Tanimoto, Y., Moriyama, K., & Marie, P. J. (2009). Fibroblast growth factor receptor 2 promotes osteogenic differentiation in mesenchymal cells via ERK1/2 and protein kinase C signaling. *The Journal of Biological Chemistry*, *284*, 4897–4904.

Miraoui, H., Ringe, J., Haupl, T., & Marie, P. J. (2010). Increased EFG- and PDGFalpha-receptor signaling by mutant FGF-receptor 2 contributes to osteoblast dysfunction in Apert craniosynostosis. *Human Molecular Genetics*, *19*, 1678–1689.

Miraoui, H., Severe, N., Vaudin, P., Pages, J. C., & Marie, P. J. (2010). Molecular silencing of Twist1 enhances osteogenic differentiation of murine mesenchymal stem cells: Implication of FGFR2 signaling. *Journal of Cellular Biochemistry*, *110*, 1147–1154.

Moenning, A., Jager, R., Egert, A., Kress, W., Wardelmann, E., & Schorle, H. (2009). Sustained platelet-derived growth factor receptor alpha signaling in osteoblasts results in craniosynostosis by overactivating the phospholipase C-gamma pathway. *Molecular and Cellular Biology*, *29*, 881–891.

Mohammadi, M., McMahon, G., Sun, L., Tang, C., Hirth, P., Yeh, B. K., et al. (1997). Structures of the tyrosine kinase domain of fibroblast growth factor receptor in complex with inhibitors. *Science*, *276*, 955–960.

Molteni, A., Modrowski, D., Hott, M., & Marie, P. J. (1999a). Alterations of matrix- and cell-associated proteoglycans inhibit osteogenesis and growth response to fibroblast growth factor-2 in cultured rat mandibular condyle and calvaria. *Cell and Tissue Research*, *295*, 523–536.

Molteni, A., Modrowski, D., Hott, M., & Marie, P. J. (1999b). Differential expression of fibroblast growth factor receptor-1, -2, and -3 and syndecan-1, -2, and -4 in neonatal rat mandibular condyle and calvaria during osteogenic differentiation in vitro. *Bone, 24*, 337–347.

Montero, A., Okada, Y., Tomita, M., Ito, M., Tsurukami, H., Nakamura, T., et al. (2000). Disruption of the fibroblast growth factor-2 gene results in decreased bone mass and bone formation. *The Journal of Clinical Investigation, 105*, 1085–1093.

Montone, R., Romanelli, M. G., Baruzzi, A., Ferrarini, F., Liboi, E., & Lievens, P. M. (2018). Mutant FGFR3 associated with SADDAN disease causes cytoskeleton disorganization through PLCgamma1/Src-mediated paxillin hyperphosphorylation. *The International Journal of Biochemistry & Cell Biology, 95*, 17–26.

Morita, J., Nakamura, M., Kobayashi, Y., Deng, C. X., Funato, N., & Moriyama, K. (2014). Soluble form of FGFR2 with S252W partially prevents craniosynostosis of the apert mouse model. *Developmental Dynamics, 243*, 560–567.

Motch Perrine, S. M., Cole, T. M., 3rd, Martinez-Abadias, N., Aldridge, K., Jabs, E. W., & Richtsmeier, J. T. (2014). Craniofacial divergence by distinct prenatal growth patterns in Fgfr2 mutant mice. *BMC Developmental Biology, 14*, 8.

Mugniery, E., Dacquin, R., Marty, C., Benoist-Lasselin, C., de Vernejoul, M. C., Jurdic, P., et al. (2012). An activating Fgfr3 mutation affects trabecular bone formation via a paracrine mechanism during growth. *Human Molecular Genetics, 21*, 2503–2513.

Murakami, S., Balmes, G., McKinney, S., Zhang, Z., Givol, D., & De Crombrugghe, B. (2004). Constitutive activation of MEK1 in chondrocytes causes Stat1-independent achondroplasia-like dwarfism and rescues the Fgfr3-deficient mouse phenotype. *Genes & Development, 18*, 290–305.

Murakami, S., Kan, M., McKeehan, W. L., & de Crombrugghe, B. (2000). Up-regulation of the chondrogenic Sox9 gene by fibroblast growth factors is mediated by the mitogen-activated protein kinase pathway. *Proceedings of the National Academy of Sciences of the United States of America, 97*, 1113–1118.

Murdoch, J. L., Walker, B. A., Hall, J. G., Abbey, H., Smith, K. K., & McKusick, V. A. (1970). Achondroplasia—a genetic and statistical survey. *Annals of Human Genetics, 33*, 227–244.

Naganawa, T., Xiao, L., Coffin, J. D., Doetschman, T., Sabbieti, M. G., Agas, D., et al. (2008). Reduced expression and function of bone morphogenetic protein-2 in bones of Fgf2 null mice. *Journal of Cellular Biochemistry, 103*, 1975–1988.

Nakamura, Y., Tensho, K., Nakaya, H., Nawata, M., Okabe, T., & Wakitani, S. (2005). Low dose fibroblast growth factor-2 (FGF-2) enhances bone morphogenetic protein-2 (BMP-2)-induced ectopic bone formation in mice. *Bone, 36*, 399–407.

Naski, M. C., Colvin, J. S., Coffin, J. D., & Ornitz, D. M. (1998). Repression of hedgehog signaling and BMP4 expression in growth plate cartilage by fibroblast growth factor receptor 3. *Development, 125*, 4977–4988.

Naski, M. C., Wang, Q., Xu, J., & Ornitz, D. M. (1996). Graded activation of fibroblast growth factor receptor 3 by mutations causing achondroplasia and thanatophoric dysplasia. *Nature Genetics, 13*, 233–237.

Neben, C. L., Idoni, B., Salva, J. E., Tuzon, C. T., Rice, J. C., Krakow, D., et al. (2014). Bent bone dysplasia syndrome reveals nucleolar activity for FGFR2 in ribosomal DNA transcription. *Human Molecular Genetics, 23*, 5659–5671.

Neben, C. L., & Merrill, A. E. (2015). Signaling pathways in craniofacial development: Insights from rare skeletal disorders. *Current Topics in Developmental Biology, 115*, 493–542.

Neben, C. L., Tuzon, C. T., Mao, X., Lay, F. D., & Merrill, A. E. (2017). FGFR2 mutations in bent bone dysplasia syndrome activate nucleolar stress and perturb cell fate determination. *Human Molecular Genetics, 26*, 3253–3270.

Niger, C., Luciotti, M. A., Buo, A. M., Hebert, C., Ma, V., & Stains, J. P. (2013). The regulation of runt-related transcription factor 2 by fibroblast growth factor-2 and connexin43 requires the inositol polyphosphate/protein kinase Cdelta cascade. *Journal of Bone and Mineral Research*, *28*, 1468–1477.

Nogami, K., Suzuki, H., Habuchi, H., Ishiguro, N., Iwata, H., & Kimata, K. (2004). Distinctive expression patterns of heparan sulfate O-sulfotransferases and regional differences in heparan sulfate structure in chick limb buds. *The Journal of Biological Chemistry*, *279*, 8219–8229.

Oh, C. D., Lu, Y., Liang, S., Mori-Akiyama, Y., Chen, D., de Crombrugghe, B., et al. (2014). SOX9 regulates multiple genes in chondrocytes, including genes encoding ECM proteins, ECM modification enzymes, receptors, and transporters. *PLoS One*, *9*, e107577.

Ohbayashi, N., Shibayama, M., Kurotaki, Y., Imanishi, M., Fujimori, T., Itoh, N., et al. (2002). FGF18 is required for normal cell proliferation and differentiation during osteogenesis and chondrogenesis. *Genes & Development*, *16*, 870–879.

Ornitz, D. M. (2000). FGFs, heparan sulfate and FGFRs: Complex interactions essential for development. *BioEssays*, *22*, 108–112.

Ornitz, D. M., & Itoh, N. (2015). The fibroblast growth factor signaling pathway. *Wiley Interdisciplinary Reviews: Developmental Biology*, *4*, 215–266.

Ornitz, D. M., & Legeai-Mallet, L. (2017). Achondroplasia: Development, pathogenesis, and therapy. *Developmental Dynamics*, *246*, 291–309.

Ornitz, D. M., & Marie, P. J. (2002). FGF signaling pathways in endochondral and intramembranous bone development and human genetic disease. *Genes & Development*, *16*, 1446–1465.

Ornitz, D. M., & Marie, P. J. (2015). Fibroblast growth factor signaling in skeletal development and disease. *Genes & Development*, *29*, 1463–1486.

Orr-Urtreger, A., Givol, D., Yayon, A., Yarden, Y., & Lonai, P. (1991). Developmental expression of two murine fibroblast growth factor receptors, flg and bek. *Development*, *113*, 1419–1434.

Pannier, S., Couloigner, V., Messaddeq, N., Elmaleh-Berges, M., Munnich, A., Romand, R., et al. (2009). Activating Fgfr3 Y367C mutation causes hearing loss and inner ear defect in a mouse model of chondrodysplasia. *Biochimica et Biophysica Acta*, *1792*, 140–147.

Park, O. J., Kim, H. J., Woo, K. M., Baek, J. H., & Ryoo, H. M. (2010). FGF2-activated ERK mitogen-activated protein kinase enhances Runx2 acetylation and stabilization. *The Journal of Biological Chemistry*, *285*, 3568–3574.

Patani, H., Bunney, T. D., Thiyagarajan, N., Norman, R. A., Ogg, D., Breed, J., et al. (2016). Landscape of activating cancer mutations in FGFR kinases and their differential responses to inhibitors in clinical use. *Oncotarget*, 7, 24252–24268.

Patel, M. D., & Filly, R. A. (1995). Homozygous achondroplasia: US distinction between homozygous, heterozygous, and unaffected fetuses in the second trimester. *Radiology*, *196*, 541–545.

Pauli, R. M., Conroy, M. M., Langer, L. O., Jr., McLone, D. G., Naidich, T., Franciosi, R., et al. (1983). Homozygous achondroplasia with survival beyond infancy. *American Journal of Medical Genetics*, *16*, 459–473.

Perlyn, C. A., Morriss-Kay, G., Darvann, T., Tenenbaum, M., & Ornitz, D. M. (2006). A model for the pharmacological treatment of crouzon syndrome. *Neurosurgery*, *59*, 210–215. discussion 210–215.

Peters, K., Ornitz, D., Werner, S., & Williams, L. (1993). Unique expression pattern of the FGF receptor 3 gene during mouse organogenesis. *Developmental Biology*, *155*, 423–430.

Peters, K. G., Werner, S., Chen, G., & Williams, L. T. (1992). Two FGF receptor genes are differentially expressed in epithelial and mesenchymal tissues during limb formation and organogenesis in the mouse. *Development*, *114*, 233–243.

Pfaff, M. J., Xue, K., Li, L., Horowitz, M. C., Steinbacher, D. M., & Eswarakumar, J. V. P. (2016). FGFR2c-mediated ERK-MAPK activity regulates coronal suture development. *Developmental Biology*, *415*, 242–250.

Priore, R., Dailey, L., & Basilico, C. (2006). Downregulation of Akt activity contributes to the growth arrest induced by FGF in chondrocytes. *Journal of Cellular Physiology*, *207*, 800–808.

Purcell, P., Joo, B. W., Hu, J. K., Tran, P. V., Calicchio, M. L., O'Connell, D. J., et al. (2009). Temporomandibular joint formation requires two distinct hedgehog-dependent steps. *Proceedings of the National Academy of Sciences of the United States of America*, *106*, 18297–18302.

Qi, H., Jin, M., Duan, Y., Du, X., Zhang, Y., Ren, F., et al. (2014). FGFR3 induces degradation of BMP type I receptor to regulate skeletal development. *Biochimica et Biophysica Acta*, *1843*, 1237–1247.

Qing, J., Du, X., Chen, Y., Chan, P., Li, H., Wu, P., et al. (2009). Antibody-based targeting of FGFR3 in bladder carcinoma and t(4,14)-positive multiple myeloma in mice. *The Journal of Clinical Investigation*, *119*, 1216–1229.

Quarto, N., Behr, B., Li, S., & Longaker, M. T. (2009). Differential FGF ligands and FGF receptors expression pattern in frontal and parietal calvarial bones. *Cells, Tissues, Organs*, *190*, 158–169.

Raucci, A., Laplantine, E., Mansukhani, A., & Basilico, C. (2004). Activation of the ERK1/2 and p38 mitogen-activated protein kinase pathways mediates fibroblast growth factor-induced growth arrest of chondrocytes. *The Journal of Biological Chemistry*, *279*, 1747–1756.

Rauchenberger, R., Borges, E., Thomassen-Wolf, E., Rom, E., Adar, R., Yaniv, Y., et al. (2003). Human combinatorial Fab library yielding specific and functional antibodies against the human fibroblast growth factor receptor 3. *The Journal of Biological Chemistry*, *278*, 38194–38205.

Reinhold, M. I., Abe, M., Kapadia, R. M., Liao, Z. X., & Naski, M. C. (2004). FGF18 represses noggin expression and is induced by calcineurin. *The Journal of Biological Chemistry*, *279*, 38209–38219.

Reinhold, M. I., & Naski, M. C. (2007). Direct interactions of Runx2 and canonical Wnt signaling induce FGF18. *The Journal of Biological Chemistry*, *282*, 3653–3663.

Rice, D. P., Aberg, T., Chan, Y., Tang, Z., Kettunen, P. J., Pakarinen, L., et al. (2000). Integration of FGF and TWIST in calvarial bone and suture development. *Development*, *127*, 1845–1855.

Rice, D. P., Rice, R., & Thesleff, I. (2003). Fgfr mRNA isoforms in craniofacial bone development. *Bone*, *33*, 14–27.

Robinson, D., Hasharoni, A., Cohen, N., Yayon, A., Moskowitz, R. M., & Nevo, Z. (1999). Fibroblast growth factor receptor-3 as a marker for precartilaginous stem cells. *Clinical Orthopaedics and Related Research*, (367 Suppl.), S163–S175.

Rousseau, F., Bonaventure, J., Legeai-Mallet, L., Pelet, A., Rozet, J. M., Maroteaux, P., et al. (1994). Mutations in the gene encoding fibroblast growth factor receptor-3 in achondroplasia. *Nature*, *371*, 252–254.

Sabbieti, M. G., Agas, D., Marchetti, L., Coffin, J. D., Xiao, L., & Hurley, M. M. (2013). BMP-2 differentially modulates FGF-2 isoform effects in osteoblasts from newborn transgenic mice. *Endocrinology*, *154*, 2723–2733.

Sabbieti, M. G., Agas, D., Xiao, L., Marchetti, L., Coffin, J. D., Doetschman, T., et al. (2009). Endogenous FGF-2 is critically important in PTH anabolic effects on bone. *Journal of Cellular Physiology*, *219*, 143–151.

Sanui, T., Tanaka, U., Fukuda, T., Toyoda, K., Taketomi, T., Atomura, R., et al. (2015). Mutation of Spry2 induces proliferation and differentiation of osteoblasts but inhibits proliferation of gingival epithelial cells. *Journal of Cellular Biochemistry*, *116*, 628–639.

Scheller, T., Hellerbrand, C., Moser, C., Schmidt, K., Kroemer, A., Brunner, S. M., et al. (2015). mTOR inhibition improves fibroblast growth factor receptor targeting in hepatocellular carcinoma. *British Journal of Cancer*, *112*, 841–850.

Schlessinger, J. (2003). Signal transduction. Autoinhibition control. *Science*, *300*, 750–752.

Segev, O., Chumakov, I., Nevo, Z., Givol, D., Madar-Shapiro, L., Sheinin, Y., et al. (2000). Restrained chondrocyte proliferation and maturation with abnormal growth plate vascularization and ossification in human FGFR-3(G380R) transgenic mice. *Human Molecular Genetics*, *9*, 249–258.

Seki, K., Fujimori, T., Savagner, P., Hata, A., Aikawa, T., Ogata, N., et al. (2003). Mouse Snail family transcription repressors regulate chondrocyte, extracellular matrix, type II collagen, and aggrecan. *The Journal of Biological Chemistry*, *278*, 41862–41870.

Senarath-Yapa, K., Chung, M. T., McArdle, A., Wong, V. W., Quarto, N., Longaker, M. T., et al. (2012). Craniosynostosis: Molecular pathways and future pharmacologic therapy. *Organogenesis*, *8*, 103–113.

Severe, N., Dieudonne, F. X., & Marie, P. J. (2013). E3 ubiquitin ligase-mediated regulation of bone formation and tumorigenesis. *Cell Death & Disease*, *4*, e463.

Shazeeb, M. S., Cox, M. K., Gupta, A., Tang, W., Singh, K., Pryce, C. T., et al. (2018). Skeletal characterization of the Fgfr3 mouse model of Achondroplasia using micro-CT and MRI volumetric imaging. *Scientific Reports*, *8*, 469.

Sheeba, C. J., Andrade, R. P., Duprez, D., & Palmeirim, I. (2010). Comprehensive analysis of fibroblast growth factor receptor expression patterns during chick forelimb development. *The International Journal of Developmental Biology*, *54*, 1517–1526.

Shiang, R., Thompson, L. M., Zhu, Y. Z., Church, D. M., Fielder, T. J., Bocian, M., et al. (1994). Mutations in the transmembrane domain of FGFR3 cause the most common genetic form of dwarfism, achondroplasia. *Cell*, *78*, 335–342.

Shimokawa, K., Kimura-Yoshida, C., Nagai, N., Mukai, K., Matsubara, K., Watanabe, H., et al. (2011). Cell surface heparan sulfate chains regulate local reception of FGF signaling in the mouse embryo. *Developmental Cell*, *21*, 257–272.

Shin, H. R., Bae, H. S., Kim, B. S., Yoon, H. I., Cho, Y. D., Kim, W. J., et al. (2018). PIN1 is a new therapeutic target of craniosynostosis. *Human Molecular Genetics*, *27*(22), 3827–3839.

Shukla, V., Coumoul, X., Wang, R. H., Kim, H. S., & Deng, C. X. (2007). RNA interference and inhibition of MEK-ERK signaling prevent abnormal skeletal phenotypes in a mouse model of craniosynostosis. *Nature Genetics*, *39*, 1145–1150.

Shung, C. Y., Ota, S., Zhou, Z. Q., Keene, D. R., & Hurlin, P. J. (2012). Disruption of a Sox9-beta-catenin circuit by mutant Fgfr3 in thanatophoric dysplasia type II. *Human Molecular Genetics*, *21*, 4628–4644.

Smith, B. N., Burton, L. J., Henderson, V., Randle, D. D., Morton, D. J., Smith, B. A., et al. (2014). Snail promotes epithelial mesenchymal transition in breast cancer cells in part via activation of nuclear ERK2. *PLoS One*, *9*, e104987.

Smith, L. B., Dally, M. R., Sainz, R. D., Rodrigue, K. L., & Oberbauer, A. M. (2006). Enhanced skeletal growth of sheep heterozygous for an inactivated fibroblast growth factor receptor 3. *Journal of Animal Science*, *84*, 2942–2949.

Song, S. J., Cool, S. M., & Nurcombe, V. (2007). Regulated expression of syndecan-4 in rat calvaria osteoblasts induced by fibroblast growth factor-2. *Journal of Cellular Biochemistry*, *100*, 402–411.

Su, N., Jin, M., & Chen, L. (2014). Role of FGF/FGFR signaling in skeletal development and homeostasis: Learning from mouse models. *Bone Research*, *2*, 14003.

Su, W. C., Kitagawa, M., Xue, N., Xie, B., Garofalo, S., Cho, J., et al. (1997). Activation of Stat1 by mutant fibroblast growth-factor receptor in thanatophoric dysplasia type II dwarfism. *Nature*, *386*, 288–292.

Su, N., Sun, Q., Li, C., Lu, X., Qi, H., Chen, S., et al. (2010). Gain-of-function mutation in FGFR3 in mice leads to decreased bone mass by affecting both osteoblastogenesis and osteoclastogenesis. *Human Molecular Genetics*, *19*, 1199 1210.

Sun, X., Lewandoski, M., Meyers, E. N., Liu, Y. H., Maxson, R. E., Jr., & Martin, G. R. (2000). Conditional inactivation of Fgf4 reveals complexity of signalling during limb bud development. *Nature Genetics*, *25*, 83–86.

Suzuki, H., Suda, N., Shiga, M., Kobayashi, Y., Nakamura, M., Iseki, S., et al. (2012). Apert syndrome mutant FGFR2 and its soluble form reciprocally alter osteogenesis of primary calvarial osteoblasts. *Journal of Cellular Physiology*, *227*, 3267–3277.

Szebenyi, G., Savage, M. P., Olwin, B. B., & Fallon, J. F. (1995). Changes in the expression of fibroblast growth factor receptors mark distinct stages of chondrogenesis in vitro and during chick limb skeletal patterning. *Developmental Dynamics*, *204*, 446–456.

Tabin, C., & Wolpert, L. (2007). Rethinking the proximodistal axis of the vertebrate limb in the molecular era. *Genes & Development*, *21*, 1433–1442.

Tan, Y., Rouse, J., Zhang, A., Cariati, S., Cohen, P., & Comb, M. J. (1996). FGF and stress regulate CREB and ATF-1 via a pathway involving p38 MAP kinase and MAPKAP kinase-2. *The EMBO Journal*, *15*, 4629–4642.

Tanimoto, Y., Yokozeki, M., Hiura, K., Matsumoto, K., Nakanishi, H., Matsumoto, T., et al. (2004). A soluble form of fibroblast growth factor receptor 2 (FGFR2) with S252W mutation acts as an efficient inhibitor for the enhanced osteoblastic differentiation caused by FGFR2 activation in Apert syndrome. *The Journal of Biological Chemistry*, *279*, 45926–45934.

Tavormina, P. L., Bellus, G. A., Webster, M. K., Bamshad, M. J., Fraley, A. E., McIntosh, I., et al. (1999). A novel skeletal dysplasia with developmental delay and acanthosis nigricans is caused by a Lys650Met mutation in the fibroblast growth factor receptor 3 gene. *American Journal of Human Genetics*, *64*, 722–731.

Tavormina, P. L., Shiang, R., Thompson, L. M., Zhu, Y. Z., Wilkin, D. J., Lachman, R. S., et al. (1995). Thanatophoric dysplasia (types I and II) caused by distinct mutations in fibroblast growth factor receptor 3. *Nature Genetics*, *9*, 321–328.

Ten Berge, D., Brugmann, S. A., Helms, J. A., & Nusse, R. (2008). Wnt and FGF signals interact to coordinate growth with cell fate specification during limb development. *Development*, *135*, 3247–3257.

Teplyuk, N. M., Haupt, L. M., Ling, L., Dombrowski, C., Mun, F. K., Nathan, S. S., et al. (2009). The osteogenic transcription factor Runx2 regulates components of the fibroblast growth factor/proteoglycan signaling axis in osteoblasts. *Journal of Cellular Biochemistry*, *107*, 144–154.

Timsah, Z., Ahmed, Z., Lin, C. C., Melo, F. A., Stagg, L. J., Leonard, P. G., et al. (2014). Competition between Grb2 and Plcgamma1 for FGFR2 regulates basal phospholipase activity and invasion. *Nature Structural & Molecular Biology*, *21*, 180–188.

Torii, S., Kusakabe, M., Yamamoto, T., Maekawa, M., & Nishida, E. (2004). Sef is a spatial regulator for Ras/MAP kinase signaling. *Developmental Cell*, *7*, 33–44.

Toydemir, R. M., Brassington, A. E., Bayrak-Toydemir, P., Krakowiak, P. A., Jorde, L. B., Whitby, F. G., et al. (2006). A novel mutation in FGFR3 causes camptodactyly, tall stature, and hearing loss (CATSHL) syndrome. *American Journal of Human Genetics*, *79*, 935–941.

Trudel, S., Li, Z. H., Wei, E., Wiesmann, M., Chang, H., Chen, C., et al. (2005). CHIR-258, a novel, multitargeted tyrosine kinase inhibitor for the potential treatment of t(4,14) multiple myeloma. *Blood*, *105*, 2941–2948.

Trudel, S., Stewart, A. K., Rom, E., Wei, E., Li, Z. H., Kotzer, S., et al. (2006). The inhibitory anti-FGFR3 antibody, PRO-001, is cytotoxic to t(4;14) multiple myeloma cells. *Blood*, *107*, 4039–4046.

Tsang, M., & Dawid, I. B. (2004). Promotion and attenuation of FGF signaling through the Ras-MAPK pathway. *Science's STKE, 2004*, pe17.

Turner, N., & Grose, R. (2010). Fibroblast growth factor signalling: From development to cancer. *Nature Reviews. Cancer, 10*, 116–129.

Twigg, S. R., Healy, C., Babbs, C., Sharpe, J. A., Wood, W. G., Sharpe, P. T., et al. (2009). Skeletal analysis of the Fgfr3(P244R) mouse, a genetic model for the Muenke craniosynostosis syndrome. *Developmental Dynamics, 238*, 331–342.

Twigg, S. R., Vorgia, E., McGowan, S. J., Peraki, I., Fenwick, A. L., Sharma, V. P., et al. (2013). Reduced dosage of ERF causes complex craniosynostosis in humans and mice and links ERK1/2 signaling to regulation of osteogenesis. *Nature Genetics, 45*, 308–313.

Vajo, Z., Francomano, C. A., & Wilkin, D. J. (2000). The molecular and genetic basis of fibroblast growth factor receptor 3 disorders: The achondroplasia family of skeletal dysplasias, Muenke craniosynostosis, and Crouzon syndrome with acanthosis nigricans. *Endocrine Reviews, 21*, 23–39.

Valverde-Franco, G., Liu, H., Davidson, D., Chai, S., Valderrama-Carvajal, H., Goltzman, D., et al. (2004). Defective bone mineralization and osteopenia in young adult FGFR3-/-mice. *Human Molecular Genetics, 13*, 271–284.

Wang, Y., Spatz, M. K., Kannan, K., Hayk, H., Avivi, A., Gorivodsky, M., et al. (1999). A mouse model for achondroplasia produced by targeting fibroblast growth factor receptor 3. *Proceedings of the National Academy of Sciences of the United States of America, 96*, 4455–4460.

Wang, Y., Sun, M., Uhlhorn, V. L., Zhou, X., Peter, I., Martinez-Abadias, N., et al. (2010). Activation of p38 MAPK pathway in the skull abnormalities of Apert syndrome Fgfr2 (+P253R) mice. *BMC Developmental Biology, 10*, 22.

Wang, Y., Xiao, R., Yang, F., Karim, B. O., Iacovelli, A. J., Cai, J., et al. (2005). Abnormalities in cartilage and bone development in the Apert syndrome FGFR2(+/S252W) mouse. *Development, 132*, 3537–3548.

Warren, S. M., Brunet, L. J., Harland, R. M., Economides, A. N., & Longaker, M. T. (2003). The BMP antagonist noggin regulates cranial suture fusion. *Nature, 422*, 625–629.

Wen, X., Li, X., Tang, Y., Tang, J., Zhou, S., Xie, Y., et al. (2016). Chondrocyte FGFR3 regulates bone mass by inhibiting osteogenesis. *The Journal of Biological Chemistry, 291*, 24912–24921.

Wendt, D. J., Dvorak-Ewell, M., Bullens, S., Lorget, F., Bell, S. M., Peng, J., et al. (2015). Neutral endopeptidase-resistant C-type natriuretic peptide variant represents a new therapeutic approach for treatment of fibroblast growth factor receptor 3-related dwarfism. *The Journal of Pharmacology and Experimental Therapeutics, 353*, 132–149.

Weng, T., Yi, L., Huang, J., Luo, F., Wen, X., Du, X., et al. (2012). Genetic inhibition of fibroblast growth factor receptor 1 in knee cartilage attenuates the degeneration of articular cartilage in adult mice. *Arthritis and Rheumatism, 64*, 3982–3992.

Wilkie, A. O. (2005). Bad bones, absent smell, selfish testes: The pleiotropic consequences of human FGF receptor mutations. *Cytokine & Growth Factor Reviews, 16*, 187–203.

Wilkie, A. O. (2007). Cancer drugs to treat birth defects. *Nature Genetics, 39*, 1057–1059.

Wilkin, D. J., Szabo, J. K., Cameron, R., Henderson, S., Bellus, G. A., Mack, M. L., et al. (1998). Mutations in fibroblast growth-factor receptor 3 in sporadic cases of achondroplasia occur exclusively on the paternally derived chromosome. *American Journal of Human Genetics, 63*, 711–716.

Wohrle, S., Henninger, C., Bonny, O., Thuery, A., Beluch, N., Hynes, N. E., et al. (2013). Pharmacological inhibition of fibroblast growth factor (FGF) receptor signaling ameliorates FGF23-mediated hypophosphatemic rickets. *Journal of Bone and Mineral Research, 28*, 899–911.

Wohrle, S., Weiss, A., Ito, M., Kauffmann, A., Murakami, M., Jagani, Z., et al. (2013). Fibroblast growth factor receptors as novel therapeutic targets in SNF5-deleted malignant rhabdoid tumors. *PLoS One, 8*, e77652.

Wu, A. L., Feng, B., Chen, M. Z., Kolumam, G., Zavala-Solorio, J., Wyatt, S. K., et al. (2013). Antibody-mediated activation of FGFR1 induces FGF23 production and hypophosphatemia. *PLoS One, 8*, e57322.

Wu, Z. L., Zhang, L., Yabe, T., Kuberan, B., Beeler, D. L., Love, A., et al. (2003). The involvement of heparan sulfate (HS) in FGF1/HS/FGFR1 signaling complex. *The Journal of Biological Chemistry, 278*, 17121–17129.

Xiao, L., Du, E., Homer-Bouthiette, C., & Hurley, M. M. (2017). Inhibition of FGFR signaling partially rescues hypophosphatemic rickets in HMWFGF2 Tg male mice. *Endocrinology, 158*, 3629–3646.

Xiao, L., Esliger, A., & Hurley, M. M. (2013). Nuclear fibroblast growth factor 2 (FGF2) isoforms inhibit bone marrow stromal cell mineralization through FGF23/FGFR/MAPK in vitro. *Journal of Bone and Mineral Research, 28*, 35–45.

Xiao, Z., Huang, J., Cao, L., Liang, Y., Han, X., & Quarles, L. D. (2014). Osteocyte-specific deletion of Fgfr1 suppresses FGF23. *PLoS One, 9*, e104154.

Xiao, G., Jiang, D., Gopalakrishnan, R., & Franceschi, R. T. (2002). Fibroblast growth factor 2 induction of the osteocalcin gene requires MAPK activity and phosphorylation of the osteoblast transcription factor, Cbfa1/Runx2. *The Journal of Biological Chemistry, 277*, 36181–36187.

Xiao, L., Liu, P., Li, X., Doetschman, T., Coffin, J. D., Drissi, H., et al. (2009). Exported 18-kDa isoform of fibroblast growth factor-2 is a critical determinant of bone mass in mice. *The Journal of Biological Chemistry, 284*, 3170–3182.

Xiao, L., Naganawa, T., Lorenzo, J., Carpenter, T. O., Coffin, J. D., & Hurley, M. M. (2010). Nuclear isoforms of fibroblast growth factor 2 are novel inducers of hypophosphatemia via modulation of FGF23 and KLOTHO. *The Journal of Biological Chemistry, 285*, 2834–2846.

Xiao, L., Sobue, T., Esliger, A., Kronenberg, M. S., Coffin, J. D., Doetschman, T., et al. (2010). Disruption of the Fgf2 gene activates the adipogenic and suppresses the osteogenic program in mesenchymal marrow stromal stem cells. *Bone, 47*, 360–370.

Xu, D., & Esko, J. D. (2014). Demystifying heparan sulfate-protein interactions. *Annual Review of Biochemistry, 83*, 129–157.

Xu, J., Lawshe, A., MacArthur, C. A., & Ornitz, D. M. (1999). Genomic structure, mapping, activity and expression of fibroblast growth factor 17. *Mechanisms of Development, 83*, 165–178.

Yan, D., Chen, D., Cool, S. M., van Wijnen, A. J., Mikecz, K., Murphy, G., et al. (2011). Fibroblast growth factor receptor 1 is principally responsible for fibroblast growth factor 2-induced catabolic activities in human articular chondrocytes. *Arthritis Research & Therapy, 13*, R130.

Yang, F., Wang, Y., Zhang, Z., Hsu, B., Jabs, E. W., & Elisseeff, J. H. (2008). The study of abnormal bone development in the Apert syndrome Fgfr2+/S252W mouse using a 3D hydrogel culture model. *Bone, 43*, 55–63.

Yasoda, A., Kitamura, H., Fujii, T., Kondo, E., Murao, N., Miura, M., et al. (2009). Systemic administration of C-type natriuretic peptide as a novel therapeutic strategy for skeletal dysplasias. *Endocrinology, 150*, 3138–3144.

Yasoda, A., Komatsu, Y., Chusho, H., Miyazawa, T., Ozasa, A., Miura, M., et al. (2004). Overexpression of CNP in chondrocytes rescues achondroplasia through a MAPK-dependent pathway. *Nature Medicine, 10*, 80–86.

Yeh, E., Atique, R., Ishiy, F. A., Fanganiello, R. D., Alonso, N., Matushita, H., et al. (2012). FGFR2 mutation confers a less drastic gain of function in mesenchymal stem cells than in fibroblasts. *Stem Cell Reviews, 8*, 685–695.

Yin, L., Du, X., Li, C., Xu, X., Chen, Z., Su, N., et al. (2008). A Pro253Arg mutation in fibroblast growth factor receptor 2 (Fgfr2) causes skeleton malformation mimicking human Apert syndrome by affecting both chondrogenesis and osteogenesis. *Bone*, *42*, 631–643.

Yin, Y., Ren, X., Smith, C., Guo, Q., Malabunga, M., Guernah, I., et al. (2016). Inhibition of fibroblast growth factor receptor 3-dependent lung adenocarcinoma with a human monoclonal antibody. *Disease Models & Mechanisms*, *9*, 563–571.

Yoon, W. J., Cho, Y. D., Kim, W. J., Bae, H. S., Islam, R., Woo, K. M., et al. (2014). Prolyl isomerase Pin1-mediated conformational change and subnuclear focal accumulation of Runx2 are crucial for fibroblast growth factor 2 (FGF2)-induced osteoblast differentiation. *The Journal of Biological Chemistry*, *289*, 8828–8838.

Yu, K., Herr, A. B., Waksman, G., & Ornitz, D. M. (2000). Loss of fibroblast growth factor receptor 2 ligand-binding specificity in Apert syndrome. *Proceedings of the National Academy of Sciences of the United States of America*, *97*, 14536–14541.

Yu, K., & Ornitz, D. M. (2001). Uncoupling fibroblast growth factor receptor 2 ligand binding specificity leads to Apert syndrome-like phenotypes. *Proceedings of the National Academy of Sciences of the United States of America*, *98*, 3641–3643.

Yu, K., & Ornitz, D. M. (2008). FGF signaling regulates mesenchymal differentiation and skeletal patterning along the limb bud proximodistal axis. *Development*, *135*, 483–491.

Yu, K., Xu, J., Liu, Z., Sosic, D., Shao, J., Olson, E. N., et al. (2003). Conditional inactivation of FGF receptor 2 reveals an essential role for FGF signaling in the regulation of osteoblast function and bone growth. *Development*, *130*, 3063–3074.

Zhang, K., Corsa, C. A., Ponik, S. M., Prior, J. L., Piwnica-Worms, D., Eliceiri, K. W., et al. (2013). The collagen receptor discoidin domain receptor 2 stabilizes SNAIL1 to facilitate breast cancer metastasis. *Nature Cell Biology*, *15*, 677–687.

Zhou, Y. X., Xu, X., Chen, L., Li, C., Brodie, S. G., & Deng, C. X. (2000). A Pro250Arg substitution in mouse Fgfr1 causes increased expression of Cbfa1 and premature fusion of calvarial sutures. *Human Molecular Genetics*, *9*, 2001–2008.

CHAPTER NINE

Wnt-signaling in skeletal development

Stefan Teufel, Christine Hartmann*
Institute of Musculoskeletal Medicine, Department Bone and Skeletal Research, Medical Faculty of the Westphalian Wilhelms University of Münster, Münster, Germany
*Corresponding author: e-mail address: hartmann@ukmuenster.de

Contents

Abstract

As the vertebrate skeleton develops it progresses from a solely cartilaginous scaffold to a mineralized bony skeleton. The cells that build up the skeleton, the chondrocytes and osteoblasts, are primarily of mesodermal origin. Yet, some facial bones, as well as the endocranium, are derived from neural crest cells. The differentiation of the mesenchymal cells to skeletal precursors as well as their subsequent differentiation and maturation along the two lineages, chondrogenic and osteogenic, is controlled by various different signaling pathways, among them Wnt-signaling. WNTs comprise a family of 19 secreted cysteine-rich glycoproteins and can signal through a variety of different intracellular pathways. Genetic loss- and gain-of-function experiments of Wnt-signaling pathway genes have helped to uncover their multiple roles in skeletogenesis, which will be discussed in this article primarily focusing on endochondral bone formation.

Current Topics in Developmental Biology, Volume 133
ISSN 0070-2153
https://doi.org/10.1016/bs.ctdb.2018.11.010

1. Introduction

The bony skeletal elements are formed during embryonic development by either a direct (intramembranous) or an indirect (endochondral) ossification process (see also "Preface" by Olsen). The flat bones of the skull are formed by intramembranous ossification. Here, mesenchymal cells condense and directly differentiate into osteoblasts. In contrast, the long bones of the limb are formed by endochondral ossification. In this case, mesenchymal cells condense into bi-potential chondro-osteoprogenitors expressing the transcription factors Sox9 and Runx2, with Sox9 being absolutely essential for chondrogenesis and Runx2 being the master regulator of osteoblastogenesis (Bi, Deng, Zhang, Behringer, & de Crombrugghe, 1999; Komori, 2006). Within these condensations the more central cells differentiate further into chondrocytes, producing an extracellular matrix rich in type II collagen and aggrecan. Eventually, the chondrocytes build up a cartilaginous template prefiguring the future skeletal element, while the loose mesenchymal surrounding cells develop into the perichondrium. From the perichondrium, precursor cells are recruited and contribute to the appositional growth controlling the width of the skeletal element. Longitudinal expansion of the skeletal element occurs by interstitial growth by dividing chondrocytes. When the skeletal element reaches a certain size, the chondrocytes within the cartilage template exit the cell cycle and mature *via* the prehypertrophic to the hypertrophic stage. After the appearance of hypertrophic chondrocytes, ossification of the long bone shaft is initiated in a discrete area within the perichondrium. This region is then referred to as the periosteum. The factor initiating osteoblastogenesis in the perichondrium, Indian hedgehog (Ihh), is produced by prehypertrophic chondrocytes (Chung, Schipani, McMahon, & Kronenberg, 2001; St-Jacques, Hammerschmidt, & McMahon, 1999). Ossification of the bone collar as such is related to the intramembranous ossification process of the flat bones in the skull, as also here mesenchymal precursors in the perichondrium directly differentiate into osteoblasts. However, it requires, in addition, the IHH signal from the chondrocytes. Osteoblasts in periosteum are all derived from the perichondrium. In contrast, the osteoblasts that build up the primary spongiosa and participate in endosteal bone formation are of dual origin. Here, the majority of preosteoblasts originate from the perichondrium. Together with blood vessels they enter the remodeling zone between the two hypertrophic zones as precursors of osteoblasts (Maes et al., 2010). The second precursor population, giving rise to about 20% of the trabecular/endosteal osteoblasts, arises by transdifferentiation from the type X collagen expressing zone of hypertrophic chondrocytes (Park, Gebhardt, et al., 2015; Tsang, Chan, & Cheah, 2015). Both precursor populations then mature into osteoblasts laying down osteoid on the calcified matrix remnants of the hypertrophic chondrocytes. In addition, osteoblasts use blood vessels as matrices to deposit the key osteoid component type I collagen, and subsequent mineral deposition (Ben Shoham et al., 2016).

In this chapter, the roles of Wnt-signaling in governing diverse steps of skeletogenesis during embryonic development and postnatal skeletal maintenance will be discussed.

1.1 Wnt-signaling

Dependent on the receptors or receptor complexes present on cells the 19 different vertebrate Wnt ligands signal either *via* the well-studied β-catenin-dependent pathway or a multitude of less well understood β-catenin-independent pathways (Acebron & Niehrs, 2016; De, 2011; MacDonald & He, 2012; MacDonald, Tamai, & He, 2009; Niehrs, 2012; Rao & Kuhl, 2010; Valenta, Hausmann, & Basler, 2012; van Amerongen, 2012; Willert & Nusse, 2012). The Wnt/β-catenin signaling pathway is named after its central component β-catenin. Activation of the pathway occurs when a Wnt ligand binds to a receptor complex composed of a member of the Frizzled (FZD) family and the low-density-lipoprotein-related proteins LRP5/6. This leads to the stabilization of the cytoplasmic β-catenin pool. Subsequently, this enables the nuclear translocation and interaction of β-catenin as a transcriptional cofactor with HMG-box transcription factors of the T cell factor/Lymphoid enhancer binding factor (TCF/LEF) family (Ramakrishnan, Sinha, Fan, & Cadigan, 2018; Schuijers, Mokry, Hatzis, Cuppen, & Clevers, 2014). In the absence of a suitable Wnt ligand, cytoplasmic β-catenin is degraded by a so-called destruction complex composed of the scaffolding protein axin, the adenomatosis polyposis coli gene product (APC), and two kinases, casein kinase 1 (CK1) and glycogen synthase kinase 3 (GSK3). These two kinases sequentially phosphorylate β-catenin on four serine/threonine residues in its amino-terminal region, resulting in its ubiquitination and subsequent proteasomal degradation (Logan & Nusse, 2004). Yet, β-catenin is a dual function protein that also serves as a structural component of adherens junctions, linking cadherins to the actin cytoskeleton (Kemler, 1993). The Wnt/β-catenin pathway has been functionally implicated in regulating essential steps during skeletogenesis. Secretion of most Wnt ligands requires the protein Wntless (Wls/Gpr177) (Das, Yu, Sakamori, Stypulkowski, & Gao, 2012). The availability of Wnt ligands for receptor binding is controlled by secreted Wnt antagonists, such as members of the family of secreted frizzled-related proteins (sFrp) and Wnt-inhibitory factor (WIF), that bind directly to Wnt ligands. Other antagonists, such as sclerostin (SOST) or Dickkopf family members (Dkks) bind to the Lrp5/6 co-receptors or Kremen, Dkk-receptors, which trigger endocytosis of Lrp receptors after binding Dkk.

The functional roles of molecules involved in the diverse β-catenin-independent signal transduction pathways have been classified based on the signaling components involved into Wnt/calcium signaling (Kohn & Moon, 2005; Kuhl, 2004), Wnt/planar cell polarity (PCP) (Gao, 2012; Lawrence & Casal, 2013), Wnt/Tsc2/mTor (Inoki et al., 2006), and Wnt/Yap/Taz signaling (Park, Kim, et al., 2015; Piccolo, Dupont, & Cordenonsi, 2014) (Fig. 1). Whether these are truly independent pathways or components of a Wnt-signaling network is a topic of long-standing debate (Arias, Brown, & Brennan, 1999; Kestler & Kuhl, 2008; Kikuchi, Yamamoto, & Sato, 2009; van Amerongen, 2012). Components of these pathways, such as members of the orphan receptor kinase family (ROR), protein kinase C (PKC) family members, Ca2+/calmodulin-dependent kinase II (CaMKII), calcineurin, and certain Wnt ligands activating these pathways have been shown to be involved in skeletogenesis on the basis of genetic mutations in humans, gain-of-function experiments in chicken, *in vitro* cell culture experiments, and genetic experiments in mice.

2. Wnt-signaling in endochondral bone formation

2.1 Roles of Wnt-signaling during the early steps of endochondral bone formation in the limbs

As mentioned already, endochondral bone formation starts with the condensation of mesenchymal cells and their subsequent differentiation into chondrocytes. It has long been known that secreted signals from the limb ectoderm can repress chondrogenesis *in vitro* (Solursh & Reiter, 1988). The likely candidates for the repressive signals are ectodermal expressed Wnts, such as Wnt3, Wnt4, Wnt6, Wnt7a, and Wnt7b, based on findings that *in vitro* or *in vivo* overexpression of certain *Wnts*, *Wnt1*, *Wnt7a*, *Wnt4*, *Wnt6*, and *Wnt9a/14* in chicken or mice suppresses chondrogenesis

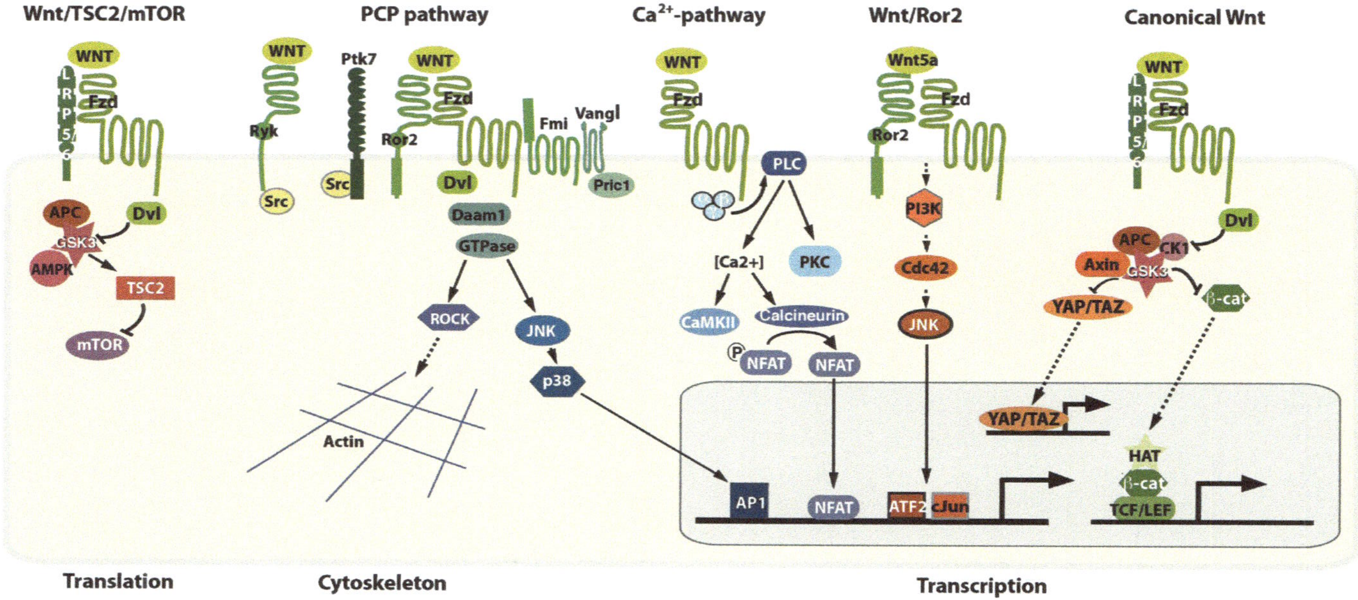

Fig. 1 Schematic overview of the different Wnt signaling pathways relevant in skeletogenesis. Abbreviations not explained in the text: AP1: Activating protein 1 complex; AMPK: AMP-activated protein kinase; ATF2: Activating transcription factor 2; Cdc42: Cell division cycle 42; c-Jun: cellular Jun proto-oncogene; Daam1: Disheveled associated activator of morphogenesis 1; Dvl: Disheveled; Fmi: Flamingo; HAT: Histone acetyl transferase; JNK: c-Jun N-terminal kinase; PLC: Phospholipase C; PI3K: Phosphoinositide-3-kinase; Pric1: Prickle 1; Ptk7: Protein tyrosine kinase 7; p38: p38 Mitogen-activated protein kinase; Rock: Rho-associated coiled-coil containing protein kinase; Ryk: Receptor-like tyrosin kinase; Src: Src kinase; Vangl: Van-Gogh-like.

(Geetha-Loganathan, Nimmagadda, Christ, Huang, & Scaal, 2010; Guo et al., 2004; Hartmann & Tabin, 2001; Loganathan, Nimmagadda, Huang, Scaal, & Christ, 2005; Rudnicki & Brown, 1997; Spater, Hill, Gruber, & Hartmann, 2006; Witte, Dokas, Neuendorf, Mundlos, & Stricker, 2009). In the case of Wnt7a the mechanism proposed is prolonged cell-cell adhesion *via* an N-cadherin/β-catenin complex interfering with chondrogenesis (Tufan & Tuan, 2001). The repressive activity of the other Wnts is probably mediated through the β-catenin-dependent Wnt-pathway (Guo et al., 2004; Hartmann & Tabin, 2000; Spater, Hill, Gruber, et al., 2006). This notion is further supported by the observation that endogenous β-catenin protein levels are low in regions undergoing chondrogenesis (Hill, Spater, Taketo, Birchmeier, & Hartmann, 2005). Furthermore, conditional inactivation of β-catenin, encoded by the *Ctnnb1* gene locus, in the entire limb mesenchyme results in an expansion of the *Sox9* expressing chondro-osteoprogenitor population, while conditional stabilization of β-catenin, mimicking active Wnt/β-catenin signaling, inhibits *Sox9* expression and overt chondrogenesis (Hill et al., 2005). Similarly, chondrogenesis was blocked by an increase in β-catenin levels through the conditional inactivation of *Apc* (Miclea et al., 2009). Taken together, during the early stages of skeletogenesis in the developing limb, ectodermal Wnts act together with fibroblast growth factors (Fgfs) to regionally inhibit chondrogenesis keeping mesenchymal cells in an undifferentiated state (ten Berge, Brugmann, Helms, & Nusse, 2008) (Fig. 2A). Interestingly, the central limb mesenchyme, undergoing chondrogenesis, also expresses the frizzled-related protein 1 (Frzb1), a secreted Wnt antagonist (Ladher et al., 2000; Wada, Kawakami, Ladher, Francis-West, & Nohno, 1999). Thereby reinforcing low levels of Wnt/β-catenin signaling in the central mesenchymal region of the limb. As such, overexpression of *Frzb1* in culture stimulates glycosaminoglycan production and neutralizes the anti-chondrogenic effect of the Wnt8 in culture (Enomoto-Iwamoto et al., 2002). In addition, a second mechanism exists that ensures low levels of β-catenin in cells differentiating into chondrocytes. Here, the master regulator of chondrogenesis, Sox9, interacts at the protein level with β-catenin and destabilizes it (Akiyama et al., 2004; Topol, Chen, Song, Day, & Yang, 2009) (Fig. 2A).

Another Wnt ligand, Wnt5a, which has been associated with β-catenin-independent Wnt-signaling, is expressed in the distal to proximal gradient in the limb mesenchyme of growing limbs and with regard to the developing skeletal elements in prehypertrophic chondrocytes and the periosteum (Gao et al., 2011; Yamaguchi, Bradley, McMahon, & Jones, 1999; Yang, Topol,

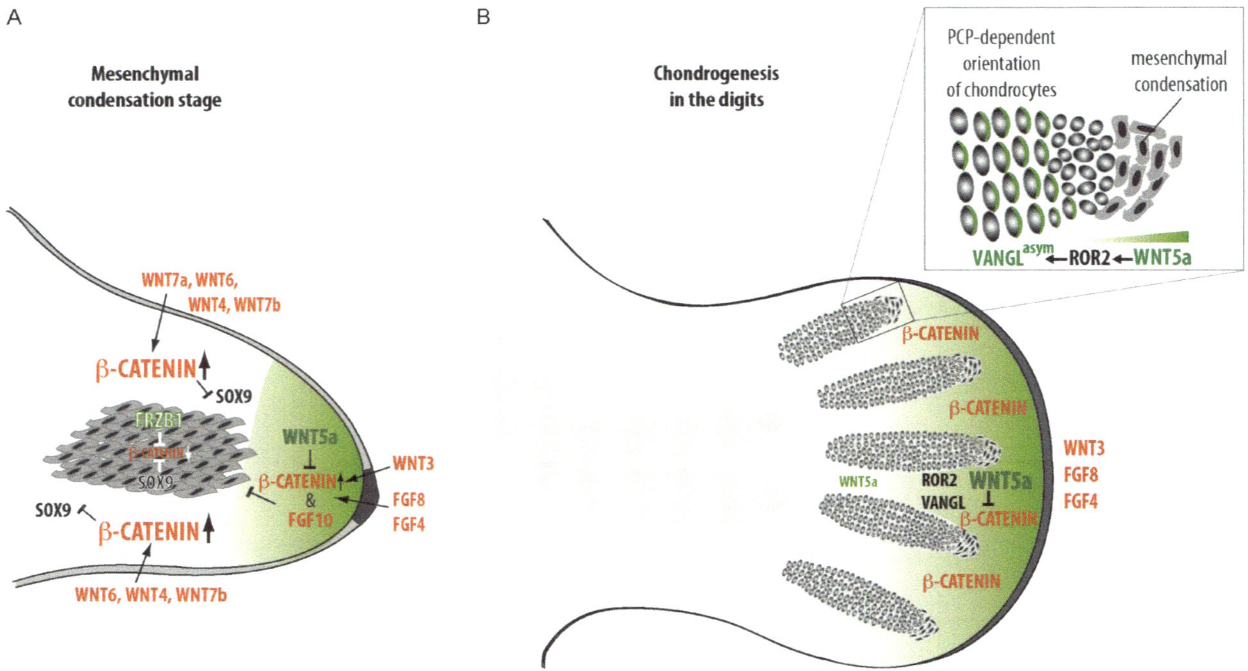

Fig. 2 Schematic overview of Wnt signaling in the mouse limb at the mesenchymal condensation stage (A) and during chondrogenesis in the digit anlagen (B). A negative influence on chondrogenesis is indicated by the red color, a positive influence by the green color. For details see text.

Lee, & Wu, 2003). *In vitro*, *Wnt5a* overexpression or treatment with recombinant Wnt5a has either no effect or slightly enhances chondrogenesis in the limb bud micromass culture system (Bradley & Drissi, 2010; Church, Nohno, Linker, Marcelle, & Francis-West, 2002; Hartmann & Tabin, 2000; Jin et al., 2006; Tufan & Tuan, 2001). Bradley and Drissi reported in their *in vitro* study differential effects upon Wnt5a treatment of undifferentiated *versus* already differentiated limb bud (Bradley & Drissi, 2010). Treatment with Wnt5a from day 3 to 21, during the differentiation phase, increased proteoglycan production but also decreased Sox9 levels. In contrast, treatment from day 14 to 21 when the cells are already differentiated decreased proteoglycan production, increased Sox9 and decreased Runx2 levels (Bradley & Drissi, 2010). Molecularly, these opposite effects were attributed to a differential use of the calcineurin/NFAT and the PI3K/Akt/Ikk/Nf-κB pathways during the early and late stage of chondrocyte differentiation, respectively (Bradley & Drissi, 2010). Wnt5a signaling is also required for the Ihh-induced degradation of the transcription factor Nkx3.2, a factor that is capable of repressing chondrocyte hypertrophy (Choi et al., 2012; Provot et al., 2006). Furthermore, experimental evidence suggests that Wnt5a signaling can, dependent on the receptor-complex present, either oppose or promote β-catenin-dependent signaling (Mikels & Nusse, 2006; Okamoto et al., 2014; Topol et al., 2003; van Amerongen, Fuerer, Mizutani, & Nusse, 2012; Yamamoto, Yoo, Nishita, Kikuchi, & Minami, 2007). To what extent these different pathways contribute to the *Wnt5a* loss-of-function phenotype is, however, currently not resolved. Genetic experiments in the mouse revealed that the loss of Wnt5a activity results in the loss of distal skeletal structures in the limbs among other effects (Yamaguchi et al., 1999). This phenotype is mimicked by the loss of the Wnt5a co-receptor Ror2 in combination with the transmembrane protein Vangl2, a component of the planar cell polarity (PCP) pathway, and results in an increase in Wnt/β-catenin signaling in the distal mesenchyme thereby blocking its differentiation into the chondrogenic digit anlagen (Gao et al., 2011) (Fig. 2B).

2.2 Effects of Wnt-signaling on proliferating chondrocytes

The chondrocytes within the chondrogenic anlagen of the skeletal elements produce an extracellular matrix rich in the proteoglycan core protein aggrecan and type II fibrillar collagen. These chondrocytes are highly proliferative and contribute to the interstitial growth of the cartilage element along its longitudinal axis. Upon activation of Wnt/β-catenin signaling in *Col2a1* expressing cells either through the expression of a constitutively active form of the transcription factor Lef1 (*CA-Lef1*), or a stabilized form of β-catenin (Δ N-β-catenin), or Wnt9a (Wnt14) expression, chondrocytes remain immature and fail to organize into growth plates (Guo et al., 2004; Tamamura et al., 2005). The skeletal elements of these transgenic mice are fused, their borders to the surrounding tissue are not very well defined, and they show abnormal expression of synovial joint interzone markers (Guo et al., 2004; Tamamura et al., 2005). Activation of the Wnt/β-catenin pathway in the long bones of chicken limbs using retroviruses expressing either a stabilized form of β-catenin, or the ligand Wnt4, or CA-Lef1 results in accelerated chondrocyte maturation and concomitantly accelerated osteoblast differentiation (Hartmann & Tabin, 2000; Kitagaki et al., 2003; Tamamura et al., 2005). The Wnt-antagonist *Frzb1/Sfrp3*, which is expressed in proliferating chondrocytes and chondrocytes in articular regions, opposes the action of Wnt/β-catenin signaling. Its retroviral mediated overexpression delays chondrocyte maturation in the chicken limb (Enomoto-Iwamoto et al., 2002). The function of the Wnt-inhibitory molecule Lrp4, a member of the low-density lipoprotein receptor-related family, which binds the Wnt-antagonists Sost and Wise, has not yet been examined *in vivo* with regard to chondrogenesis. Nevertheless, *Lrp4* is upregulated during chondrocyte differentiation and *in vitro* experiments suggest that it opposes Wnt/β-catenin signaling during chondrogenesis promoting matrix production and chondrocyte differentiation (Asai et al., 2014).

The serine/threonine kinase, protein kinase C (PKC), is a downstream component of the Wnt/Calcium pathway (Kuhl, 2004). PKC is actually a multigene family with 12 different mammalian isoforms, which can be subdivided into four classes depending on their co-factor requirement and structural aspects: classical PKCs (cPKCs) that require diacylglycerol (DAG) and Ca^{2+} for their activation, novel PKCs (nPKCs) that require only DAG, atypical PKCs (aPKCs) and the PKN family, which both require either DAG or Ca^{2+} (Rosse et al., 2010). In the chicken limb mesenchyme at least four PKCs (cPKCα, nPKCε, aPKCζ and aPKCλ/ι) are expressed (Choi, Chun, Lee, Sonn, & Kang, 1995; Nicolin et al., 2004; Yang et al., 1998). In micromass cultures, the PKC activity increases during the first 3 days (Sonn & Solursh, 1993). Numerous *in vitro* experiments have been performed using PKC agonist and antagonists, the results of which are difficult to interpret and sometimes conflicting (reviewed in Matta &

Mobasheri, 2014). Knockout mice for cPKCδ display structural alterations of the articular cartilage, the intra-articular, and the subchondral compartments, accompanied *in vitro* by a decrease in GAG production of their chondrocytes (Yang et al., 2015). Knock-down of nPKCε in chondrocytes isolated from osteoarthritic knees exaggerated the hypertrophic phenotype suggesting that nPKCε is a negative regulator of hypertrophy (Queirolo et al., 2016).

The kinase CaMKII is another component of the non-canonical Wnt/Calcium pathways (Kuhl, 2004). Endogenously, CaMKII is activated in the prehypertrophic zone (Li, Ahrens, Wu, Liu, & Dudley, 2011). Surprisingly, the pCaMKII signal is downregulated in limbs infected with a *Wnt5a* expressing virus, while it is ectopically present in cells infected with a dominant-negative form of Fzd7 (Li et al., 2011). Overexpression of a constitutively active form of CaMKII in chicken results in an overgrowth of the infected skeletal element (Taschner, Rafigh, Lampert, Schnaiter, & Hartmann, 2008). The underlying mechanisms are complex; on the one side, a broad downregulation of the expression of *Cyclin A* and *Cyclin D1* and of the AP1 family members, *cFos* and *cJun* are observed (Taschner et al., 2008). On the other side, a subset of the infected cells prematurely expresses the maturation markers *Ihh* and *Col10a1* outside their endogenous domain (Li et al., 2011; Taschner et al., 2008). Both effects occur in a cell-autonomous manner, yet, the ectopic maturation occurs in a much more restricted manner. The transcription factors Runx2 and Mef2c are important regulators of chondrocyte hypertrophy (Arnold et al., 2007; Yoshida & Komori, 2005). CaMKII activity may act through these two transcription factors, as it synergizes with the transcription factor Runx2 promoting its nuclear import, and it potentiates the transcriptional activity of Mef2c *in vitro* (Amara, Fabritius, Houben, Wolff, & Hartmann, 2017; Li et al., 2011). Endogenously, the activity of CaMKII is probably counteracted by some kind of inhibitory gradient in the skeletal elements and only when a certain threshold is reached can it exert its stimulatory function on hypertrophy (Li et al., 2011). The nature of this inhibitory gradient is still elusive.

Increased intracellular calcium levels also activate the calcium/calmodulin-dependent serine/threonine phosphatase calcineurin (Aramburu, Rao, & Klee, 2000). Treatment of micromass cultures with the calcineurin inhibitor cyclosporin A inhibits chondrogenesis, while the agonist ionomycin promotes it (Tomita, Reinhold, Molkentin, & Naski, 2002). The transcription factors of the NFAT family are dephosphorylated by calcineurin resulting in their nuclear translocation (Crabtree, 2001). *In vitro*, a constitutively active form of NFAT4 stimulates glycosaminoglycan production along with the transcription of the core protein aggrecan (Acan) (Tomita et al., 2002). These prochondrogenic effects are probably mediated by BMP-signaling (Tomita et al., 2002). Another family member, NFATc2/NFATp, has been shown to act as a repressor of chondrogenesis during postnatal development in connective tissue precursor cells in the vicinity of synovial joints (Ranger et al., 2000).

2.3 Wnt-signaling and growth plate functions

The growth plate of endochondral bones is an important structure involved in long bone morphogenesis and embryonic and postnatal skeletal growth. In contrast to postnatal development, where the growth plate is sandwiched between the secondary ossification center and the primary spongiosa, the organization of the embryonic growth plate is less of a plate and more reflected by the columnar organization of the differentiated chondrocyte types also present in the postnatal growth plate. In the embryonic growth plate, the resting chondrocytes, which contrary to their name are proliferating and metabolically active, are round in shape and are located in the epiphysis near the articulations. This zone is also referred to as zone I of undifferentiated chondrocytes (Yang et al., 2003). It is followed by the zone of columnar proliferating chondrocytes, with a flattened and stacked appearance, also referred to as zone II of undifferentiated chondrocytes (Dodds, 1930; Yang et al., 2003). Chondrocyte proliferation within these two zones is differentially influenced by the activity of two highly related Wnt molecules, Wnt5a and Wnt5b. Wnt5a positively regulates the proliferation of zone I chondrocytes *via* P130 and of zone II chondrocytes *via* Ihh-mediated Cyclin D1 regulation. In addition, excessive Wnt5a inhibits the transition of chondrocytes from zone I into zone II and thus delays their further maturation to prehypertrophic and hypertrophic chondrocytes (Yang et al., 2003). In contrast, excessive Wnt5b appears to promote the transition of chondrocytes from zone I into zone II and prevents chondrocytes in zone II to undergo cell cycle withdrawal, thereby delaying chondrocyte maturation (Yang et al., 2003).

The columnar zone elongates internally by a combined process of cell division followed by intercalation movements (Ahrens, Li, Jiang, & Dudley, 2009; Aszodi, Hunziker, Brakebusch, & Fassler, 2003; Dodds, 1930). Evidence from the chick system suggests that signaling through non-canonical Frizzleds regulates the polarity of the columnar chondrocytes (Li & Dudley, 2009). This is supported by studies in mice, showing that a Ror2/Vangl2 complex senses a Wnt5a gradient, thereby regulating chondrocyte polarity at least in the autopod region of the limb (Gao et al., 2011). The chondrocytes within the carpal elements of *Wnt5a* mutant mice as well as the surrounding perichondrial cells also display a cell polarity defect (Kuss et al., 2014). Interestingly, activation of the PCP pathway can support the columnar organization of chondrocytes in pellet cultures *in vitro* (Randall, Shao, Wang, & Ballock, 2012). In a recent publication using time-lapse confocal imaging, it has been shown that the daughter cells in the columnar zone maintain an intimate contact after cell division that lasts until the end of the rotation process (Romereim, Conoan, Chen, & Dudley, 2014). A cadherin-β-catenin adhesion complex has been proposed to maintain the cell-cell contacts during this rotational event, as both proteins are enriched at the common daughter cell interface (Romereim et al., 2014). Nevertheless, cell arrangement and tissue architecture are normal in *Ctnnb1*-deficient growth plates (Ahrens, Romereim, & Dudley, 2011).

Ror2 mutant mice, like *Wnt5a* mutants, have shortened long bones (DeChiara et al., 2000; Oishi et al., 2003; Schwabe et al., 2004; Takeuchi et al., 2000). Yet, *Ror2* mutants, in contrast to *Wnt5a*, display a mesomelic limb shortening phenotype with the skeletal elements of the zeugopod (radius and ulna in the forelimb) being more affected than the stylopod (humerus in the forelimb) or autopod elements (digits) (DeChiara et al., 2000; Oishi et al., 2003; Schwabe et al., 2004; Takeuchi et al., 2000). In elements of the zeugopod, such as the ulna, the columnar organization of chondrocytes fails to occur and the prehypertrophic marker, *Ihh*, and the hypertrophic marker, *Col10a1*, are not expressed at embryonic stage E15.5 (Schwabe et al., 2004). In comparison, stylopod elements at that stage show hypertrophic chondrocyte maturation, which, however, is delayed. No differences in proliferation, according to the percentage of BrdU positive cycling cells, have been reported for the two proliferative zones at least not within the stylopod elements (Schwabe et al., 2004). Yet, the distribution of zone-specific markers such as P130 and Cyclin D1 has not been examined in the *Ror2* mutant skeletal elements. *Ror2* is expressed in all chondrocytes, except the hypertrophic chondrocytes, and in cells of the perichondrium (DeChiara et al., 2000). Hence, it is currently unclear why the skeletal elements of the zeugopod are more affected and to what extent the chondrogenic *versus* the perichondrial expression of Ror2 contribute to the *Ror2* mutant phenotype. Mice carrying a hypomorphic allele of the closely related receptor *Ror1* are viable and display alterations in their growth plates leading to retardation of postnatal growth (Lyashenko et al., 2010).

Wnt9a mutant mice also have slightly shorter skeletal elements (Spater, Hill, O'Sullivan, et al., 2006). In these mutants, the central regulator of skeletogenesis, Ihh (Kronenberg, 2003), is temporarily downregulated during the onset of endochondral ossification (E12.5–E13.5), when the chondrocytes begin to hypertrophy (Spater, Hill, O'Sullivan, et al., 2006). The regulatory effect of Wnt9a on *Ihh* expression is mediated through the Wnt/β-catenin pathway and is likely to be direct as β-catenin and Lef1 both bind to the *Ihh* promoter (Spater, Hill, O'Sullivan, et al., 2006). Furthermore, exogenous application of Wnt9a promotes hypertrophy *in vitro* and stimulates chondrocyte maturation and osteoblastogenesis *in vivo* (Day, Guo, Garrett-Beal, & Yang, 2005; Dong, Soung do, Schwarz, O'Keefe, & Drissi, 2006). A similar *in vitro* effect is also seen upon stimulation with Wnt8c and is probably mediated *via* the β-catenin pathway acting directly on the *Runx2* promotor upregulating the expression of *Runx2* (Dong et al., 2006). Whether this mechanism is relevant *in vivo* remains to be addressed, at least no obvious downregulation of *Runx2* was observed in RNA-*in situ* hybridizations performed on material from *Ctnnb1*- and *Wnt9a*-deficient limbs or limbs from mice expressing *Wnt9a* under the control of the *Col2a1*-promoter in chondrocytes (Day et al., 2005; Hill et al., 2005; Hu et al., 2005; Spater, Hill, O'Sullivan, et al., 2006).

Wnt4, which is expressed in prehypertrophic chondrocytes, acts redundantly with *Wnt9a* on chondrocyte maturation in a temporally restricted fashion. In *Wnt4* single mutants, a slight delay in chondrocyte maturation is observed around the onset of chondrocyte maturation (Spater, Hill, O'Sullivan, et al., 2006). Interestingly, overexpression of *Wnt4* in *Col2a1*-expressing chondrocytes results in a delay of the formation of the primary ossification center and delayed vascular invasion during embryonic skeletal development (Lee & Behringer, 2007). This phenotype is similar, although less pronounced, to the phenotype of β-catenin stabilization in hypertrophic chondrocytes (Houben et al., 2016). Surprisingly, shortly after birth, no size differences are noted in *Wnt4* transgenic mice. However, starting around postnatal day 10, the transgenic mice display a growth retardation phenotype associated with alterations of their growth plates (Lee & Behringer, 2007). Here, the hypertrophic *Col10a1*-expressing zone is expanded at the expense of the proliferative *Col2a1*-expressing chondrocytes. The delay in the turnover of hypertrophic chondrocytes is also apparent during the formation of the secondary ossification center as this process is also delayed in the *Wnt4* transgenic mice (Lee & Behringer, 2007).

Loss of β-catenin activity in the mesenchyme or chondrocytes also results in a delay of chondrocyte maturation (Ahrens et al., 2011; Akiyama et al., 2004; Day et al., 2005; Hill et al., 2005; Hu et al., 2005). This phenotype is recapitulated by the combined loss of the two co-receptors Lrp5/6 (Joeng, Schumacher, Zylstra-Diegel, Long, & Williams, 2011; Schumacher, Joiner, Less, Drewry, & Williams, 2016). Functionally, this is probably in part mediated by alterations in Sox9 activity due to increased transcriptional levels or stabilization at the protein level, as Sox9 interferes with chondrocyte hypertrophy (Akiyama et al., 2004). Loss of the antagonist sFrp1 slightly accelerates chondrocyte hypertrophy (Gaur et al., 2006). Exogenous application of Wnt antagonists, such as Frzb1, Dkk1, or a secreted Fzd8, inhibits hypertrophic differentiation *in vitro* (Narcisi et al., 2015; Zhong et al., 2016). Postnatal loss of β-catenin activity in chondrocytes results also in a reduction of the hypertrophic zone (Wang et al., 2014). In addition, slow cycling cells present in the reserve zone of the growth plate require β-catenin activity (Candela et al., 2014). Activation of β-catenin signaling in the postnatal growth plate leads to closure of the growth plate which could be associated with an exhaustion of these slow cycling cells (Yuasa et al., 2009). In essence, within the growth plate, the Wnt/β-catenin pathway appears to promote primarily chondrocyte maturation and maintains a slow cycling population of chondrocytes.

β-Catenin activity in chondrocytes, or even more specifically, in hypertrophic chondrocytes influences osteoclastogenesis locally at the chondro-osseous border, primarily *via* repression of *Rankl* expression (Golovchenko et al., 2013; Houben et al., 2016; Wang et al., 2014, 2017). As a potential mechanism, it has been proposed that the repression of *Rankl* by Wnt/β-catenin in hypertrophic chondrocytes is mediated by a glucocorticoid receptor-dependent mechanism (Wang et al., 2014). However, mice mutant for glucocorticoid receptor show no apparent embryonic or postnatal cartilage or bone phenotype (Tu et al., 2014). Based on studies interfering with Wnt/β-catenin signaling by expressing a protein inhibitor (ICAT) of β-catenin and TCF/LEF under the control of the *Col2a1* promoter, *Mmp13* and *Vegfa* have been identified as positively regulated potential targets in hypertrophic chondrocytes (Chen et al., 2008). Yet, the effects upon deletion of *Ctnnb1* in hypertrophic chondrocytes on their expression levels are variable and in the case of *Mmp13* even opposite results have been reported (Golovchenko et al., 2013; Houben et al., 2016). Postnatally, the *Col2a1*-ICAT mice show reduced chondrocyte proliferation and differentiation, and increased chondrocyte apoptosis leading to an overall decrease in growth plate width (Chen et al., 2008).

Inactivation of *Wls/Gpr177*, required for the secretion of Wnts in mesenchymal progenitors or chondrocytes, leads to defects in chondrocyte maturation associated with greatly reduced Wnt/β-catenin signaling (Maruyama, Jiang, & Hsu, 2013). These defects are less pronounced when *Wls* is deleted in *Col2a1* expressing chondrocytes (Maruyama et al., 2013). Consistent with the proposal that ectodermal Wnts are responsible for the inhibition of chondrogenesis in the non-chondrogenic mesenchyme, inactivation of *Wls* in the mesenchyme using the *Dermo1*-Cre line does not fully recapitulate the phenotype observed in *Prx1*-Cre *Ctnnb1* deficient limbs (Hill et al., 2005; Maruyama et al., 2013). Unfortunately, the analysis upon conditional *Wls* inactivation is not sufficiently detailed to conclude whether or not the phenotype fully recapitulates that of *Wnt5a* mutants.

The transcriptional co-activators Yap and Taz, originally identified as downstream components of the HIPPO pathway, can also be induced by Wnt5a/5b and Wnt3a (Park, Kim, et al., 2015). Based on genetic experiments using transgenic and conditional knockouts, Yap1 promotes early chondrocyte proliferation but interferes with chondrocyte hypertrophy (Deng et al., 2016). This is in agreement with its endogenous distribution being lowest in hypertrophic chondrocytes, where the inactive phosphorylated form of Yap1 is most prominent (Deng et al., 2016). The promotive effect on chondrocyte proliferation is mediated through the regulation of *Sox6* transcription in a TEAD-dependent manner, while the inhibitory effect on chondrocyte hypertrophy is mediated through physical interaction with Runx2 affecting its transcriptional activity. Paradoxically, *in vitro* data point toward an opposite function for Taz in chondrogenesis (Deng et al., 2016). Nevertheless, mice lacking one copy of *Yap* and both copies of *Taz* in the *Prx1*-Cre derived cells have a slightly enlarged hypertrophic zone (Xiong, Almeida, & O'Brien, 2018). Yet, the situation with Yap/Taz may even be more complex as they have also been shown to interact with Smad2/3 downstream of Tgfβ-signaling, to interfere with the nuclear translocation of Disheveled (Dvl), a scaffolding protein that acts between Fzd and the destruction complex, and to participate in the β-catenin destruction complex itself (Azzolin et al., 2014; Piersma, Bank, Boersema, 2015; Varelas et al., 2010). Furthermore, in mesenchymal stem cells, Yap has been shown to inhibit chondrogenic differentiation (Karystinou et al., 2015).

In conclusion, Wnt-signaling modulates chondrogenesis and chondrocyte differentiation at multiple steps during development, but also postnatally. A schematic overview of the different actions of Wnt ligands and signaling components in chondrogenesis is shown in Fig. 3.

3. Role of Wnt signaling in osteoblast differentiation and osteoblast function

Osteoblasts are, like chondrocytes, of mesenchymal origin and share a common progenitor, the osteo-chondroprogenitor. In humans, numerous susceptibility loci for osteoporosis have been identified on the basis of reduced bone mineral density in genome-wide association studies (GWAS) (Hsu & Kiel, 2012; Liu, Zhang, Papasian, & Deng, 2014). Some of these loci map to genes encoding Wnt-signaling associated proteins, such as the ligands Wnt4 and Wnt16, the co-receptor Lrp5, intracellular components Axin1 and Ctnnb1, Gpr177, and the antagonist sclerostin (Sost) (Liu et al., 2014). *Wnt4* mutant mice die at birth due to kidney defects (Stark, Vainio, Vassileva, & McMahon, 1994). As mentioned above, chondrocyte maturation is slightly delayed in the *Wnt4* mutants (Spater, Hill, O'Sullivan, et al., 2006). Yet, so far, no defects in osteoblastogenesis or bone formation have been reported in *Wnt4* mutants. Mouse studies revealed that Wnt16 is, at least in female mice, a critical regulator of cortical bone thickness and homeostasis, while not affecting trabecular bone (Gori, Lerner, Ohlsson, & Baron, 2015; Moverare-Skrtic et al., 2014; Ohlsson et al., 2018; Wergedal, Kesavan, Brommage, Das, & Mohan, 2015; Zheng et al., 2012). Mechanical loading of bones normally results in a significantly increased bone formation rate of the cortical bone. Yet, this response does not occur in the *Wnt16* knockout mice (Wergedal et al., 2015). Interestingly, another study reported that Wnt16 levels are not controlled by mechanical loading; instead, they are influenced by estrogen-receptor signaling (Todd et al., 2015). However, the bone-sparing effect of Wnt16 occurs independently of estrogen signaling (Moverare-Skrtic et al., 2015; Todd et al., 2015). Conditional postnatal inactivation of *Wnt16* in 10-week-old, sexual-mature female mice recapitulates the cortical bone phenotype, suggesting that Wnt16 activity is required also during adulthood for cortical bone thickness and strength (Ohlsson et al., 2018). Wnt16 has a strong anti-resorptive activity and acts both on osteoblasts and osteoclasts (Gori et al., 2015; Moverare-Skrtic et al., 2014). Its effect on osteoclasts is of dual nature, one being indirect—mediated by an increase

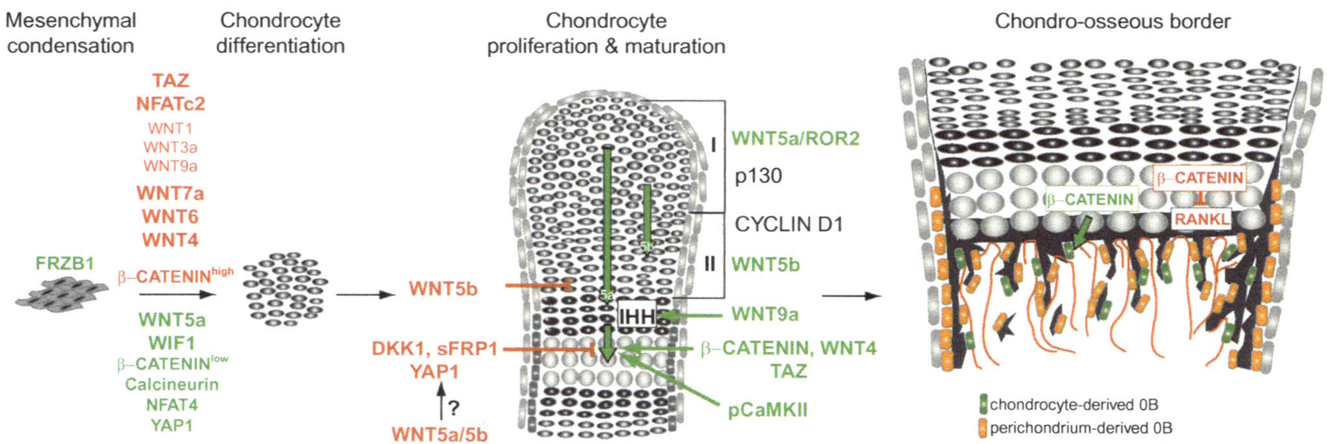

Fig. 3 Schematic overview of the different steps of chondrogenesis and the influences of Wnt signaling molecules therein. A negative influence is indicated by the red color, a positive influence by the green color. For details see text.

in the expression of osteoprotegerin (*Opg*), a negative regulator of osteoclastogenesis, in osteoblasts (Moverare-Skrtic et al., 2014). Opg is acting as a decoy receptor for the pro-osteoclastic factor Rank-ligand (Rankl) (Simonet et al., 1997). The other one is a direct effect, whereby Wnt16 interferes with the early steps of Rank signaling, probably *via* Jnk activation (Moverare-Skrtic et al., 2014). In support of this, Wnt16 can still inhibit osteoclastogenesis in osteoclast precursor/osteoblast co-cultures using cells derived from *Opg*-deficient mice (Kobayashi et al., 2015). Interestingly, osteoblast-specific Wnt16 overexpression under the control of the rat procollagen type I α1 promoter has a more pronounced effect on trabecular than cortical bone formation (Alam et al., 2016; Moverare-Skrtic et al., 2015). This may be due to the fact that endogenous *Wnt16* is only moderately expressed in trabecular but highly expressed in cortical bone (Moverare-Skrtic et al., 2015). What is the signaling pathway downstream of Wnt16? In the literature, there is evidence that Wnt16 can signal through the Wnt/β-catenin pathway as well as a Wnt/Jnk pathway not involving Ror2 (Jiang, Von den Hoff, Torensma, Meng, & Bian, 2014; Moverare-Skrtic et al., 2014). The latter may be responsible for the Wnt/β-catenin pathway antagonizing effects of Wnt16 (Nalesso et al., 2017).

In numerous studies, Wnt ligand expression in osteoblasts has been analyzed giving different results regarding their relative expression levels and revealing differences in their expression in calvarial and trabecular bone (Andrade, Nilsson, Barnes, & Baron, 2007; Martineau, Abed, Martel-Pelletier, Pelletier, & Lajeunesse, 2017; Tan et al., 2014; Wan et al., 2013). One ligand found consistently in different studies is Wnt5a. Interestingly, *Wnt5a* is expressed by osteoblasts, while its receptor Ror2 is present in osteoclasts, suggesting that the Wnt5a/Ror2 pathway primarily affects osteoclastogenesis (Maeda et al., 2012). Yet, loss of Wnt5a activity also affects osteoblastogenesis and increased adipogenesis in calvarial cultures (Maeda et al., 2012; Takada et al., 2007). This effect of Wnt5a on osteoblasts occurs in a Ror2-independent way. The work by Okamoto and colleagues suggested that Wnt5a in osteoblasts supports Wnt/β-catenin signaling by maintaining the Lrp5/6 levels (Okamoto et al., 2014). These observations are consistent with the finding that postnatal loss of Wnt/β-catenin signaling in preosteoblasts shifts their fate toward the adipogenic lineage (Song et al., 2012).

Another Wnt involved in bone is Wnt3a. Polymorphisms in the *WNT3A* gene are associated with bone mineral density variations (Velazquez-Cruz et al., 2014). *Wnt3a* heterozygous mice have a low bone mass phenotype (Takada et al., 2007). *In vitro*, Wnt3a stimulates osteoblastogenesis. Here, two alternative mechanisms have been proposed: one showing that Wnt3a signals through the $G\alpha_q/11$ subunit of G-proteins and eventually activates Pkcδ in the murine bone marrow-derived ST2 cell line (Tu et al., 2007), while the other mechanism requires the interaction of the transcription factor Lef1 with the transcriptional coactivator Taz (Kida et al., 2018). Heterozygous *Lef1*-deficient mice also show, in a gender and age-dependent manner, a decrease in bone volume that is associated with reduced osteoblast activity (Noh et al., 2009). Yap/Taz have also been implicated to act downstream of Wnt4 in osteoblasts. Wnt4 stimulates their nuclear accumulation and their depletion interferes with the pro-osteoblastic activity of Wnt4 (Park, Kim, et al., 2015). In addition, osteoblast-specific Taz overexpression increases the bone mass *in vivo* and lentiviral-mediated Taz expression alleviates ovariectomy-induced bone loss (Yang et al., 2013; Zhang et al., 2016). Combined conditional loss of Yap and Taz using Cre-lines active during different stages of osteoblastogenesis suggests that they oppose osteoblast lineage differentiation at the progenitor stage but promote bone formation and inhibit bone resorption in mature osteoblasts and osteocytes (Xiong et al., 2018).

Two other Wnt ligands, Wnt7b and Wnt10b, positively affect osteoblastogenesis. *Wnt7b* is expressed in osteoblasts and has been proposed to stimulate osteoblastogenesis *in vitro* through Pkcδ (Hu et al., 2005; Tu et al., 2007). In line with this, *Pkcδ*-deficient embryos display alterations in bone formation primarily in neural crest-derived craniofacial elements. Their long bones show only minor changes, which may be secondary as chondrocyte maturation is also altered in these mice (Tu et al., 2007). A similar phenotype is observed upon conditional deletion of *Wnt7b* in mesenchymal cells using the *Dermo1*-Cre line (Tu et al., 2007). More recently, Wnt7b has been proposed to activate mTorc1 primarily on the basis of transgenic experiments expressing Wnt7b in osteoblasts (Chen et al., 2014). Wnt7b overexpression leads to a massive increase in bone mass due to an increased osteoblast number. This effect can be partially reversed deleting a component of the mTorc1 complex (Chen et al., 2014). Wnt signaling through Lrp5 also activates mTorc2 to stimulate glycolysis in osteoblasts (Esen et al., 2013). Transgenic overexpression of Wnt10b under the control of the Fabp4 regulatory region stimulates osteoblast differentiation at the expense of adipocytes (Bennett et al., 2005). Concomitantly, osteoblastogenesis is accelerated *in vitro* upon Wnt10b treatment (Bennett et al., 2005). Osteoblast-specific overexpression of Wnt10b under the control of the osteocalcin promoter also results in increased trabecular bone formation without affecting osteoclastogenesis (Bennett et al., 2007). *Wnt10b*-deficient mice, in contrast, display a reduction in their trabecular bone volume progressing with age due to a reduction in osteoprogenitor cells (Bennett et al., 2005; Stevens et al., 2010).

In embryonic development, β-catenin is required to suppress the chondrogenic potential of the uncommitted bi-potential osteo-chondroprogenitors in the perichondrium/periosteum (Day & Yang, 2008; Hill et al., 2005; Rodda & McMahon, 2006). During osteoblastogenesis, β-catenin acts functionally downstream of Ihh-signaling (Hu et al., 2005; Rodda & McMahon, 2006). Stabilization of β-catenin in *Osx* expressing osteoblast precursors results in their expansion and premature mineralization, while on the other hand interfering with their terminal differentiation into osteocalcin-producing cells (Rodda & McMahon, 2006). Hence, all experimental data point toward the concept that β-catenin acts at multiple steps and that its levels have to be tightly modulated during osteoblastogenesis to allow the maturation of functional osteoblasts. This has also been observed in an *in vitro* model (van der Horst et al., 2005). In differentiated perichondrial osteoblasts, β-catenin directly regulates the expression of *Opg via* Tcf1 and/or a transcriptional complex of Ebf-2 with Lef1, thereby negatively influencing osteoclastogenesis (Boyce, Xing, & Chen, 2005; Colnot et al., 2005; Glass et al., 2005; Holmen et al., 2005; Kieslinger et al., 2005). In addition, it has been shown that β-catenin activity in osteoblasts negatively regulates the expression of the pro-osteoclastic factor, Rankl (Holmen et al., 2005; Sato, Nakashima, Nashimoto, Yawaka, & Tamura, 2009; Spencer, Utting, Etheridge, Arnett, & Genever, 2006). Osteoblasts are of dual origin, differentiating from precursors within the perichondrium and the hypertrophic zone (Manolagas, 2014; Park, Gebhardt, et al., 2015; Tsang et al., 2015). Conditional inactivation of β-catenin in chondrocytes, or even more specifically in hypertrophic chondrocytes, results in an almost complete inhibition of chondrocyte-derived osteoblastogenesis (Houben et al., 2016; Jing et al., 2018). In contrast, β-catenin stabilization in either *Acan*-Cre expressing chondrocytes or *Col10a1*-Cre expressing hypertrophic chondrocytes potentiated chondrocyte-derived osteoblastogenesis (Houben et al., 2016; Jing et al., 2018). The underlying mechanism is so far not known, but could potentially be a non-cell autonomous effect (Houben et al., 2016).

The combined loss of Lrp5 and Lrp6 activity recapitulates the osteoblast-lineage differentiation defects observed upon loss of β-catenin. Also here, ectopic chondrocyte differentiation in the perichondrium and anlagen of the skull bones are observed (Joeng et al., 2011). Mutations in *LRP5* affect bone mass in humans. Loss-of-function mutations are found in patients with Osteoporosis-Pseudoglioma Syndrome (OPPG), while patients with gain-of-function mutations display an autosomal-dominant high bone mass (Williams & Insogna, 2009). *Lrp5* null mutant mice or those lacking Lrp5 activity specifically in osteocytes are osteopenic, faithfully recapitulating the human loss-of-function phenotype (Clement-Lacroix et al., 2005; Cui et al., 2011; Kato et al., 2002). Conversely, mice expressing *Lrp5* mutant alleles mimicking the human gain-of-function mutations in the first β-propeller in osteocytes develop a high bone mass phenotype (Cui et al., 2011). The concept that Lrp5 is cell-autonomously acting in osteoblasts/osteocytes has been challenged by the reports of Yadav and colleagues showing that serotonin synthesis is increased in *Lrp5* mutant mice and that mice with a *Lrp5* deletion in mesenchymal cells display no skeletal abnormalities (Yadav, Arantes, Barros, Lazaretti-Castro, & Ducy, 2010; Yadav et al., 2008). This lead to an ongoing debate in the field, whether or not the altered bone mass phenotype in *Lrp5* mutant mice is due to a cell-autonomous function of Lrp5 in bone or mediated by the proposed gut/serotonin axis (Cui et al., 2014; Kode et al., 2014; Yadav & Ducy, 2010). Mice carrying a hypomorphic mutation for Lrp6 (the *Ringelschwanz* (*rs*) allele) are also osteopenic but due to increased bone resorption and not a defect in osteoblastogenesis (Kubota et al., 2008).

The Wnt-antagonist Dkk binds to a single transmembrane receptor of the Kremen1/2 family. This leads to the internalization of a complex containing also the Wnt co-receptor Lrp5 or Lrp6, thereby interfering with Wnt/β-catenin signaling (Mao et al., 2002). Yet, they are not absolutely required for Dkk-mediated inhibition of Wnt/β-catenin signaling, as Dkk1 can also directly bind to Lrp6 (Semenov et al., 2001). Osteoblast-specific overexpression of the Dkk-receptor Kremen2 (Krm2) under the control of the *Col1a1*-promoter perturbs osteoblast maturation and stimulates osteoclastogenesis *via* downregulation of *Opg* (Schulze et al., 2010). Mice deficient for *Krm2* display a late onset high bone mass phenotype. In 24-week-old female *Krm2*$^{-/-}$ mice the trabecular bone volume and number are elevated and the bone formation rate was even threefold increased compared to wildtype (Schulze et al., 2010). This is not observed in 12-week-old *Krm2*-deficient animals (Ellwanger et al., 2008). Yet, osteoclastogenesis is not affected in the aged *Krm2*$^{-/-}$ animals, suggesting that the downregulation of *Opg* observed in the transgenic mice is likely to be the consequence of the defective osteoblastogenesis. A sex-independent increase in bone mass is already observed in 12-week-old mice lacking Krm1 and Krm2. Again, the phenotype is due to an increase in osteoblast numbers and activity, while bone resorption is unaffected (Ellwanger et al., 2008). The Wnt antagonists sFrp1, sFrp2, Dkk1 have also been implicated in negatively regulating osteoblastogenesis (Bodine et al., 2009, 2004; Gaur et al., 2009; Morvan et al., 2006; Sathi et al., 2009). Surprisingly, mice lacking *Dkk2* are osteopenic with increased but poorly mineralized osteoid. As such Dkk2 has probably a role in late osteoblasts controlling proper matrix mineralization, which may, however, not be entirely mediated by its Wnt antagonistic activity (Li, Liu, et al., 2005).

The effect of the loss of all Wnt ligands produced by osteoblasts has been addressed by deleting *Wls/Gpr177* from *Osx*-expressing cells. This results in shortened skeletal elements with aberrant ossification (Maruyama et al., 2013). Deletion of *Wls* at later stages of osteoblast differentiation using an *Osteocalcin*-Cre or a *Col1a1*-Cre line dramatically decreases cortical and trabecular bone due to decreased matrix formation and increased bone resorption (Wan et al., 2013; Zhong et al., 2012). Embryonic skeletal development in these mice is largely normal with the exception of a slight delay in chondrocyte hypertrophy in the *Col1a1*-Cre *Wls*-deficient mice (Wan et al., 2013). Interestingly, the block in osteoclastogenesis present in mice carrying the *Col1a1*-Cre $Ctnnb1^{ex3}$ gain-of-function allele is alleviated upon additional deletion of *Wls* without changes in Opg or Rankl (Wan et al., 2013). This is probably due to the loss of a Wnt ligand, which is produced by osteoblasts and inhibits osteoclastogenesis directly in a paracrine fashion. This Wnt ligand cannot be Wnt5a because it has been shown to stimulate osteoclastogenesis (Maeda et al., 2012). Yet, it could be a Wnt ligand signaling through Fzd8 repressing osteoclastogenesis in an Opg-independent manner, possibly Wnt16, as it fulfills this criterium (Albers et al., 2013; Kobayashi, Uehara, Udagawa, & Takahashi, 2016).

Members of the R-spondin protein family binding to Lrg receptors are considered to potentiate Wnt/β-catenin signaling (de Lau, Peng, Gros, & Clevers, 2014; de Lau, Snel, & Clevers, 2012). R-spondins are regulators of bone metabolism (Shi, Mao, Zheng, & Jiang, 2016). R-spondin 1 promotes osteoblast differentiation and bone formation while blocking osteoclastogenesis (Kronke et al., 2010). R-spondin 2 also supports osteoblastogenesis acting through binding to Lrg4 promoting Wnt/β-catenin signaling (Zhu et al., 2016). The osteoblastogenesis-promoting effect of Wnt11 requires R-spondin 2 (Friedman, Oyserman, & Hankenson, 2009).

3.1 Wnt-signaling and osteocytes

The terminal fate of an osteoblast is to become an osteocyte (Noble, 2008). Historically, osteocytes where considered to be retired osteoblasts that became entrapped in the bone matrix during bone formation. Osteocytes account for the majority of bone cells (90–95%). They reside in lacunae within the compact bone and are interconnected by a dense network of dendritic processes projecting through a canalicular system. Osteocytes act as mechanosensory cells within the bone (Nguyen & Jacobs, 2013; Pead, Suswillo, Skerry, Vedi, & Lanyon, 1988). Several inhibitors of the Wnt/β-catenin pathway are expressed by osteocytes, such as Dickkopf 1 (Dkk1), secreted frizzled-related protein 1 (sFrp1), and sclerostin (Sost). The latter was originally identified as an inhibitor of bone morphogenetic protein (BMP) (Winkler et al., 2003). Yet, Sost also binds to the Wnt co-receptor Lrp5 interfering with the binding of Wnt ligands (Li, Zhang, et al., 2005). In addition, Sost, as well as Dkk1 and Wise, can bind to Lrp4, which is also present on the surface of osteoblasts and osteocytes (Choi, Dieckmann, Herz, & Niemeier, 2009; Leupin et al., 2011). Once secreted into the osteocyte lacuna, Sost is thought to be diffusing through the canalicular network. Mechanical stimulation leads to increased bone formation associated with a decrease in Sost and Dkk1 production (Robling et al., 2008). The mechanostimulated anabolic bone response is lost in *Lrp5* deficient mice and diminished in mice lacking *Lrp5* specifically in osteocytes (Sawakami et al., 2006; Zhao, Shim, Dodge, Robling, & Yokota, 2013). Mice expressing a truncated extracellular form of the receptor Lrp4 ($Lrp4^{ECD}$) show reduced skeletal growth and increased bone turnover (Choi et al., 2009). In contrast, mice homozygous for *Lrp4* loss-of-function alleles or with an osteoblast-lineage-specific *Lrp4* deletion or carrying the sclerosteosis allele Lrp4R1170Q display an increase in bone mass along with an increase in serum Sost levels (Boudin et al., 2017; Chang et al., 2014; Xiong et al., 2015). These phenotypic changes can be in part recapitulated *in vivo* using antibodies that block the interaction between Lrp4 and Sost (Chang et al., 2014). Functionally, Lrp4 is required primarily on osteoblasts and osteocytes, but not on osteoclasts (Xiong et al., 2015).

Osteocytes also produce Wnt1 (Laine et al., 2013). Cell-type specific inactivation of *Wnt1* in osteocytes is responsible for the impaired bone formation, low bone mass, and spontaneous fracture phenotype observed in the *Swaying* mouse harboring a spontaneous *Wnt1* mutation (Joeng et al., 2014, 2017). Wnt1 overexpression in osteocytes leads to markedly increased bone formation in both trabecular and cortical bone (Joeng et al., 2017). Joeng and colleagues also provided evidence that Wnt1-signaling is mediated *via* an mTor (mechanistic target of rapamycin) pathway and not *via* β-catenin. Interestingly, anti-sclerostin antibody treatment of the *Swaying* mouse results in a marked improvement of its bone-related phenotypic features, suggesting that other Wnt ligands, signaling through the Lrp5 receptor, can partially compensate for the absence of Wnt1 (Joeng et al., 2017). The likely candidates are Wnt7b and Wnt10b, which both interact with Lrp5 and enhance bone formation as mentioned above (Chen et al., 2014; Esen et al., 2013).

Similarly to the deletion of *Ctnnb1* in osteoblasts, alterations in bone homeostasis are observed upon conditional deletion of *Ctnnb1* in osteocytes using the *Dmp1*-Cre line. The cancellous trabecular bone and the cortical bone are severely reduced in these mice at 2 months of age due to an increased osteoclastic bone resorption, with the bone-loss phenotype being more dramatic in female than male mice. Molecularly, this is associated with a downregulation of *Opg* expression, an increase in the Rankl serum levels but not with alterations in the *Rankl* expression levels (Kramer et al., 2010). Mechanical load activates β-catenin signaling in osteocytes (Lara-Castillo et al., 2015). Yet, regarding the response to mechanical load in mice with reduced β-catenin levels in osteocytes, conflicting reports have been published. One reports the abolishment of the anabolic response, while the other reports that, at least, periosteal load-induced bone formation is not affected upon postnatally induced deletion of *Ctnnb1* in osteocytes (Javaheri et al., 2014; Kang, Hong, & Robling, 2016).

Taken together, most of the experimental evidence regarding the role of β-catenin-mediated signaling in osteoblasts/osteocytes points toward a regulatory role in the catabolic branch of bone homeostasis (Fig. 4). In contrast, the anabolic Wnt activities appear to be mediated by non-canonical Wnt signaling pathways.

4. Wnt signaling and osteoclastogenesis

In vitro, Wnt3a, Wnt4, and Wnt16 exert an anti-osteoclastic effect in co-cultures of osteoclast precursors and osteoblasts (Kobayashi et al., 2015). Here, the effects of Wnt3a and Wnt4 appear to be Opg-dependent, while Wnt16 acts in an Opg-independent fashion. In contrast, Wnt5a signaling through Ror2, present on osteoclasts, exhibits a pro-osteoclastic effect by increasing *Rank* expression and counteracts the inhibitory effect of Wnt16 on osteoclastogenesis (Kobayashi et al., 2015; Maeda et al., 2012). Alternatively, Wnt5a and Wnt5b may signal through the Fzd-receptor-like

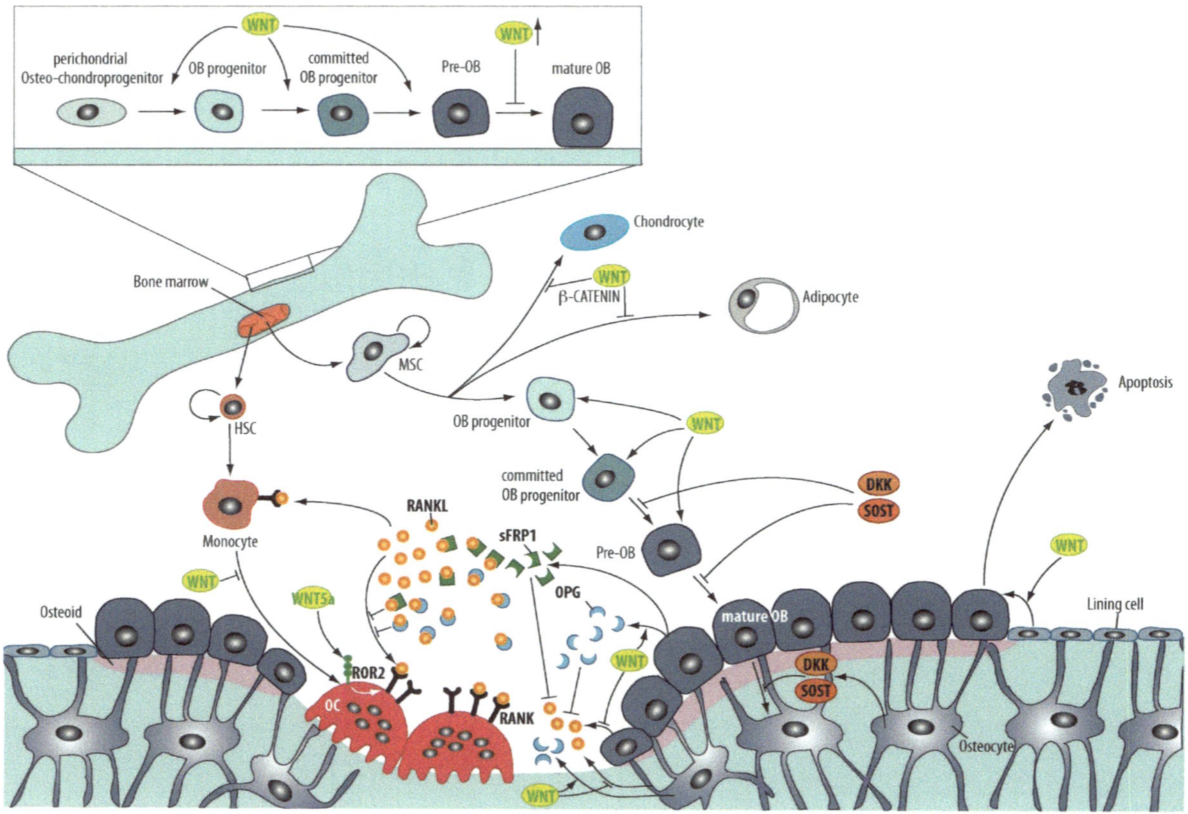

Fig. 4 Schematic representation of the origins of osteoblasts and osteoclasts along with their differentiation steps and the influence of Wnt signaling (primarily Wnt/β-catenin signaling) on the various differentiation steps. For further details see text.

tyrosine kinase (Ryk) receptor to stimulate osteoclastogenesis (Santiago, Oguma, Brown, & Laurence, 2012). Wnt3a has been reported to suppress Rankl-mediated *Nfatc1* expression in osteoclast precursors by activating a cAmp/Pka pathway in addition to canonical Wnt/β-catenin (Weivoda et al., 2016). This has been conceptually challenged (Martin, 2015). *Fzd8* is expressed on osteoblasts and osteoclasts and its loss results in an osteopenic phenotype due to increased osteoclastogenesis without affecting parameters of bone formation (Albers et al., 2013). This anti-osteoclastic effect is not associated with alterations in the level of Opg or Rankl. Deletion of *Ctnnb1* in the monocyte lineage, giving rise to macrophages, neutrophils, and osteoclasts, using the *LysM*-Cre line, or even more specifically in the osteoclasts themselves, using the *CathepsinK*-Cre line, results in an increase in osteoclasts, suggesting that the Wnt/β-catenin pathway plays a cell-autonomous inhibitory role in osteoclastogenesis (Albers et al., 2013; Otero et al., 2012; Ruiz et al., 2016; Wei et al., 2011). Interestingly, the Wnt antagonist sFrp1, which is expressed in osteoblasts, has been shown to inhibit osteoclastogenesis by binding directly to Rankl (Hausler et al., 2004). Finally, Wnts are bidirectionally involved in the coupling of osteoclasts with osteoblasts, as the expression of the pro-osteogenic *Wnt10b* in osteoclasts is under the control of Tgfβ, at least in female mice (Ota et al., 2013; Pederson, Ruan, Westendorf, Khosla, & Oursler, 2008).

5. Roles of Wnt-signaling in intramembranous bone formation

The Wnt/β-catenin pathway is active in the skull bones and given its importance for osteoblastogenesis it is, of course, playing a functional role in intermembranous bone formation (Day et al., 2005; Yu et al., 2005). Loss of β-catenin activity in the head mesenchyme or *in vitro* in osteoblast precursor cells, results in the differentiation of these cells into chondrocytes (Day et al., 2005; Hill et al., 2005). Similarly, loss of *Wnt9a*, which is expressed in the developing sagittal suture, leads to ectopic differentiation of mesenchymal cells into chondrocytes (Spater, Hill, O'Sullivan, et al., 2006). Hence, one important role of the Wnt/β-catenin pathway in intramembranous ossification, as well as bone formation during endochondral skeleton formation, is to repress the chondrogenic potential of the precursor cells, thereby enabling their differentiation along the osteoblastic lineage. A phenotype similar to the one observed in *Wnt9a* mutants is also detected upon deletion of *Axin2* in combination with the fibroblast growth factor receptor *Fgfr1*, whereby reduced Fgf-signaling counteracts the inhibitory chondrogenesis effects of the elevated β-catenin levels (Maruyama, Mirando, Deng, & Hsu, 2010). This is reminiscent of the observation, mentioned earlier, that Fgf-signaling in combination with Wnt/β-catenin signaling represses chondrogenesis in the distal limb bud mesenchyme (ten Berge et al., 2008). Axin2 is a negative regulator of the Wnt/β-catenin pathway and is expressed in cells at the osteogenic fronts and the periosteum of the developing sutures except for the posterior frontal suture (Yu et al., 2005). More recently, it was shown that the Axin2 positive population of cells also contains the suture stem cell population (Maruyama, Jeong, Sheu, & Hsu, 2016). *Axin2* mutants display a craniosynostosis-like phenotype with premature fusion of sutures between neural crest-derived flat bones, while mesenchymal-derived bones such as the parietal bones are unaffected. The osteoprogenitor population in neural crest-derived bones is expanded in *Axin2* mutants and ossification is accelerated (Yu et al., 2005). Complementary *in vitro* studies confirmed differences in the proliferation and differentiation behavior of neural crest *versus* mesenchymal-derived *Axin2*$^{-/-}$ osteoblasts. The lack of Axin2 stimulates, on the one hand, β-catenin signaling in osteoblast precursors, while on the other hand, it promotes Cadherin/β-catenin-mediated cell adhesion in mature osteoblasts (Liu, Yu, & Hsu, 2007; Yu et al., 2005). The latter is known to promote osteogenesis (Stains & Civitelli, 2005). Yet, the underlying mechanism for this regional difference is not known so far. In contrast to the local effects in the *Axin2*-deficient mice, global conditional stabilization of β-catenin in head mesenchymal cells, using the *Prx1*-Cre line, inhibits the formation of mesenchymal-derived bone (Hill et al., 2005). The effect of conditional inactivation of *Wls/Gpr177* in neural crest cells resembles the craniofacial phenotype of *Ctnnb1* inactivation in neural crest cells as well as that of *Wnt1/Wnt3a* double mutant mice lacking neural crest-derived skeletal elements (Brault et al., 2001; Fu, Ivy Yu, Maruyama, Mirando, & Hsu, 2011; Ikeya, Lee, Johnson, McMahon, & Takada, 1997). In the case of the neural crest-specific *Ctnnb1* inactivation this phenotype has been attributed to increased apoptosis in the region where mesenchymal condensations take place (Brault et al., 2001).

6. Wnt signaling in joint development, maintenance, and degeneration

Chondrocytes and cells contributing to the joint interzone of the synovial joints in the limb share a common progenitor. Wnt signaling has been originally implicated in joint development on the basis of Wnt9a

(Wnt14) retroviral misexpression experiments in the chicken (Hartmann & Tabin, 2001). Here, Wnt9a is capable of suppressing chondrogenic differentiation and inducing markers of the joint interzone. Similar observations have been made using a transgenic approach in mice, revealing in addition the requirement of β-catenin in this process (Guo et al., 2004). Yet, all joints are present in *Wnt9a* mutant mice. In these mutants, fibrous and synovial cells in the elbow joint develop a chondroid metaplasia phenotype (Spater, Hill, O'Sullivan, et al., 2006). In addition to Wnt9a, Wnt4 and Wnt16 are also expressed in the developing joints (Guo et al., 2004; Hartmann & Tabin, 2001). Wnt4 appears to act partially redundant with Wnt9a, as additional joints display the chondroid metaplasia phenotype, with ectopic differentiation of chondrocytes within the synovial or joint capsule, in the *Wnt9a/Wnt4* double mutants. In addition, the carpal and tarsal elements fuse secondarily in the double mutants (Spater, Hill, O'Sullivan, et al., 2006). Mice lacking β-catenin activity in the limb mesenchyme have multiple limb defects, among them, lack of a distinct joint interzone formation. The molecular analysis suggests that the joint program is initiated, however, as chondrogenesis is not suppressed in the developing joint interzone due to the absence of β-catenin activity. Instead, the cells are stuck in an early uncommitted stage, expressing both chondrogenic and joint interzone markers (Spater, Hill, Gruber, et al., 2006). Similarly, overexpression of the antagonist Frzb1 affects joint development in the chicken limb (Enomoto-Iwamoto et al., 2002). Immobilization of the limbs also affects the formation of a joint interzone and is associated with downregulation of Wnt/β-catenin signaling (Kahn et al., 2009). Wnt signaling interacts with Ihh in the control of joint formation (Mak, Chen, Day, Chuang, & Yang, 2006).

Wnt signaling has been implicated in pathogenesis of osteoarthritis (OA) (Blom et al., 2009; Corr, 2008; Zhang et al., 2018). Several studies reported an upregulation of Wnt signaling in OA patients and in a surgical model of OA in mice, caused by destabilization of the medial meniscus (DMM). This is associated with inhibition of chondrogenesis and stimulation of catabolic activities leading to matrix degradation (Deshmukh et al., 2018; Ma et al., 2014; Yuasa, Otani, Koike, Iwamoto, & Enomoto-Iwamoto, 2008; Zhu et al., 2009). Treatment with XAV-939, a small-molecule inhibitor of Wnt/β-catenin signaling that stabilizes Axin2, reduces cartilage degeneration and synovitis in the DMM model (Lietman et al., 2018). In agreement with findings that increased Wnt signaling play a role in OA, modulation of Wnt antagonists, such as Dkk1, Frzb1, sFrp1, and Sost, in mice affects OA progression (Bouaziz et al., 2015; Chang et al., 2018;

Funck-Brentano et al., 2014; Lories et al., 2007; Oh, Chun, & Chun, 2012; Thysen, Luyten, & Lories, 2015). Interestingly, downregulation of β-catenin activity in articular chondrocytes also results in articular cartilage destruction (Zhu et al., 2008). *Ctnnb1* inactivation in articular chondrocytes causes loss of Prg4-producing superficial cells and a strong increase in *Acn* and *Col10a1* expression, the latter being a trait of hypertrophic chondrocytes (Yasuhara et al., 2011). The ligand, Wnt16, is upregulated after injury of human articular cartilage (Dell'accio, De Bari, Eltawil, Vanhummelen, & Pitzalis, 2008). Yet, *Wnt16* mutant mice show no defects in joint development or maintenance (Todd et al., 2015; Zheng et al., 2012). Nevertheless, Wnt16 is proposed to act as a "buffering" agent in articular chondrocytes by controlling Wnt/β-catenin signaling, as it can dampen an excessive activation of this pathway (Nalesso et al., 2017). Hence, it appears that the levels of Wnt/β-catenin signaling need to be tightly controlled to maintain a healthy joint. Drugs modulating Wnt signaling are currently being investigated as an OA treatment strategy (Miyamoto et al., 2017; Yazici et al., 2017, 2018).

7. Defects in Wnt-signaling associated with skeletal diseases

The importance of Wnt-signaling for skeletogenesis is reflected by the findings that mutations in humans are associated with alterations of the skeleton. Mutations in *WNT5a* and *ROR2* are associated with the Robinow syndrome (OMIM #268310; #164975) and/or Brachydactyly type B1 (OMIM # 113000) (Patton & Afzal, 2002; Person et al., 2010; Roifman, Brunner, Lohr, Mazzeu, & Chitayat, 2015). Numerous Wnt-pathway molecules were also identified in genome-wide association studies (GWAS) related to skeletal phenotypes (Hsu & Kiel, 2012). The first indications for a critical role of Wnt/β-catenin signaling in osteoblastogenesis were based on the findings that human loss-of-function mutations in the Wnt receptor *LRP5* cause Osteoporosis-Pseudoglioma Syndrome (OMIM #259770) (Gong et al., 2001; Lara-Castillo & Johnson, 2015). In contrast, gain-of-function mutations in the *LRP5* b-propeller 1 encoded region are responsible for an autosomal-dominant high bone mass phenotype (OMIM #601884). More recently, it was discovered that mutations in the *WNT1* gene cause Osteogenesis imperfecta type XV and an autosomal-dominant form of susceptibility to early onset of osteoporosis (OMIM #615220, #615221) (Keupp et al., 2013; Laine et al., 2013; Pyott et al., 2013). Furthermore, mutations in *WNT16* are associated with cortical bone porosity as well as increased fracture risk (Gori et al., 2015; Zheng et al., 2012). Wnt16 has also been found to be upregulated in osteoarthritic articular chondrocytes (Dell'accio et al., 2008). A region on chromosome 1 encompassing the *WNT3a* and *WNT9a* loci has been identified in a recent GWAS study as a novel locus associated with OA endophenotypes affecting the thumb (Boer et al., 2018).

8. Conclusions and implications

Genetic studies in mice revealed that during embryonic development the canonical Wnt/β-catenin pathway plays an important role as a permissive signal for the differentiation of osteoblasts and joint interzone cells from common precursors during skeletal lineage differentiation. This knowledge can be applied to tissue engineering. In the absence of Wnt/β-catenin signaling, chondrogenesis appears to be the default route during embryonic skeletogenesis. Yet, this pathway has additional functions in chondrocytes and osteoblasts. In chondrocytes, it modulates, on the one hand, the pace of maturation *via* the regulation of *Ihh* and, on the other the hand, the transdifferentiation of hypertrophic chondrocytes to osteoblasts. In mature osteoblasts and hypertrophic chondrocytes, it regulates the expression of pro- and anti-osteoclastic factors thereby influencing bone homeostasis. As such, the pathway is key to almost all aspects of embryonic skeletogenesis, including mechanosensation during joint development. Yet, in many cases, the Wnt ligands responsible for the activation of the pathway remain to be uncovered. As β-catenin is a dual function protein that is also implicated in cell adhesion, it remains to be shown if all phenotypic aspects are truly based on its function as a transcriptional co-activator in the Wnt-pathway. Furthermore, genetic studies resemble either only the on or the off state. Yet, for many purposes it would be more interesting if the level of signaling could be fine-tuned to unravel the relevant thresholds. Such fine-tuning mechanisms exist physiologically, as activation of the Wnt/β-catenin pathway is regulated at the extracellular and the intracellular levels through the presence or absence of certain co-receptors, the availability of the ligands, the presence of endogenous antagonists and feedback loops leading to their upregulation upon ligand-binding, etc. Other pathways, including the non-canonical Wnt-pathway, are also implicated in fine-tuning β-catenin mediated signaling, but have additional roles that remain to be elucidated further. Future research will probably also focus on uncovering cross-talks with other pathways during skeletogenesis.

The discovery of mutations in Wnt-signaling pathway components associated by skeletal diseases in humans has substantiated the relevance of Wnt-signaling for mammalian skeletogenesis. Therapeutic exploration of the knowledge gained from mouse models and human mutations at the level of the Wnt ligands is hampered, as it has not been possible to produce all ligands in their active form. As such, blocking antibodies of secreted agonists are currently tested as bone modeling agents.

Acknowledgments

The authors acknowledge the contributions of all researchers in the field of Wnt-signaling and skeletogenesis and hope that they have neither miscited nor overlooked important contributions to the field. Research by the authors is supported by grants from the Deutsche Forschungsgemeinschaft (HA 4767/2-1, HA 4767/4-2, HA 4767/5-1), from the Federal Ministry of Education and Science (01EC1408E) as part of the OVERLOAD-Prev-OP consortium. S.T. has previously been supported by a grant from the Interdisciplinary Center of Clinical Research (IZKF, Har2/002/14).

References

Acebron, S. P., & Niehrs, C. (2016). Beta-catenin-independent roles of Wnt/LRP6 Signaling. *Trends in Cell Biology*, *26*, 956–967.

Ahrens, M. J., Li, Y., Jiang, H., & Dudley, A. T. (2009). Convergent extension movements in growth plate chondrocytes require gpi-anchored cell surface proteins. *Development*, *136*, 3463–3474.

Ahrens, M. J., Romereim, S., & Dudley, A. T. (2011). A re-evaluation of two key reagents for in vivo studies of Wnt signaling. *Developmental Dynamics*, *240*, 2060–2068.

Akiyama, H., Lyons, J. P., Mori-Akiyama, Y., Yang, X., Zhang, R., Zhang, Z., et al. (2004). Interactions between Sox9 and beta-catenin control chondrocyte differentiation. *Genes & Development*, *18*, 1072–1087.

Alam, I., Alkhouli, M., Gerard-O'Riley, R. L., Wright, W. B., Acton, D., Gray, A. K., et al. (2016). Osteoblast-specific overexpression of human WNT16 increases both cortical and trabecular bone mass and structure in mice. *Endocrinology*, *157*, 722–736.

Albers, J., Keller, J., Baranowsky, A., Beil, F. T., Catala-Lehnen, P., Schulze, J., et al. (2013). Canonical Wnt signaling inhibits osteoclastogenesis independent of osteoprotegerin. *The Journal of Cell Biology*, *200*, 537–549.

Amara, C. S., Fabritius, C., Houben, A., Wolff, L. I., & Hartmann, C. (2017). CaMKII signaling stimulates Mef2c activity in vitro but only minimally affects murine long bone development in vivo. *Frontiers in Cell and Development Biology*, *5*, 20.

Andrade, A. C., Nilsson, O., Barnes, K. M., & Baron, J. (2007). Wnt gene expression in the post-natal growth plate: Regulation with chondrocyte differentiation. *Bone*, *40*, 1361–1369.

Aramburu, J., Rao, A., & Klee, C. B. (2000). Calcineurin: From structure to function. *Current Topics in Cellular Regulation*, *36*, 237–295.

Arias, A. M., Brown, A. M., & Brennan, K. (1999). Wnt signalling: Pathway or network? *Current Opinion in Genetics & Development*, *9*, 447–454.

Arnold, M. A., Kim, Y., Czubryt, M. P., Phan, D., McAnally, J., Qi, X., et al. (2007). MEF2C transcription factor controls chondrocyte hypertrophy and bone development. *Developmental Cell*, *12*, 377–389.

Asai, N., Ohkawara, B., Ito, M., Masuda, A., Ishiguro, N., & Ohno, K. (2014). LRP4 induces extracellular matrix productions and facilitates chondrocyte differentiation. *Biochemical and Biophysical Research Communications*, *451*, 302–307.

Aszodi, A., Hunziker, E. B., Brakebusch, C., & Fassler, R. (2003). Beta1 integrins regulate chondrocyte rotation, G1 progression, and cytokinesis. *Genes & Development*, *17*, 2465–2479.

Azzolin, L., Panciera, T., Soligo, S., Enzo, E., Bicciato, S., Dupont, S., et al. (2014). YAP/TAZ incorporation in the beta-catenin destruction complex orchestrates the Wnt response. *Cell*, *158*, 157–170.

Ben Shoham, A., Rot, C., Stern, T., Krief, S., Akiva, A., Dadosh, T., et al. (2016). Deposition of collagen type I onto skeletal endothelium reveals a new role for blood vessels in regulating bone morphology. *Development*, *143*, 3933–3943.

Bennett, C. N., Longo, K. A., Wright, W. S., Suva, L. J., Lane, T. F., Hankenson, K. D., et al. (2005). Regulation of osteoblastogenesis and bone mass by Wnt10b. *PNAS*, *102*, 3324–3329.

Bennett, C. N., Ouyang, H., Ma, Y. L., Zeng, Q., Gerin, I., Sousa, K. M., et al. (2007). Wnt10b increases postnatal bone formation by enhancing osteoblast differentiation. *Journal of Bone and Mineral Research*, *22*, 1924–1932.

Bi, W., Deng, J. M., Zhang, Z., Behringer, R. R., & de Crombrugghe, B. (1999). Sox9 is required for cartilage formation. *Nature Genetics*, *22*, 85–89.

Blom, A. B., Brockbank, S. M., van Lent, P. L., van Beuningen, H. M., Geurts, J., Takahashi, N., et al. (2009). Involvement of the Wnt signaling pathway in experimental and human osteoarthritis: Prominent role of Wnt-induced signaling protein 1. *Arthritis and Rheumatism*, *60*, 501–512.

Bodine, P. V., Stauffer, B., Ponce-de-Leon, H., Bhat, R. A., Mangine, A., Seestaller-Wehr, L. M., et al. (2009). A small molecule inhibitor of the Wnt antagonist secreted frizzled-related protein-1 stimulates bone formation. *Bone*, *44*, 1063–1068.

Bodine, P. V., Zhao, W., Kharode, Y. P., Bex, F. J., Lambert, A. J., Goad, M. B., et al. (2004). The Wnt antagonist secreted frizzled-related protein-1 is a negative regulator of trabecular bone formation in adult mice. *Molecular Endocrinology*, *18*, 1222–1237.

Boer, C. G., Broer, L., Schiphof, D., Runhaar, J., Oei, E., Waarsing, E., et al. (2018). Genetic exploration of osteoarthritis endophenotypes identifies new biological pathways in osteoarthritis. 2018 OARSI world congress on osteoarthritis: Promoting clinical and basic research in osteoarthritis. *Osteoarthritis and Cartilage*, *26*, S30.

Bouaziz, W., Funck-Brentano, T., Lin, H., Marty, C., Ea, H. K., Hay, E., et al. (2015). Loss of sclerostin promotes osteoarthritis in mice via beta-catenin-dependent and -independent Wnt pathways. *Arthritis Research & Therapy*, *17*, 24.

Boudin, E., Yorgan, T., Fijalkowski, I., Sonntag, S., Steenackers, E., Hendrickx, G., et al. (2017). The Lrp4R1170Q homozygous knock-in mouse recapitulates the bone phenotype of Sclerosteosis in humans. *Journal of Bone and Mineral Research*, *32*, 1739–1749.

Boyce, B. F., Xing, L., & Chen, D. (2005). Osteoprotegerin, the bone protector, is a surprising target for beta-catenin signaling. *Cell Metabolism*, *2*, 344–345.

Bradley, E. W., & Drissi, M. H. (2010). WNT5A regulates chondrocyte differentiation through differential use of the CaN/NFAT and IKK/NF-kappaB pathways. *Molecular Endocrinology*, *24*, 1581–1593.

Brault, V., Moore, R., Kutsch, S., Ishibashi, M., Rowitch, D. H., McMahon, A. P., et al. (2001). Inactivation of the beta-catenin gene by Wnt1-Cre-mediated deletion results in dramatic brain malformation and failure of craniofacial development. *Development*, *128*, 1253–1264.

Candela, M. E., Cantley, L., Yasuaha, R., Iwamoto, M., Pacifici, M., & Enomoto-Iwamoto, M. (2014). Distribution of slow-cycling cells in epiphyseal cartilage and requirement of beta-catenin signaling for their maintenance in growth plate. *Journal of Orthopaedic Research*, *32*, 661–668.

Chang, J. C., Christiansen, B. A., Murugesh, D. K., Sebastian, A., Hum, N. R., Collette, N. M., et al. (2018). SOST/Sclerostin improves posttraumatic osteoarthritis and inhibits MMP2/3 expression after injury. *Journal of Bone and Mineral Research, 33*, 1105–1113.

Chang, M. K., Kramer, I., Huber, T., Kinzel, B., Guth-Gundel, S., Leupin, O., et al. (2014). Disruption of Lrp4 function by genetic deletion or pharmacological blockade increases bone mass and serum sclerostin levels. *PNAS, 111*, E5187–E5195.

Chen, J., Tu, X., Esen, E., Joeng, K. S., Lin, C., Arbeit, J. M., et al. (2014). WNT7B promotes bone formation in part through mTORC1. *PLoS Genetics, 10*, e1004145.

Chen, M., Zhu, M., Awad, H., Li, T. F., Sheu, T. J., Boyce, B. F., et al. (2008). Inhibition of beta-catenin signaling causes defects in postnatal cartilage development. *Journal of Cell Science, 121*, 1455–1465.

Choi, B., Chun, J. S., Lee, Y. S., Sonn, J. K., & Kang, S. S. (1995). Expression of protein kinase C isozymes that are required for chondrogenesis of chick limb bud mesenchymal cells. *Biochemical and Biophysical Research Communications, 216*, 1034–1040.

Choi, H. Y., Dieckmann, M., Herz, J., & Niemeier, A. (2009). Lrp4, a novel receptor for Dickkopf 1 and sclerostin, is expressed by osteoblasts and regulates bone growth and turnover in vivo. *PLoS One, 4*, e7930.

Choi, S. W., Jeong, D. U., Kim, J. A., Lee, B., Joeng, K. S., Long, F., et al. (2012). Indian hedgehog signalling triggers Nkx3.2 protein degradation during chondrocyte maturation. *The Biochemical Journal, 443*, 789–798.

Chung, U. I., Schipani, E., McMahon, A. P., & Kronenberg, H. M. (2001). Indian hedgehog couples chondrogenesis to osteogenesis in endochondral bone development. *Journal of Clinical Investigation, 107*, 295–304.

Church, V., Nohno, T., Linker, C., Marcelle, C., & Francis-West, P. (2002). Wnt regulation of chondrocyte differentiation. *Journal of Cell Science, 115*, 4809–4818.

Clement-Lacroix, P., Ai, M., Morvan, F., Roman-Roman, S., Vayssiere, B., Belleville, C., et al. (2005). Lrp5-independent activation of Wnt signaling by lithium chloride increases bone formation and bone mass in mice. *PNAS, 102*, 17406–17411.

Colnot, C., de la Fuente, L., Huang, S., Hu, D., Lu, C., St-Jacques, B., et al. (2005). Indian hedgehog synchronizes skeletal angiogenesis and perichondrial maturation with cartilage development. *Development, 132*, 1057–1067.

Corr, M. (2008). Wnt-beta-catenin signaling in the pathogenesis of osteoarthritis. *Nature Clinical Practice. Rheumatology, 4*, 550–556.

Crabtree, G. R. (2001). Calcium, calcineurin, and the control of transcription. *Journal of Biological Chemistry, 276*, 2313–2316.

Cui, Y., Niziolek, P. J., MacDonald, B. T., Alenina, N., Matthes, S., Jacobsen, C. M., et al. (2014). Reply to Lrp5 regulation of bone mass and gut serotonin synthesis. *Nature Medicine, 20*, 1229–1230.

Cui, Y., Niziolek, P. J., MacDonald, B. T., Zylstra, C. R., Alenina, N., Robinson, D. R., et al. (2011). Lrp5 functions in bone to regulate bone mass. *Nature Medicine, 17*, 684–691.

Das, S., Yu, S., Sakamori, R., Stypulkowski, E., & Gao, N. (2012). Wntless in Wnt secretion: Molecular, cellular and genetic aspects. *Frontiers in Biology, 7*, 587–593.

Day, T. F., Guo, X., Garrett-Beal, L., & Yang, Y. (2005). Wnt/beta-catenin signaling in mesenchymal progenitors controls osteoblast and chondrocyte differentiation during vertebrate skeletogenesis. *Developmental Cell, 8*, 739–750.

Day, T. F., & Yang, Y. (2008). Wnt and hedgehog signaling pathways in bone development. *The Journal of Bone and Joint Surgery, 90*(Suppl. 1), 19–24.

De, A. (2011). Wnt/Ca2+ signaling pathway: A brief overview. *Acta Biochimica et Biophysica Sinica, 43*, 745–756.

de Lau, W., Peng, W. C., Gros, P., & Clevers, H. (2014). The R-spondin/Lgr5/Rnf43 module: Regulator of Wnt signal strength. *Genes & Development, 28*, 305–316.

de Lau, W. B., Snel, B., & Clevers, H. C. (2012). The R-spondin protein family. *Genome Biology*, *13*, 242.

DeChiara, T. M., Kimble, R. B., Poueymirou, W. T., Rojas, J., Masiakowski, P., Valenzuela, D. M., et al. (2000). Ror2, encoding a receptor-like tyrosine kinase, is required for cartilage and growth plate development. *Nature Genetics*, *24*, 271–274.

Dell'accio, F., De Bari, C., Eltawil, N. M., Vanhummelen, P., & Pitzalis, C. (2008). Identification of the molecular response of articular cartilage to injury, by microarray screening: Wnt-16 expression and signaling after injury and in osteoarthritis. *Arthritis and Rheumatism*, *58*, 1410–1421.

Deng, Y., Wu, A., Li, P., Li, G., Qin, L., Song, H., et al. (2016). Yap1 regulates multiple steps of chondrocyte differentiation during skeletal development and bone repair. *Cell Reports*, *14*, 2224–2237.

Deshmukh, V., Hu, H., Barroga, C., Bossard, C., Kc, S., Dellamary, L., et al. (2018). A small-molecule inhibitor of the Wnt pathway (SM04690) as a potential disease modifying agent for the treatment of osteoarthritis of the knee. *Osteoarthritis and Cartilage*, *26*, 18–27.

Dodds, G. S. (1930). Row formation and other types of arrangement of cartilage cells in endochondral ossification. *The Anatomical Record Banner*, *46*, 385–399.

Dong, Y. F., Soung do, Y., Schwarz, E. M., O'Keefe, R. J., & Drissi, H. (2006). Wnt induction of chondrocyte hypertrophy through the Runx2 transcription factor. *Journal of Cellular Physiology*, *208*, 77–86.

Ellwanger, K., Saito, H., Clement-Lacroix, P., Maltry, N., Niedermeyer, J., Lee, W. K., et al. (2008). Targeted disruption of the Wnt regulator Kremen induces limb defects and high bone density. *Molecular and Cellular Biology*, *28*, 4875–4882.

Enomoto-Iwamoto, M., Kitagaki, J., Koyama, E., Tamamura, Y., Wu, C., Kanatani, N., et al. (2002). The Wnt antagonist Frzb-1 regulates chondrocyte maturation and long bone development during limb skeletogenesis. *Developmental Biology*, *251*, 142–156.

Esen, E., Chen, J., Karner, C. M., Okunade, A. L., Patterson, B. W., & Long, F. (2013). WNT-LRP5 signaling induces Warburg effect through mTORC2 activation during osteoblast differentiation. *Cell Metabolism*, *17*, 745–755.

Friedman, M. S., Oyserman, S. M., & Hankenson, K. D. (2009). Wnt11 promotes osteoblast maturation and mineralization through R-spondin 2. *Journal of Biological Chemistry*, *284*, 14117–14125.

Fu, J., Ivy Yu, H. M., Maruyama, T., Mirando, A. J., & Hsu, W. (2011). Gpr177/mouse Wntless is essential for Wnt-mediated craniofacial and brain development. *Developmental Dynamics*, *240*, 365–371.

Funck-Brentano, T., Bouaziz, W., Marty, C., Geoffroy, V., Hay, E., & Cohen-Solal, M. (2014). Dkk-1-mediated inhibition of Wnt signaling in bone ameliorates osteoarthritis in mice. *Arthritis & Rheumatology*, *66*, 3028–3039.

Gao, B. (2012). Wnt regulation of planar cell polarity (PCP). *Current Topics in Developmental Biology*, *101*, 263–295.

Gao, B., Song, H., Bishop, K., Elliot, G., Garrett, L., English, M. A., et al. (2011). Wnt signaling gradients establish planar cell polarity by inducing Vangl2 phosphorylation through Ror2. *Developmental Cell*, *20*, 163–176.

Gaur, T., Rich, L., Lengner, C. J., Hussain, S., Trevant, B., Ayers, D., et al. (2006). Secreted frizzled related protein 1 regulates Wnt signaling for BMP2 induced chondrocyte differentiation. *Journal of Cellular Physiology*, *208*, 87–96.

Gaur, T., Wixted, J. J., Hussain, S., O'Connell, S. L., Morgan, E. F., Ayers, D. C., et al. (2009). Secreted frizzled related protein 1 is a target to improve fracture healing. *Journal of Cellular Physiology*, *220*, 174–181.

Geetha-Loganathan, P., Nimmagadda, S., Christ, B., Huang, R., & Scaal, M. (2010). Ectodermal Wnt6 is an early negative regulator of limb chondrogenesis in the chicken embryo. *BMC Developmental Biology*, *10*, 32.

Glass, D. A., 2nd, Bialek, P., Ahn, J. D., Starbuck, M., Patel, M. S., Clevers, H., et al. (2005). Canonical Wnt signaling in differentiated osteoblasts controls osteoclast differentiation. *Developmental Cell*, *8*, 751–764.

Golovchenko, S., Hattori, T., Hartmann, C., Gebhardt, M., Gebhard, S., Hess, A., et al. (2013). Deletion of beta catenin in hypertrophic growth plate chondrocytes impairs trabecular bone formation. *Bone*, *55*, 102–112.

Gong, Y., Slee, R. B., Fukai, N., Rawadi, G., Roman-Roman, S., Reginato, A. M., et al. (2001). LDL receptor-related protein 5 (LRP5) affects bone accrual and eye development. *Cell*, *107*, 513–523.

Gori, F., Lerner, U., Ohlsson, C., & Baron, R. (2015). A new WNT on the bone: WNT16, cortical bone thickness, porosity and fractures. *Bonekey Reports*, *4*, 669.

Guo, X., Day, T. F., Jiang, X., Garrett-Beal, L., Topol, L., & Yang, Y. (2004). Wnt/beta-catenin signaling is sufficient and necessary for synovial joint formation. *Genes & Development*, *18*, 2404–2417.

Hartmann, C., & Tabin, C. J. (2000). Dual roles of Wnt signaling during chondrogenesis in the chicken limb. *Development*, *127*, 3141–3159.

Hartmann, C., & Tabin, C. J. (2001). Wnt-14 plays a pivotal role in inducing synovial joint formation in the developing appendicular skeleton. *Cell*, *104*, 341–351.

Hausler, K. D., Horwood, N. J., Chuman, Y., Fisher, J. L., Ellis, J., Martin, T. J., et al. (2004). Secreted frizzled-related protein-1 inhibits RANKL-dependent osteoclast formation. *Journal of Bone and Mineral Research*, *19*, 1873–1881.

Hill, T. P., Spater, D., Taketo, M. M., Birchmeier, W., & Hartmann, C. (2005). Canonical Wnt/beta-catenin signaling prevents osteoblasts from differentiating into chondrocytes. *Developmental Cell*, *8*, 727–738.

Holmen, S. L., Zylstra, C. R., Mukherjee, A., Sigler, R. E., Faugere, M. C., Bouxsein, M. L., et al. (2005). Essential role of beta-catenin in postnatal bone acquisition. *Journal of Biological Chemistry*, *280*, 21162–21168.

Houben, A., Kostanova-Poliakova, D., Weissenbock, M., Graf, J., Teufel, S., von der Mark, K., et al. (2016). Beta-catenin activity in late hypertrophic chondrocytes locally orchestrates osteoblastogenesis and osteoclastogenesis. *Development*, *143*, 3826–3838.

Hsu, Y. H., & Kiel, D. P. (2012). Clinical review: Genome-wide association studies of skeletal phenotypes: What we have learned and where we are headed. *The Journal of Clinical Endocrinology and Metabolism*, *97*, E1958–E1977.

Hu, H., Hilton, M. J., Tu, X., Yu, K., Ornitz, D. M., & Long, F. (2005). Sequential roles of hedgehog and Wnt signaling in osteoblast development. *Development*, *132*, 49–60.

Ikeya, M., Lee, S. M., Johnson, J. E., McMahon, A. P., & Takada, S. (1997). Wnt signalling required for expansion of neural crest and CNS progenitors. *Nature*, *389*, 966–970.

Inoki, K., Ouyang, H., Zhu, T., Lindvall, C., Wang, Y., Zhang, X., et al. (2006). TSC2 integrates Wnt and energy signals via a coordinated phosphorylation by AMPK and GSK3 to regulate cell growth. *Cell*, *126*, 955–968.

Javaheri, B., Stern, A. R., Lara, N., Dallas, M., Zhao, H., Liu, Y., et al. (2014). Deletion of a single beta-catenin allele in osteocytes abolishes the bone anabolic response to loading. *Journal of Bone and Mineral Research*, *29*, 705–715.

Jiang, Z., Von den Hoff, J. W., Torensma, R., Meng, L., & Bian, Z. (2014). Wnt16 is involved in intramembranous ossification and suppresses osteoblast differentiation through the Wnt/beta-catenin pathway. *Journal of Cellular Physiology*, *229*, 384–392.

Jin, E. J., Park, J. H., Lee, S. Y., Chun, J. S., Bang, O. S., & Kang, S. S. (2006). Wnt-5a is involved in TGF-beta3-stimulated chondrogenic differentiation of chick wing bud mesenchymal cells. *The International Journal of Biochemistry & Cell Biology*, *38*, 183–195.

Jing, Y., Jing, J., Wang, K., Chan, K., Harris, S. E., Hinton, R. J., et al. (2018). Vital roles of beta-catenin in trans-differentiation of chondrocytes to bone cells. *International Journal of Biological Sciences*, *14*, 1–9.

Joeng, K. S., Lee, Y. C., Jiang, M. M., Bertin, T. K., Chen, Y., Abraham, A. M., et al. (2014). The swaying mouse as a model of osteogenesis imperfecta caused by WNT1 mutations. *Human Molecular Genetics*, *23*, 4035–4042.

Joeng, K. S., Lee, Y. C., Lim, J., Chen, Y., Jiang, M. M., Munivez, E., et al. (2017). Osteocyte-specific WNT1 regulates osteoblast function during bone homeostasis. *Journal of Clinical Investigation*, *127*, 2678–2688.

Joeng, K. S., Schumacher, C. A., Zylstra-Diegel, C. R., Long, F., & Williams, B. O. (2011). Lrp5 and Lrp6 redundantly control skeletal development in the mouse embryo. *Developmental Biology*, *359*, 222–229.

Kahn, J., Shwartz, Y., Blitz, E., Krief, S., Sharir, A., Breitel, D. A., et al. (2009). Muscle contraction is necessary to maintain joint progenitor cell fate. *Developmental Cell*, *16*, 734–743.

Kang, K. S., Hong, J. M., & Robling, A. G. (2016). Postnatal beta-catenin deletion from Dmp1-expressing osteocytes/osteoblasts reduces structural adaptation to loading, but not periosteal load-induced bone formation. *Bone*, *88*, 138–145.

Karystinou, A., Roelofs, A. J., Neve, A., Cantatore, F. P., Wackerhage, H., & De Bari, C. (2015). Yes-associated protein (YAP) is a negative regulator of chondrogenesis in mesenchymal stem cells. *Arthritis Research & Therapy*, *17*, 147.

Kato, M., Patel, M. S., Levasseur, R., Lobov, I., Chang, B. H., Glass, D. A., 2nd, et al. (2002). Cbfa1-independent decrease in osteoblast proliferation, osteopenia, and persistent embryonic eye vascularization in mice deficient in Lrp5, a Wnt coreceptor. *The Journal of Cell Biology*, *157*, 303–314.

Kemler, R. (1993). From cadherins to catenins: Cytoplasmic protein interactions and regulation of cell adhesion. *Trends in Genetics*, *9*, 317–321.

Kestler, H. A., & Kuhl, M. (2008). From individual Wnt pathways towards a Wnt signalling network. *Philosophical Transactions of the Royal Society of London. Series B, Biological Sciences*, *363*, 1333–1347.

Keupp, K., Beleggia, F., Kayserili, H., Barnes, A. M., Steiner, M., Semler, O., et al. (2013). Mutations in WNT1 cause different forms of bone fragility. *American Journal of Human Genetics*, *92*, 565–574.

Kida, J., Hata, K., Nakamura, E., Yagi, H., Takahata, Y., Murakami, T., et al. (2018). Interaction of LEF1 with TAZ is necessary for the osteoblastogenic activity of Wnt3a. *Scientific Reports*, *8*, 10375.

Kieslinger, M., Folberth, S., Dobreva, G., Dorn, T., Croci, L., Erben, R., et al. (2005). EBF2 regulates osteoblast-dependent differentiation of osteoclasts. *Developmental Cell*, *9*, 757–767.

Kikuchi, A., Yamamoto, H., & Sato, A. (2009). Selective activation mechanisms of Wnt signaling pathways. *Trends in Cell Biology*, *19*, 119–129.

Kitagaki, J., Iwamoto, M., Liu, J. G., Tamamura, Y., Pacifci, M., & Enomoto-Iwamoto, M. (2003). Activation of beta-catenin-LEF/TCF signal pathway in chondrocytes stimulates ectopic endochondral ossification. *Osteoarthritis and Cartilage*, *11*, 36–43.

Kobayashi, Y., Thirukonda, G. J., Nakamura, Y., Koide, M., Yamashita, T., Uehara, S., et al. (2015). Wnt16 regulates osteoclast differentiation in conjunction with Wnt5a. *Biochemical and Biophysical Research Communications*, *463*, 1278–1283.

Kobayashi, Y., Uehara, S., Udagawa, N., & Takahashi, N. (2016). Regulation of bone metabolism by Wnt signals. *Journal of Biochemistry*, *159*, 387–392.

Kode, A., Obri, A., Paone, R., Kousteni, S., Ducy, P., & Karsenty, G. (2014). Lrp5 regulation of bone mass and serotonin synthesis in the gut. *Nature Medicine*, *20*, 1228–1229.

Kohn, A. D., & Moon, R. T. (2005). Wnt and calcium signaling: Beta-catenin-independent pathways. *Cell Calcium*, *38*, 439–446.

Komori, T. (2006). Regulation of osteoblast differentiation by transcription factors. *Journal of Cellular Biochemistry*, *99*, 1233–1239.

Kramer, I., Halleux, C., Keller, H., Pegurri, M., Gooi, J. H., Weber, P. B., et al. (2010). Osteocyte Wnt/beta-catenin signaling is required for normal bone homeostasis. *Molecular and Cellular Biology, 30*, 3071–3085.

Kronenberg, H. M. (2003). Developmental regulation of the growth plate. *Nature, 423*, 332–336.

Kronke, G., Uderhardt, S., Kim, K. A., Stock, M., Scholtysek, C., Zaiss, M. M., et al. (2010). R-spondin 1 protects against inflammatory bone damage during murine arthritis by modulating the Wnt pathway. *Arthritis and Rheumatism, 62*, 2303–2312.

Kubota, T., Michigami, T., Sakaguchi, N., Kokubu, C., Suzuki, A., Namba, N., et al. (2008). Lrp6 hypomorphic mutation affects bone mass through bone resorption in mice and impairs interaction with Mesd. *Journal of Bone and Mineral Research, 23*, 1661–1671.

Kuhl, M. (2004). The WNT/calcium pathway: Biochemical mediators, tools and future requirements. *Frontiers in Bioscience, 9*, 967–974.

Kuss, P., Kraft, K., Stumm, J., Ibrahim, D., Vallecillo-Garcia, P., Mundlos, S., et al. (2014). Regulation of cell polarity in the cartilage growth plate and perichondrium of metacarpal elements by HOXD13 and WNT5A. *Developmental Biology, 385*, 83–93.

Ladher, R. K., Church, V. L., Allen, S., Robson, L., Abdelfattah, A., Brown, N. A., et al. (2000). Cloning and expression of the Wnt antagonists Sfrp-2 and Frzb during chick development. *Developmental Biology, 218*, 183–198.

Laine, C. M., Joeng, K. S., Campeau, P. M., Kiviranta, R., Tarkkonen, K., Grover, M., et al. (2013). WNT1 mutations in early-onset osteoporosis and osteogenesis imperfecta. *The New England Journal of Medicine, 368*, 1809–1816.

Lara-Castillo, N., & Johnson, M. L. (2015). LRP receptor family member associated bone disease. *Reviews in Endocrine & Metabolic Disorders, 16*, 141–148.

Lara-Castillo, N., Kim-Weroha, N. A., Kamel, M. A., Javaheri, B., Ellies, D. L., Krumlauf, R. E., et al. (2015). In vivo mechanical loading rapidly activates beta-catenin signaling in osteocytes through a prostaglandin mediated mechanism. *Bone, 76*, 58–66.

Lawrence, P. A., & Casal, J. (2013). The mechanisms of planar cell polarity, growth and the hippo pathway: Some known unknowns. *Developmental Biology, 377*, 1–8.

Lee, H. H., & Behringer, R. R. (2007). Conditional expression of Wnt4 during chondrogenesis leads to dwarfism in mice. *PLoS One, 2*, e450.

Leupin, O., Piters, E., Halleux, C., Hu, S., Kramer, I., Morvan, F., et al. (2011). Bone overgrowth-associated mutations in the LRP4 gene impair sclerostin facilitator function. *Journal of Biological Chemistry, 286*, 19489–19500.

Li, Y., Ahrens, M. J., Wu, A., Liu, J., & Dudley, A. T. (2011). Calcium/calmodulin-dependent protein kinase II activity regulates the proliferative potential of growth plate chondrocytes. *Development, 138*, 359–370.

Li, Y., & Dudley, A. T. (2009). Noncanonical frizzled signaling regulates cell polarity of growth plate chondrocytes. *Development, 136*, 1083–1092.

Li, X., Liu, P., Liu, W., Maye, P., Zhang, J., Zhang, Y., et al. (2005). Dkk2 has a role in terminal osteoblast differentiation and mineralized matrix formation. *Nature Genetics, 37*, 945–952.

Li, X., Zhang, Y., Kang, H., Liu, W., Liu, P., Zhang, J., et al. (2005). Sclerostin binds to LRP5/6 and antagonizes canonical Wnt signaling. *Journal of Biological Chemistry, 280*, 19883–19887.

Lietman, C., Wu, B., Lechner, S., Shinar, A., Sehgal, M., Rossomacha, E., et al. (2018). Inhibition of Wnt/beta-catenin signaling ameliorates osteoarthritis in a murine model of experimental osteoarthritis. *JCI Insight, 3*(3), pii: 96308, https://doi.org/10.1172/jci.insight.96308, [Epub ahead of print].

Liu, B., Yu, H. M., & Hsu, W. (2007). Craniosynostosis caused by Axin2 deficiency is mediated through distinct functions of beta-catenin in proliferation and differentiation. *Developmental Biology, 301*, 298–308.

Liu, Y. J., Zhang, L., Papasian, C. J., & Deng, H. W. (2014). Genome-wide association studies for osteoporosis: A 2013 update. *Journal of Bone Metabolism*, *21*, 99–116.
Logan, C. Y., & Nusse, R. (2004). The Wnt signaling pathway in development and disease. *Annual Review of Cell and Developmental Biology*, *20*, 781–810.
Loganathan, P. G., Nimmagadda, S., Huang, R., Scaal, M., & Christ, B. (2005). Comparative analysis of the expression patterns of Wnts during chick limb development. *Histochemistry and Cell Biology*, *123*, 195–201.
Lories, R. J., Peeters, J., Bakker, A., Tylzanowski, P., Derese, I., Schrooten, J., et al. (2007). Articular cartilage and biomechanical properties of the long bones in Frzb-knockout mice. *Arthritis and Rheumatism*, *56*, 4095–4103.
Lyashenko, N., Weissenbock, M., Sharir, A., Erben, R. G., Minami, Y., & Hartmann, C. (2010). Mice lacking the orphan receptor ror1 have distinct skeletal abnormalities and are growth retarded. *Developmental Dynamics*, *239*, 2266–2277.
Ma, C. H., Lv, Q., Cao, Y., Wang, Q., Zhou, X. K., Ye, B. W., et al. (2014). Genes relevant with osteoarthritis by comparison gene expression profiles of synovial membrane of osteoarthritis patients at different stages. *European Review for Medical and Pharmacological Sciences*, *18*, 431–439.
MacDonald, B. T., & He, X. (2012). Frizzled and LRP5/6 receptors for Wnt/beta-catenin signaling. *Cold Spring Harbor Perspectives in Biology*, *4*(12), pii: a007880, https://doi.org/10.1101/cshperspect.a007880.
MacDonald, B. T., Tamai, K., & He, X. (2009). Wnt/beta-catenin signaling: Components, mechanisms, and diseases. *Developmental Cell*, *17*, 9–26.
Maeda, K., Kobayashi, Y., Udagawa, N., Uehara, S., Ishihara, A., Mizoguchi, T., et al. (2012). Wnt5a-Ror2 signaling between osteoblast-lineage cells and osteoclast precursors enhances osteoclastogenesis. *Nature Medicine*, *18*, 405–412.
Maes, C., Kobayashi, T., Selig, M. K., Torrekens, S., Roth, S. I., Mackem, S., et al. (2010). Osteoblast precursors, but not mature osteoblasts, move into developing and fractured bones along with invading blood vessels. *Developmental Cell*, *19*, 329–344.
Mak, K. K., Chen, M. H., Day, T. F., Chuang, P. T., & Yang, Y. (2006). Wnt/beta-catenin signaling interacts differentially with Ihh signaling in controlling endochondral bone and synovial joint formation. *Development*, *133*, 3695–3707.
Manolagas, S. C. (2014). Wnt signaling and osteoporosis. *Maturitas*, *78*, 233–237.
Mao, B., Wu, W., Davidson, G., Marhold, J., Li, M., Mechler, B. M., et al. (2002). Kremen proteins are Dickkopf receptors that regulate Wnt/beta-catenin signalling. *Nature*, *417*, 664–667.
Martin, T. J. (2015). Comment on: Wnt signaling inhibits osteoclast differentiation by activating canonical and non-canonical cAMP/PKA pathways. *Journal of Bone and Mineral Research*, *30*, 2133–2134.
Martineau, X., Abed, E., Martel-Pelletier, J., Pelletier, J. P., & Lajeunesse, D. (2017). Alteration of Wnt5a expression and of the non-canonical Wnt/PCP and Wnt/PKC-Ca2+ pathways in human osteoarthritis osteoblasts. *PLoS One*, *12*, e0180711.
Maruyama, T., Jeong, J., Sheu, T. J., & Hsu, W. (2016). Stem cells of the suture mesenchyme in craniofacial bone development, repair and regeneration. *Nature Communications*, *7*, 10526.
Maruyama, T., Jiang, M., & Hsu, W. (2013). Gpr177, a novel locus for bone mineral density and osteoporosis, regulates osteogenesis and chondrogenesis in skeletal development. *Journal of Bone and Mineral Research*, *28*, 1150–1159.
Maruyama, T., Mirando, A. J., Deng, C. X., & Hsu, W. (2010). The balance of WNT and FGF signaling influences mesenchymal stem cell fate during skeletal development. *Science Signaling*, *3*, ra40.
Matta, C., & Mobasheri, A. (2014). Regulation of chondrogenesis by protein kinase C: Emerging new roles in calcium signalling. *Cellular Signalling*, *26*, 979–1000.

Miclea, R. L., Karperien, M., Bosch, C. A., van der Horst, G., van der Valk, M. A., Kobayashi, T., et al. (2009). Adenomatous polyposis coli-mediated control of beta-catenin is essential for both chondrogenic and osteogenic differentiation of skeletal precursors. *BMC Developmental Biology*, *9*, 26.

Mikels, A. J., & Nusse, R. (2006). Purified Wnt5a protein activates or inhibits beta-catenin-TCF signaling depending on receptor context. *PLoS Biology*, *4*, e115.

Miyamoto, K., Ohkawara, B., Ito, M., Masuda, A., Hirakawa, A., Sakai, T., et al. (2017). Fluoxetine ameliorates cartilage degradation in osteoarthritis by inhibiting Wnt/beta-catenin signaling. *PLoS One*, *12*, e0184388.

Morvan, F., Boulukos, K., Clement-Lacroix, P., Roman Roman, S., Suc-Royer, I., Vayssiere, B., et al. (2006). Deletion of a single allele of the Dkk1 gene leads to an increase in bone formation and bone mass. *Journal of Bone and Mineral Research*, *21*, 934–945.

Moverare-Skrtic, S., Henning, P., Liu, X., Nagano, K., Saito, H., Borjesson, A. E., et al. (2014). Osteoblast-derived WNT16 represses osteoclastogenesis and prevents cortical bone fragility fractures. *Nature Medicine*, *20*, 1279–1288.

Moverare-Skrtic, S., Wu, J., Henning, P., Gustafsson, K. L., Sjogren, K., Windahl, S. H., et al. (2015). The bone-sparing effects of estrogen and WNT16 are independent of each other. *PNAS*, *112*, 14972–14977.

Nalesso, G., Thomas, B. L., Sherwood, J. C., Yu, J., Addimanda, O., Eldridge, S. E., et al. (2017). WNT16 antagonises excessive canonical WNT activation and protects cartilage in osteoarthritis. *Annals of the Rheumatic Diseases*, *76*, 218–226.

Narcisi, R., Cleary, M. A., Brama, P. A., Hoogduijn, M. J., Tuysuz, N., ten Berge, D., et al. (2015). Long-term expansion, enhanced chondrogenic potential, and suppression of endochondral ossification of adult human MSCs via WNT signaling modulation. *Stem Cell Reports*, *4*, 459–472.

Nguyen, A. M., & Jacobs, C. R. (2013). Emerging role of primary cilia as mechanosensors in osteocytes. *Bone*, *54*, 196–204.

Nicolin, V., Sandrucci, M. A., Basa, M., Bareggi, R., Martelli, A. M., Narducci, P., et al. (2004). Expression of protein kinase C (PKC) alpha, delta, epsilon, zeta in primary chick chondrocyte cultures: Immunocytochemical study. *Italian Journal of Anatomy and Embryology*, *109*, 55–65.

Niehrs, C. (2012). The complex world of WNT receptor signalling. *Nature Reviews. Molecular Cell Biology*, *13*, 767–779.

Noble, B. S. (2008). The osteocyte lineage. *Archives of Biochemistry and Biophysics*, *473*, 106–111.

Noh, T., Gabet, Y., Cogan, J., Shi, Y., Tank, A., et al. (2009). Lef1 haploinsufficient mice display a low turnover and low bone mass phenotype in a gender- and age-specific manner. *PLoS One*, *4*, e5438.

Oh, H., Chun, C. H., & Chun, J. S. (2012). Dkk-1 expression in chondrocytes inhibits experimental osteoarthritic cartilage destruction in mice. *Arthritis and Rheumatism*, *64*, 2568–2578.

Ohlsson, C., Henning, P., Nilsson, K. H., Wu, J., Gustafsson, K. L., & Sjogren, K. (2018). Inducible Wnt16 inactivation: WNT16 regulates cortical bone thickness in adult mice. *The Journal of Endocrinology*, *237*, 113–122.

Oishi, I., Suzuki, H., Onishi, N., Takada, R., Kani, S., Ohkawara, B., et al. (2003). The receptor tyrosine kinase Ror2 is involved in non-canonical Wnt5a/JNK signalling pathway. *Genes to Cells*, *8*, 645–654.

Okamoto, M., Udagawa, N., Uehara, S., Maeda, K., Yamashita, T., Nakamichi, Y., et al. (2014). Noncanonical Wnt5a enhances Wnt/beta-catenin signaling during osteoblastogenesis. *Scientific Reports*, *4*, 4493.

Ota, K., Quint, P., Ruan, M., Pederson, L., Westendorf, J. J., Khosla, S., et al. (2013). TGF-beta induces Wnt10b in osteoclasts from female mice to enhance coupling to osteoblasts. *Endocrinology*, *154*, 3745–3752.

Otero, K., Shinohara, M., Zhao, H., Cella, M., Gilfillan, S., & Colucci, A. (2012). TREM2 and beta-catenin regulate bone homeostasis by controlling the rate of osteoclastogenesis. *Journal of Immunology*, *188*, 2612–2621.

Park, J., Gebhardt, M., Golovchenko, S., Branguli, F., Hattori, T., Hartmann, C., et al. (2015). Dual pathways to endochondral osteoblasts: A novel chondrocyte-derived osteoprogenitor cell identified in hypertrophic cartilage. *Biology Open*, *4*, 608–621.

Park, H. W., Kim, Y. C., Yu, B., Moroishi, T., Mo, J. S., Plouffe, S. W., et al. (2015). Alternative Wnt Signaling activates YAP/TAZ. *Cell*, *162*, 780–794.

Patton, M. A., & Afzal, A. R. (2002). Robinow syndrome. *Journal of Medical Genetics*, *39*, 305–310.

Pead, M. J., Suswillo, R., Skerry, T. M., Vedi, S., & Lanyon, L. E. (1988). Increased 3H-uridine levels in osteocytes following a single short period of dynamic bone loading in vivo. *Calcified Tissue International*, *43*, 92–96.

Pederson, L., Ruan, M., Westendorf, J. J., Khosla, S., & Oursler, M. J. (2008). Regulation of bone formation by osteoclasts involves Wnt/BMP signaling and the chemokine sphingosine-1-phosphate. *PNAS*, *105*, 20764–20769.

Person, A. D., Beiraghi, S., Sieben, C. M., Hermanson, S., Neumann, A. N., Robu, M. E., et al. (2010). WNT5A mutations in patients with autosomal dominant Robinow syndrome. *Developmental Dynamics*, *239*, 327–337.

Piccolo, S., Dupont, S., & Cordenonsi, M. (2014). The biology of YAP/TAZ: Hippo signaling and beyond. *Physiological Reviews*, *94*, 1287–1312.

Piersma, B., Bank, R. A., & Boersema, M. (2015). Signaling in fibrosis: TGF-beta, WNT, and YAP/TAZ converge. *Frontiers in Medicine*, *2*, 59.

Provot, S., Kempf, H., Murtaugh, L. C., Chung, U. I., Kim, D. W., Chyung, J., et al. (2006). Nkx3.2/Bapx1 acts as a negative regulator of chondrocyte maturation. *Development*, *133*, 651–662.

Pyott, S. M., Tran, T. T., Leistritz, D. F., Pepin, M. G., Mendelsohn, N. J., Temme, R. T., et al. (2013). WNT1 mutations in families affected by moderately severe and progressive recessive osteogenesis imperfecta. *American Journal of Human Genetics*, *92*, 590–597.

Queirolo, V., Galli, D., Masselli, E., Borzi, R. M., Martini, S., Vitale, F., et al. (2016). PKCepsilon is a regulator of hypertrophic differentiation of chondrocytes in osteoarthritis. *Osteoarthritis and Cartilage*, *24*, 1451–1460.

Ramakrishnan, A. B., Sinha, A., Fan, V. B., & Cadigan, K. M. (2018). The Wnt transcriptional switch: TLE removal or inactivation? *BioEssays: News and Reviews in Molecular, Cellular and Developmental Biology*, *40*(2), Review, PMID: 29250807, https://doi.org/10.1002/bies.201700162, [Epub 2017 Dec 18].

Randall, R. M., Shao, Y. Y., Wang, L., & Ballock, R. T. (2012). Activation of Wnt planar cell polarity (PCP) signaling promotes growth plate column formation in vitro. *Journal of Orthopaedic Research*, *30*, 1906–1914.

Ranger, A. M., Gerstenfeld, L. C., Wang, J., Kon, T., Bae, H., Gravallese, E. M., et al. (2000). The nuclear factor of activated T cells (NFAT) transcription factor NFATp (NFATc2) is a repressor of chondrogenesis. *The Journal of Experimental Medicine*, *191*, 9–22.

Rao, T. P., & Kuhl, M. (2010). An updated overview on Wnt signaling pathways: A prelude for more. *Circulation Research*, *106*, 1798–1806.

Robling, A. G., Niziolek, P. J., Baldridge, L. A., Condon, K. W., Allen, M. R., Alam, I., et al. (2008). Mechanical stimulation of bone in vivo reduces osteocyte expression of Sost/sclerostin. *Journal of Biological Chemistry*, *283*, 5866–5875.

Rodda, S. J., & McMahon, A. P. (2006). Distinct roles for hedgehog and canonical Wnt signaling in specification, differentiation and maintenance of osteoblast progenitors. *Development*, *133*, 3231–3244.

Roifman, M., Brunner, H. G., Lohr, J. L., Mazzeu, J. F., & Chitayat, D. (2015). In M. P. Adam, H. H. Ardinger, R. A. Pagon, S. E. Wallace, L. J. H. Bean, K. Stephens, & A. Amemiya (Eds.), *Autosomal dominant robinow syndrome*. Seattle (WA): GeneReviews [Internet], PMID: 25577943.

Romereim, S. M., Conoan, N. H., Chen, B., & Dudley, A. T. (2014). A dynamic cell adhesion surface regulates tissue architecture in growth plate cartilage. *Development, 141*, 2085–2095.

Rosse, C., Linch, M., Kermorgant, S., Cameron, A. J., Boeckeler, K., & Parker, P. J. (2010). PKC and the control of localized signal dynamics. *Nature Reviews. Molecular Cell Biology, 11*, 103–112.

Rudnicki, J. A., & Brown, A. M. (1997). Inhibition of chondrogenesis by Wnt gene expression in vivo and in vitro. *Developmental Biology, 185*, 104–118.

Ruiz, P., Martin-Millan, M., Gonzalez-Martin, M. C., Almeida, M., Gonzalez-Macias, J., & Ros, M. A. (2016). CathepsinKCre mediated deletion of betacatenin results in dramatic loss of bone mass by targeting both osteoclasts and osteoblastic cells. *Scientific Reports, 6*, 36201.

Santiago, F., Oguma, J., Brown, A. M., & Laurence, J. (2012). Noncanonical Wnt signaling promotes osteoclast differentiation and is facilitated by the human immunodeficiency virus protease inhibitor ritonavir. *Biochemical and Biophysical Research Communications, 417*, 223–230.

Sathi, G. A., Inoue, M., Harada, H., Rodriguez, A. P., Tamamura, R., Tsujigiwa, H., et al. (2009). Secreted frizzled related protein (sFRP)-2 inhibits bone formation and promotes cell proliferation in ameloblastoma. *Oral Oncology, 45*, 856–860.

Sato, M. M., Nakashima, A., Nashimoto, M., Yawaka, Y., & Tamura, M. (2009). Bone morphogenetic protein-2 enhances Wnt/beta-catenin signaling-induced osteoprotegerin expression. *Genes to Cells, 14*, 141–153.

Sawakami, K., Robling, A. G., Ai, M., Pitner, N. D., Liu, D., Warden, S. J., et al. (2006). The Wnt co-receptor LRP5 is essential for skeletal mechanotransduction but not for the anabolic bone response to parathyroid hormone treatment. *Journal of Biological Chemistry, 281*, 23698–23711.

Schuijers, J., Mokry, M., Hatzis, P., Cuppen, E., & Clevers, H. (2014). Wnt-induced transcriptional activation is exclusively mediated by TCF/LEF. *EMBO Journal, 33*, 146–156.

Schulze, J., Seitz, S., Saito, H., Schneebauer, M., Marshall, R. P., Baranowsky, A., et al. (2010). Negative regulation of bone formation by the transmembrane Wnt antagonist Kremen-2. *PLoS One, 5*, e10309.

Schumacher, C. A., Joiner, D. M., Less, K. D., Drewry, M. O., & Williams, B. O. (2016). Characterization of genetically engineered mouse models carrying Col2a1-cre-induced deletions of Lrp5 and/or Lrp6. *Bone Research, 4*, 15042.

Schwabe, G. C., Trepczik, B., Suring, K., Brieske, N., Tucker, A. S., Sharpe, P. T., et al. (2004). Ror2 knockout mouse as a model for the developmental pathology of autosomal recessive Robinow syndrome. *Developmental Dynamics, 229*, 400–410.

Semenov, M. V., Tamai, K., Brott, B. K., Kuhl, M., Sokol, S., & He, X. (2001). Head inducer Dickkopf-1 is a ligand for Wnt coreceptor LRP6. *Current Biology, 11*, 951–961.

Shi, G. X., Mao, W. W., Zheng, X. F., & Jiang, L. S. (2016). The role of R-spondins and their receptors in bone metabolism. *Progress in Biophysics and Molecular Biology, 122*, 93–100.

Simonet, W. S., Lacey, D. L., Dunstan, C. R., Kelley, M., Chang, M. S., Luthy, R., et al. (1997). Osteoprotegerin: A novel secreted protein involved in the regulation of bone density. *Cell, 89*, 309–319.

Solursh, M., & Reiter, R. S. (1988). Inhibitory and stimulatory effects of limb ectoderm on in vitro chondrogenesis. *Journal of Experimental Zoology, 248*, 147–154.

Song, L., Liu, M., Ono, N., Bringhurst, F. R., Kronenberg, H. M., & Guo, J. (2012). Loss of wnt/beta-catenin signaling causes cell fate shift of preosteoblasts from osteoblasts to adipocytes. *Journal of Bone and Mineral Research*, *27*, 2344–2358.

Sonn, J. K., & Solursh, M. (1993). Activity of protein kinase C during the differentiation of chick limb bud mesenchymal cells. *Differentiation*, *53*, 155–162.

Spater, D., Hill, T. P., Gruber, M., & Hartmann, C. (2006). Role of canonical Wnt-signalling in joint formation. *European Cells & Materials*, *12*, 71–80.

Spater, D., Hill, T. P., O'Sullivan, R. J., Gruber, M., Conner, D. A., & Hartmann, C. (2006). Wnt9a signaling is required for joint integrity and regulation of Ihh during chondrogenesis. *Development*, *133*, 3039–3049.

Spencer, G. J., Utting, J. C., Etheridge, S. L., Arnett, T. R., & Genever, P. G. (2006). Wnt signalling in osteoblasts regulates expression of the receptor activator of NFkappaB ligand and inhibits osteoclastogenesis in vitro. *Journal of Cell Science*, *119*, 1283–1296.

Stains, J. P., & Civitelli, R. (2005). Cell-cell interactions in regulating osteogenesis and osteoblast function. *Birth Defects Research. Part C, Embryo Today*, *75*, 72–80.

Stark, K., Vainio, S., Vassileva, G., & McMahon, A. P. (1994). Epithelial transformation of metanephric mesenchyme in the developing kidney regulated by Wnt-4. *Nature*, *372*, 679–683.

Stevens, J. R., Miranda-Carboni, G. A., Singer, M. A., Brugger, S. M., Lyons, K. M., & Lane, T. F. (2010). Wnt10b deficiency results in age-dependent loss of bone mass and progressive reduction of mesenchymal progenitor cells. *Journal of Bone and Mineral Research*, *25*, 2138–2147.

St-Jacques, B., Hammerschmidt, M., & McMahon, A. P. (1999). Indian hedgehog signaling regulates proliferation and differentiation of chondrocytes and is essential for bone formation. *Genes & Development*, *13*, 2072–2086.

Takada, I., Mihara, M., Suzawa, M., Ohtake, F., Kobayashi, S., Igarashi, M., et al. (2007). A histone lysine methyltransferase activated by non-canonical Wnt signalling suppresses PPAR-gamma transactivation. *Nature Cell Biology*, *9*, 1273–1285.

Takeuchi, S., Takeda, K., Oishi, I., Nomi, M., Ikeya, M., Itoh, K., et al. (2000). Mouse Ror2 receptor tyrosine kinase is required for the heart development and limb formation. *Genes to Cells*, *5*, 71–78.

Tamamura, Y., Otani, T., Kanatani, N., Koyama, E., Kitagaki, J., Komori, T., et al. (2005). Developmental regulation of Wnt/beta-catenin signals is required for growth plate assembly, cartilage integrity, and endochondral ossification. *Journal of Biological Chemistry*, *280*, 19185–19195.

Tan, S. H., Senarath-Yapa, K., Chung, M. T., Longaker, M. T., Wu, J. Y., & Nusse, R. (2014). Wnts produced by Osterix-expressing osteolineage cells regulate their proliferation and differentiation. *PNAS*, *111*, E5262–E5271.

Taschner, M. J., Rafigh, M., Lampert, F., Schnaiter, S., & Hartmann, C. (2008). Ca2+/Calmodulin-dependent kinase II signaling causes skeletal overgrowth and premature chondrocyte maturation. *Developmental Biology*, *317*, 132–146.

ten Berge, D., Brugmann, S. A., Helms, J. A., & Nusse, R. (2008). Wnt and FGF signals interact to coordinate growth with cell fate specification during limb development. *Development*, *135*, 3247–3257.

Thysen, S., Luyten, F. P., & Lories, R. J. (2015). Loss of Frzb and Sfrp1 differentially affects joint homeostasis in instability-induced osteoarthritis. *Osteoarthritis and Cartilage*, *23*, 275–279.

Todd, H., Galea, G. L., Meakin, L. B., Delisser, P. J., Lanyon, L. E., Windahl, S. H., et al. (2015). Wnt16 is associated with age-related bone loss and estrogen withdrawal in murine bone. *PLoS One*, *10*, e0140260.

Tomita, M., Reinhold, M. I., Molkentin, J. D., & Naski, M. C. (2002). Calcineurin and NFAT4 induce chondrogenesis. *Journal of Biological Chemistry*, *277*, 42214–42218.

Topol, L., Chen, W., Song, H., Day, T. F., & Yang, Y. (2009). Sox9 inhibits Wnt signaling by promoting beta-catenin phosphorylation in the nucleus. *Journal of Biological Chemistry, 284*, 3323–3333.

Topol, L., Jiang, X., Choi, H., Garrett-Beal, L., Carolan, P. J., & Yang, Y. (2003). Wnt-5a inhibits the canonical Wnt pathway by promoting GSK-3-independent beta-catenin degradation. *The Journal of Cell Biology, 162*, 899–908.

Tsang, K. Y., Chan, D., & Cheah, K. S. (2015). Fate of growth plate hypertrophic chondrocytes: Death or lineage extension? *Development, Growth & Differentiation, 57*, 179–192.

Tu, J., Henneicke, H., Zhang, Y., Stoner, S., Cheng, T. L., Schindeler, A., et al. (2014). Disruption of glucocorticoid signaling in chondrocytes delays metaphyseal fracture healing but does not affect normal cartilage and bone development. *Bone, 69*, 12–22.

Tu, X., Joeng, K. S., Nakayama, K. I., Nakayama, K., Rajagopal, J., Carroll, T. J., et al. (2007). Noncanonical Wnt signaling through G protein-linked PKCdelta activation promotes bone formation. *Developmental Cell, 12*, 113–127.

Tufan, A. C., & Tuan, R. S. (2001). Wnt regulation of limb mesenchymal chondrogenesis is accompanied by altered N-cadherin-related functions. *FASEB Journal, 15*, 1436–1438.

Valenta, T., Hausmann, G., & Basler, K. (2012). The many faces and functions of beta-catenin. *EMBO Journal, 31*, 2714–2736.

van Amerongen, R. (2012). Alternative Wnt pathways and receptors. *Cold Spring Harbor Perspectives in Biology, 4*.

van Amerongen, R., Fuerer, C., Mizutani, M., & Nusse, R. (2012). Wnt5a can both activate and repress Wnt/beta-catenin signaling during mouse embryonic development. *Developmental Biology, 369*, 101–114.

van der Horst, G., van der Werf, S. M., Farih-Sips, H., van Bezooijen, R. L., Lowik, C. W., & Karperien, M. (2005). Downregulation of Wnt signaling by increased expression of Dickkopf-1 and -2 is a prerequisite for late-stage osteoblast differentiation of KS483 cells. *Journal of Bone and Mineral Research, 20*, 1867–1877.

Varelas, X., Miller, B. W., Sopko, R., Song, S., Gregorieff, A., Fellouse, F. A., et al. (2010). The hippo pathway regulates Wnt/beta-catenin signaling. *Developmental Cell, 18*, 579–591.

Velazquez-Cruz, R., Garcia-Ortiz, H., Castillejos-Lopez, M., Quiterio, M., Valdes-Flores, M., Orozco, L., et al. (2014). WNT3A gene polymorphisms are associated with bone mineral density variation in postmenopausal mestizo women of an urban Mexican population: Findings of a pathway-based high-density single nucleotide screening. *Age (Dordrecht, Netherlands), 36*, 9635.

Wada, N., Kawakami, Y., Ladher, R., Francis-West, P. H., & Nohno, T. (1999). Involvement of Frzb-1 in mesenchymal condensation and cartilage differentiation in the chick limb bud. *The International Journal of Developmental Biology, 43*, 495–500.

Wan, Y., Lu, C., Cao, J., Zhou, R., Yao, Y., Yu, J., et al. (2013). Osteoblastic Wnts differentially regulate bone remodeling and the maintenance of bone marrow mesenchymal stem cells. *Bone, 55*, 258–267.

Wang, B., Jin, H., Zhu, M., Li, J., Zhao, L., Zhang, Y., et al. (2014). Chondrocyte beta-catenin signaling regulates postnatal bone remodeling through modulation of osteoclast formation in a murine model. *Arthritis & Rheumatology, 66*, 107–120.

Wang, T., Li, J., Zhou, G. Q., Ma, P., Zhao, Y., Wang, B., et al. (2017). Specific deletion of beta-catenin in Col2-expressing cells leads to defects in epiphyseal bone. *International Journal of Biological Sciences, 13*, 1540–1546.

Wei, W., Zeve, D., Suh, J. M., Wang, X., Du, Y., Zerwekh, J. E., et al. (2011). Biphasic and dosage-dependent regulation of osteoclastogenesis by beta-catenin. *Molecular and Cellular Biology, 31*, 4706–4719.

Weivoda, M. M., Ruan, M., Hachfeld, C. M., Pederson, L., Howe, A., Davey, R. A., et al. (2016). Wnt Signaling inhibits osteoclast differentiation by activating canonical and non-canonical cAMP/PKA pathways. *Journal of Bone and Mineral Research*, *31*, 65–75.
Wergedal, J. E., Kesavan, C., Brommage, R., Das, S., & Mohan, S. (2015). Role of WNT16 in the regulation of periosteal bone formation in female mice. *Endocrinology*, *156*, 1023–1032.
Willert, K., & Nusse, R. (2012). Wnt proteins. *Cold Spring Harbor Perspectives in Biology*, *4*, a007864.
Williams, B. O., & Insogna, K. L. (2009). Where Wnts went: The exploding field of Lrp5 and Lrp6 signaling in bone. *Journal of Bone and Mineral Research*, *24*, 171–178.
Winkler, D. G., Sutherland, M. K., Geoghegan, J. C., Yu, C., Hayes, T., et al. (2003). Osteocyte control of bone formation via sclerostin, a novel BMP antagonist. *EMBO Journal*, *22*, 6267–6276.
Witte, F., Dokas, J., Neuendorf, F., Mundlos, S., & Stricker, S. (2009). Comprehensive expression analysis of all Wnt genes and their major secreted antagonists during mouse limb development and cartilage differentiation. *Gene Expression Patterns*, *9*, 215–223.
Xiong, J., Almeida, M., & O'Brien, C. A. (2018). The YAP/TAZ transcriptional co-activators have opposing effects at different stages of osteoblast differentiation. *Bone*, *112*(1–9).
Xiong, L., Jung, J. U., Wu, H., Xia, W. F., Pan, J. X., Shen, C., et al. (2015). Lrp4 in osteoblasts suppresses bone formation and promotes osteoclastogenesis and bone resorption. *PNAS*, *112*, 3487–3492.
Yadav, V. K., Arantes, H. P., Barros, E. R., Lazaretti-Castro, M., & Ducy, P. (2010). Genetic analysis of Lrp5 function in osteoblast progenitors. *Calcified Tissue International*, *86*, 382–388.
Yadav, V. K., & Ducy, P. (2010). Lrp5 and bone formation: A serotonin-dependent pathway. *Annals of the New York Academy of Sciences*, *1192*, 103–109.
Yadav, V. K., Ryu, J. H., Suda, N., Tanaka, K. F., Gingrich, J. A., Schutz, G., et al. (2008). Lrp5 controls bone formation by inhibiting serotonin synthesis in the duodenum. *Cell*, *135*, 825–837.
Yamaguchi, T. P., Bradley, A., McMahon, A. P., & Jones, S. (1999). A Wnt5a pathway underlies outgrowth of multiple structures in the vertebrate embryo. *Development*, *126*, 1211–1223.
Yamamoto, H., Yoo, S. K., Nishita, M., Kikuchi, A., & Minami, Y. (2007). Wnt5a modulates glycogen synthase kinase 3 to induce phosphorylation of receptor tyrosine kinase Ror2. *Genes to Cells*, *12*, 1215–1223.
Yang, M. S., Chang, S. H., Sonn, J. K., Lee, Y. S., Kang, S. S., Park, T. K., et al. (1998). Regulation of chondrogenic differentiation of mesenchymes by protein kinase C alpha. *Molecules and Cells*, *8*, 266–271.
Yang, J. Y., Cho, S. W., An, J. H., Jung, J. Y., Kim, S. W., Kim, S. Y., et al. (2013). Osteoblast-targeted overexpression of TAZ increases bone mass in vivo. *PLoS One*, *8*, e56585.
Yang, X., Teguh, D., Wu, J. P., He, B., Kirk, T. B., Qin, S., et al. (2015). Protein kinase C delta null mice exhibit structural alterations in articular surface, intra-articular and subchondral compartments. *Arthritis Research & Therapy*, *17*, 210.
Yang, Y., Topol, L., Lee, H., & Wu, J. (2003). Wnt5a and Wnt5b exhibit distinct activities in coordinating chondrocyte proliferation and differentiation. *Development*, *130*, 1003–1015.
Yasuhara, R., Ohta, Y., Yuasa, T., Kondo, N., Hoang, T., Addya, S., et al. (2011). Roles of beta-catenin signaling in phenotypic expression and proliferation of articular cartilage superficial zone cells. *Laboratory Investigation: A Journal of Technical Methods and Pathology*, *91*, 1739–1752, ISSN 1530-0307.

Yazici, Y., McAlindon, T. E., Fleischmann, R., Gibofsky, A., Lane, N. E., & Kivitz, A. J. (2017). A novel Wnt pathway inhibitor, SM04690, for the treatment of moderate to severe osteoarthritis of the knee: Results of a 24-week, randomized, controlled, phase 1 study. *Osteoarthritis and Cartilage, 25*, 1598–1606.

Yazici, Y., McAlindon, T. E., Gibofsky, A., Lane, N. E., Clauw, D. J., Jones, M. H., et al. (2018). Results from a 52-week randomized, double-blind, placebo-controlled, phase 2 study of a novel, intra-articular wnt pathway inhibitor (SM04690) for the treatment of knee osteoarthritis. *Osteoarthritis and Cartilage, 26*, S293–S294.

Yoshida, C. A., & Komori, T. (2005). Role of Runx proteins in chondrogenesis. *Critical Reviews in Eukaryotic Gene Expression, 15*, 243–254.

Yu, H. M., Jerchow, B., Sheu, T. J., Liu, B., Costantini, F., Puzas, J. E., et al. (2005). The role of Axin2 in calvarial morphogenesis and craniosynostosis. *Development, 132*, 1995–2005.

Yuasa, T., Kondo, N., Yasuhara, R., Shimono, K., Mackem, S., Pacifici, M., et al. (2009). Transient activation of Wnt/{beta}-catenin signaling induces abnormal growth plate closure and articular cartilage thickening in postnatal mice. *The American Journal of Pathology, 175*, 1993–2003.

Yuasa, T., Otani, T., Koike, T., Iwamoto, M., & Enomoto-Iwamoto, M. (2008). Wnt/beta-catenin signaling stimulates matrix catabolic genes and activity in articular chondrocytes: Its possible role in joint degeneration. *Laboratory Investigation; A Journal of Technical Methods and Pathology, 88*, 264–274.

Zhang, X., Bu, Y., Zhu, B., Zhao, Q., Lv, Z., Li, B., et al. (2018). Global transcriptome analysis to identify critical genes involved in the pathology of osteoarthritis. *Bone & Joint Research, 7*, 298–307.

Zhang, Y., Wang, Z., Ding, L., Damaolar, A., Li, Z., Qiu, Y., et al. (2016). Lentivirus-TAZ administration alleviates osteoporotic phenotypes in the femoral neck of Ovariectomized rats. *Cellular Physiology and Biochemistry, 38*, 283–294.

Zhao, L., Shim, J. W., Dodge, T. R., Robling, A. G., & Yokota, H. (2013). Inactivation of Lrp5 in osteocytes reduces young's modulus and responsiveness to the mechanical loading. *Bone, 54*, 35–43.

Zheng, H. F., Tobias, J. H., Duncan, E., Evans, D. M., Eriksson, J., Paternoster, L., et al. (2012). WNT16 influences bone mineral density, cortical bone thickness, bone strength, and osteoporotic fracture risk. *PLoS Genetics, 8*, e1002745.

Zhong, L., Huang, X., Rodrigues, E. D., Leijten, J. C., Verrips, T., El Khattabi, M., et al. (2016). Endogenous DKK1 and FRZB regulate chondrogenesis and hypertrophy in three-dimensional cultures of human chondrocytes and human Mesenchymal stem cells. *Stem Cells and Development, 25*, 1808–1817.

Zhong, Z., Zylstra-Diegel, C. R., Schumacher, C. A., Baker, J. J., Carpenter, A. C., Rao, S., et al. (2012). Wntless functions in mature osteoblasts to regulate bone mass. *PNAS, 109*, E2197–E2204.

Zhu, M., Chen, M., Zuscik, M., Wu, Q., Wang, Y. J., Rosier, R. N., et al. (2008). Inhibition of beta-catenin signaling in articular chondrocytes results in articular cartilage destruction. *Arthritis and Rheumatism, 58*, 2053–2064.

Zhu, M., Tang, D., Wu, Q., Hao, S., Chen, M., Xie, C., et al. (2009). Activation of beta-catenin signaling in articular chondrocytes leads to osteoarthritis-like phenotype in adult beta-catenin conditional activation mice. *Journal of Bone and Mineral Research, 24*, 12–21.

Zhu, C., Zheng, X. F., Yang, Y. H., Li, B., Wang, Y. R., Jiang, S. D., et al. (2016). LGR4 acts as a key receptor for R-spondin 2 to promote osteogenesis through Wnt signaling pathway. *Cellular Signalling, 28*, 989–1000.

CHAPTER TEN

$G\alpha_s$ signaling in skeletal development, homeostasis and diseases

Qian Cong[a,†], Ruoshi Xu[b,†], Yingzi Yang[a,*]
[a]Department of Developmental Biology, Harvard School of Dental Medicine, Boston, MA, United States
[b]State Key Laboratory of Oral Diseases, National Clinical Research Center for Oral Diseases, Department of Cariology and Endodontology, West China Hospital of Stomatology, Sichuan University, Chengdu, China
*Corresponding author: e-mail address: yingzi_yang@hsdm.harvard.edu

Contents

Abstract

Skeletal development is exquisitely controlled both spatially and temporally by cell signaling networks. $G\alpha_s$ is the stimulatory α-subunit in a heterotrimeric G protein complex transducing the signaling of G-protein-coupled receptors (GPCRs), responsible for controlling both skeletal development and homeostasis. $G\alpha_s$, encoded by the *GNAS* gene in

† These authors contributed equally to this review.

Current Topics in Developmental Biology, Volume 133
ISSN 0070-2153
https://doi.org/10.1016/bs.ctdb.2018.11.019

humans, plays critical roles in skeletal development and homeostasis by regulating commitment, differentiation and maturation of skeletal cells. $G\alpha_s$-mediated signaling interacts with the Wnt and Hedgehog signaling pathways, both crucial regulators of skeletal development, remodeling and injury repair. Genetic mutations that disrupt $G\alpha_s$ functions cause human disorders with severe skeletal defects, such as fibrous dysplasia of bone and heterotopic bone formation. This chapter focuses on the crucial roles of $G\alpha_s$ signaling during skeletal development and homeostasis, and the pathological mechanisms underlying skeletal diseases caused by GNAS mutations.

1. Introduction

G-protein-coupled receptors (GPCRs) are the largest family of cellular surface receptors that are crucial for the control of a diverse array of developmental and physiological processes, disruptions of which cause diseases (O'Hayre et al., 2013; Pierce, Premont, & Lefkowitz, 2002). Heterotrimeric G-proteins are composed of three subunits, $G\alpha$, $G\beta$ and $G\gamma$ (Morris & Malbon, 1999; O'Hayre et al., 2013). There are four subgroups $G\alpha$ subunits, $G\alpha_s$, $G\alpha_{i/o}$, $G\alpha_q$ and $G\alpha_{12}$ (Oldham & Hamm, 2008; Simon, Strathmann, & Gautam, 1991). $G\alpha_s$, encoded by the *GNAS* gene in humans, is the stimulatory α-subunit (Weinstein et al., 1990). Upon agonist binding to GPCRs, $G\alpha_s$ dissociates from $G\beta$ and $G\gamma$ and activates downstream effectors, a major one of which is adenylate cyclase that produces cAMP to regulate downstream signaling events, including activation of gene expression (Neves, Ram, & Iyengar, 2002).

Skeletal development and homeostasis require proper coordination of cell fate determination, proliferation and maturation of several key cell types, including chondrocytes and osteoblasts for cartilage and bone formation, as well as osteoclasts for bone remodeling. $G\alpha_s$ signaling controls skeletal development by regulating chondrocyte proliferation and hypertrophy (Bastepe et al., 2004; Chagin et al., 2014; Jin et al., 2018), mesenchymal progenitor cell commitment to osteoblasts(Wu et al., 2011), osteoblast differentiation (Sakamoto et al., 2005; Sinha et al., 2014), and mineralization (Khan et al., 2018; Zhao et al., 2018). The fundamental functions of $G\alpha_s$ in skeletal development and homeostasis are underlined by severe skeletal defects in human diseases caused by altered $G\alpha_s$ signaling due to mutations in *GNAS* or changes in other interacting factors (Kaplan, Hahn, & Zasloff, 1994; Riminucci, Robey, Saggio, & Bianco, 2010).

Activating mutations in *GNAS* cause fibrous dysplasia (FD) (OMIM# 174800) in bone; clinically characterized by bone marrow fibrosis, lack of hematopoietic tissue and marrow fat as well as abnormal trabecular bone, resulting in deformity, fracture and pain in the affected bone (Riminucci et al., 1997, 1999; Robinson, Collins, & Boyce, 2016). Conversely, inactivation of *GNAS* causes progressive osseous heteroplasia (POH, OMIM#166350) and Albright hereditary osteodystrophy (AHO, OMIM#103580) (Eddy et al., 2000; Kaplan, Hahn, et al., 1994). Clinically, POH and AHO are characterized by extraskeletal or heterotopic ossification (HO) (Shore & Kaplan, 2010), predominantly through the mechanism of intramembranous ossification (Eddy et al., 2000; Kaplan, Hahn, et al., 1994). As an important player in regulating skeletal development and physiological functions, $G\alpha_s$ signaling is also integrated in a central signaling network by interacting with the Wnt and Hedgehog signaling pathways; both fundamentally important in regulating various aspects of skeletal development. Here we focus on the role of $G\alpha_s$ in bone development and homeostasis and new mechanistic understanding of skeletal diseases caused by genetic alterations of *GNAS*.

2. $G\alpha_s$ signaling in human skeletal development and homeostasis

Formation and maintenance of bone is orchestrated by a genetically controlled time course and spatial rules. Accumulating evidence indicates that $G\alpha_s$ signaling plays fundamentally important roles during skeletal development and homeostasis by regulating various processes of cell proliferation and differentiation, such as chondrocyte hypertrophy and osteoblast differentiation of mesenchymal stem and progenitor cells (Regard et al., 2011; Riminucci et al., 2010; Wu et al., 2011). The critical roles of $G\alpha_s$ signaling in skeletal development are illustrated by the skeletal defects in human patients carrying either activating or inactivating mutations in *GNAS*.

2.1 Skeletal diseases caused by activating mutations in the *GNAS* gene

Fibrous dysplasia (FD) of bone (OMIM#174800) is a rare skeletal disorder caused by mosaic activating mutations (R201H or R201C) of $G\alpha_s$ (Schwindinger, Francomano, & Levine, 1992; Shenker, Weinstein, Sweet, & Spiegel, 1994; Weinstein et al., 1991). The mutations cause $G\alpha_s$ to lose its inherent GTPase activity so that it remains in a constitutively active form and stimulates excessive cAMP production (Lania et al., 1998). FD is characterized by a bone marrow space devoid of both hematopoietic cells and adipocytes and filled with fibrotic tissue. FD bone also exhibits an abnormal architecture ("Chinese writing" pattern), structure and mineral content of bone trabeculae (Riminucci et al., 1997, 1999; Robinson et al., 2016). These complex changes result in a mechanically incompetent, brittle and fracture prone bone that may necessitate wheel chair confinement of severely affected individuals. The presentations of FD may be monostotic or polyostotic, depending on whether one or more bones are involved. Monostotic fibrous dysplasia is much more common than the polyostotic type and it usually involves the femur or the bones of the craniofacial skeleton (Waldron, 1993). The lesions are composed of immature osteoblastic progenitor cells derived from the mesenchymal precursors, spicules of immature woven bone and, in some cases, nests of hyaline cartilage (Weinstein, 2006). In most cases, FD lesions progress from the medullary cavity to the neighboring cortical bone and contain a rim of osteoclasts. Although FD is often asymptomatic and discovered fortuitously, some patients present with bone pain, deformities, most notably in craniofacial forms, or pathological fractures, which often constitute the first clinical manifestation (Chapurlat & Orcel, 2008). In 80% of patients, the first symptoms occur before 15 years of age (Chapurlat & Orcel, 2008). In addition to fractures, FD complications include exophthalmos, dental abnormalities, and leontiasis ossea.

FD may occur in isolation or along with other clinical features such as skin pigmentation and endocrine dysfunction in McCune–Albright syndrome (MAS) (Albright, Butler, Hampton, & Smith, 1937; Boyce et al., 1993–2018). Endocrine system abnormalities in MAS may consist of hyperthyroidism, excessive growth hormone production, hypercorticism, and hyperprolactinemia. Renal dysfunction is a feature in about 50% of patients with MAS. Non-endocrine abnormalities reported in MAS include intramuscular myxoma, cholestasis, cardiac involvement, myelofibrosis, and gastrointestinal disorders (Weinstein, 2006).

There is no report of disease inheritance, therefore, the activating mutations of *GNAS* are thought to occur postzygotically, resulting in a somatic mosaic state (Bianco et al., 1998; Shenker et al., 1993, 1994; Weinstein et al., 1991). The lack of inheritance of the disease is due to embryonic lethality caused by germline-transmitted activating *GNAS* mutations, which can only be recovered through mosaicism (Happle, 1986). The phenotype may depend on the extent and distribution of the mutated cells, which in turn depend on the time at which the mutation occurred during embryonic development. A somatic mutation that occurs later during development may lead to a less severe phenotype (e.g., monostotic fibrous dysplasia) (Weinstein, 2006).

In FD and MAS, *GNAS* mutations cause gain-of-function of the $G\alpha_s$ protein. The amino acid substitutions due to the mutations occur at sites that are critical to the activities of the GTPase, which normally inactivates $G\alpha_s$. Therefore, FD or MAS mutations in *GNAS* result in prolonged $G\alpha_s$ protein activation that induces adenyl cyclase to overproduce cyclic AMP (cAMP) even in the absence of stimulating hormone (Weinstein, 2006). Activation of $G\alpha_s$/cAMP also leads to increased pigment production by the melanocytes via the induction of tyrosinase expression and to cardiac hypertrophy via activation of the MAP kinase pathway (Weinstein, 2006). In bone, activation of the Gs/protein kinase A signaling pathway represses the transcription factor Runx2 and activation of Wnt/β-catenin signaling (Khan et al., 2018; Regard et al., 2011) may cause the abnormalities in osteoblast differentiation. Gs/protein activation increases the secretion of interleukin-6, which may cause the osteolytic lesions (Weinstein, 2006).

2.2 Skeletal diseases caused by inactivating mutations in the GNAS gene

Heterozygous inactivating mutations of GNAS cause progressive osseous heteroplasia (POH, OMIM#166350), Albright hereditary osteodystrophy (AHO, OMIM#103580), Pseudo-pseudohypoparathyroidism (PPHP) and osteoma cutis (OC), which share the common features of heterotopic ossification (HO) through the mechanisms of intramembrane ossification (Bastepe & Juppner, 2005; Farfel, Bourne, & Iiri, 1999; Kumagai et al., 2008; Plagge, Kelsey, & Germain-Lee, 2008; Shore et al., 2002; Weinstein, Chen, Xie, & Liu, 2006). HO is a pathological condition of extraskeletal bone formation that can be devastating and even life threatening.

POH or AHO represents one of the two illustrative inherited genetic forms of HO. The other one is fibrodysplasia ossificans progressiva (FOP, OMIM#135100) caused by heterozygous activating mutation in the type I bone morphogenetic protein (BMP) receptor known as activin receptor-like kinase 2 (ALK2), or Activin A receptor, type I (ACVR1) (Chakkalakal & Shore, 2019; Xu, Hu, Zhou, & Yang, 2018). HO in FOP is regulated by the endochondral ossification mechanism. Heterozygous inactivating mutations within one of 13 GNAS exons encoding Gsα are responsible for POH or AHO (Patten et al., 1990; Weinstein et al., 1990). Albright et al. described originally a constellation of physical features including obesity, short stature, brachydactyly, cognitive impairment, and superficial heterotopic ossification. The patients show a state of resistance to parathyroid hormone (PTH), thyroid stimulating hormone (TSH), gonadotropins, and growth hormone releasing hormone, developing not only hypocalcemia, but also hyperphosphatemia (Bastepe, 2007; Plagge et al., 2008; Tashjian, Frantz, & Lee, 1966; Turan & Bastepe, 2015; Weinstein, Liu, Sakamoto, Xie, & Chen, 2004; Weinstein, Yu, Warner, & Liu, 2001). This disorder of AHO with multihormone resistance is referred to as pseudohypoparathyroidism type-Ia (PHP-Ia) (Bastepe, 2018). POH is a rare genetic disorder of progressive ectopic ossification that was first described by Kaplan, Craver, et al. (1994). Clinically, POH is characterized by dermal ossification during infancy with progressive heterotopic ossification of subcutaneous, and deep connective tissue during childhood, including muscle and fascia, in the absence of multiple features of AHO or hormone resistance (Kaplan & Shore, 2000). Most cases of POH are caused by heterozygous inactivating mutations of GNAS, the gene encoding the alpha subunit of the G-stimulatory protein of adenylyl cyclase (Shore et al., 2002).

POH can be distinguished from FOP by the presence of cutaneous ossification, superficial to deep progression of HO, the mosaic distribution of lesions, the absence of congenital skeletal malformations or inflammatory tumor-like swellings, and the predominance of intramembranous (in contrast to endochondral) ossification (Kaplan & Shore, 2000; Shore & Kaplan, 2008) (Table 1). Importantly, POH can be distinguished from AHO by the progression of heterotopic ossification from skin and subcutaneous tissue into skeletal muscle, the absence of a distinctive habitus associated with AHO, and the presence of normal endocrine function respectively (Adegbite, Xu, Kaplan, Shore, & Pignolo, 2008; Kaplan & Shore, 2000; Kaplan, Craver, et al., 1994; Shore & Kaplan, 2008) (Table 1). As with many newly described and extremely rare conditions, POH is probably underdiagnosed. Careful consideration of clinical and radiographic signs is usually enough to recognize the disorder and to differentiate it from FOP and AHO (Adegbite et al., 2008).

Taken together, both activating and inactivating mutations of GNAS alter osteoblastic differentiation of mesenchymal stem or progenitor cells, leading to bone marrow fibrosis or HO, respectively.

Table 1 Features of heterotopic ossification in POH, FOP, and AHO (Schimmel et al., 2010; Xu, Hu, et al., 2018).

Feature	POH	FOP	AHO
Genetic transmission	Autosomal dominant	Autosomal dominant	Autosomal dominant
Temporal occurrence	Bony lesion occurs during infancy, followed by progressive ossification	Appear normal at birth, preosseous swellings occur, and then bony lesion occur in different location	LBonyLesion occurs during infancy
Cutaneous ossification	+	−	±
Subcutaneous ossification	+	−	±
Muscle ossification	+	+	−
Superficial to deep progression of ossification	+	−	−
Severe limitation of mobility	+	+	−
Ectopic ossification after intramuscular injections	−	+	−
Ectopic ossification after trauma	±	+	−
Inheritance	Paternal or maternal allele	Paternal or maternal allele	Maternal allele
Predominant mechanism of ossification	Intramembranous	Endochondral	Intramembranous
Hematopoietic marrow in ectopic bone	±	+	−
PTH resistance	−	−	+
Hypocalcemia and hyperphosphatemia	−	−	+
Genetic mutations	Heterozygous inactivating mutations of GNAS	Activating mutation of the gene encoding the BMP type I receptor ACVR1/ALK2	Heterozygous inactivating mutations of GNAS

3. Regulation of osteoblast differentiation by $G\alpha_s$ signaling

3.1 $G\alpha_s$ in osteochondral progenitor cells

Skeletal development starts from mesenchymal condensations, in which osteochondral progenitor cells differentiate into either osteoblasts or chondrocytes (Carroll & Ravid, 2013; Kozhemyakina, Lassar, & Zelzer, 2015; Toosi, Behravan, & Behravan, 2018). The human skeletal defects caused by *GNAS* mutations indicate that $G\alpha_s$ signaling plays essential roles in osteochondral progenitor cells with strong impacts on the entire process of skeletal development and homeostasis. In mouse models, $G\alpha_s$ signaling has been altered in osteochondral progenitor cells by several *Cre* lines. The Cre recombinase in the *Prrx1-Cre* recombinase is expressed in limb bud mesenchymal progenitor cells and in a subset of craniofacial mesenchyme (Logan et al., 2002); *Dermo1-Cre* is expressed in mesenchymal condensations giving rise to both osteoblast and chondrocyte lineages (Yu et al., 2003); and *Ap2-Cre* is expressed in mesenchymal progenitor cells in the frontonasal processes and the limb (Nelson & Williams, 2004).

To investigate the tissue-specific role of *Gnas*, a conditional knock-in allele of *Gnas* gain-of-function mutation ($Gnas^{f(R201H)}$) was recently generated (Khan et al., 2018). When *Gnas* is constitutively activated in osteochondral progenitor cells at an early embryonic stage by crossing the $Gnas^{f(R201H)}$ line with the *Prrx1-Cre line*, chondrocyte hypertrophy is severely delayed and bone formation is disrupted (Khan et al., 2018). As a result of delay in chondrocyte hypertrophy at P0 (postnatal day 0), formation of trabecular and cortical bone, bone marrow cavity or cartilaginous epiphyses, cannot be found (Khan et al., 2018). Throughout adulthood, growth plate is irregular and bone marrow cavity is filled with trabecular bones and fibrotic tissues with no cortical bone (Khan et al., 2018), all of which closely resemble the pathological findings in human FD (Riminucci et al., 1997, 1999; Robinson et al., 2016). Conversely, when *Gnas* is deleted in osteochondral progenitor cells using *Prrx1Cre*, *Dermo1Cre* or *AP2-Cre*, ossification is accelerated and extensive ectopic ossification is found at E16.5 (Regard et al., 2013; Wu et al., 2011). Taken together, *Gnas* is required in mesenchymal progenitor cells to initiate proper bone development and homeostasis.

3.2 Gα$_s$ in the osteoblast lineage

Osterix (Osx) is expressed in the committed osteoblastic cells and *Osx-Cre* has been generated to investigate gene functions in the osteoblast lineage (Rodda & McMahon, 2006). Expression of the activating *Gnas*^{R201H} mutant using the *Osx-Cre* line also led to early perinatal lethality and strong FD phenotypes. Conversely, conditional deletion of *Gnas* in *Osx*-expressing cells caused early postnatal mortality and severe osteoporosis (Chen et al., 2005). At E15.5, no obvious morphology difference in long bones could be found in the mutant and early markers of osteoblast differentiation were similarly expressed in the mutants compared to the controls (Wu et al., 2011). At birth, fractures of ribs and long bones were identified in mutants, which was associated with a low expression level of osteocalcin, a terminal differentiation marker of osteoblasts (Wu et al., 2011). By 1–2 weeks postnatally, decreased trabecular bone and thinned cortical bone were observed in long bones that contained disorganized woven bone and insufficient bone mass (Wu et al., 2011). Immature woven bone formation in osteoporosis was caused by decreased osteoblast numbers and accelerated differentiation of mature osteoblasts with altered bone matrix instead of increased bone resorption (Wu et al., 2011). Many craniofacial bones are formed by intramembranous ossification, during which cells in the mesenchymal condensation differentiate into osteoblasts without a cartilaginous template. Ablation of Gα$_s$ in the osteoblast lineage also displayed woven cranial bone with patchy mineralization in a reticular pattern, misshapen osteocyte lacunae and irregular collagen fibril deposition at birth. These data indicate that Gα$_s$ plays a crucial role in regulating osteoblast maturation during bone formation (Wu et al., 2011).

Besides osteoporosis, abundant adipocytes were found in bone marrow of *Osx-Cre;Gnas* mutant mice at 2 weeks of age (Sinha et al., 2014), suggesting that loss of Gα$_s$ in the osteoblast lineage also regulates the commitment of mesenchymal progenitors into osteoblasts versus the adipocyte lineage. In long bones, adipocytes were increasingly evident as the *Osx-Cre; Gnas* mutant mice got older so that bone marrow in 6-month old mutant mice was almost completely replaced by adipocytes. Deletion of Gα$_s$ in the osteoblast lineage increased expression of Sclerostin (SOST) and DKK1 and reduced Wnt signaling in osteoblasts; these changes contribute to the alteration of osteoblastic cell fate in favor of adipocytes versus osteoblasts (Sinha et al., 2014). However, Gα$_s$ is not required for Wnt-mediated inhibition of adipogenesis *in vitro* (Sinha et al., 2014).

Osteoblasts are known to support hematopoietic stem cells (Arai et al., 2004; Calvi et al., 2003) and have been confirmed as components of the B lymphocyte niche in a $G\alpha_s$-dependent way (Wu et al., 2008). Removal of osteoblast-specific $G\alpha_s$ caused general reduction of $B220^+$ committed B lymphocyte lineage cells and impaired B lymphopoiesis from pro-B to pre-B cell transition in the bone marrow (Wu et al., 2008). Transplantation of $B220^+$ cells into a wild type environment completely rescued bone marrow cellularity and B lymphopoiesis (Wu et al., 2008). $G\alpha_s$ deficient osteoblastic cells showed reduced IL-7 expression and administration of IL-7 rescued pro-B cell production in mutant mice to the wildtype level, suggesting that IL-7 mediates B lymphopoiesis regulated by $G\alpha_s$ in osteoblasts (Wu et al., 2008).

3.3 $G\alpha_s$ in osteocyte lineage

Osteocytes, differentiated from osteoblasts and embedded in bone matrix, are the most abundant bone cells in adult animals (Franz-Odendaal, Hall, & Witten, 2006). Osteocytes control bone homeostasis by regulating both osteoblasts and osteoclasts. Removal of $G\alpha_s$ in differentiated osteoblasts using the collagen type I alpha1 promoter driven-*Cre* transgenic mice (*Col1-Cre*) leads to reduced bone turnover (Sakamoto et al., 2005). Mutant mice showed reduced trabecular bone volume, defective formation of primary spongiosa and reduced immature bone formation with decreased expression of the late osteoblastic differentiation markers osteopontin and osteocalcin (Sakamoto et al., 2005). In addition, osteoclast numbers were decreased on the endosteal surface of cortical bone, leading to thickened cortical bone and narrowing bone marrow space (Sakamoto et al., 2005). However, transgenic mice expressing a rat $G\alpha_s^{R201C}$ cDNA selectively in mature osteoblasts, using the 2.3kb *Col1a1* promoter (Remoli et al., 2015), showed no significant skeletal abnormalities at birth by X-ray examination or histological analysis. Increased bone density was detected, and progressively increased as excessive bone mass, starting from 3 weeks of age (Remoli et al., 2015).

Ablation of $G\alpha_s$ in osteocytes (OCY-$G\alpha_s$KO), using the *DMP-Cre* mice (Lu et al., 2007), led to osteopenia, characterized by decreased bone mineral density in trabecular and cortical bone (Fulzele et al., 2013). In the mutant mice, osteocyte density was increased but the lacunar-canalicular network was disorganized (Fulzele et al., 2013). In addition, myeloid cells were increased in the bone marrow, spleen and peripheral blood of the osteocyte-specific $G\alpha_s$ mutant mice as a consequence of the disrupted bone marrow microenvironment by $G\alpha_s$ regulated osteocyte-derived G-CSF (Fulzele et al., 2013).

3.4 $G\alpha_s$ in osteoclastogenesis

$G\alpha_s$ also regulates osteoclast differentiation during bone development and homeostasis (Ramaswamy et al., 2017). Mice with paternal allele inactivation of *Gnas* ($Gnas^{+/p-}$) showed more defective cortical bone than mice with maternal mutations during early development and adulthood. This was associated with increased endosteal osteoclast differentiation and bone resorption without significant effects on osteoblast numbers and function (Ramaswamy et al., 2017). Mechanically, it is suggested that reduced β-catenin and the downstream cyclin D1 result in increased levels of Wnt inhibitors and enhanced osteoclast differentiation in the $Gnas^{+/p-}$ mice (Ramaswamy et al., 2017). Together with data from a previous study (Weivoda et al., 2016), it suggests that crosstalk between $G\alpha_s$/cAMP/PKA and Wnt/β-catenin signaling pathways is important for osteoclastogenesis (Ramaswamy et al., 2017). This is further supported by data of another study in which $G\alpha_s$ depletion, independent of embryonic skeletal development, was induced by Tamoxifen treatment of adult *CreERT2;Gnas*$^{fl/fl}$ mice (Ramaswamy et al., 2018). Trabecular bone volume, trabecular number and thickness were shown to be reduced based on μCT analysis, and this was associated with increased numbers of multi-nucleated trabecular osteoclasts in homozygous mutant mice (Ramaswamy et al., 2018).

To further test if *Gnas* has a cell-autonomous role in osteoclastic regulation of bone remodeling, a mouse model was generated in which homozygous removal of *Gnas* in macrophage and osteoclast lineages was induced by *LysM-Cre* (Clausen, Burkhardt, Reith, Renkawitz, & Forster, 1999). Mutant mice displayed decreased trabecular bone quality, marginally reduced microarchitecture parameters, and a trend towards increased osteoclast numbers. Although these effects were milder than what was observed in *CreERT2;Gnas*$^{fl/fl}$ mice (Ramaswamy et al., 2018), further evidence that $G\alpha_s$ can regulate skeletal homeostasis by regulating osteoclast number and activity, is suggested by *in vitro* experiments that the role of $G\alpha_s$ is cell-autonomous in osteoclasts (Ramaswamy et al., 2017). Taken together, $G\alpha_s$ regulates bone formation and homeostasis by facilitating the commitment of mesenchymal progenitors to the osteoblast lineage, and at the same time restraining, via an effect on osteoclasts, the committed osteoblasts to enable production of bone of optimal mass, quality and strength.

4. Cross talk of $G\alpha_s$ signaling with other signaling pathways in the skeletal system

4.1 $G\alpha_s$ is an inhibitor of Hedgehog signaling

Hedgehog (Hh) signaling is required to induce the osteogenic program. Indian hedgehog (Ihh) is expressed by prehypertrophic chondrocytes and important in coordinating chondrocyte proliferation with hypertrophy in the developing long bones (Karsenty, 2003; Kronenberg, 2003). Ihh also signals directly to perichondrial mesenchymal progenitor cells to induce osteoblast differentiation (Karsenty, 2003; Kronenberg, 2003). Removal of Ihh completely abolished osteoblast differentiation during endochondral ossification (Long et al., 2004; St-Jacques, Hammerschmidt, & McMahon, 1999). $G\alpha_s$ signaling has been reported to regulate skeletal development and genetic bone disorders through Hh signaling (Regard et al., 2013; Xu, Khan, et al., 2018).

In the developing long bone cartilage, loss of $G\alpha_s$ signaling in the *Prrx1-cre; Gnas*$^{fl/-}$ limb led to upregulated and ectopic expression of Hedgehog target genes, indicating that loss of *Gnas* leads to activation of Hh signaling (Regard et al., 2013). Loss of $G\alpha_s$ signaling in the subcutaneous mesenchymal progenitor cells (SMPs) also activated Hh signaling. Such Hh signaling activation is both necessary and sufficient to induce increased and ectopic osteoblast differentiation (Regard et al., 2013). Mechanistically, $G\alpha_s$ inhibits Hh signaling by activating protein kinase A (PKA) through cAMP production. PKA inhibits Gli transcription factors by phosphorylating Gli3 and Gli2. Gli3 phosphorylation generates cleaved Gli3 repressor form $Gli3^R$ while Gli2 phosphorylation destabilizes it. $G\alpha_s$ regulation of Hh signaling is downstream of the Hh receptor Smoothened (Smo), as Cyclopamine, an inhibitor of Smo, was unable to suppress Hh target gene expression in *Gnas*$^{-/-}$ cells, while the Gli inhibitors arsenic trioxide and GANT-58 could, indicating that $G\alpha_s$ acts downstream of Smo and upstream of Gli transcription factors to suppress Hh signaling (Regard et al., 2013). Such regulation is not specific to the skeletal system and has general regulatory implications. Besides bone tissue, loss of human $G\alpha_s$ in the brain causes medulloblastoma also by activating Hh signaling through the $G\alpha_s$-controlled cAMP-PKA-GLI signaling axis (He et al., 2014).

4.2 $G\alpha_s$ signaling regulates bone through Wnt/β-catenin signaling

Wnt/β-catenin signaling is also required to induce osteoblast cell fate (Hill, Spater, Taketo, Birchmeier, & Hartmann, 2005) and regulate bone mass later after bone formation (Rodda & McMahon, 2006) and $G\alpha_s$ signaling has been reported to regulate skeletal development and homeostasis through Wnt/β-catenin signaling (Xu, Khan, et al., 2018). Activating mutations in human low-density lipoprotein receptor-related protein 5 (LRP5), a co-receptor for the Wnt ligands in the Wnt/β-catenin signaling pathways, cause increased bone mass while inactivating mutations lead to decreased bone mass (Boyden et al., 2002; Gong et al., 2001). Bone marrow stromal cells from FD patients showed elevated expression of both *AXIN2* and *TCF1* (Regard et al., 2011). FD cells showed higher Wnt/β-catenin signaling activity both *in vivo* and *in vitro*. Activated $G\alpha_s$ sensitizes BMSCs to Wnt signaling. Activation of $G\alpha_s$ proteins alone is not sufficient to activate Wnt/β-catenin pathway de novo but can augment Wnt/β-catenin signaling activity by binding to Axin to increase its association with Lrp5/6 (Regard et al., 2011).

In the FD mouse models generated with *Prrx1-Cre* or *Sox9-CreER* lines, Wnt/β-catenin signaling was highly upregulated (Khan et al., 2018). In *Osx-Cre*-driven mutants, the FD phenotypes were partially rescued in long bones as shown by rescued bone marrow space and appearance of cortical bone by removing one copy of *Lrp6*. Like Lrp5, Lrp6 is also a Wnt co-receptor which transmits Wnt/β-catenin signaling (Pinson, Brennan, Monkley, Avery, & Skarnes, 2000). In addition, osteoblast maturation inhibited by $Gnas^{R201H}$ expression in BMSC *in vitro* was improved by treatment with LGK-974 (Khan et al., 2018), a potent small-molecule inhibitor of Wnt secretion (Tammela et al., 2017).

4.3 $G\alpha_s$ signaling and Hippo signaling

The Hippo/YAP signaling pathway has emerged as a major regulator of multiple developmental and oncogenic processes (Barron & Kagey, 2014; Cox et al., 2018; Mach et al., 2018; Pan, 2010; Piccolo, Dupont, & Cordenonsi, 2014). Yes-associated protein (YAP) is a transcription factor that is negatively regulated by Hippo kinases. In addition to being an integrator for cell proliferation, differentiation and survival regulated by diverse intrinsic and extrinsic cues (Mo, Park, & Guan, 2014; Zhao, Li, Lei, & Guan, 2010), Yap serves as a co-regulator for bone-related transcription factors, such as phospho-smad1/5/8 (Huang et al., 2016), RUNX2

(Zaidi et al., 2004), and β-catenin (Azzolin et al., 2014), suggesting a function in osteogenesis.

Several studies have shown that $G\alpha_s$ may regulate YAP activities in the context of basal cell carcinomas, Schwann cells and intraductal papillary mucinous neoplasms (IPMNs).

In a basal cell carcinoma (BCC) mouse model, generated by K14CreER-driven loss of *Gnas* in the epidermal stem cell compartment (Vasioukhin, Degenstein, Wise, & Fuchs, 1999), Yap1 transcriptional activity was significantly increased, suggesting that Yap1 activity in the epidermis might be regulated by *Gnas* (Iglesias-Bartolome et al., 2015), consistent with previous literature indicating that the Hippo pathway is tightly regulated by GPCRs (Yu et al., 2012). A constitutively active form of *Gnas* was expressed in pancreatic tumor cells resembling human IPMNs (Ideno et al., 2018) by generating the *p48-Cre; LSL-KrasG12D; Rosa26RLSL-rtTA-TetO-GnasR201C (Kras;Gnas)* mice. Pancreatic ductal adenocarcinomas (PDACs) isolated from the *Kras;Gnas* mice exhibited activation of Hippo signaling and phosphorylated YAP1 was sequestered in the cytoplasm, similar to what happens in human IPMNs with GNAS mutations (Ideno et al., 2018). Taken together, the data indicate that $G\alpha_s$ signaling regulates skeletal development and control bone-related genetic disorders through Hh signaling and Wnt/β-catenin signaling, while its interaction with Hippo signaling is less investigated.

5. Mouse models of skeletal diseases caused by GNAS mutations

5.1 Mouse models of FD

To understand the cellular and molecular mechanisms underlying FD and develop therapeutic approaches to treat FD, several mouse models have been generated to mimic bone defects caused by activated $G\alpha_s$ in humans. The first mouse model of human FD was made by constitutive expression of $G\alpha_s^{R201C}$, one of the FD-causing mutations (Saggio et al., 2014). The transgenic line was generated by lentiviral constructs with rat $G\alpha_s^{R201C}$ cDNA (Piersanti et al., 2010), under the control of the promoters of human elongation factor 1a, EF1a, or the human phosphoglycerate kinase, PGK, both ubiquitously expressed (Conget & Minguell, 2000; Saggio et al., 2014). The transgene was expressed in embryos and germline-inherited.

However, the transgenic mice did not show bone defects until 2 months after birth. In a stepwise and polyostotic pattern, bone lesions firstly appeared in tail vertebrae, then femur, followed by tibia, spine, humerus, acropodial short bones, cranium, ribs, and pelvis (Saggio et al., 2014). The earliest histological changes included endosteal thickening and excessive trabeculae formation (Saggio et al., 2014). Then narrowing of marrow cavity and large gaps in cortical bone regions were found (Saggio et al., 2014). At late stages (more than 10 months after birth), a full-blown FD-like lesion was found with "Chinese writing" patterns of defective trabeculae, abundant collagen in inter-trabecular regions, woven bone and prominent fibrosis (Saggio et al., 2014).

To investigate the role of $G\alpha_s$ in skeletal cells specifically, a transgenic mouse line expressing the rat $G\alpha_s^{R201C}$ cDNA under the control of 2.3 kb *Col1a1* promoter was genenerated (Remoli et al., 2015). Although high bone mass was observed in mutant mice, other FD-features, such as marrow fibrosis or osteolytic changes, were not reproduced in this model, suggesting that additional cells in the bone environment other than differentiated osteoblasts are responsible for the formation of FD bone lesions (Remoli et al., 2015). A tet-off transgenic mouse was generated by breeding the ColI(2.3)-tTA (Peng et al., 2008) with TetO-R_s1 transgenic mice to express R_s1 cDNA, an activated G_s-coupled GPCR, in mature osteoblasts when doxycycline is absent (Hsiao et al., 2008). When the ColI(2.3)-tTA; TetO-R_s1 mice were maintained in the absence of doxycycline, R_s1 GPCR expression caused age-dependent increase of trabecular bone. The ColI (2.3)-tTA;TetO-R_s1 mice were phenotypically indistinguishable from the wild type control mice at birth. However, profound enlargement of the skeleton started at 3 weeks postnatally (Hsiao et al., 2008). At 9 weeks postnatally, bone volume was dramatically increased, cortical bone was almost completely disappeared, normal bone marrow spaces were substantially narrowed and scattered as small islands among increased amounts of scattered bone trabeculae. All these defects share features of human FD conditions (Hsiao et al., 2008).

More recently, inducible triple tet-on transgenic mice were generated by breeding *Prrx-1cre* with the *ROSA26-LSL-rtTA-IRES-GFP; Tet-GNAS*^{R201C} mice (Iglesias-Bartolome et al., 2015) to express human *GNAS*^{R201C} cDNA under the control of reverse tetracycline transactivator (rtTA) after *Cre* recombination in *Prrx1*$^{+}$ osteochondral progenitors and their descendants when doxycycline (Dox) is administrated

(Zhao et al., 2018). When Dox was administered from embryonic day 4.5, dysmorphic limbs were found at P0 in mutant pups (Zhao et al., 2018). When Dox was administered at weaning age, FD-like bone lesions were observed by histological analysis after 2 weeks of dox-feeding (Zhao et al., 2018).

To better model FD and other phenotypes caused by activating mutations in human *GNAS*, a conditional knock-in allele of *Gnas*^{R201H} was generated to express the corresponding mouse mutant *Gnas* from its original genomic locus when a *Cre* driver is available (Khan et al., 2018). Indeed, germline expression of *Gnas*^{R201H} was embryonic lethal, consistent with findings in FD patients that the FD mutations are not germline-inherited and patients only survive with somatic mosaicism. Mice expressing *Gnas*^{R201H} in osteochondral progenitor cells, using *Prrx1-Cre,* exhibited FD features and mice expressing *GNAS*^{R201H} in bone marrow mesenchymal stem cells showed impaired osteoblastic differentiation (Khan et al., 2018). Interestingly, mosaic expression of *Gnas*^{R201H} *in vivo* by crossing the *Gnas*$^{f(R201H)/+}$ mice with the *Sox9CreER;Rosa26TdTomato* mice showed prominent features of FD, including increased bone volume, delayed chondrocyte hypertrophy with disorganized growth plate, and extensive marrow fibrosis (Khan et al., 2018). Interestingly, mosaic expression of *Gnas*^{R201H} caused FD-like fibrotic lesions in a non-cell-autonomous manner (Khan et al., 2018).

5.1.1 Current treatment

Currently, there are no medical therapies of proven efficacy to treat FD. The orthopedic goals for managing skeletal FD lesions include optimizing function and decreasing morbidity with regard to bone pain, fractures, and deformity. Surgical management is challenging, and there are few data that provide information about appropriate indications and techniques. Approaches such as curettage, grafting, plates and screws, and other external fixation devices are frequently ineffective and should generally be avoided (Leet et al., 2016; Stanton et al., 2012). Conservative techniques are recommended, particularly in the pediatric population. Approximately 40% of patients with craniofacial FD present with pain, which is often treated with bisphosphonates (Collins et al., 2005; Kelly, Brillante, & Collins, 2008). Bisphosphonates are medications that decrease bone turnover by inhibiting the activity of bone-resorbing osteoclasts, and are commonly used to treat osteoporosis and skeletal metastases. Bisphosphonates have been advocated as a potential treatment for FD based upon the high levels of osteoclastogenesis present in some FD lesions (Riminucci et al., 2003). Of concern, recent evidence establishes that patients with FD are at risk of osteonecrosis of the jaw (ONJ) as a consequence of bisphosphonate therapy (Metwally, Burke, Tsai, Collins, & Boyce, 2016). Based on the current literature, intravenous bisphosphonates likely have utility in the treatment for FD-related bone pain; however, they should be used at the lowest dose and at intervals so as to minimize the risk of ONJ.

One approach that holds promise for the future is skeletal stem cell/bone marrow stromal cell transplantation; however, more research is needed before such applications can be implemented (Bianco et al., 2013). The report of denosumab treatment in patients with FD (Benhamou, Gensburger, & Chapurlat, 2014; Boyce et al., 2012) has questioned the role of RANKL inhibition in the treatment of FD. Denosumab, a humanized monoclonal antibody to RANKL, is currently approved in the treatment of osteoporosis, giant cell tumors of the long bones, and prevention of skeletal-related events from bone metastases (Thomas et al., 2010). Denosumab inhibits RANKL binding to its receptor, RANK, thereby preventing the activation and function of osteoclasts (Xu, Adams, Yu, & Xu, 2013). The overexpression of RANKL has been reported in FD-like bone cells (Piersanti et al., 2010) and FD tissue (Wang et al., 2014), which suggests that it may play a role in the pathophysiology of FD. Recent findings of overactivated Wnt/β-catenin signaling in FD mouse models suggest that Wnt inhibitors may be therapeutic candidates.

5.2 POH mouse models

It is reported that mice with heterozygous *Gnas* deletion of the paternal allele form bony spicules in the subcutaneous region at 9 months of age (Pignolo et al., 2011). Adipose stromal cells isolated from this mouse model show accelerated osteogenic differentiation *in vitro* (Pignolo et al., 2011). A more robust mouse model that better model POH is the *Gnas* conditional knock-out mouse driven by *Prrx1Cre* (Regard et al., 2013). Ectopic bone formation was initiated at E16.5–17.5 and progressed quickly after birth (Regard et al., 2013). Beside prenatal inactivation of *Gnas*, locally induced *Gnas* deficiency in the subcutaneous mesenchymal tissues in adult mice is sufficient to induce progressive HO, similar to that found in POH patients (Regard et al., 2013). In addition, the *Gnas* knock-out mouse, driven by *renin-Cre*, shows heterotopic ossification in basal cells of hair follicles and subcutaneous and deep muscle areas (Castrop et al., 2007).

5.2.1 Current treatment options

Nonsteroidal anti-inflammatory drugs (NSAIDs) have been used in the prevention of HO (Legosz, Drela, Pulik, Sarzynska, & Maldyk, 2018).

However, these drugs show relatively low efficacy when HO is already developed (Teasell et al., 2010) and their molecular and cellular targets remain to be elucidated. The main mechanism for NSAIDs is blockage of an enzyme responsible for an inflammatory reaction. The use of NSAIDs in combination with radiotherapy is the standard treatment for HO prophylaxis and shows the highest effectiveness (Pakos et al., 2009; Pavlou, Kyrkos, Tsialogiannis, Korres, & Tsiridis, 2012). Radiation therapy has been successfully used to prevent or treat HO. It inhibits osteogenic differentiation of mesenchymal stem cells; thus, preventing bone repair and abnormal bone growth. In 1981, the efficacy of postoperative irradiation in preventing HO after total hip arthroplasty in 48 high-risk patients was reported (Coventry & Scanlon, 1981). After comparing the efficacy of radiotherapy, there is no statistically significant difference in the effectiveness of pre- and postoperative intervention (Winkler et al., 2015). However, increasing the time after surgery before radiotherapy can reduce its effectiveness, and radiotherapy should be applied as soon as possible (Mourad et al., 2012). In addition, diphosphonates can also be used to prevent HO (Thomas & Amstutz, 1985). Surgical resection of HO is an effective treatment to increase joint mobility or alter limb position (Freed, Hahn, Menter, & Dillon, 1982). However, it is not recommended if HO has reached maturity, as trauma can increase the size of the lesion and induce more ectopic ossification (Kaplan et al., 1993).

Based on findings of the molecular and cellular mechanisms underlying HO in POH, Hh signaling activation is both necessary and sufficient to induce osteoblast differentiation from mesenchymal progenitor cells. Indeed, Stoeger et al. demonstrated the role of HH signaling in BMP2-induced ectopic bone formation as critical for chondrocyte differentiation in HO (Stoeger et al., 2002). Collectively, these observations suggest that HH signaling plays a significant role in various forms of HO, but the exact mechanism of the crosstalk between HH and BMP or other factors is not fully known (Legosz et al., 2018). Consistent with this idea, the observations suggest that HH inhibitors or Gli inhibitors could be used for POH prevention and treatment. Gli inhibitors, such as arsenic trioxide (ATO) (Kim, Lee, Kim, Gardner, & Beachy, 2010) or GANT58 (Lauth, Bergstrom, Shimokawa, & Toftgard, 2007), have been found to reduce HO in the POH mouse models (Regard et al., 2013). Because activation of Hh signaling is oncogenic in many tissues, it is foreseeable that some of the approved small-molecule tumor inhibitors targeting Hh signaling can be repurposed to inhibit HO in POH patients or even other non-genetic forms of HO (Pignolo, Ramaswamy, Fong, Shore, & Kaplan, 2015).

6. Conclusions and implications

$G\alpha_s$ mediates a critical signaling arm of the classical heterotrimeric G-proteins in transducing GPCR signaling, which regulates a diverse array of developmental and homeostatic processes in the skeletal and other systems. The crucial functions of $G\alpha_s$ in the signaling networks regulating the skeletal system are first manifested by severe human skeletal diseases caused by both activating and inactivating mutations in the *GNAS* gene. Establishment of proper mouse models that recapitulate human skeletal diseases resulted from *GNAS* mutations both genetically and phenotypically have provided invaluable tools to uncover the molecular and cellular mechanisms whereby $G\alpha_s$ signaling regulates commitment, differentiation and maturation of skeletal cells. From these mechanistic studies, it is evident that $G\alpha_s$ signaling executes many of its functions by interacting with the Wnt and Hh signaling pathways, both of which are fundamentally important in skeletal development and homeostasis. While these findings advanced previous understandings of signaling regulation of skeletal development, more questions have been raised. Among these, how $G\alpha_s$ signaling exerts non-cell autonomous roles in regulating skeletal stem cell proliferation and differentiation has top significance as few mutant cells in both FD and POH patients are able to cause severe and broad tissue defects by altering wild type cells. In addition, from a pharmacological point of view, further understanding the role of $G\alpha_s$ signaling in GPCR-dependent and independent pathways will open a door for the development of more pathway-specific therapeutics. GPCRs have been one of the primary therapeutic targets for numerous diseases. However, GPCRs can signal to multiple $G\alpha$ pathways and there are also other GPCR-interacting proteins that mediate G-protein-independent signals downstream of GPCRs. In this regard, treatment using GPCR agonists, antagonists and inverse agonists for the diseases primarily caused by alterations of $G\alpha_s$ signaling must be selected to reduce potential side effects. Furthermore, more research efforts are required to test whether drugs targeting effectors of $G\alpha_s$ signaling, such as the Wnt and Hh pathways, can be potentially repurposed to treat diseases caused by alterations of $G\alpha_s$ or GPCRs.

Acknowledgments

Qian Cong is supported by the NIH grant R01DE025866 from NIDCR and the MDBR-18-115-FD/MAS from the Penn Center for Musculoskeletal Disorders. Ruoshi Xu is supported by the 111 Project, MOE (B14038), China. Yingzi Yang is supported by R01DE025866 from NIDCR and R01AR070877 from NIAMS. We thank Dr. Bjorn Olsen for careful editing of the chapter.

References

Adegbite, N. S., Xu, M., Kaplan, F. S., Shore, E. M., & Pignolo, R. J. (2008). Diagnostic and mutational spectrum of progressive osseous heteroplasia (POH) and other forms of GNAS-based heterotopic ossification. *American Journal of Medical Genetics Part A, 146A*, 1788–1796. https://doi.org/10.1002/ajmg.a.32346.

Albright, F., Butler, A. M., Hampton, A. O., & Smith, P. (1937). Syndrome characterized by osteitis fibrosa disseminata, areas of pigmentation and endocrine dysfunction with precocious puberty in females. *The New England Journal of Medicine, 216*, 727–746.

Arai, F., et al. (2004). Tie2/angiopoietin-1 signaling regulates hematopoietic stem cell quiescence in the bone marrow niche. *Cell, 118*, 149–161. https://doi.org/10.1016/j.cell.2004.07.004.

Azzolin, L., et al. (2014). YAP/TAZ incorporation in the beta-catenin destruction complex orchestrates the Wnt response. *Cell, 158*, 157–170. https://doi.org/10.1016/j.cell.2014.06.013.

Barron, D. A., & Kagey, J. D. (2014). The role of the Hippo pathway in human disease and tumorigenesis. *Clinical and Translational Medicine, 3*, 25. https://doi.org/10.1186/2001-1326-3-25.

Bastepe, M. (2007). The GNAS locus: Quintessential complex gene encoding Gsalpha, XLalphas, and other imprinted transcripts. *Current Genomics, 8*, 398–414. https://doi.org/10.2174/138920207783406488.

Bastepe, M. (2018). GNAS mutations and heterotopic ossification. *Bone, 109*, 80–85. https://doi.org/10.1016/j.bone.2017.09.002.

Bastepe, M., & Juppner, H. (2005). GNAS locus and pseudohypoparathyroidism. *Hormone Research, 63*, 65–74. https://doi.org/10.1159/000083895.

Bastepe, M., et al. (2004). Stimulatory G protein directly regulates hypertrophic differentiation of growth plate cartilage in vivo. *Proceedings of the National Academy of Sciences of the United States of America, 101*, 14794–14799. https://doi.org/10.1073/pnas.0405091101.

Benhamou, J., Gensburger, D., & Chapurlat, R. (2014). Transient improvement of severe pain from fibrous dysplasia of bone with denosumab treatment. *Joint, Bone, Spine, 81*, 549–550. https://doi.org/10.1016/j.jbspin.2014.04.013.

Bianco, P., et al. (1998). Reproduction of human fibrous dysplasia of bone in immunocompromised mice by transplanted mosaics of normal and Gsalpha-mutated skeletal progenitor cells. *The Journal of Clinical Investigation, 101*, 1737–1744. https://doi.org/10.1172/JCI2361.

Bianco, P., et al. (2013). Regulation of stem cell therapies under attack in Europe: For whom the bell tolls. *The EMBO Journal, 32*, 1489–1495. https://doi.org/10.1038/emboj.2013.114.

Boyce, A. M., Florenzano, P., de Castro, L. F., et al. (1993–2018). Fibrous dysplasia/McCune-Albright syndrome. In M. P. Adam, H. H. Ardinger, & R. A. Pagon et al., (Eds.), GeneReviews®. Seattle, WA: University of Washington, Seattle Available from: https://www.ncbi.nlm.nih.gov/books/NBK274564/. (Updated 2018 Aug 16).

Boyce, A. M., et al. (2012). Denosumab treatment for fibrous dysplasia. *Journal of Bone and Mineral Research, 27*, 1462–1470. https://doi.org/10.1002/jbmr.1603.

Boyden, L. M., et al. (2002). High bone density due to a mutation in LDL-receptor-related protein 5. *The New England Journal of Medicine, 346*, 1513–1521. https://doi.org/10.1056/NEJMoa013444.

Calvi, L. M., et al. (2003). Osteoblastic cells regulate the haematopoietic stem cell niche. *Nature, 425*, 841–846. https://doi.org/10.1038/nature02040.

Carroll, S. H., & Ravid, K. (2013). Differentiation of mesenchymal stem cells to osteoblasts and chondrocytes: A focus on adenosine receptors. *Expert Reviews in Molecular Medicine, 15*, e1. https://doi.org/10.1017/erm.2013.2.

Castrop, H., et al. (2007). Skeletal abnormalities and extra-skeletal ossification in mice with restricted Gsalpha deletion caused by a renin promoter-Cre transgene. *Cell and Tissue Research*, *330*, 487–501. https://doi.org/10.1007/s00441-007-0491-6.

Chagin, A. S., et al. (2014). G-protein stimulatory subunit alpha and Gq/11alpha G-proteins are both required to maintain quiescent stem-like chondrocytes. *Nature Communications*, *5*, 3673. https://doi.org/10.1038/ncomms4673.

Chakkalakal, S. A., & Shore, E. M. (2019). Heterotopic ossification in mouse models of fibrodysplasia ossificans progressiva. *Methods in Molecular Biology*, *1891*, 247–255. https://doi.org/10.1007/978-1-4939-8904-1_18.

Chapurlat, R. D., & Orcel, P. (2008). Fibrous dysplasia of bone and McCune-Albright syndrome. *Best Practice & Research. Clinical Rheumatology*, *22*, 55–69. https://doi.org/10.1016/j.berh.2007.11.004.

Chen, M., et al. (2005). Increased glucose tolerance and reduced adiposity in the absence of fasting hypoglycemia in mice with liver-specific Gsalpha deficiency. *The Journal of Clinical Investigation*, *115*, 3217–3227. https://doi.org/10.1172/JCI24196.

Clausen, B. E., Burkhardt, C., Reith, W., Renkawitz, R., & Forster, I. (1999). Conditional gene targeting in macrophages and granulocytes using LysMcre mice. *Transgenic Research*, *8*, 265–277.

Collins, M. T., et al. (2005). An instrument to measure skeletal burden and predict functional outcome in fibrous dysplasia of bone. *Journal of Bone and Mineral Research*, *20*, 219–226. https://doi.org/10.1359/JBMR.041111.

Conget, P. A., & Minguell, J. J. (2000). Adenoviral-mediated gene transfer into ex vivo expanded human bone marrow mesenchymal progenitor cells. *Experimental Hematology*, *28*, 382–390.

Coventry, M. B., & Scanlon, P. W. (1981). The use of radiation to discourage ectopic bone. A nine-year study in surgery about the hip. *The Journal of Bone and Joint Surgery. American Volume*, *63*, 201–208.

Cox, A. G., Tsomides, A., Yimlamai, D., Hwang, K. L., Miesfeld, J., Galli, G. G., et al. (2018). Yap regulates glucose utilization and sustains nucleotide synthesis to enable organ growth. *The EMBO Journal*, *37*(22), e100294. https://doi.org/10.15252/embj.2018100294.

Eddy, M. C., et al. (2000). Deficiency of the alpha-subunit of the stimulatory G protein and severe extraskeletal ossification. *Journal of Bone and Mineral Research*, *15*, 2074–2083. https://doi.org/10.1359/jbmr.2000.15.11.2074.

Farfel, Z., Bourne, H. R., & Iiri, T. (1999). The expanding spectrum of G protein diseases. *The New England Journal of Medicine*, *340*, 1012–1020. https://doi.org/10.1056/NEJM199904013401306.

Franz-Odendaal, T. A., Hall, B. K., & Witten, P. E. (2006). Buried alive: How osteoblasts become osteocytes. *Developmental Dynamics*, *235*, 176–190. https://doi.org/10.1002/dvdy.20603.

Freed, J. H., Hahn, H., Menter, R., & Dillon, T. (1982). The use of the three-phase bone scan in the early diagnosis of heterotopic ossification (HO) and in the evaluation of Didronel therapy. *Paraplegia*, *20*, 208–216. https://doi.org/10.1038/sc.1982.39.

Fulzele, K., et al. (2013). Myelopoiesis is regulated by osteocytes through Gsalpha-dependent signaling. *Blood*, *121*, 930–939. https://doi.org/10.1182/blood-2012-06-437160.

Gong, Y., et al. (2001). LDL receptor-related protein 5 (LRP5) affects bone accrual and eye development. *Cell*, *107*, 513–523.

Happle, R. (1986). The McCune-Albright syndrome: A lethal gene surviving by mosaicism. *Clinical Genetics*, *29*, 321–324.

He, X., et al. (2014). The G protein alpha subunit Galphas is a tumor suppressor in sonic hedgehog-driven medulloblastoma. *Nature Medicine*, *20*, 1035–1042. https://doi.org/10.1038/nm.3666.

Hill, T. P., Spater, D., Taketo, M. M., Birchmeier, W., & Hartmann, C. (2005). Canonical Wnt/beta-catenin signaling prevents osteoblasts from differentiating into chondrocytes. *Developmental Cell*, *8*, 727–738. https://doi.org/10.1016/j.devcel.2005.02.013.

Hsiao, E. C., et al. (2008). Osteoblast expression of an engineered Gs-coupled receptor dramatically increases bone mass. *Proceedings of the National Academy of Sciences of the United States of America*, *105*, 1209–1214. https://doi.org/10.1073/pnas.0707457105.

Huang, Z., et al. (2016). YAP stabilizes SMAD1 and promotes BMP2-induced neocortical astrocytic differentiation. *Development*, *143*, 2398–2409. https://doi.org/10.1242/dev.130658.

Ideno, N., et al. (2018). GNAS(R201C) induces pancreatic cystic neoplasms in mice that express activated KRAS by inhibiting YAP1 signaling. *Gastroenterology*, *155*, 1593–1607 e1512. https://doi.org/10.1053/j.gastro.2018.08.006.

Iglesias-Bartolome, R., et al. (2015). Inactivation of a Galpha(s)-PKA tumour suppressor pathway in skin stem cells initiates basal-cell carcinogenesis. *Nature Cell Biology*, *17*, 793–803. https://doi.org/10.1038/ncb3164.

Jin, Y., Cong, Q., Gvozdenovic-Jeremic, J., Hu, J., Zhang, Y., Terkeltaub, R., et al. (2018). Enpp1 inhibits ectopic joint calcification and maintains articular chondrocytes by repressing hedgehog signaling. *Development*, *145*(18), dev164830. https://doi.org/10.1242/dev.164830.

Kaplan, F. S., Craver, R., MacEwen, G. D., Gannon, F. H., Finkel, G., Hahn, G., et al. (1994). Progressive osseous heteroplasia: A distinct developmental disorder of heterotopic ossification. Two new case reports and follow-up of three previously reported cases. *The Journal of Bone and Joint Surgery. American Volume*, *76*, 425–436.

Kaplan, F. S., Hahn, G. V., & Zasloff, M. A. (1994). Heterotopic ossification: Two rare forms and what they can teach us. *The Journal of the American Academy of Orthopaedic Surgeons*, *2*, 288–296.

Kaplan, F. S., & Shore, E. M. (2000). Progressive osseous heteroplasia. *Journal of Bone and Mineral Research*, *15*, 2084–2094. https://doi.org/10.1359/jbmr.2000.15.11.2084.

Kaplan, F. S., et al. (1993). The histopathology of fibrodysplasia ossificans progressiva. An endochondral process. *The Journal of Bone and Joint Surgery. American Volume*, *75*, 220–230.

Karsenty, G. (2003). The complexities of skeletal biology. *Nature*, *423*, 316–318. https://doi.org/10.1038/nature01654.

Kelly, M. H., Brillante, B., & Collins, M. T. (2008). Pain in fibrous dysplasia of bone: Age-related changes and the anatomical distribution of skeletal lesions. *Osteoporosis International*, *19*, 57–63. https://doi.org/10.1007/s00198-007-0425-x.

Khan, S. K., et al. (2018). Induced Gnas(R201H) expression from the endogenous Gnas locus causes fibrous dysplasia by up-regulating Wnt/beta-catenin signaling. *Proceedings of the National Academy of Sciences of the United States of America*, *115*, E418–E427. https://doi.org/10.1073/pnas.1714313114.

Kim, J., Lee, J. J., Kim, J., Gardner, D., & Beachy, P. A. (2010). Arsenic antagonizes the Hedgehog pathway by preventing ciliary accumulation and reducing stability of the Gli2 transcriptional effector. *Proceedings of the National Academy of Sciences of the United States of America*, *107*, 13432–13437. https://doi.org/10.1073/pnas.1006822107.

Kozhemyakina, E., Lassar, A. B., & Zelzer, E. (2015). A pathway to bone: Signaling molecules and transcription factors involved in chondrocyte development and maturation. *Development*, *142*, 817–831. https://doi.org/10.1242/dev.105536.

Kronenberg, H. M. (2003). Developmental regulation of the growth plate. *Nature*, *423*, 332–336. https://doi.org/10.1038/nature01657.

Kumagai, K., et al. (2008). A case of progressive osseous heteroplasia: A first case in Japan. *Skeletal Radiology*, *37*, 563–567. https://doi.org/10.1007/s00256-008-0469-9.

Lania, A., et al. (1998). Constitutively active Gs alpha is associated with an increased phosphodiesterase activity in human growth hormone-secreting adenomas. *The Journal of Clinical Endocrinology and Metabolism, 83*, 1624–1628. https://doi.org/10.1210/jcem.83.5.4814.

Lauth, M., Bergstrom, A., Shimokawa, T., & Toftgard, R. (2007). Inhibition of GLI-mediated transcription and tumor cell growth by small-molecule antagonists. *Proceedings of the National Academy of Sciences of the United States of America, 104*, 8455–8460. https://doi.org/10.1073/pnas.0609699104.

Leet, A. I., et al. (2016). Bone-grafting in polyostotic fibrous dysplasia. *The Journal of Bone and Joint Surgery. American Volume, 98*, 211–219. https://doi.org/10.2106/JBJS.O.00547.

Legosz, P., Drela, K., Pulik, L., Sarzynska, S., & Maldyk, P. (2018). Challenges of heterotopic ossification-molecular background and current treatment strategies. *Clinical and Experimental Pharmacology & Physiology, 45*, 1229–1235. https://doi.org/10.1111/1440-1681.13025.

Logan, M., et al. (2002). Expression of Cre Recombinase in the developing mouse limb bud driven by a Prxl enhancer. *Genesis, 33*, 77–80. https://doi.org/10.1002/gene.10092.

Long, F., et al. (2004). Ihh signaling is directly required for the osteoblast lineage in the endochondral skeleton. *Development, 131*, 1309–1318. https://doi.org/10.1242/dev.01006.

Lu, Y., et al. (2007). DMP1-targeted Cre expression in odontoblasts and osteocytes. *Journal of Dental Research, 86*, 320–325. https://doi.org/10.1177/154405910708600404.

Mach, J., et al. (2018). Modulation of the Hippo pathway and organ growth by RNA processing proteins. *Proceedings of the National Academy of Sciences of the United States of America, 115*, 10684–10689. https://doi.org/10.1073/pnas.1807325115.

Metwally, T., Burke, A., Tsai, J. Y., Collins, M. T., & Boyce, A. M. (2016). Fibrous dysplasia and medication-related osteonecrosis of the jaw. *Journal of Oral and Maxillofacial Surgery, 74*, 1983–1999. https://doi.org/10.1016/j.joms.2016.04.001.

Mo, J. S., Park, H. W., & Guan, K. L. (2014). The Hippo signaling pathway in stem cell biology and cancer. *EMBO Reports, 15*, 642–656. https://doi.org/10.15252/embr.201438638.

Morris, A. J., & Malbon, C. C. (1999). Physiological regulation of G protein-linked signaling. *Physiological Reviews, 79*, 1373–1430. https://doi.org/10.1152/physrev.1999.79.4.1373.

Mourad, W. F., et al. (2012). A prolonged time interval between trauma and prophylactic radiation therapy significantly increases the risk of heterotopic ossification. *International Journal of Radiation Oncology, Biology, Physics, 82*, e339–e344. https://doi.org/10.1016/j.ijrobp.2011.06.1981.

Nelson, D. K., & Williams, T. (2004). Frontonasal process-specific disruption of AP-2alpha results in postnatal midfacial hypoplasia, vascular anomalies, and nasal cavity defects. *Developmental Biology, 267*, 72–92. https://doi.org/10.1016/j.ydbio.2003.10.033.

Neves, S. R., Ram, P. T., & Iyengar, R. (2002). G protein pathways. *Science, 296*, 1636–1639. https://doi.org/10.1126/science.1071550.

O'Hayre, M., et al. (2013). The emerging mutational landscape of G proteins and G-protein-coupled receptors in cancer. *Nature Reviews. Cancer, 13*, 412–424. https://doi.org/10.1038/nrc3521.

Oldham, W. M., & Hamm, H. E. (2008). Heterotrimeric G protein activation by G-protein-coupled receptors. *Nature Reviews. Molecular Cell Biology, 9*, 60–71. https://doi.org/10.1038/nrm2299.

Pakos, E. E., et al. (2009). Combined radiotherapy and indomethacin for the prevention of heterotopic ossification after total hip arthroplasty. *Strahlentherapie und Onkologie, 185*, 500–505. https://doi.org/10.1007/s00066-009-1954-3.

Pan, D. (2010). The Hippo signaling pathway in development and cancer. *Developmental Cell, 19*, 491–505. https://doi.org/10.1016/j.devcel.2010.09.011.

Patten, J. L., et al. (1990). Mutation in the gene encoding the stimulatory G protein of adenylate cyclase in Albright's hereditary osteodystrophy. *The New England Journal of Medicine, 322*, 1412–1419. https://doi.org/10.1056/NEJM199005173222002.

Pavlou, G., Kyrkos, M., Tsialogiannis, E., Korres, N., & Tsiridis, E. (2012). Pharmacological treatment of heterotopic ossification following hip surgery: An update. *Expert Opinion on Pharmacotherapy*, *13*, 619–622. https://doi.org/10.1517/14656566.2012.662342.

Peng, J., et al. (2008). Conditional expression of a Gi-coupled receptor in osteoblasts results in trabecular osteopenia. *Endocrinology*, *149*, 1329–1337. https://doi.org/10.1210/en.2007-0235.

Piccolo, S., Dupont, S., & Cordenonsi, M. (2014). The biology of YAP/TAZ: Hippo signaling and beyond. *Physiological Reviews*, *94*, 1287–1312. https://doi.org/10.1152/physrev.00005.2014.

Pierce, K. L., Premont, R. T., & Lefkowitz, R. J. (2002). Seven-transmembrane receptors. *Nature Reviews. Molecular Cell Biology*, *3*, 639–650. https://doi.org/10.1038/nrm908.

Piersanti, S., et al. (2010). Transfer, analysis, and reversion of the fibrous dysplasia cellular phenotype in human skeletal progenitors. *Journal of Bone and Mineral Research*, *25*, 1103–1116. https://doi.org/10.1359/jbmr.091036.

Pignolo, R. J., Ramaswamy, G., Fong, J. T., Shore, E. M., & Kaplan, F. S. (2015). Progressive osseous heteroplasia: Diagnosis, treatment, and prognosis. *The Application of Clinical Genetics*, *8*, 37–48. https://doi.org/10.2147/TACG.S51064.

Pignolo, R. J., et al. (2011). Heterozygous inactivation of Gnas in adipose-derived mesenchymal progenitor cells enhances osteoblast differentiation and promotes heterotopic ossification. *Journal of Bone and Mineral Research*, *26*, 2647–2655. https://doi.org/10.1002/jbmr.481.

Pinson, K. I., Brennan, J., Monkley, S., Avery, B. J., & Skarnes, W. C. (2000). An LDL-receptor-related protein mediates Wnt signalling in mice. *Nature*, *407*, 535–538. https://doi.org/10.1038/35035124.

Plagge, A., Kelsey, G., & Germain-Lee, E. L. (2008). Physiological functions of the imprinted Gnas locus and its protein variants Galpha(s) and XLalpha(s) in human and mouse. *The Journal of Endocrinology*, *196*, 193–214. https://doi.org/10.1677/JOE-07-0544.

Ramaswamy, G., et al. (2017). Gsalpha controls cortical bone quality by regulating osteoclast differentiation via cAMP/PKA and beta-catenin pathways. *Scientific Reports*, 7, 45140. https://doi.org/10.1038/srep45140.

Ramaswamy, G., et al. (2018). Ablation of Gsalpha signaling in osteoclast progenitor cells adversely affects skeletal bone maintenance. *Bone*, *109*, 86–90. https://doi.org/10.1016/j.bone.2017.11.019.

Regard, J. B., et al. (2011). Wnt/beta-catenin signaling is differentially regulated by Galpha proteins and contributes to fibrous dysplasia. *Proceedings of the National Academy of Sciences of the United States of America*, *108*, 20101–20106. https://doi.org/10.1073/pnas.1114656108.

Regard, J. B., et al. (2013). Activation of Hedgehog signaling by loss of GNAS causes heterotopic ossification. *Nature Medicine*, *19*, 1505–1512. https://doi.org/10.1038/nm.3314.

Remoli, C., et al. (2015). Osteoblast-specific expression of the fibrous dysplasia (FD)-causing mutation Gsalpha(R201C) produces a high bone mass phenotype but does not reproduce FD in the mouse. *Journal of Bone and Mineral Research*, *30*, 1030–1043. https://doi.org/10.1002/jbmr.2425.

Riminucci, M., Robey, P. G., Saggio, I., & Bianco, P. (2010). Skeletal progenitors and the GNAS gene: Fibrous dysplasia of bone read through stem cells. *Journal of Molecular Endocrinology*, *45*, 355–364. https://doi.org/10.1677/JME-10-0097.

Riminucci, M., et al. (1997). Fibrous dysplasia of bone in the McCune-Albright syndrome: Abnormalities in bone formation. *The American Journal of Pathology*, *151*, 1587–1600.

Riminucci, M., et al. (1999). The histopathology of fibrous dysplasia of bone in patients with activating mutations of the Gs alpha gene: Site-specific patterns and recurrent histological hallmarks. *The Journal of Pathology*, *187*, 249–258. https://doi.org/10.1002/(SICI)1096-9896(199901)187:2<249::AID-PATH222>3.0.CO;2-J.

Riminucci, M., et al. (2003). Osteoclastogenesis in fibrous dysplasia of bone: In situ and in vitro analysis of IL-6 expression. *Bone*, *33*, 434–442.
Robinson, C., Collins, M. T., & Boyce, A. M. (2016). Fibrous dysplasia/McCune-Albright syndrome: Clinical and translational perspectives. *Current Osteoporosis Reports*, *14*, 178–186. https://doi.org/10.1007/s11914-016-0317-0.
Rodda, S. J., & McMahon, A. P. (2006). Distinct roles for Hedgehog and canonical Wnt signaling in specification, differentiation and maintenance of osteoblast progenitors. *Development*, *133*, 3231–3244. https://doi.org/10.1242/dev.02480.
Saggio, I., et al. (2014). Constitutive expression of Gsalpha(R201C) in mice produces a heritable, direct replica of human fibrous dysplasia bone pathology and demonstrates its natural history. *Journal of Bone and Mineral Research*, *29*, 2357–2368. https://doi.org/10.1002/jbmr.2267.
Sakamoto, A., et al. (2005). Deficiency of the G-protein alpha-subunit G(s)alpha in osteoblasts leads to differential effects on trabecular and cortical bone. *The Journal of Biological Chemistry*, *280*, 21369–21375. https://doi.org/10.1074/jbc.M500346200.
Schimmel, R. J., et al. (2010). GNAS-associated disorders of cutaneous ossification: Two different clinical presentations. *Bone*, *46*, 868–872. https://doi.org/10.1016/j.bone.2009.11.001.
Schwindinger, W. F., Francomano, C. A., & Levine, M. A. (1992). Identification of a mutation in the gene encoding the alpha subunit of the stimulatory G protein of adenylyl cyclase in McCune-Albright syndrome. *Proceedings of the National Academy of Sciences of the United States of America*, *89*, 5152–5156.
Shenker, A., Weinstein, L. S., Sweet, D. E., & Spiegel, A. M. (1994). An activating Gs alpha mutation is present in fibrous dysplasia of bone in the McCune-Albright syndrome. *The Journal of Clinical Endocrinology and Metabolism*, *79*, 750–755. https://doi.org/10.1210/jcem.79.3.8077356.
Shenker, A., et al. (1993). Severe endocrine and nonendocrine manifestations of the McCune-Albright syndrome associated with activating mutations of stimulatory G protein GS. *The Journal of Pediatrics*, *123*, 509–518.
Shore, E. M., & Kaplan, F. S. (2008). Insights from a rare genetic disorder of extra-skeletal bone formation, fibrodysplasia ossificans progressiva (FOP). *Bone*, *43*, 427–433. https://doi.org/10.1016/j.bone.2008.05.013.
Shore, E. M., & Kaplan, F. S. (2010). Inherited human diseases of heterotopic bone formation. *Nature Reviews. Rheumatology*, *6*, 518–527. https://doi.org/10.1038/nrrheum.2010.122.
Shore, E. M., et al. (2002). Paternally inherited inactivating mutations of the GNAS1 gene in progressive osseous heteroplasia. *The New England Journal of Medicine*, *346*, 99–106. https://doi.org/10.1056/NEJMoa011262.
Simon, M. I., Strathmann, M. P., & Gautam, N. (1991). Diversity of G proteins in signal transduction. *Science*, *252*, 802–808.
Sinha, P., et al. (2014). Loss of Gsalpha early in the osteoblast lineage favors adipogenic differentiation of mesenchymal progenitors and committed osteoblast precursors. *Journal of Bone and Mineral Research*, *29*, 2414–2426. https://doi.org/10.1002/jbmr.2270.
Stanton, R. P., et al. (2012). The surgical management of fibrous dysplasia of bone. *Orphanet Journal of Rare Diseases*, 7(Suppl. 1), S1. https://doi.org/10.1186/1750-1172-7-S1-S1.
St-Jacques, B., Hammerschmidt, M., & McMahon, A. P. (1999). Indian hedgehog signaling regulates proliferation and differentiation of chondrocytes and is essential for bone formation. *Genes & Development*, *13*, 2072–2086.
Stoeger, T., et al. (2002). In situ gene expression analysis during BMP2-induced ectopic bone formation in mice shows simultaneous endochondral and intramembranous ossification. *Growth Factors*, *20*, 197–210.
Tammela, T., et al. (2017). A Wnt-producing niche drives proliferative potential and progression in lung adenocarcinoma. *Nature*, *545*, 355–359. https://doi.org/10.1038/nature22334.

Tashjian, A. H., Jr., Frantz, A. G., & Lee, J. B. (1966). Pseudohypoparathyroidism: Assays of parathyroid hormone and thyrocalcitonin. *Proceedings of the National Academy of Sciences of the United States of America, 56*, 1138–1142.

Teasell, R. W., et al. (2010). A systematic review of the therapeutic interventions for heterotopic ossification after spinal cord injury. *Spinal Cord, 48*, 512–521. https://doi.org/10.1038/sc.2009.175.

Thomas, B. J., & Amstutz, H. C. (1985). Results of the administration of diphosphonate for the prevention of heterotopic ossification after total hip arthroplasty. *The Journal of Bone and Joint Surgery. American Volume, 67*, 400–403.

Thomas, D., et al. (2010). Denosumab in patients with giant-cell tumour of bone: An open-label, phase 2 study. *The Lancet Oncology, 11*, 275–280. https://doi.org/10.1016/S1470-2045(10)70010-3.

Toosi, S., Behravan, N., & Behravan, J. (2018). Nonunion fractures, mesenchymal stem cells and bone tissue engineering. *Journal of Biomedical Materials Research. Part A, 106*, 2552–2562. https://doi.org/10.1002/jbm.a.36433.

Turan, S., & Bastepe, M. (2015). GNAS spectrum of disorders. *Current Osteoporosis Reports, 13*, 146–158. https://doi.org/10.1007/s11914-015-0268-x.

Vasioukhin, V., Degenstein, L., Wise, B., & Fuchs, E. (1999). The magical touch: Genome targeting in epidermal stem cells induced by tamoxifen application to mouse skin. *Proceedings of the National Academy of Sciences of the United States of America, 96*, 8551–8556.

Waldron, C. A. (1993). Fibro-osseous lesions of the jaws. *Journal of Oral and Maxillofacial Surgery, 51*, 828–835.

Wang, H. D., et al. (2014). Effects of denosumab treatment and discontinuation on human growth plates. *The Journal of Clinical Endocrinology and Metabolism, 99*, 891–897. https://doi.org/10.1210/jc.2013-3081.

Weinstein, L. S. (2006). G(s)alpha mutations in fibrous dysplasia and McCune-Albright syndrome. *Journal of Bone and Mineral Research, 21*(Suppl. 2), P120–P124. https://doi.org/10.1359/jbmr.06s223.

Weinstein, L. S., Chen, M., Xie, T., & Liu, J. (2006). Genetic diseases associated with heterotrimeric G proteins. *Trends in Pharmacological Sciences, 27*, 260–266. https://doi.org/10.1016/j.tips.2006.03.005.

Weinstein, L. S., Liu, J., Sakamoto, A., Xie, T., & Chen, M. (2004). Minireview: GNAS: Normal and abnormal functions. *Endocrinology, 145*, 5459–5464. https://doi.org/10.1210/en.2004-0865.

Weinstein, L. S., Yu, S., Warner, D. R., & Liu, J. (2001). Endocrine manifestations of stimulatory G protein alpha-subunit mutations and the role of genomic imprinting. *Endocrine Reviews, 22*, 675–705. https://doi.org/10.1210/edrv.22.5.0439.

Weinstein, L. S., et al. (1990). Mutations of the Gs alpha-subunit gene in Albright hereditary osteodystrophy detected by denaturing gradient gel electrophoresis. *Proceedings of the National Academy of Sciences of the United States of America, 87*, 8287–8290.

Weinstein, L. S., et al. (1991). Activating mutations of the stimulatory G protein in the McCune-Albright syndrome. *The New England Journal of Medicine, 325*, 1688–1695. https://doi.org/10.1056/NEJM199112123252403.

Weivoda, M. M., et al. (2016). Wnt signaling inhibits osteoclast differentiation by activating canonical and noncanonical cAMP/PKA pathways. *Journal of Bone and Mineral Research, 31*, 65–75. https://doi.org/10.1002/jbmr.2599.

Winkler, S., et al. (2015). Current therapeutic strategies of heterotopic ossification—A survey amongst orthopaedic and trauma departments in Germany. *BMC Musculoskeletal Disorders, 16*, 313. https://doi.org/10.1186/s12891-015-0764-2.

Wu, J. Y., et al. (2008). Osteoblastic regulation of B lymphopoiesis is mediated by Gs{alpha}-dependent signaling pathways. *Proceedings of the National Academy of Sciences of the United States of America, 105*, 16976–16981. https://doi.org/10.1073/pnas.0802898105.

Wu, J. Y., et al. (2011). Gsalpha enhances commitment of mesenchymal progenitors to the osteoblast lineage but restrains osteoblast differentiation in mice. *The Journal of Clinical Investigation, 121*, 3492–3504. https://doi.org/10.1172/JCI46406.

Xu, S. F., Adams, B., Yu, X. C., & Xu, M. (2013). Denosumab and giant cell tumour of bone-a review and future management considerations. *Current Oncology, 20*, e442–e447. https://doi.org/10.3747/co.20.1497.

Xu, R., Hu, J., Zhou, X., & Yang, Y. (2018). Heterotopic ossification: Mechanistic insights and clinical challenges. *Bone, 109*, 134–142. https://doi.org/10.1016/j.bone.2017.08.025.

Xu, R., Khan, S. K., Zhou, T., Gao, B., Zhou, Y., Zhou, X., et al. (2018). Gαs signaling controls intramembranous ossification during cranial bone development by regulating both Hedgehog and Wnt/β-catenin signaling. *Bone Research, 6*, 33.

Yu, K., et al. (2003). Conditional inactivation of FGF receptor 2 reveals an essential role for FGF signaling in the regulation of osteoblast function and bone growth. *Development, 130*, 3063–3074.

Yu, F. X., et al. (2012). Regulation of the Hippo-YAP pathway by G-protein-coupled receptor signaling. *Cell, 150*, 780–791. https://doi.org/10.1016/j.cell.2012.06.037.

Zaidi, S. K., et al. (2004). Tyrosine phosphorylation controls Runx2-mediated subnuclear targeting of YAP to repress transcription. *The EMBO Journal, 23*, 790–799. https://doi.org/10.1038/sj.emboj.7600073.

Zhao, B., Li, L., Lei, Q., & Guan, K. L. (2010). The Hippo-YAP pathway in organ size control and tumorigenesis: An updated version. *Genes & Development, 24*, 862–874. https://doi.org/10.1101/gad.1909210.

Zhao, X., et al. (2018). Expression of an active Galphas mutant in skeletal stem cells is sufficient and necessary for fibrous dysplasia initiation and maintenance. *Proceedings of the National Academy of Sciences of the United States of America, 115*, E428–E437. https://doi.org/10.1073/pnas.1713710115.

CHAPTER ELEVEN

Importance of the circadian clock in tendon development

Ching-Yan Chloé Yeung[a,b,*], Karl E. Kadler[c]
[a]Institute of Sports Medicine Copenhagen, Bispebjerg Hospital, Copenhagen, Denmark
[b]Center for Healthy Aging, University of Copenhagen, Copenhagen, Denmark
[c]Wellcome Centre for Cell-Matrix Research, Faculty of Biology, Medicine & Health, University of Manchester, Manchester Academic Health Science Centre, Manchester, United Kingdom
*Corresponding author: e-mail address: chloe.yeung@gmail.com

Contents

Abstract

Tendons are remarkable tissues that transmit force from muscle to bone during joint movement. They are remarkable because they withstand tensile forces that are orders of magnitude greater than can be withstood by isolated cells. The ability of the cells to survive is directly attributable to the stress shielding properties of the collagen-rich

ISSN 0070-2153
https://doi.org/10.1016/bs.ctdb.2018.11.004

extracellular matrix of the tissue. A further remarkable feature is that the vast majority (>98%) of the collagen is never turned over; it is synthesized during embryonic through early adult development and persists for the lifetime of the person. How the collagen is synthesized, and importantly, how it is protected from fatigue failure for decades of countless loading cycles, remains a mystery. A recent discovery is that tendons are peripheral circadian clock tissues in which the expression of ~5% of the transcriptome is rhythmic during 24h. Evidence is emerging that a fraction of the total amount of collagen is synthesized and removed on a daily basis without being incorporated into the lifelong permanent collagen. This review provides some of the background, and summarizes the findings, of these latest discoveries. Detailed descriptions of tendon development, collagen synthesis and collagen fibrillogenesis can be found in excellent reviews (cited here) and will not be a major part of this review.

1. Introduction

Tendons transmit forces from muscle to bone and their ability to perform this function is directly attributable to the organization and composition of their extracellular matrix (ECM). Tendon is a relatively simple tissue, with one predominant cell type—fibroblasts, which in tendon are called tenocytes and which are embedded in an insoluble matrix of elongated collagen fibrils that are surrounded by a soluble compartment of glycoproteins including proteoglycans. The collagen fibrils are arranged parallel in bundles that enable tendon to withstand high tensile forces (for extensive reviews on the tendon ECM and other cell types in tendon, please refer to reviews by Kjaer, 2004; Screen, Berk, Kadler, Ramirez, & Young, 2015). Tendon tissue development begins at embryonic day 12.5 in the mouse embryo. Force-transmitting and intermuscular tendons are derived from a population of scleraxis (encoded by *Scx*)-expressing progenitors derived from the syndetome (Brent, Schweitzer, & Tabin, 2003; Murchison et al., 2007). Mohawk (*Mkx*) expression in these progenitors regulates tendon differentiation by suppressing the expression of genes that drive chondrogenesis and osteogenesis (Ito et al., 2010; Liu et al., 2010; Suzuki et al., 2016). Expression of *Scx*, *Mkx* and early growth response 1 and 2 (*Egr1*, *Egr2*) promotes expression of tendon matrix ECM proteins, including type I collagen (*Col1a1*, *Col1a2*) and tenomodulin (*Tnmd*) (Guerquin et al., 2013; Lejard et al., 2011). However, formation of tendons in *Scx*−/−, *Mkx*−/−, *Egr1*−/− and *Egr2*−/− mice demonstrates that these factors are dispensable for tendon progenitor cell specification and collagen-I deposition, suggesting that other genes are required but these are currently unknown (reviewed by Subramanian & Schilling, 2015). Through specialized membrane structures called fibripositors tenocytes play an active role in the synthesis and assembly of the highly organized collagen type I-rich ECM containing uniform diameter (~30nm) fibrils (Canty et al., 2004, 2006; Kalson et al., 2013). After birth, cells release the fibrils that then begin the second phase of growth in length and width, taking on a bimodal distribution of diameters (Parry, Barnes, & Craig, 1978). Tenocytes that were organized on top of one another in the embryonic tendon increase their surface area through lateral protrusions. These protrusions connect to adjacent cells and are aligned longitudinally, forming channels that maintain the parallel alignment of fibril bundles (Kalson et al., 2015). This organized ECM undergoes repeated cycles of mechanical loading daily, which can be up to 70MPa in human tendons (Magnusson, Langberg, & Kjaer, 2010). Overloading of tendons is a key factor that leads to injuries and tendinopathies (Scott, Backman, & Speed, 2015); therefore, maintenance of the ECM in adult tendons is essential for tissue homeostasis in postnatal tendon development.

Tissue homeostasis is defined as the process of the maintenance of an internal steady state within a defined tissue of an organism, which includes control of cell numbers through regulating proliferation and cell death, and maintenance of ECM composition and turnover. Disruption to the balance of matrix synthesis and degradation in postnatal tendon may lead to deregulation of the cell and development of pathological conditions including fibrosis, ectopic calcification and impaired wound healing. Within the last few years, it has become apparent that ECM-rich tissues, including tendon and cartilage, are peripheral circadian clocks. And it is becoming evident that their endogenous 24-h rhythms play a major role in tissue-specific homeostasis and that aging and circadian disruptions increase the risk of musculoskeletal disorders (Bunger et al., 2005; Dudek et al., 2016; Gossan et al., 2013; Kondratov, Kondratova, Gorbacheva, Vykhovanets, & Antoch, 2006; Yeung et al., 2014). The goal of this chapter is to provide an overview of the mammalian circadian clock network and discuss what is currently known about the role of the molecular clock in tendon tissue homeostasis.

2. Mammalian circadian clock

The circadian clock is an evolutionarily conserved time-keeping mechanism that all but a few living organisms on Earth have developed to anticipate changes in physiological demands during a 24-h day. The features of the clock are to sustain a sufficient oscillation amplitude through the circadian cycle, compose a phase that is properly aligned with the light-dark cycle, and be entrain able by light to maintain a ~24-h period (reviewed by Bass & Takahashi, 2010; Welsh, Takahashi, & Kay, 2010). In mammals, the circadian rhythm regulates crucial homeostatic processes, including feeding, metabolism, sleep and arousal, hormone secretion, body temperature, and waste elimination. The identification of "clock genes" enabled the subsequent discovery of cell autonomous clocks in peripheral tissues, which are entrained by rhythmic signals, or "zeitgebers," that include feeding, temperature, and social cues.

2.1 "Master" clock

In mammals the circadian rhythm is driven by a highly conserved and specialized region of the brain located in the anterior hypothalamus called the suprachiasmatic nuclei (SCN) (Cassone, Speh, Card, & Moore, 1988). The SCN is an essential timekeeper for behavioral rhythmicity and is the most robust molecular clock in the body. The robustness (large amplitude and ability to sustain a rhythm over a long period of time) of the SCN rhythm allows animals to preserve their endogenous behavioral rhythm in the absence of environmental light cues. The very first circadian studies showed that rodents were able to maintain near 24-h behavioral and gene expression rhythms when kept in total darkness for long periods (Ebihara, Tsuji, & Kondo, 1978; Stephan, 1983). The importance of the SCN in driving circadian rhythms was established in studies where surgical ablation of the SCN led to loss of behavioral rhythms (Ibuka, Inouye, & Kawamura, 1977; Ibuka, Nihonmatsu, & Sekiguchi, 1980; Mosko & Moore, 1979; Welsh, Richardson, & Dement, 1988), which were restored by SCN transplantation, where the circadian characteristics of the donor dictated period length (Lehman et al., 1987; Ralph, Foster, Davis, & Menaker, 1990).

The SCN receives light information from intrinsically photosensitive ganglion cells of the retina via the retino-hypothalamic tract, which is essential for light-entrainment of the SCN (Brzezinski et al., 2005; Guler et al., 2008). There are approximately 20,000 neurons of the SCN and each contains an autonomous clock mechanism. Coupling of SCN neurons enable the SCN to produce a high amplitude circadian rhythm that is synchronous across the whole tissue; however, the mechanisms of neuronal coupling are still unclear. SCN neurons are electrically coupled together via connexin

36-containing gap junctions, which when knocked out disrupted electrical synapses and caused disruption to circadian behavioral rhythms in juvenile mice (17- to 23-days old) (Long, Jutras, Connors, & Burwell, 2005). Conversely, a very recent study demonstrated that, in fact, connexin 36-null SCNs could maintain protein oscillations but they exhibited a longer period, and that connexin 36 was not necessary for behavioral rhythms in older knockout mice (9- to 30-weeks old) (Diemer et al., 2017). Therefore, other mechanisms of coupling (e.g., synaptic communication) may be responsible for the synchrony of SCN neurons (reviewed by Welsh et al., 2010).

2.2 Cell autonomous molecular oscillator

The cell-autonomous clock mechanism is an auto-regulatory transcription-translation feedback loop (TTFL) (Fig. 1). All the molecular components of the circadian clock need to be expressed in the correct phasing to produce a period near 24 h (Mirsky, Liu, Welsh, Kay, & Doyle, 2009). The TTFL is driven by two transcription factors, BMAL1 (brain and muscle ARNT-like 1, encoded by *Arntl1*) and CLOCK (circadian locomotor output cycles kaput, encoded by *Clock*) that dimerize to initiate the expression of *Period* (*Per1*, *Per2* and *Per3*) and *Cryptochrome* (*Cry1*, *Cry2*). PER and CRY proteins accumulate and assemble into heterodimers, become phosphorylated, and then translocate into the nucleus to bind and inhibit BMAL1/CLOCK (Yagita et al., 2000). Proteosomal degradation of ubiquitinated PER and CRY then allows the TTFL to begin again (Keesler et al., 2000; Lowrey et al., 2000; Yoo et al., 2013). A stabilizing loop controls the temporal expression of *Bmal1* and *Clock*. ROR (α, β and γ, encoded by *Nr1f1*, *Nr1f2* and *Nr1f3*) positively activates *Bmal1* and *Clock* via RORE elements in their promoters and REV-ERB (α and β, encoded by *Nr1d1* and *Nr1d2*) competes for RORE element binding to repress *Bmal1* and *Clock* expression (Bell-Pedersen et al., 2005; Harding & Lazar, 1993; Zhang et al., 2015). There is an additional negative feedback loop that involves CHRONO (ChIP-derived repressor of network oscillator or computationally highlighted repressor of network oscillator; encoded by *Chrono* aka *Gm129*) (Anafi et al., 2014; Goriki et al., 2014). CHRONO binds to histone deacetylase and behaves as a transcriptional repressor of circadian gene promoters (Goriki et al., 2014). The completion of the TTFL takes ~24 h and this period is tightly controlled by post-translational and epigenetic modifications of the core clock components.

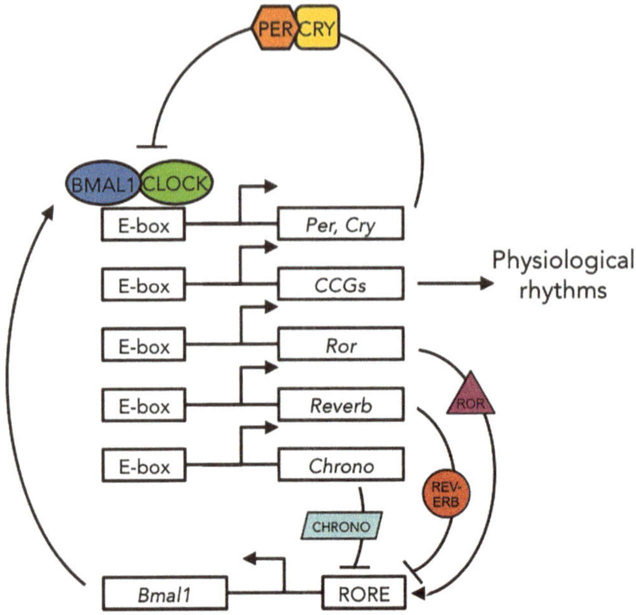

Fig. 1 The circadian transcription-translation feedback loop. The molecular clock is auto-regulated by a transcription-translation feedback loop (TTFL) that takes ~24 h to complete. BMAL1/CLOCK activates the transcription of genes containing E-box motifs, which include *Per* and *Cry* genes. PER/CRY complexes then translocate to the nucleus to inhibit BMAL1/CLOCK, forming the core TTFL. Other E-box genes that are also activated by BMAL1/CLOCK and of these, ROR and REVERB regulate *Bmal1* activation via RORE elements in promoter, which forms the stabilizing loop. CHRONO was recently discovered as a negative regulator of *Bmal1* by regulating DNA conformation of its promoter. *Adapted from Husse, J., Eichele, G., & Oster, H. (2015). Synchronization of the mammalian circadian timing system: Light can control peripheral clocks independently of the SCN clock: Alternate routes of entrainment optimize the alignment of the body's circadian clock network with external time.* BioEssays: News and Reviews in Molecular, Cellular and Developmental Biology, 37, *1119–1128.*

3. Peripheral clocks

The output of the molecular pacemaker is the activation of E-box motif- or E-box motif-like-containing genes, termed "clock-controlled genes" (CCGs) (Munoz & Baler, 2003). The E-box sequence CACGTG is a core *cis*-element for the circadian regulation of transcription and is recognized by the basic helix-loop-helix family of transcription factors, which include BMAL1 and CLOCK. While the molecular mechanism of the circadian clock in SCN neurons is shared among tissues (Dibner, Schibler, &

Albrecht, 2010), the rhythmic output of CCGs is highly tissue-specific but the processes that govern this specificity are not well characterized (Korencic et al., 2014; Storch et al., 2002; Yan, Wang, Liu, & Shao, 2008; Zhang, Lahens, Ballance, Hughes, & Hogenesch, 2014). Tissue-specific clock outputs are likely a result of interaction with tissue-specific factors (e.g., transcription factors). In this section we will describe what is known about the mechanisms regulating tissue-specific clock outputs, how peripheral clock entrainment is mediated and what the consequences of aging on the peripheral clock are.

3.1 Tissue-specificity of peripheral clocks

Clock outputs make up to ~15% of all transcripts in a tissue (Zhang et al., 2014). Expression of CCGs occurs in phases rather than "all on" in the day and "all off" at night, supporting the idea that additional mechanisms are involved in activation of CCGs (Fang et al., 2014). For example, chromatin confirmation of CCGs and the activity of tissue-specific transcription factors that recruit core clock components to regulatory sequences of the CCGs are known to regulate tissue-specificity of clock outputs (Koike et al., 2012; Yeung, Mermet, et al., 2018). In addition to differences in which CCGs are expressed, there is also variation in the amplitude of CCG expression across different tissues. For example, metabolically active tissues (liver and muscle) have roughly 100 transcripts that exhibit peak-trough differences between 2- and 10-folds, whereas in brain, there are no CCG transcripts with peak-trough differences greater than 4-folds (Yeung, Mermet, et al., 2018). It is unclear why these differences in CCG amplitudes exist.

The functional and relevant output of gene expression is the protein and accordingly transcriptomics only reveals a portion of circadian-regulated processes. First, there is post-transcriptional regulation, where the rhythmicity of CCG transcripts is driven not only by rhythmic transcription but also rhythmic degradation. For example, in liver, 20% of rhythmic transcripts were found to be regulated by rhythmic degradation alone (Wang et al., 2018). Interestingly this rhythmic transcription and degradation of CCG transcripts in liver does not require BMAL1 but is dependent on the tissue's own zeitgeber—timed feeding (Wang et al., 2018). These data suggest that tissue-specific entrainment signals contribute to tissue-specificity of circadian outputs at various post-transcriptional levels. Second, phase and half-life of protein oscillations are highly regulated by the circadian clock and is highly tissue-specific. For example, timed feeding is also responsible for protein oscillations in liver, which is mediated via regulation of translation efficiency of CCG transcripts (Atger et al., 2015). Further, differences in translational efficiencies between circadian transcriptomes of different peripheral clock tissues have also been demonstrated to contribute to the identities, the phases and the levels of rhythmic protein biosynthesis (Castelo-Szekely, Arpat, Janich, & Gatfield, 2017). Third, there are proteins that oscillate with a 24-h period that do not have corresponding oscillations in mRNA expression, e.g., collagen-I in tendon (discussed below). Comparison of the circadian transcriptome and circadian proteome in liver revealed that half of all oscillating proteins do not have rhythmic transcripts (Reddy et al., 2006; Robles, Cox, & Mann, 2014). The mechanisms that regulate the rhythmic synthesis of this subset of oscillating proteins are not entirely understood but rhythmic degradation of mRNA and protein of these non-rhythmic transcripts is known to contribute to the phase and amplitude of protein oscillations (Luck, Thurley, Thaben, & Westermark, 2014). These data highlight the complexity in peripheral clock outputs and the need for more in-depth studies to elucidate the mechanisms underlying tissue-specificity of these outputs at both the mRNA and protein levels.

3.2 Peripheral clock entrainment

SCN outputs, including neurotransmitter waves, melatonin and glucocorticoid release help to synchronize metabolic processes in peripheral tissues. Therefore, the "master clock" was assumed to coordinate the peripheral clocks in a hierarchical manner, whereby only the SCN receives zeitgeber (light) and is responsible for the alignment of all peripheral tissues clocks (Dibner et al., 2010; Hastings, O'Neill, & Maywood, 2007; McNamara et al., 2001). However, it is now clear that non-light zeitgebers, including timed feeding and exercise, can override the SCN and synchronize peripheral clocks. This phenomenon is most obvious during jet lag, which causes a temporary disruption of the sleep-wake cycle. The hierarchical model of peripheral clock entrainment came from mouse studies that used *Per1* promoter-driven expression of luciferase (*Per1-LUC*) or green fluorescent protein (*Per1-GFP*) reporters, which showed that the SCN was able to maintain robust, self-sustained oscillations whereas peripheral tissues showed dampening after 2–7 cycles (Abe et al., 2002; Kuhlman, Quintero, & McMahon, 2000; Yamaguchi et al., 2000; Yamazaki et al., 2000). Luciferase expression driven by a *Per* promoter only reports the transcription activation of the *Per* gene. A few years later, Yoo and colleagues created a new circadian reporter mouse where the *Luciferase* gene was fused in-frame to the 3′ end of the *Per2* gene, encoding a PER2::LUC fusion protein (2004). This fusion protein is able to undergo endogenous post-translational modifications that regulate protein stability essential for producing a correct period length, so this reporter permits the monitoring of endogenous PER2 protein oscillations (Nishii et al., 2006). Using the *PER2::LUC* mouse model, the authors observed that some peripheral tissues were actually able to sustain a persistent circadian rhythm for more than 20 days (Yoo et al., 2004), which was controversial to the hierarchical model.

Studies performed on mice with SCN lesions or *Bmal1*-deficient SCNs showed that peripheral clocks maintained circadian rhythms independent of entrainment signals from a functional master clock (Husse, Leliavski, Tsang, Oster, & Eichele, 2014; Saini et al., 2013). Entrainment of different peripheral clocks requires specific zeitgebers, for example, scheduled exercise entrains skeletal muscle and lung clocks (Sasaki et al., 2016; Wolff & Esser, 2012), and timed feeding entrains liver and kidney clocks (Damiola et al., 2000). However, the SCN is not entirely dispensable; it is required for stabilizing the phase of peripheral clocks and is important for the amplitude of tissue oscillations (Yeung, Mermet, et al., 2018). It is now understood that the role of non-light zeitgebers is to help integrate complex periodic changes from the organism's environment into the circadian system, making the network more robust. Consequently, this integration of external non-light zeitgebers causes the overall circadian network slow to adapt to changes such as jet lag, but it allows for noise in the environment and prevents unwanted phase shifts. This system also allows for adaptation in sustained zeitgebers (e.g., food availability) to elicit appropriate tissue responses (reviewed by Husse, Eichele, & Oster, 2015). There is a real need in circadian research to better understand the connectivity between tissue clocks because misalignment of peripheral clocks (e.g., in aging and circadian disorders) is a key to unraveling the mechanisms linking the circadian clock and health (Bass, 2017; Roenneberg & Merrow, 2016).

3.3 Aging of peripheral clocks

Mouse models for mutations in core clock genes have premature aging phenotypes, suggesting that age-related tissue homeostasis insufficiencies are related to a decline in the circadian clock (reviewed by Yu & Weaver, 2011). Interestingly, aged mice are more susceptible to circadian challenges (e.g., jet lag) and the circadian rhythms and outputs of peripheral clocks, including ECM-rich tissues, are dampened (Davidson et al., 2006; Sellix et al., 2012). It is unclear why this dampening occurs but possible factors could be reduced endocrine function, peripheral clock phase alignment and sleep (Gibson, Williams, & Kriegsfeld, 2009; Scarbrough, Losee-Olson, Wallen, & Turek, 1997; Valentinuzzi, Scarbrough, Takahashi, & Turek, 1997).

The first suspect of causing circadian decline would be the SCN because in spite of maintaining a robust circadian rhythm, there is a decline in SCN outputs, including gene expression, electrical activity and neuropeptide synthesis (reviewed by Bedont & Blackshaw, 2015). As discussed above, the precise mechanisms that link peripheral clocks to the SCN remain largely obscure but there are data to suggest that peripheral clock entrainment by the SCN is impaired in aging. Similar to SCN ablation studies where entrainment of peripheral clocks to non-light zeitgebers was faster than controls (Husse et al., 2014; Saini et al., 2013), timed feeding entrainment of peripheral clocks was faster in aged mice than in young mice, and peripheral clocks were more susceptible to a 6-h interval feeding schedule, which superseded 12-h light/12-h dark (LD) entrainment and abolished circadian rhythms in aged tissues (Tahara et al., 2017). When food was available *ad libitum* in LD there were no differences in amplitude and phase alignment in peripheral clocks (kidney, liver and submandibular gland) in older mice (>18 months) compared to young mice (3–6 months), but aged mice did show weaker entrainment to stress- and exercise-induced entrainment signals that act via glucocorticoid receptors (Tahara et al., 2017). These data suggest that in aging there is uncoupling between SCN-dependent entrainment and some non-light zeitgeber signals in the circadian network, both of which are required for alignment of peripheral clocks.

4. Circadian clock regulation of tendon homeostasis

The tendon ECM is predominantly made up of type I collagen but also comprises other collagens (including type III, V, XI, XII, and XIV), proteoglycans (including biglycan, fibromodulin, lumican, decorin, versican, and aggrecan), glycoproteins (tenomodulin, lubricin, tenascin-C, cartilage oligomeric matrix protein), elastic fibers (elastin, fibrillin, fibulin), growth factors and proteinases including BMP1/tolloids, ADAMTSs (a disintegrin and metalloproteinase with thrombospondin motifs) and matrix metalloproteinases (MMPs). Collagen biosynthesis is tightly regulated and disruption to any step of its biogenesis can impact on the mechanical and biochemical properties of the ECM, which is most obvious in aging (reviewed by Phillip, Aifuwa, Walston, & Wirtz, 2015). Research into the tendon circadian clock is still in its infancy. Recent works on the tendon clock have revealed a role in ectopic calcification (Yeung et al., 2014), endoplasmic reticulum (ER) homeostasis (Pickard et al., 2018 preprint), and collagen-I synthesis and secretion (Yeung, Garva, et al., 2018 preprint) and therefore this section will largely focus on these topics. Fig. 2 summarizes what is currently known in the regulation of tissue homeostasis by the tendon clock.

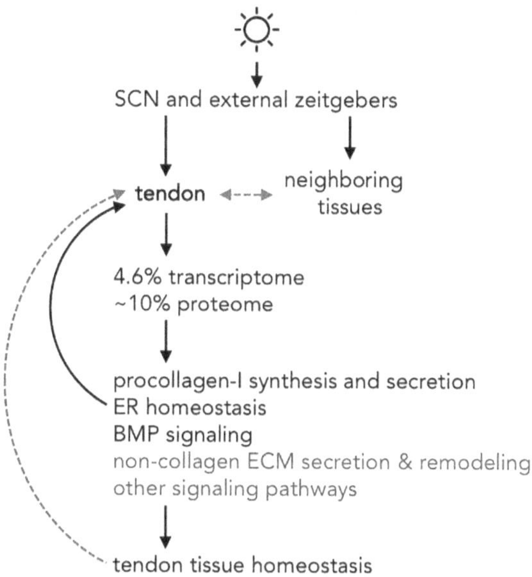

Fig. 2 The circadian control of tissue homeostasis in postnatal tendon. A combination of possible SCN outputs and extra-SCN zeitgebers (feeding, physical activity, glucocorticoid release), and potential inter-tissue signals entrains the tendon clock. The combination of autonomous circadian rhythm and tissue-specific factors results in the rhythmic oscillation of 4.6% of the tendon transcriptome. In addition to gene expression, the circadian clock regulates post-transcriptional and post-translational events that produce oscillations in ~10% of the tendon proteome. These tendon clock outputs coordinate cellular processes including procollagen-I synthesis and secretion (Yeung, Garva, et al., 2018 preprint), which may also impact on secretion of other ECM molecules, ECM remodeling, signaling via the BMP pathway (Yeung et al., 2014) and potentially through other pathways, and ER homeostasis (Pickard et al., 2018 preprint) with the rest-activity cycle and drives tissue homeostasis. ER homeostasis feedbacks on the tendon clock, the extracellular environment can also potentially modulate tendon clock outputs.

4.1 Tendon circadian transcriptome

Using circadian reporter mice, *PER2::LUC*, *Per2-YFP* (yellow fluorescent protein), circadian rhythms were observed in tail and Achilles tendons of wild type mice for over a week *ex vivo* but not in tendons from *ClockΔ19* mice (Lande-Diner, Stewart-Ornstein, Weitz, & Lahav, 2015; Yeung et al., 2014; see Movie 1 in the online version at https://doi.org/10.1016/bs.ctdb.2018.11.004). Endogenous circadian rhythms were also demonstrated in primary human tendon cell cultures via *Per2-LUC* and *Bmal1-LUC* reporters (Yeung et al., 2014). Similar to other peripheral tissue clocks, circadian rhythms in tendons *ex vivo* dampen over time. Analysis of single cells from *Per2-YFP* tail tendons showed that the amplitude of individual oscillators was preserved and its variability did not increase with time (Lande-Diner et al., 2015), and dampened tendon circadian rhythm *ex vivo* could be reinitiated by glucocorticoid treatment (dexamethasone) (Yeung et al., 2014). Therefore in the absence of daily systemic entrainment signals, the dampening of the circadian rhythm in *ex vivo* tendons is due to uncoupling of the individual cells in the population rather than a loss in cell intrinsic rhythms.

The use of microarrays for high-throughput gene expression analysis has allowed the identification of tissue-specific circadian transcriptomes. Constant darkness, where animals are housed in 12-h dark/12-h dark (DD) cycles with free access to food and water rather than the artificial laboratory setting of 12-h light/12-h dark cycles (LD), is considered a "free-running" circadian state that is free from any zeitgebers and usually the period of a free-run circadian rhythm is slightly deviated from exactly 24 h. A 3′ microarray performed on tail tendons harvested from free-running mice, every 4 h for 48 h revealed 4.6% of the tendon transcriptome (745 genes) to be rhythmically expressed with a ~24-h period (Yeung et al., 2014). With the knowledge that there are genes that oscillate at the protein level, independent of mRNA oscillations, a proteomics approach was used in a second study, which identified 141 proteins (10% of proteins identified) as robustly rhythmic with a ~24-h period in tendons of mice kept in LD (Yeung, Garva, et al., 2018 preprint). When compared to CCGs of other musculoskeletal tissues and other ECM-rich tissues there was very little overlap except for core clock genes, which confirmed that circadian regulation of gene expression in tendon is highly tissue-specific (Dudek & Meng, 2014; Dudek et al., 2017).

4.2 Collagen synthesis

Patellar tendons in humans were found to be stiffer in the morning and able to elongate more in the evening, which affected muscle force generation (Pearson & Onambele, 2005, 2006), suggesting there are diurnal variations in the collagen-rich ECM. Collagen synthesis shows diurnal rhythms in some ECM-rich tissues, including the growth plate (Igarashia, Saekia, & Shinodab, 2013), and bone (Hassager, Risteli, Risteli, Jensen, & Christiansen, 1992; Russell, Walker, Fenster, & Simmons, 1985), and oscillations in *Col1a1* gene expression in osteoblasts *in vitro* is responsive to glucocorticoid synchronization (Fujihara, Kondo, Noguchi, & Togari, 2014; Komoto, Kondo, Fukuta, & Togari, 2012). Expression of type II collagen (*Col2a1*) is also rhythmic in growth plates and in rib cartilage in rats (Honda et al., 2013), in mouse hip cartilage (Dudek et al., 2016) and xiphoid cartilage (Gossan et al., 2013). In mouse tendons transcription of collagen genes was not rhythmic but peptides corresponding to the collagen α1(I) and α2(I) chains exhibited prominent diurnal rhythms that peaked in the rest phase (7–11 h into the light of LD cycle) (Yeung, Garva, et al., 2018 preprint). The levels of C-propeptide of the pro-collagen α1(I) chain also oscillated with a ~24-h period and peaked ~4 h prior to the α1(I) and α2 (I) collagen peptides, suggesting collagen-I production is regulated by the tendon clock at a post-transcriptional level. In line with this idea, targeting collagen transcripts to the ER for translation was found to be regulated by the tendon clock. The 5′ untranslated region of *Col1a1* and *Col1a2* transcripts bind to LARP6 (La ribonucleoprotein domain family member 6), which targets them to SEC61 ER membrane translocator protein complex for transport into the ER (Stefanovic, Longo, Zhang, & Stefanovic, 2014). *Sec61a2* encodes one of the subunits of SEC61 and its mRNA and protein expression is rhythmic in tendon and is required for procollagen-I entry into the ER and secretion by tendon fibroblasts (Yeung, Garva, et al., 2018 preprint).

4.3 Collagen post-translational modification, folding and secretion

The correct folding of the three polypeptide chains of procollagen into a thermally stable triple helix capable of assembling into fibrils is fundamental to tissue function. Folding of procollagen-I begins with association of the

C-propeptides of two pro-α1(I) chains and one pro-α2(I) chain and proceeds in a zipper-like action toward the N-terminus of the molecule (reviewed by Canty & Kadler, 2002). The ER chaperones that are required for folding of the globular C-propeptide are calnexin, calreticulin, protein disulfide isomerase, GRP94 (heat shock protein 90 kDa member1, encoded by *Hsp80b1*) and BiP (binding immunoglobulin protein aka GRP78, encoded by *Hspa5*). There are also collagen-specific chaperones that are unique among molecular chaperones in that they bind preferentially to the folded triple helix. These include collagen prolyl-4-hydroxylase, heat shock protein 47 (HSP47), FKBP56 (65-kDa FK506-binding protein), and the P3H1 (prolyl-3-hydroxylase 1)/CRTAP (cartilage-associated protein)/CYBP (calcyclin-binding protein) complex, which stabilize the collagen triple helix (reviewed by Makareeva, Aviles, & Leikin, 2011). Protein folding machinery in the mouse liver is regulated in circadian manner with many chaperones called heat shock proteins (HSP90, HSP110, HSP70, HSP40) that oscillate at the protein level with a 24-h rhythm (Robles et al., 2014). In tendon *Hsp70* expression but no other transcripts encoding HSPs are rhythmic (Yeung et al., 2014).

Accumulation of misfolded collagen leads to ER stress (reviewed by Boot-Handford & Briggs, 2010; Lamande & Bateman, 1999). ER stress is regulated by the ER-resident chaperone BiP that is closely related to HSP70 (Wooden & Lee, 1992). BiP senses misfolded proteins in the ER and triggers the unfolded protein response (Lee, 2005). BiP protein levels are rhythmic in tendon and mouse embryonic fibroblasts, and peak preemptively ahead of the peak in procollagen-I protein levels (Pickard et al., 2018 preprint). Drug-induced ER stress blocked procollagen-I secretion and surprisingly, dampened PER2::LUC reporter rhythms. Short pulse treatments with ER stress inducers or inhibitors of protein secretion disrupted circadian oscillations and were able to inhibit dexamethasone-induced collagen secretion (Pickard et al., 2018 preprint). These data suggest that a circadian rhythm is required for collagen secretion and that induction of ER stress via protein retention negatively feeds back onto the tendon molecular clock.

HSP47 is a collagen-specific chaperone that transiently associates with the folded collagen triple helix from the ER to the *cis*-Golgi or to the ER to Golgi intermediate compartment (ERGIC) (Ishida & Nagata, 2011; Makareeva & Leikin, 2007). HSP47 contains an ER retrieval sequence that is able to activate KDEL receptors (endoplasmic reticulum protein retention receptor 1) (Satoh, Hirayoshi, Yokota, Hosokawa, &

Nagata, 1996). These are located in the Golgi and upon activation of phosphodiesterases (PDE) mediate the transport of HSP47 back to the ER (Cancino et al., 2014). In tendon, *Pde4d* is rhythmic and its deletion in fibroblasts caused HSP47 retention in the Golgi and inhibited procollagen-I secretion (Yeung, Garva, et al., 2018 preprint).

It has been suggested that procollagen is too large to fit into conventional 60–90 nm diameter coat protein complex II (COPII) vesicles and requires TANGO1 (transport and Golgi organization protein 1, encoded by *Mia3*) to assist in the targeting of procollagen to large ER exit sites (Malhotra & Erlmann, 2011). Very recently, it was shown in primary human fibroblasts that procollagen-I transfer from the ER to the *cis*-Golgi can occur independent of vesicle trafficking and instead, procollagen-I at ER exit sites are coated with COPII and matures to form the ERGIC (McCaughey, Stevenson, Cross, & Stephens, 2018 preprint). TANGO1 knockout mice have impairment in the efficient secretion of all collagens type I, II, III, IV, VII, and IX (Saito et al., 2009; Wilson et al., 2011). TANGO1 interacts with HSP47 enabling it to mediate the packaging of all collagens (Ishikawa, Ito, Nagata, Sakai, & Bachinger, 2016). *Mia3* transcript is rhythmic in tendon, and unsurprisingly knockdown of *Mia3* in tendon fibroblasts impaired procollagen-I secretion (Yeung, Garva, et al., 2018 preprint). The importance of tendon clock-regulated TANGO1 may also extend to the secretion of other ECM molecules. Small and larger ECM molecules, including cartilage oligomeric matrix protein (COMP), can piggy back on to the collagen-containing COPII vesicles (Ishikawa et al., 2016; Rios-Barrera, Sigurbjornsdottir, Baer, & Leptin, 2017). TANGO1 also plays a role in ER homeostasis, where loss of TANGO1 perturbs ER-Golgi morphology independent of large cargo, and induces ER stress in the presence of bulky cargo (collagen) (Maiers et al., 2017; Rios-Barrera et al., 2017). In tendon peak TANGO1 levels coincide with peak levels of pro-collagen $\alpha 1$(I) peptides, and TANGO1 may act in concert with BiP to regulate procollagen-I trafficking and ER homeostasis.

Post-translational modification of procollagen-I appears to also be regulated by the circadian clock in tendons. VPS33B (late endosome and lysosome associated, encoded by *Vps33b*) and VIPAR (VPS33B interacting protein, apical-basolateral polarity regulator) regulate the trafficking of lysyl hydroxylase 3, which is essential for collagen crosslinking and homeostasis (Banushi et al., 2016; Sricholpech et al., 2012). Inducible deletion of either *Vps33b* or *Vipas39* in mice resulted in abnormal collagen fibril structure (Banushi et al., 2016). *Vps33b* is rhythmic in tendon and

CRISPR/Cas9-mediated knockout in tendon fibroblasts prevented the rhythmic secretion of procollagen-I (Yeung, Garva, et al., 2018 preprint).

The procollagen-I molecule then undergoes processing whereby the N- and C-terminal flanking propeptides are cleaved by N- and C-proteinases producing a 300 nm-long collagen molecule. Collagen molecules then assemble into fibrils and undergo crosslinking mediated by lysyl oxidases (LOXs) that stabilize inter-molecular interactions and is essential for the development of mechanical properties of collagen-rich matrices (reviewed by Eyre, Paz, & Gallop, 1984). *Loxl4*, which encodes lysyl oxidase-like 4, a LOX homologue, is rhythmic in tendon, suggesting that the circadian clock might also regulate this step of post-translational modification of collagen. Taken together, these data strongly implicate the tendon circadian clock in regulating the rhythmic production and secretion of procollagen-I that may also cause secretion of other ECM molecules to be rhythmic, and that this process is tightly regulated to prevent ER stress, which can disrupt the endogenous circadian rhythm.

4.4 ECM remodeling

Tendon undergoes dynamic turnover especially after loading, where there is synthesis of new ECM and degradation of old or damaged ECM (reviewed by Magnusson et al., 2010). Tendons from mice show time-of-day differences in viscoelastic properties during cyclic loading suggesting that there are diurnal variations in the biochemical properties of the ECM, which may be optimized to the rest-activity cycle of the mouse (Yeung, Garva, et al., 2018 preprint). Fibroblasts use ECM receptors called integrins and the actin-myosin machinery to assemble or physically pull on and remodel fibrillar ECM (reviewed by Humphrey, Dufresne, & Schwartz, 2014; Kadler, Hill, & Canty-Laird, 2008; Schwartz, 2010). The intrinsic clock in dermal fibroblasts controls actin dynamics and generates a circadian rhythm in the ratio between filamentous and globular actin (Hoyle et al., 2017). Fibroblasts isolated from tissues retain circadian oscillations in culture and can be synchronized by dexamethasone treatment, serum shock or temperature entrainment (Balsalobre et al., 2000; Balsalobre, Damiola, & Schibler, 1998; Brown, Zumbrunn, Fleury-Olela, Preitner, & Schibler, 2002). In scratch assays, dermal fibroblasts wounded at various time points after synchronization exhibited varying extents of wound closure, where the wounding time for efficient healing was 20–24, or 44–48 h post-synchronization, coinciding with peak PER2::LUC expression time (Hoyle et al., 2017). Time-dependent optimal healing was also observed *in vivo*, where full thickness wounds made in mouse skin during their active phase showed greater collagen deposition when examined after 14 days than wounds made during the rest period. The data from this study demonstrate that adaptation of peripheral clock alignment with the rest-activity cycle is crucial and that challenges that are not time-optimized may have long-term detrimental effects.

Deletion of *Clock* does not affect the circadian clock due to functional substitution from NPAS2 (neuronal PAS domain protein 2) (Debruyne et al., 2006; DeBruyne, Weaver, & Reppert, 2007). The mutant *ClockΔ19* allele harbor a deletion in exon 19 of the *Clock* gene producing a dominant negative mutant protein and *ClockΔ19* mice do not display circadian or behavioral rhythms (Vitaterna et al., 1994). Tendons from *ClockΔ19* mice and tendon-specific, *Scx*-driven *Bmal1* knockout mice were fibrotic and had abnormal collagen fibril structures, diameter distributions and poorer mechanical properties compared to wild type (Yeung, Garva, et al., 2018 preprint). These data unequivocally demonstrate that the tendon clock plays a critical role in regulating tendon ECM and the mechanisms involve regulation of collagen secretion (discussed above), but also ECM degradation. Cornea is another collagen-rich peripheral clock tissue containing an ECM of very narrow fibrils arranged in orthogonal layers, which forms a tough, light-permissible tissue. Similar to tendons, loss of *Bmal1* caused thickening of cornea tissue (Baba & Tosini, 2018; Kondratov et al., 2006; Yang et al., 2016), further supporting the clock's role in collagen turnover. Time-series microarray analyses of collagen-II-rich cartilage tissues revealed oscillations in many extracellular proteases and their regulators, which are involved in ECM turnover (*Adamts4*, *Adamts9*, *Mmp14*, *Timp4*) (Dudek et al., 2016; Gossan et al., 2013). Transporters of the solute carrier (SLC) family are implicated in the transport of degraded ECM, a proposed mechanism by which tumor cells utilize the collagen in the ECM as a source of amino acids (Olivares et al., 2017). In tendon tissue transcripts for *Mmp11*, *Mmp14*, *Adam17*, *Adam19*, *Adamts4* and *Adamts20*, and 15 different members of the SLC family are rhythmic (Yeung et al., 2014), and therefore tendon fibrosis caused by an absence of a functional circadian clock could be a result of deregulated proteinases preventing efficient turnover of the ECM.

4.5 Ectopic calcification

The circadian clock was first suspected of regulating tendon homeostasis because *Bmal1*-null mice had ectopic calcification in tendons (Bunger et al., 2005), which was also present when *Bmal1* expression was rescued in the SCN or muscle, suggesting that there was a tissue-specific function for BMAL1 or the circadian clock in tendon (McDearmon et al., 2006). Evidence of ectopic tendon calcification in both *ClockΔ19* and *Scx-Cre;Bmal1*$^{f/f}$ mice as early as 18 weeks of age confirmed that calcification was the result of a deregulated tendon clock (Yeung, Garva, et al., 2018 preprint; Yeung et al., 2014). The osteogenic potential of tendon-derived cells, including stem cells, is well characterized (Agarwal et al., 2017; Bi et al., 2007; Cadby, Buehler, Godbout, van Weeren, & Snedeker, 2014; de Mos et al., 2007; Rui et al., 2010; Salingcarnboriboon et al., 2003). A bone morphogenetic protein (BMP) inhibitor under the regulation of the tendon clock called gremlin-2 (encoded by *Grem2*) was found to regulate BMP-SMAD1/5 signaling in mouse tendons, where phosphorylation of SMAD1/5 and activation of BMP target genes were gated during the rest phase. Addition of recombinant gremlin-2 to primary human tendon fibroblasts cultures reduced calcium deposition induced by osteogenic medium and reduced the extent of SMAD1/5 phosphorylation induced by recombinant BMP2 (Yeung et al., 2014). BMP signaling plays a physiological role in regulating tendon ECM and is required for the early phases of tendon healing (Chhabra et al., 2003; Clark et al., 2001; Mikic, Schalet, Clark, Gaschen, & Hunziker, 2001). It is plausible that tendon clock-regulated gating of BMP signaling could be important in attenuating its signal transduction in tendon cells or in aligning the signaling pathway with time of activity to prevent BMP-induced osteogenesis.

4.6 mTOR signaling

In addition to BMP signaling, postnatal tendon development is regulated by a number of signaling pathways but these are not well understood. The mTOR (mechanistic target of the rapamycin, encoded by *Mtor*) pathway is a master regulator of cell metabolism and cell growth and its expression is under circadian control in tendon (Yeung et al., 2014). Inhibition of mTOR complex 1 (mTORC1) by rapamycin has been shown to attenuate the aging effects in tendon (increased stiffness, calcification, cell density, fibrocartilage development) (Wilkinson et al., 2012; Zaseck, Miller, & Brooks, 2016). Data from *Scx-Cre*-driven knockout mouse models of *Rptor* (raptor) and *Tsc1* (tuberous sclerosis complex 1), which inhibits and activates mTOR signaling, respectively, suggested that the regulation of postnatal tendon maturation by mTOR is more complex. Both activation and inactivation of the pathway prevented lateral growth of collagen fibrils, which was evident in 1-month-old mice. *Tsc1*-deficient tendons had increased vascularization and increased proliferation; however, *Rptor*-deficient tendons were thinner, contained smaller bundles and ectopic fibrocartilage (Lim et al., 2017). A caveat of these mouse models is that mTOR signaling was modulated from embryonic tendon development onward, so it is difficult to dissect the phenotypes resulting exclusively from mTOR signaling in postnatal tendon. Interestingly, mTOR signaling is required for a robust circadian rhythm in the SCN and peripheral clocks via regulating protein expression of core clock genes (Cao, Li, Cho, Lee, & Obrietan, 2010; Ramanathan et al., 2018). Together, these data suggest that rhythmic *Mtor* expression in tendon has the potential to regulate tissue homeostasis via mTOR signaling and/or via its role as a circadian timekeeper.

4.7 TGFβ signaling

Transforming growth factor β (TGFβ) signaling plays an important role in embryonic tendon development by promoting the *Scx*-expressing tendon progenitor cell fate, but how it regulates postnatal tendon is largely unknown (Havis et al., 2016; Kuo, Petersen, & Tuan, 2008; Pryce et al., 2009). TGFβ signaling is mediated by three isoform ligands TGFβ1, 2 and 3 and two receptors TGFβR1 and 2, and its downstream activation is mediated via SMAD2/3 and non-canonically via mitogen-activated protein kinases (MAPKs) (Massague, 2012). TGFβ ligands are synthesized as precursors that homodimerize and become covalently associated with latent TGFβ-binding protein (LTBP, of which there are four isoforms 1–4) and is then secreted and sequestered in the ECM by transglutaminase crosslinking (Nunes, Gleizes, Metz, & Rifkin, 1997; Rifkin, 2005). TGFβ release from LTBP can be proteolytic (e.g., MMPs) and non-proteolytic (e.g., pH changes, mechanical forces) or a combination of both (see reviews by Rifkin, 2005; Subramanian & Schilling, 2015). Activation of TGFβ signaling in fibroblasts leads to a pro-fibrotic response, activating ECM genes, including *Col1a1*, via the TGFβ target gene, *Ccn2* (also known as connective tissue growth factor) (Duncan et al., 1999; Lin et al., 2013; Tall, Bernstein, Oliver, Gray, & Masur, 2010). CCN2 also regulates a population of tendon progenitors that express CD146 and has been demonstrated to improve tendon healing *in vivo* (Lee et al., 2015; Tarafder et al., 2017). Disrupted TGFβ signaling was observed in cartilage-specific (*Col2a1-Cre*-driven) knockout of *Bmal1* articular cartilage tissues, along with progressive cartilage degeneration in mice at only 2 months of age (Dudek et al., 2016). In tendon, *Tgfbr3* and *Ccn2* are both rhythmic transcripts in tendon and peak during the active phase of the mouse (Yeung et al., 2014), suggesting that TGFβ signaling may regulate tendon progenitors and tendon health.

5. Chronotherapy for tendinopathy treatment

Deficiency in tendon homeostasis results in tendinopathy, which affects 1 in 4 persons over 40 years of age and is the second most common musculoskeletal disorder behind osteoarthritis (Bevan et al., 2009). Tendinopathy is an umbrella term used to describe non-rupture injuries, where there is swelling of the tendon tissue that may or may not be accompanied by inflammation, and that is exacerbated by mechanical loading (reviewed by Scott et al., 2015). It is unclear how tendinopathy develops, but one popular model is chronic overloading of the tissue beyond its physiological capacity (Gross, 1992; Magnusson et al., 2010). Tendinopathic tendons are usually thicker, have reduced mechanical properties, disorganized collagen-I fibers, increase in proteoglycan content that leads to swelling, increase in collagen-III content, hyper-vascularization and nerve growth (Helland et al., 2013; Magnusson et al., 2010; Scott et al., 2015, 2008). Some of these characteristics have also been described for arrhythmic tendons (Yeung, Garva, et al., 2018 preprint). As discussed above, age-related peripheral clock dampening could explain the decline in tissue homeostasis, and further understanding of processes regulated by the tendon clock, tissue-specific entrainment signals and how these signals are integrated with the systemic circadian system are needed in order to exploit peripheral clocks as therapeutic targets for treatment of tendinopathies (Fig. 2).

5.1 Aging of tendon clock

Aging of mouse tendon is accompanied by ~40% reduction in PER2::LUC amplitude and a 6-h delay in phase relative to the SCN (Yeung et al., 2014). The overall effect on tendon clock outputs is diminished time differences in core clock gene expression and potential uncoupling of CCG expression with the rest-activity cycle. The rapid dampening of circadian oscillation in older tendon could be due to reduced inter-cellular communication in aged tissues, much like when cells are released from tissues. The delay in phase in older tendon clocks is indicative of impairment in the entrainment from the central clock or a shift in the post-translation modification mechanisms that regulate phase of protein expression, as discussed above

(Castelo-Szekely et al., 2017). At present we do not know what the entrainment signals for tendon clock are, but the endogenous tendon clock is sensitive to synchronization using glucocorticoids (Yeung et al., 2014), and mechanisms of inter-cellular and inter-tissue synchronization may also be involved (reviewed by Husse et al., 2015).

5.2 Possible methods of tendon clock entrainment

An obvious possible entrainment mechanism for tendons is exercise, which is known to induce glucocorticoid release (Sasaki et al., 2016; Tahara et al., 2015). In muscle, exercise upregulates core clock gene expression and this is reflected in the temporal expression pattern of muscle CCGs, with activation mostly occurring during the active phase (McCarthy et al., 2007; Zambon et al., 2003). In tendon, CCG activation is highest at the night-day transition (Yeung et al., 2014), suggesting that signals downstream of exercise are able to entrain the tendon clock. Energy storing tendons, including the Achilles in humans and mice or the superficial digital flexor tendon (SDFT) in horses, generate heat during cycles of loading (Ker, 1981; Riemersma & Schamhardt, 1985). For example, temperatures of SDFTs can increase from 37 to 45 °C when a horse is galloping (Wilson & Goodship, 1994). The circadian TTFL is temperature-compensated so that although it can be entrained by temperature changes, the period remains unaffected (Pittendrigh, 1954). The fluctuation in body temperatures during a 24-h day was found to be an entrainment signal for the cartilage (Gossan et al., 2013) and it is possible that exercise-induced temperature changes in the tendon could act as an additional entrainment signal to optimize phase alignment of the tendon clock to the rest-activity cycle and surrounding musculoskeletal tissues.

Cells released from tissues quickly lose their rhythm (Yoo et al., 2004), suggesting that short range signaling required for synchronizing the cell population requires three-dimensional interaction with the ECM. The impact of the extracellular environment on circadian rhythm is very relevant for ECM-rich peripheral clocks. A recent study elegantly demonstrated that endogenous circadian rhythms are modulated by the extracellular environment. Primary mammary epithelial cells cultured in a three-dimensional environment exhibited a larger amplitude of PER2::LUC oscillation and clock gene expression than two-dimensional cultures (Yang et al., 2017). The authors also showed that the dampened circadian oscillation in aged mammary tissues was due to the changes in the aged ECM and was not cell intrinsic. However, fibroblast circadian oscillations appear to be regulated differently. Skin explants of young and aged rats showed no difference in Per1::LUC rhythms but cultures of dermal fibroblasts released from old tissue showed that *Per1* activation was reduced compared to young fibroblasts (Sandu et al., 2015). Indeed, circadian clocks of fibroblasts and epithelial cells from the same tissues are inversely regulated by their ECM (Williams et al., 2018). However, whether the tendon fibroblast circadian oscillation could be affected by the ECM and whether age-related changes to the matrix in ECM-rich peripheral clock tissues contributes to dampening of the circadian rhythm require further investigation.

A yet unexplored mechanism of entrainment for the tendon clock could be inter-tissue synchronization, whereby neighboring tissues clocks have to be synchronized with each other, to maintain systemic internal synchrony (Husse et al., 2015). Tendon, muscle, and bone of the same joint may require their circadian rhythms be synchronous with each other and with neighboring tissues of the joint (cartilage, ligament) (Fig. 2). Inter-tissue signaling between tendon and muscle underlies tendon development in the embryo (reviewed by Schweitzer, Zelzer, & Volk, 2010; Subramanian & Schilling, 2015). During development, signaling between muscle and tendon and tendon and bone is mediated by many signaling pathways, including fibroblast growth factor (FGF), TGFβ, BMP pathways (Schweitzer et al., 2010). As discussed above, these pathways may be regulated by the tendon clock, so crosstalk between cells of the neighboring tissues (myotendinous junction and enthesis) in the adult tendon could potentially exist.

5.3 Implications for around-the-clock tendon care

Time-of-day differences in ECM homeostasis in ECM-rich tissues ultimately affect the tissue's mechanical properties. In tendon, this is reflected in changes in viscoelastic properties *in vivo* and *ex vivo* (Pearson & Onambele, 2005, 2006; Yeung, Garva, et al., 2018 preprint) and these may contribute to the function of the musculoskeletal system, and the phase alignment of peripheral clocks will determine time of peak performance. In fact, human chronotypes (early risers or so-called larks versus late-night "owls") show variations in peak athletic performance time. When analyzed as a function of time since awakening, early and intermediate chronotypes showed the highest average performance at ~6h but this was significantly delayed in late chronotypes, who reached average peak performance after 11h (Facer-Childs & Brandstaetter, 2015). Chronic sleep deprivation is a risk factor in sports injuries. Less than 8 h of sleep per night was the biggest predictor of sports injury in adolescent athletes (12–18 years old), followed by increase in age being the second biggest risk factor (Milewski et al., 2014). These studies further highlight the importance of synchronizing individual's circadian rhythms (chronotypes) and age to their rest-activity cycle and suggest that personalized chronotherapies should be taken into consideration.

Chronic exposure to jet lag causes misalignment of circadian clocks and increases mortality in aged mice (Davidson et al., 2006). It is now also clear that chronic disruptions to circadian clocks in humans (e.g., sleep disorders, evening screen time, shift work) can severely affect health, increasing risks of cancer, metabolic diseases, cardiovascular disease, weight gain and addiction to nicotine or alcohol (reviewed by Roenneberg & Merrow, 2016). The oncogenic mechanism of a disrupted clock was thought to be a result of reduced scavenging of reactive oxygen species by lowered melatonin levels (Hansen, 2001). It is now understood that circadian misalignment caused by modern life style cues, exposure to light at night and shift work is the key factor in these pathologies. Sleep deprivation disorders, where advanced or delayed sleep phase are two or more hours earlier or later, respectively, relative to desired or socially customary sleep times, affect approximately 25% of the population and treatments for sleep deprivation including light therapy or timed melatonin administration have little evidence to support their uses (Auger et al., 2015; Roenneberg, Allebrandt, Merrow, & Vetter, 2012). Again, more work into how the circadian clock in peripheral tissues is synchronized to the environment and other tissues is required because these light and melatonin treatments may not be sufficient for entrainment of all peripheral tissues.

6. Conclusions and implications

Recent discoveries of autonomous circadian oscillations in peripheral tissues have created an exciting new research field for exploring its role in tissue-specific homeostasis and in temporal orchestration of known biological processes. The goal of this chapter was to provide an overview of the mammalian circadian clock network and discuss what is currently known about the role of the molecular clock in tendon tissue homeostasis. Data from very recent research on the tendon clock show that it regulates BMP signaling, procollagen-I synthesis and secretion, and ER homeostasis, the latter of which can feed back onto the circadian pathway. Tendons from aged mice also exhibit ectopic calcification and have dampened and phase-delayed circadian oscillations. Further research to identify the mechanisms of tendon clock entrainment and alignment with the rest-activity cycle is critical for our understanding of how tissue homeostasis becomes insufficient in aging.

Acknowledgments

C.-Y.C.Y. is supported by a postdoctoral fellowship from Region Hovedstaden Bispebjerg and Frederiksberg Hospital, the Nordea Foundation (to the Center for Healthy Aging), and a Lundbeck Foundation Grant (R198-2015-207 awarded to Michael Kjær). The research in Karl Kadler's laboratory is supported by Wellcome Trust Investigator and Wellcome Centre Core awards to K.E.K. (110126/Z/15/Z and 203128/Z/16/Z).

References

Abe, M., Herzog, E. D., Yamazaki, S., Straume, M., Tei, H., Sakaki, Y., et al. (2002). Circadian rhythms in isolated brain regions. *The Journal of Neuroscience: The Official Journal of the Society for Neuroscience*, *22*, 350–356.

Agarwal, S., Loder, S. J., Cholok, D., Peterson, J., Li, J., Breuler, C., et al. (2017). Scleraxis-lineage cells contribute to ectopic bone formation in muscle and tendon. *Stem Cells*, *35*, 705–710.

Anafi, R. C., Lee, Y., Sato, T. K., Venkataraman, A., Ramanathan, C., Kavakli, I. H., et al. (2014). Machine learning helps identify CHRONO as a circadian clock component. *PLoS Biology*, *12*, e1001840.

Atger, F., Gobet, C., Marquis, J., Martin, E., Wang, J., Weger, B., et al. (2015). Circadian and feeding rhythms differentially affect rhythmic mRNA transcription and translation in mouse liver. *Proceedings of the National Academy of Sciences of the United States of America*, *112*, E6579–E6588.

Auger, R. R., Burgess, H. J., Emens, J. S., Deriy, L. V., Thomas, S. M., & Sharkey, K. M. (2015). Clinical practice guideline for the treatment of intrinsic circadian rhythm sleep-wake disorders: Advanced sleep-wake phase disorder (ASWPD), delayed sleep-wake phase disorder (DSWPD), Non-24-hour sleep-wake rhythm disorder (N24SWD), and irregular sleep-wake rhythm disorder (ISWRD). An update for 2015: An American Academy of Sleep Medicine Clinical Practice Guideline. *Journal of Clinical Sleep Medicine: JCSM: Official Publication of the American Academy of Sleep Medicine*, *11*, 1199–1236.

Baba, K., & Tosini, G. (2018). Aging alters circadian rhythms in the mouse eye. *Journal of Biological Rhythms*, *33*(4), 441–445. https://doi.org/10.1177/0748730418783648. Epub 2018 Jun 25.

Balsalobre, A., Brown, S. A., Marcacci, L., Tronche, F., Kellendonk, C., Reichardt, H. M., et al. (2000). Resetting of circadian time in peripheral tissues by glucocorticoid signaling. *Science*, *289*, 2344–2347.

Balsalobre, A., Damiola, F., & Schibler, U. (1998). A serum shock induces circadian gene expression in mammalian tissue culture cells. *Cell*, *93*, 929–937.

Banushi, B., Forneris, F., Straatman-Iwanowska, A., Strange, A., Lyne, A. M., Rogerson, C., et al. (2016). Regulation of post-Golgi LH3 trafficking is essential for collagen homeostasis. *Nature Communications*, *7*, 12111.

Bass, J. T. (2017). The circadian clock system's influence in health and disease. *Genome Medicine*, *9*, 94.

Bass, J., & Takahashi, J. S. (2010). Circadian integration of metabolism and energetics. *Science, 330*, 1349–1354.

Bedont, J. L., & Blackshaw, S. (2015). Constructing the suprachiasmatic nucleus: A watchmaker's perspective on the central clockworks. *Frontiers in Systems Neuroscience, 9*, 74.

Bell-Pedersen, D., Cassone, V. M., Earnest, D. J., Golden, S. S., Hardin, P. E., Thomas, T. L., et al. (2005). Circadian rhythms from multiple oscillators: Lessons from diverse organisms. *Nature Reviews Genetics, 6*, 544–556.

Bevan, S., Quadrell, T., McGee, R., Mahdon, M., Vavrovsky, A., & Barham, L. (2009). *Fit for work? Musculoskeletal disorders in the European workforce*. London: The Work Foundation.

Bi, Y., Ehirchiou, D., Kilts, T. M., Inkson, C. A., Embree, M. C., Sonoyama, W., et al. (2007). Identification of tendon stem/progenitor cells and the role of the extracellular matrix in their niche. *Nature Medicine, 13*, 1219–1227.

Boot-Handford, R. P., & Briggs, M. D. (2010). The unfolded protein response and its relevance to connective tissue diseases. *Cell and Tissue Research, 339*, 197–211.

Brent, A. E., Schweitzer, R., & Tabin, C. J. (2003). A somitic compartment of tendon progenitors. *Cell, 113*, 235–248.

Brown, S. A., Zumbrunn, G., Fleury-Olela, F., Preitner, N., & Schibler, U. (2002). Rhythms of mammalian body temperature can sustain peripheral circadian clocks. *Current Biology, 12*, 1574–1583.

Brzezinski, J. A. t., Brown, N. L., Tanikawa, A., Bush, R. A., Sieving, P. A., Vitaterna, M. H., et al. (2005). Loss of circadian photoentrainment and abnormal retinal electrophysiology in Math5 mutant mice. *Investigative Ophthalmology & Visual Science, 46*, 2540–2551.

Bunger, M. K., Walisser, J. A., Sullivan, R., Manley, P. A., Moran, S. M., Kalscheur, V. L., et al. (2005). Progressive arthropathy in mice with a targeted disruption of the Mop3/Bmal-1 locus. *Genesis, 41*, 122–132.

Cadby, J. A., Buehler, E., Godbout, C., van Weeren, P. R., & Snedeker, J. G. (2014). Differences between the cell populations from the peritenon and the tendon core with regard to their potential implication in tendon repair. *PLoS One, 9*, e92474.

Cancino, J., Capalbo, A., Di Campli, A., Giannotta, M., Rizzo, R., Jung, J. E., et al. (2014). Control systems of membrane transport at the interface between the endoplasmic reticulum and the Golgi. *Developmental Cell, 30*, 280–294.

Canty, E. G., & Kadler, K. E. (2002). Collagen fibril biosynthesis in tendon: A review and recent insights. *Comparative Biochemistry and Physiology. Part A, Molecular & Integrative Physiology, 133*, 979–985.

Canty, E. G., Lu, Y., Meadows, R. S., Shaw, M. K., Holmes, D. F., & Kadler, K. E. (2004). Coalignment of plasma membrane channels and protrusions (fibripositors) specifies the parallelism of tendon. *The Journal of Cell Biology, 165*, 553–563.

Canty, E. G., Starborg, T., Lu, Y., Humphries, S. M., Holmes, D. F., Meadows, R. S., et al. (2006). Actin filaments are required for fibripositor-mediated collagen fibril alignment in tendon. *The Journal of Biological Chemistry, 281*, 38592–38598.

Cao, R., Li, A., Cho, H. Y., Lee, B., & Obrietan, K. (2010). Mammalian target of rapamycin signaling modulates photic entrainment of the suprachiasmatic circadian clock. *The Journal of Neuroscience: The Official Journal of the Society for Neuroscience, 30*, 6302–6314.

Cassone, V. M., Speh, J. C., Card, J. P., & Moore, R. Y. (1988). Comparative anatomy of the mammalian hypothalamic suprachiasmatic nucleus. *Journal of Biological Rhythms, 3*, 71–91.

Castelo-Szekely, V., Arpat, A. B., Janich, P., & Gatfield, D. (2017). Translational contributions to tissue specificity in rhythmic and constitutive gene expression. *Genome Biology, 18*, 116.

Chhabra, A., Tsou, D., Clark, R. T., Gaschen, V., Hunziker, E. B., & Mikic, B. (2003). GDF-5 deficiency in mice delays Achilles tendon healing. *Journal of Orthopaedic Research: Official Publication of the Orthopaedic Research Society*, *21*, 826–835.

Clark, R. T., Johnson, T. L., Schalet, B. J., Davis, L., Gaschen, V., Hunziker, E. B., et al. (2001). GDF-5 deficiency in mice leads to disruption of tail tendon form and function. *Connective Tissue Research*, *42*, 175–186.

Damiola, F., Le Minh, N., Preitner, N., Kornmann, B., Fleury-Olela, F., & Schibler, U. (2000). Restricted feeding uncouples circadian oscillators in peripheral tissues from the central pacemaker in the suprachiasmatic nucleus. *Genes & Development*, *14*, 2950–2961.

Davidson, A. J., Sellix, M. T., Daniel, J., Yamazaki, S., Menaker, M., & Block, G. D. (2006). Chronic jet-lag increases mortality in aged mice. *Current Biology*, *16*, R914–R916.

de Mos, M., Koevoet, W. J., Jahr, H., Verstegen, M. M., Heijboer, M. P., Kops, N., et al. (2007). Intrinsic differentiation potential of adolescent human tendon tissue: An in-vitro cell differentiation study. *BMC Musculoskeletal Disorders*, *8*, 16.

Debruyne, J. P., Noton, E., Lambert, C. M., Maywood, E. S., Weaver, D. R., & Reppert, S. M. (2006). A clock shock: Mouse CLOCK is not required for circadian oscillator function. *Neuron*, *50*, 465–477.

DeBruyne, J. P., Weaver, D. R., & Reppert, S. M. (2007). CLOCK and NPAS2 have overlapping roles in the suprachiasmatic circadian clock. *Nature Neuroscience*, *10*, 543–545.

Dibner, C., Schibler, U., & Albrecht, U. (2010). The mammalian circadian timing system: Organization and coordination of central and peripheral clocks. *Annual Review of Physiology*, *72*, 517–549.

Diemer, T., Landgraf, D., Noguchi, T., Pan, H., Moreno, J. L., & Welsh, D. K. (2017). Cellular circadian oscillators in the suprachiasmatic nucleus remain coupled in the absence of connexin-36. *Neuroscience*, *357*, 1–11.

Dudek, M., Gossan, N., Yang, N., Im, H. J., Ruckshanthi, J. P., Yoshitane, H., et al. (2016). The chondrocyte clock gene Bmal1 controls cartilage homeostasis and integrity. *The Journal of Clinical Investigation*, *126*, 365–376.

Dudek, M., & Meng, Q. J. (2014). Running on time: The role of circadian clocks in the musculoskeletal system. *The Biochemical Journal*, *463*, 1–8.

Dudek, M., Yang, N., Ruckshanthi, J. P., Williams, J., Borysiewicz, E., Wang, P., et al. (2017). The intervertebral disc contains intrinsic circadian clocks that are regulated by age and cytokines and linked to degeneration. *Annals of the Rheumatic Diseases*, *76*, 576–584.

Duncan, M. R., Frazier, K. S., Abramson, S., Williams, S., Klapper, H., Huang, X., et al. (1999). Connective tissue growth factor mediates transforming growth factor beta-induced collagen synthesis: Down-regulation by cAMP. *FASEB Journal: Official Publication of the Federation of American Societies for Experimental Biology*, *13*, 1774–1786.

Ebihara, S., Tsuji, K., & Kondo, K. (1978). Strain differences of the mouse's free-running circadian rhythm in continuous darkness. *Physiology & Behavior*, *20*, 795–799.

Eyre, D. R., Paz, M. A., & Gallop, P. M. (1984). Cross-linking in collagen and elastin. *Annual Review of Biochemistry*, *53*, 717–748.

Facer-Childs, E., & Brandstaetter, R. (2015). The impact of circadian phenotype and time since awakening on diurnal performance in athletes. *Current Biology*, *25*, 518–522.

Fang, B., Everett, L. J., Jager, J., Briggs, E., Armour, S. M., Feng, D., et al. (2014). Circadian enhancers coordinate multiple phases of rhythmic gene transcription in vivo. *Cell*, *159*, 1140–1152.

Fujihara, Y., Kondo, H., Noguchi, T., & Togari, A. (2014). Glucocorticoids mediate circadian timing in peripheral osteoclasts resulting in the circadian expression rhythm of osteoclast-related genes. *Bone*, *61*, 1–9.

Gibson, E. M., Williams, W. P., 3rd, & Kriegsfeld, L. J. (2009). Aging in the circadian system: Considerations for health, disease prevention and longevity. *Experimental Gerontology, 44*, 51–56.
Goriki, A., Hatanaka, F., Myung, J., Kim, J. K., Yoritaka, T., Tanoue, S., et al. (2014). A novel protein, CHRONO, functions as a core component of the mammalian circadian clock. *PLoS Biology, 12*, e1001839.
Gossan, N., Zeef, L., Hensman, J., Hughes, A., Bateman, J. F., Rowley, L., et al. (2013). The circadian clock in murine chondrocytes regulates genes controlling key aspects of cartilage homeostasis. *Arthritis and Rheumatism, 65*, 2334–2345.
Gross, M. T. (1992). Chronic tendinitis: Pathomechanics of injury, factors affecting the healing response, and treatment. *The Journal of Orthopaedic and Sports Physical Therapy, 16*, 248–261.
Guerquin, M. J., Charvet, B., Nourissat, G., Havis, E., Ronsin, O., Bonnin, M. A., et al. (2013). Transcription factor EGR1 directs tendon differentiation and promotes tendon repair. *The Journal of Clinical Investigation, 123*, 3564–3576.
Guler, A. D., Ecker, J. L., Lall, G. S., Haq, S., Altimus, C. M., Liao, H. W., et al. (2008). Melanopsin cells are the principal conduits for rod-cone input to non-image-forming vision. *Nature, 453*, 102–105.
Hansen, J. (2001). Light at night, shiftwork, and breast cancer risk. *Journal of the National Cancer Institute, 93*, 1513–1515.
Harding, H. P., & Lazar, M. A. (1993). The orphan receptor Rev-ErbA alpha activates transcription via a novel response element. *Molecular and Cellular Biology, 13*, 3113–3121.
Hassager, C., Risteli, J., Risteli, L., Jensen, S. B., & Christiansen, C. (1992). Diurnal variation in serum markers of type I collagen synthesis and degradation in healthy premenopausal women. *Journal of Bone and Mineral Research: The Official Journal of the American Society for Bone and Mineral Research*, 7, 1307–1311.
Hastings, M., O'Neill, J. S., & Maywood, E. S. (2007). Circadian clocks: Regulators of endocrine and metabolic rhythms. *The Journal of Endocrinology, 195*, 187–198.
Havis, E., Bonnin, M. A., Esteves de Lima, J., Charvet, B., Milet, C., & Duprez, D. (2016). TGFbeta and FGF promote tendon progenitor fate and act downstream of muscle contraction to regulate tendon differentiation during chick limb development. *Development, 143*, 3839–3851.
Helland, C., Bojsen-Moller, J., Raastad, T., Seynnes, O. R., Moltubakk, M. M., Jakobsen, V., et al. (2013). Mechanical properties of the patellar tendon in elite volleyball players with and without patellar tendinopathy. *British Journal of Sports Medicine, 47*, 862–868.
Honda, K. K., Kawamoto, T., Ueda, H. R., Nakashima, A., Ueshima, T., Yamada, R. G., et al. (2013). Different circadian expression of major matrix-related genes in various types of cartilage: Modulation by light-dark conditions. *Journal of Biochemistry, 154*, 373–381.
Hoyle, N. P., Seinkmane, E., Putker, M., Feeney, K. A., Krogager, T. P., Chesham, J. E., et al. (2017). Circadian actin dynamics drive rhythmic fibroblast mobilization during wound healing. *Science Translational Medicine, 9*(415), eaal2774.
Humphrey, J. D., Dufresne, E. R., & Schwartz, M. A. (2014). Mechanotransduction and extracellular matrix homeostasis. *Nature Reviews. Molecular Cell Biology, 15*, 802–812.
Husse, J., Eichele, G., & Oster, H. (2015). Synchronization of the mammalian circadian timing system: Light can control peripheral clocks independently of the SCN clock: Alternate routes of entrainment optimize the alignment of the body's circadian clock network with external time. *BioEssays: News and Reviews in Molecular, Cellular and Developmental Biology, 37*, 1119–1128.
Husse, J., Leliavski, A., Tsang, A. H., Oster, H., & Eichele, G. (2014). The light-dark cycle controls peripheral rhythmicity in mice with a genetically ablated suprachiasmatic

nucleus clock. *FASEB Journal: Official Publication of the Federation of American Societies for Experimental Biology*, *28*, 4950–4960.

Ibuka, N., Inouye, S. I., & Kawamura, H. (1977). Analysis of sleep-wakefulness rhythms in male rats after suprachiasmatic nucleus lesions and ocular enucleation. *Brain Research*, *122*, 33–47.

Ibuka, N., Nihonmatsu, I., & Sekiguchi, S. (1980). Sleep-wakefulness rhythms in mice after suprachiasmatic nucleus lesions. *Waking and Sleeping*, *4*, 167–173.

Igarashia, K., Saekia, S., & Shinodab, H. (2013). Diurnal rhythms in the incorporation and secretion of 3H-proline and 3H-galactose by cartilage cells and osteoblasts in various bone-forming sites in growing rats. *Orthodontic Waves*, *72*, 11–15.

Ishida, Y., & Nagata, K. (2011). Hsp47 as a collagen-specific molecular chaperone. *Methods in Enzymology*, *499*, 167–182.

Ishikawa, Y., Ito, S., Nagata, K., Sakai, L. Y., & Bachinger, H. P. (2016). Intracellular mechanisms of molecular recognition and sorting for transport of large extracellular matrix molecules. *Proceedings of the National Academy of Sciences of the United States of America*, *113*, E6036–E6044.

Ito, Y., Toriuchi, N., Yoshitaka, T., Ueno-Kudoh, H., Sato, T., Yokoyama, S., et al. (2010). The Mohawk homeobox gene is a critical regulator of tendon differentiation. *Proceedings of the National Academy of Sciences of the United States of America*, *107*, 10538–10542.

Kadler, K. E., Hill, A., & Canty-Laird, E. G. (2008). Collagen fibrillogenesis: Fibronectin, integrins, and minor collagens as organizers and nucleators. *Current Opinion in Cell Biology*, *20*, 495–501.

Kalson, N. S., Lu, Y., Taylor, S. H., Starborg, T., Holmes, D. F., & Kadler, K. E. (2015). A structure-based extracellular matrix expansion mechanism of fibrous tissue growth. *eLife*, *4*, e05958.

Kalson, N. S., Starborg, T., Lu, Y., Mironov, A., Humphries, S. M., Holmes, D. F., et al. (2013). Nonmuscle myosin II powered transport of newly formed collagen fibrils at the plasma membrane. *Proceedings of the National Academy of Sciences of the United States of America*, *110*, E4743–E4752.

Keesler, G. A., Camacho, F., Guo, Y., Virshup, D., Mondadori, C., & Yao, Z. (2000). Phosphorylation and destabilization of human period I clock protein by human casein kinase I epsilon. *Neuroreport*, *11*, 951–955.

Ker, R. F. (1981). Dynamic tensile properties of the plantaris tendon of sheep (Ovis aries). *The Journal of Experimental Biology*, *93*, 283–302.

Kjaer, M. (2004). Role of extracellular matrix in adaptation of tendon and skeletal muscle to mechanical loading. *Physiological Reviews*, *84*, 649–698.

Koike, N., Yoo, S. H., Huang, H. C., Kumar, V., Lee, C., Kim, T. K., et al. (2012). Transcriptional architecture and chromatin landscape of the core circadian clock in mammals. *Science*, *338*, 349–354.

Komoto, S., Kondo, H., Fukuta, O., & Togari, A. (2012). Comparison of beta-adrenergic and glucocorticoid signaling on clock gene and osteoblast-related gene expressions in human osteoblast. *Chronobiology International*, *29*, 66–74.

Kondratov, R. V., Kondratova, A. A., Gorbacheva, V. Y., Vykhovanets, O. V., & Antoch, M. P. (2006). Early aging and age-related pathologies in mice deficient in BMAL1, the core componentof the circadian clock. *Genes & Development*, *20*, 1868–1873.

Korencic, A., Kosir, R., Bordyugov, G., Lehmann, R., Rozman, D., & Herzel, H. (2014). Timing of circadian genes in mammalian tissues. *Scientific Reports*, *4*, 5782.

Kuhlman, S. J., Quintero, J. E., & McMahon, D. G. (2000). GFP fluorescence reports period 1 circadian gene regulation in the mammalian biological clock. *Neuroreport*, *11*, 1479–1482.

Kuo, C. K., Petersen, B. C., & Tuan, R. S. (2008). Spatiotemporal protein distribution of TGF-betas, their receptors, and extracellular matrix molecules during embryonic tendon development. *Developmental Dynamics: An Official Publication of the American Association of Anatomists, 237*, 1477–1489.

Lamande, S. R., & Bateman, J. F. (1999). Procollagen folding and assembly: The role of endoplasmic reticulum enzymes and molecular chaperones. *Seminars in Cell & Developmental Biology, 10*, 455–464.

Lande-Diner, L., Stewart-Ornstein, J., Weitz, C. J., & Lahav, G. (2015). Single-cell analysis of circadian dynamics in tissue explants. *Molecular Biology of the Cell, 26*, 3940–3945.

Lee, A. S. (2005). The ER chaperone and signaling regulator GRP78/BiP as a monitor of endoplasmic reticulum stress. *Methods, 35*, 373–381.

Lee, C. H., Lee, F. Y., Tarafder, S., Kao, K., Jun, Y., Yang, G., et al. (2015). Harnessing endogenous stem/progenitor cells for tendon regeneration. *The Journal of Clinical Investigation, 125*, 2690–2701.

Lehman, M. N., Silver, R., Gladstone, W. R., Kahn, R. M., Gibson, M., & Bittman, E. L. (1987). Circadian rhythmicity restored by neural transplant. Immunocytochemical characterization of the graft and its integration with the host brain. *The Journal of Neuroscience: The Official Journal of the Society for Neuroscience*, 7, 1626–1638.

Lejard, V., Blais, F., Guerquin, M. J., Bonnet, A., Bonnin, M. A., Havis, E., et al. (2011). EGR1 and EGR2 involvement in vertebrate tendon differentiation. *The Journal of Biological Chemistry, 286*, 5855–5867.

Lim, J., Munivez, E., Jiang, M. M., Song, I. W., Gannon, F., Keene, D. R., et al. (2017). mTORC1 signaling is a critical regulator of postnatal tendon development. *Scientific Reports*, 7, 17175.

Lin, C. H., Yu, M. C., Tung, W. H., Chen, T. T., Yu, C. C., Weng, C. M., et al. (2013). Connective tissue growth factor induces collagen I expression in human lung fibroblasts through the Rac1/MLK3/JNK/AP-1 pathway. *Biochimica et Biophysica Acta, 1833*, 2823–2833.

Liu, W., Watson, S. S., Lan, Y., Keene, D. R., Ovitt, C. E., Liu, H., et al. (2010). The atypical homeodomain transcription factor Mohawk controls tendon morphogenesis. *Molecular and Cellular Biology, 30*, 4797–4807.

Long, M. A., Jutras, M. J., Connors, B. W., & Burwell, R. D. (2005). Electrical synapses coordinate activity in the suprachiasmatic nucleus. *Nature Neuroscience, 8*, 61–66.

Lowrey, P. L., Shimomura, K., Antoch, M. P., Yamazaki, S., Zemenides, P. D., Ralph, M. R., et al. (2000). Positional syntenic cloning and functional characterization of the mammalian circadian mutation tau. *Science, 288*, 483–492.

Luck, S., Thurley, K., Thaben, P. F., & Westermark, P. O. (2014). Rhythmic degradation explains and unifies circadian transcriptome and proteome data. *Cell Reports, 9*, 741–751.

Magnusson, S. P., Langberg, H., & Kjaer, M. (2010). The pathogenesis of tendinopathy: Balancing the response to loading. *Nature Reviews. Rheumatology, 6*, 262–268.

Maiers, J. L., Kostallari, E., Mushref, M., deAssuncao, T. M., Li, H., Jalan-Sakrikar, N., et al. (2017). The unfolded protein response mediates fibrogenesis and collagen I secretion through regulating TANGO1 in mice. *Hepatology, 65*, 983–998.

Makareeva, E., Aviles, N. A., & Leikin, S. (2011). Chaperoning osteogenesis: New protein-folding disease paradigms. *Trends in Cell Biology, 21*, 168–176.

Makareeva, E., & Leikin, S. (2007). Procollagen triple helix assembly: An unconventional chaperone-assisted folding paradigm. *PLoS One, 2*, e1029.

Malhotra, V., & Erlmann, P. (2011). Protein export at the ER: Loading big collagens into COPII carriers. *The EMBO Journal, 30*, 3475–3480.

Massague, J. (2012). TGFbeta signalling in context. *Nature Reviews. Molecular Cell Biology, 13*, 616–630.

McCarthy, J. J., Andrews, J. L., McDearmon, E. L., Campbell, K. S., Barber, B. K., Miller, B. H., et al. (2007). Identification of the circadian transcriptome in adult mouse skeletal muscle. *Physiological Genomics*, *31*, 86–95.
McCaughey, J., Stevenson, N., Cross, S., & Stephens, D. (2018). ER-to-Golgi trafficking of procollagen in the absence of large carriers. *bioRxiv*. https://doi.org/10.1101/339804. preprint.
McDearmon, E. L., Patel, K. N., Ko, C. H., Walisser, J. A., Schook, A. C., Chong, J. L., et al. (2006). Dissecting the functions of the mammalian clock protein BMAL1 by tissue-specific rescue in mice. *Science*, *314*, 1304–1308.
McNamara, P., Seo, S. B., Rudic, R. D., Sehgal, A., Chakravarti, D., & FitzGerald, G. A. (2001). Regulation of CLOCK and MOP4 by nuclear hormone receptors in the vasculature: A humoral mechanism to reset a peripheral clock. *Cell*, *105*, 877–889.
Mikic, B., Schalet, B. J., Clark, R. T., Gaschen, V., & Hunziker, E. B. (2001). GDF-5 deficiency in mice alters the ultrastructure, mechanical properties and composition of the Achilles tendon. *Journal of Orthopaedic Research: Official Publication of the Orthopaedic Research Society*, *19*, 365–371.
Milewski, M. D., Skaggs, D. L., Bishop, G. A., Pace, J. L., Ibrahim, D. A., Wren, T. A., et al. (2014). Chronic lack of sleep is associated with increased sports injuries in adolescent athletes. *Journal of Pediatric Orthopedics*, *34*, 129–133.
Mirsky, H. P., Liu, A. C., Welsh, D. K., Kay, S. A., & Doyle, F. J., 3rd. (2009). A model of the cell-autonomous mammalian circadian clock. *Proceedings of the National Academy of Sciences of the United States of America*, *106*, 11107–11112.
Mosko, S. S., & Moore, R. Y. (1979). Neonatal suprachiasmatic nucleus lesions: Effects on the development of circadian rhythms in the rat. *Brain Research*, *164*, 17–38.
Munoz, E., & Baler, R. (2003). The circadian E-box: When perfect is not good enough. *Chronobiology International*, *20*, 371–388.
Murchison, N. D., Price, B. A., Conner, D. A., Keene, D. R., Olson, E. N., Tabin, C. J., et al. (2007). Regulation of tendon differentiation by scleraxis distinguishes force-transmitting tendons from muscle-anchoring tendons. *Development*, *134*, 2697–2708.
Nishii, K., Yamanaka, I., Yasuda, M., Kiyohara, Y. B., Kitayama, Y., Kondo, T., et al. (2006). Rhythmic post-transcriptional regulation of the circadian clock protein mPER2 in mammalian cells: A real-time analysis. *Neuroscience Letters*, *401*, 44–48.
Nunes, I., Gleizes, P. E., Metz, C. N., & Rifkin, D. B. (1997). Latent transforming growth factor-beta binding protein domains involved in activation and transglutaminase-dependent cross-linking of latent transforming growth factor-beta. *The Journal of Cell Biology*, *136*, 1151–1163.
Olivares, O., Mayers, J. R., Gouirand, V., Torrence, M. E., Gicquel, T., Borge, L., et al. (2017). Collagen-derived proline promotes pancreatic ductal adenocarcinoma cell survival under nutrient limited conditions. *Nature Communications*, *8*, 16031.
Parry, D. A., Barnes, G. R., & Craig, A. S. (1978). A comparison of the size distribution of collagen fibrils in connective tissues as a function of age and a possible relation between fibril size distribution and mechanical properties. *Proceedings of the Royal Society of London. Series B, Biological Sciences*, *203*, 305–321.
Pearson, S. J., & Onambele, G. N. (2005). Acute changes in knee-extensors torque, fiber pennation, and tendon characteristics. *Chronobiology International*, *22*, 1013–1027.
Pearson, S. J., & Onambele, G. N. (2006). Influence of time of day on tendon compliance and estimations of voluntary activation levels. *Muscle & Nerve*, *33*, 792–800.
Phillip, J. M., Aifuwa, I., Walston, J., & Wirtz, D. (2015). The mechanobiology of aging. *Annual Review of Biomedical Engineering*, *17*, 113–141.
Pickard, A., Chang, J., Alachkar, N., Calverley, B., Garva, R., Arvan, P., et al. (2018). Protection of circadian rhythms by the protein folding chaperone, BiP. *bioRxiv*. https://doi.org/10.1101/348078. preprint.

Pittendrigh, C. S. (1954). On temperature Independence in the clock system controlling emergence time in Drosophila. *Proceedings of the National Academy of Sciences of the United States of America*, *40*, 1018–1029.

Pryce, B. A., Watson, S. S., Murchison, N. D., Staverosky, J. A., Dunker, N., & Schweitzer, R. (2009). Recruitment and maintenance of tendon progenitors by TGFbeta signaling are essential for tendon formation. *Development*, *136*, 1351–1361.

Ralph, M. R., Foster, R. G., Davis, F. C., & Menaker, M. (1990). Transplanted suprachiasmatic nucleus determines circadian period. *Science*, *247*, 975–978.

Ramanathan, C., Kathale, N. D., Liu, D., Lee, C., Freeman, D. A., Hogenesch, J. B., et al. (2018). mTOR signaling regulates central and peripheral circadian clock function. *PLoS Genetics*, *14*, e1007369.

Reddy, A. B., Karp, N. A., Maywood, E. S., Sage, E. A., Deery, M., O'Neill, J. S., et al. (2006). Circadian orchestration of the hepatic proteome. *Current Biology: CB*, *16*, 1107–1115.

Riemersma, D. J., & Schamhardt, H. C. (1985). In vitro mechanical properties of equine tendons in relation to cross-sectional area and collagen content. *Research in Veterinary Science*, *39*, 263–270.

Rifkin, D. B. (2005). Latent transforming growth factor-beta (TGF-beta) binding proteins: Orchestrators of TGF-beta availability. *The Journal of Biological Chemistry*, *280*, 7409–7412.

Rios-Barrera, L. D., Sigurbjornsdottir, S., Baer, M., & Leptin, M. (2017). Dual function for Tango1 in secretion of bulky cargo and in ER-Golgi morphology. *Proceedings of the National Academy of Sciences of the United States of America*, *114*, E10389–E10398.

Robles, M. S., Cox, J., & Mann, M. (2014). In-vivo quantitative proteomics reveals a key contribution of post-transcriptional mechanisms to the circadian regulation of liver metabolism. *PLoS Genetics*, *10*, e1004047.

Roenneberg, T., Allebrandt, K. V., Merrow, M., & Vetter, C. (2012). Social jetlag and obesity. *Current Biology*, *22*, 939–943.

Roenneberg, T., & Merrow, M. (2016). The circadian clock and human health. *Current Biology*, *26*, R432–R443.

Rui, Y. F., Lui, P. P., Li, G., Fu, S. C., Lee, Y. W., & Chan, K. M. (2010). Isolation and characterization of multipotent rat tendon-derived stem cells. *Tissue Engineering. Part A*, *16*, 1549–1558.

Russell, J. E., Walker, W. V., Fenster, R. J., & Simmons, D. J. (1985). In vitro evaluation of circadian patterns of bone collagen formation. *Proceedings of the Society for Experimental Biology and Medicine. Society for Experimental Biology and Medicine (New York, N.Y.)*, *180*, 375–381.

Saini, C., Liani, A., Curie, T., Gos, P., Kreppel, F., Emmenegger, Y., et al. (2013). Real-time recording of circadian liver gene expression in freely moving mice reveals the phase-setting behavior of hepatocyte clocks. *Genes & Development*, *27*, 1526–1536.

Saito, K., Chen, M., Bard, F., Chen, S., Zhou, H., Woodley, D., et al. (2009). TANGO1 facilitates cargo loading at endoplasmic reticulum exit sites. *Cell*, *136*, 891–902.

Salingcarnboriboon, R., Yoshitake, H., Tsuji, K., Obinata, M., Amagasa, T., Nifuji, A., et al. (2003). Establishment of tendon-derived cell lines exhibiting pluripotent mesenchymal stem cell-like property. *Experimental Cell Research*, *287*, 289–300.

Sandu, C., Liu, T., Malan, A., Challet, E., Pevet, P., & Felder-Schmittbuhl, M. P. (2015). Circadian clocks in rat skin and dermal fibroblasts: Differential effects of aging, temperature and melatonin. *Cellular and Molecular Life Sciences: CMLS*, *72*, 2237–2248.

Sasaki, H., Hattori, Y., Ikeda, Y., Kamagata, M., Iwami, S., Yasuda, S., et al. (2016). Forced rather than voluntary exercise entrains peripheral clocks via a corticosterone/noradrenaline increase in PER2::LUC mice. *Scientific Reports*, *6*, 27607.

Satoh, M., Hirayoshi, K., Yokota, S., Hosokawa, N., & Nagata, K. (1996). Intracellular interaction of collagen-specific stress protein HSP47 with newly synthesized procollagen. *The Journal of Cell Biology*, *133*, 469–483.

Scarbrough, K., Losee-Olson, S., Wallen, E. P., & Turek, F. W. (1997). Aging and photoperiod affect entrainment and quantitative aspects of locomotor behavior in Syrian hamsters. *The American Journal of Physiology*, *272*, R1219–R1225.

Schwartz, M. A. (2010). Integrins and extracellular matrix in mechanotransduction. *Cold Spring Harbor Perspectives in Biology*, *2*, a005066.

Schweitzer, R., Zelzer, E., & Volk, T. (2010). Connecting muscles to tendons: Tendons and musculoskeletal development in flies and vertebrates. *Development*, *137*, 2807–2817.

Scott, A., Backman, L. J., & Speed, C. (2015). Tendinopathy: Update on pathophysiology. *The Journal of Orthopaedic and Sports Physical Therapy*, *45*, 833–841.

Scott, A., Lian, O., Roberts, C. R., Cook, J. L., Handley, C. J., Bahr, R., et al. (2008). Increased versican content is associated with tendinosis pathology in the patellar tendon of athletes with jumper's knee. *Scandinavian Journal of Medicine & Science in Sports*, *18*, 427–435.

Screen, H. R., Berk, D. E., Kadler, K. E., Ramirez, F., & Young, M. F. (2015). Tendon functional extracellular matrix. *Journal of Orthopaedic Research: Official Publication of the Orthopaedic Research Society*, *33*, 793–799.

Sellix, M. T., Evans, J. A., Leise, T. L., Castanon-Cervantes, O., Hill, D. D., DeLisser, P., et al. (2012). Aging differentially affects the re-entrainment response of central and peripheral circadian oscillators. *The Journal of Neuroscience: The Official Journal of the Society for Neuroscience*, *32*, 16193–16202.

Sricholpech, M., Perdivara, I., Yokoyama, M., Nagaoka, H., Terajima, M., Tomer, K. B., et al. (2012). Lysyl hydroxylase 3-mediated glucosylation in type I collagen: Molecular loci and biological significance. *The Journal of Biological Chemistry*, *287*, 22998–23009.

Stefanovic, L., Longo, L., Zhang, Y., & Stefanovic, B. (2014). Characterization of binding of LARP6 to the 5′ stem-loop of collagen mRNAs: Implications for synthesis of type I collagen. *RNA Biology*, *11*, 1386–1401.

Stephan, F. K. (1983). Circadian rhythms in the rat: Constant darkness, entrainment to T cycles and to skeleton photoperiods. *Physiology & Behavior*, *30*, 451–462.

Storch, K. F., Lipan, O., Leykin, I., Viswanathan, N., Davis, F. C., Wong, W. H., et al. (2002). Extensive and divergent circadian gene expression in liver and heart. *Nature*, *417*, 78–83.

Subramanian, A., & Schilling, T. F. (2015). Tendon development and musculoskeletal assembly: Emerging roles for the extracellular matrix. *Development*, *142*, 4191–4204.

Suzuki, H., Ito, Y., Shinohara, M., Yamashita, S., Ichinose, S., Kishida, A., et al. (2016). Gene targeting of the transcription factor Mohawk in rats causes heterotopic ossification of Achilles tendon via failed tenogenesis. *Proceedings of the National Academy of Sciences of the United States of America*, *113*, 7840–7845.

Tahara, Y., Shiraishi, T., Kikuchi, Y., Haraguchi, A., Kuriki, D., Sasaki, H., et al. (2015). Entrainment of the mouse circadian clock by sub-acute physical and psychological stress. *Scientific Reports*, *5*, 11417.

Tahara, Y., Takatsu, Y., Shiraishi, T., Kikuchi, Y., Yamazaki, M., Motohashi, H., et al. (2017). Age-related circadian disorganization caused by sympathetic dysfunction in peripheral clock regulation. *NPJ Aging and Mechanisms of Disease*, *3*, 16030.

Tall, E. G., Bernstein, A. M., Oliver, N., Gray, J. L., & Masur, S. K. (2010). TGF-beta-stimulated CTGF production enhanced by collagen and associated with biogenesis of a novel 31-kDa CTGF form in human corneal fibroblasts. *Investigative Ophthalmology & Visual Science*, *51*, 5002–5011.

Tarafder, S., Chen, E., Jun, Y., Kao, K., Sim, K. H., Back, J., et al. (2017). Tendon stem/progenitor cells regulate inflammation in tendon healing via JNK and STAT3 signaling.

FASEB Journal: Official Publication of the Federation of American Societies for Experimental Biology, *31*, 3991–3998.

Valentinuzzi, V. S., Scarbrough, K., Takahashi, J. S., & Turek, F. W. (1997). Effects of aging on the circadian rhythm of wheel-running activity in C57BL/6 mice. *The American Journal of Physiology*, *273*, R1957–R1964.

Vitaterna, M. H., King, D. P., Chang, A. M., Kornhauser, J. M., Lowrey, P. L., McDonald, J. D., et al. (1994). Mutagenesis and mapping of a mouse gene, clock, essential for circadian behavior. *Science*, *264*, 719–725.

Wang, J., Symul, L., Yeung, J., Gobet, C., Sobel, J., Luck, S., et al. (2018). Circadian clock-dependent and -independent posttranscriptional regulation underlies temporal mRNA accumulation in mouse liver. *Proceedings of the National Academy of Sciences of the United States of America*, *115*, E1916–E1925.

Welsh, D., Richardson, G. S., & Dement, W. C. (1988). Effect of running wheel availability on circadian patterns of sleep and wakefulness in mice. *Physiology & Behavior*, *43*, 771–777.

Welsh, D. K., Takahashi, J. S., & Kay, S. A. (2010). Suprachiasmatic nucleus: Cell autonomy and network properties. *Annual Review of Physiology*, *72*, 551–577.

Wilkinson, J. E., Burmeister, L., Brooks, S. V., Chan, C. C., Friedline, S., Harrison, D. E., et al. (2012). Rapamycin slows aging in mice. *Aging Cell*, *11*, 675–682.

Williams, J., Yang, N., Wood, A., Zindy, E., Meng, Q. J., & Streuli, C. H. (2018). Epithelial and stromal circadian clocks are inversely regulated by their mechano-matrix environment. *Journal of Cell Science*, *131*, jcs208223.

Wilson, A. M., & Goodship, A. E. (1994). Exercise-induced hyperthermia as a possible mechanism for tendon degeneration. *Journal of Biomechanics*, *27*, 899–905.

Wilson, D. G., Phamluong, K., Li, L., Sun, M., Cao, T. C., Liu, P. S., et al. (2011). Global defects in collagen secretion in a Mia3/TANGO1 knockout mouse. *The Journal of Cell Biology*, *193*, 935–951.

Wolff, G., & Esser, K. A. (2012). Scheduled exercise phase shifts the circadian clock in skeletal muscle. *Medicine and Science in Sports and Exercise*, *44*, 1663–1670.

Wooden, S. K., & Lee, A. S. (1992). Comparison of the genomic organizations of the rat grp78 and hsc73 gene and their evolutionary implications. *DNA Sequence: The Journal of DNA Sequencing and Mapping*, *3*, 41–48.

Yagita, K., Yamaguchi, S., Tamanini, F., van Der Horst, G. T., Hoeijmakers, J. H., Yasui, A., et al. (2000). Dimerization and nuclear entry of mPER proteins in mammalian cells. *Genes & Development*, *14*, 1353–1363.

Yamaguchi, S., Mitsui, S., Miyake, S., Yan, L., Onishi, H., Yagita, K., et al. (2000). The 5′ upstream region of mPer1 gene contains two promoters and is responsible for circadian oscillation. *Current Biology*, *10*, 873–876.

Yamazaki, S., Numano, R., Abe, M., Hida, A., Takahashi, R., Ueda, M., et al. (2000). Resetting central and peripheral circadian oscillators in transgenic rats. *Science*, *288*, 682–685.

Yan, J., Wang, H., Liu, Y., & Shao, C. (2008). Analysis of gene regulatory networks in the mammalian circadian rhythm. *PLoS Computational Biology*, *4*, e1000193.

Yang, G., Chen, L., Grant, G. R., Paschos, G., Song, W. L., Musiek, E. S., et al. (2016). Timing of expression of the core clock gene Bmal1 influences its effects on aging and survival. *Science Translational Medicine*, *8*, 324ra16.

Yang, N., Williams, J., Pekovic-Vaughan, V., Wang, P., Olabi, S., McConnell, J., et al. (2017). Cellular mechano-environment regulates the mammary circadian clock. *Nature Communications*, *8*, 14287.

Yeung, C. Y. C., Garva, R., Pickard, A., Chang, J., Holmes, D. F., Lu, Y., et al. (2018). Circadian clock regulation of the secretory pathway. *bioRxiv*. https://doi.org/10.1101/304014. preprint.

Yeung, C. Y. C., Gossan, N., Lu, Y., Hughes, A., Hensman, J. J., Bayer, M. L., et al. (2014). Gremlin-2 is a BMP antagonist that is regulated by the circadian clock. *Scientific Reports*, *4*, 5183.

Yeung, J., Mermet, J., Jouffe, C., Marquis, J., Charpagne, A., Gachon, F., et al. (2018). Transcription factor activity rhythms and tissue-specific chromatin interactions explain circadian gene expression across organs. *Genome Research*, *28*, 182–191.

Yoo, S. H., Mohawk, J. A., Siepka, S. M., Shan, Y., Huh, S. K., Hong, H. K., et al. (2013). Competing E3 ubiquitin ligases govern circadian periodicity by degradation of CRY in nucleus and cytoplasm. *Cell*, *152*, 1091–1105.

Yoo, S. H., Yamazaki, S., Lowrey, P. L., Shimomura, K., Ko, C. H., Buhr, E. D., et al. (2004). PERIOD2::LUCIFERASE real-time reporting of circadian dynamics reveals persistent circadian oscillations in mouse peripheral tissues. *Proceedings of the National Academy of Sciences of the United States of America*, *101*, 5339–5346.

Yu, E. A., & Weaver, D. R. (2011). Disrupting the circadian clock: Gene-specific effects on aging, cancer, and other phenotypes. *Aging*, *3*, 479–493.

Zambon, A. C., McDearmon, E. L., Salomonis, N., Vranizan, K. M., Johansen, K. L., Adey, D., et al. (2003). Time- and exercise-dependent gene regulation in human skeletal muscle. *Genome Biology*, *4*, R61.

Zaseck, L. W., Miller, R. A., & Brooks, S. V. (2016). Rapamycin attenuates age-associated changes in tibialis anterior tendon viscoelastic properties. *The Journals of Gerontology. Series A, Biological Sciences and Medical Sciences*, *71*, 858–865.

Zhang, Y., Fang, B., Emmett, M. J., Damle, M., Sun, Z., Feng, D., et al. (2015). Gene regulation. Discrete functions of nuclear receptor Rev-erbalpha couple metabolism to the clock. *Science*, *348*, 1488–1492.

Zhang, R., Lahens, N. F., Ballance, H. I., Hughes, M. E., & Hogenesch, J. B. (2014). A circadian gene expression atlas in mammals: Implications for biology and medicine. *Proceedings of the National Academy of Sciences of the United States of America*, *111*, 16219–16224.

CHAPTER TWELVE

Mechanistic insights into skeletal development gained from genetic disorders

Raymond K.H. Yip†, Danny Chan, Kathryn S.E. Cheah*
School of Biomedical Sciences, The University of Hong Kong, Pok Fu Lam, Hong Kong
*Corresponding author: e-mail address: kathycheah@hku.hk

Contents

Abstract

A complex cascade of highly regulated processes of cell fate determination, differentiation, proliferation and transdifferentiation dictate the patterning, morphogenesis and growth of the vertebrate skeleton, perturbation of which results in malformation. In humans over 450 different dysplasias involving the skeletal system constitute a significant fraction of documented Mendelian disorders. The combination of clinical, phenotypic characterization of rare human skeletal dysmorphologies, the discovery of causative mutations and functional validation in animal models has contributed enormously to the understanding of molecular control of skeletal development. These studies revealed a myriad of genes and pathways, such as WNT, Hedgehog (HH), planar cell polarity and primary cilia, as key regulators for skeletal patterning, growth and

† *Current address*: ACRF Stem Cells and Cancer Division, The Walter and Eliza Hall Institute of Medical Research, Parkville, VIC, Australia; Department of Medical Biology, University of Melbourne, Parkville, VIC, Australia.

Current Topics in Developmental Biology, Volume 133
ISSN 0070-2153
https://doi.org/10.1016/bs.ctdb.2019.02.002

homeostasis. The generation of mouse models recapitulating human congenital skeletal dysplasia has provided mechanistic insights into the diverse pathologies caused by single gene mutations, integrated action of developmental pathways such as WNT and HH and the role of stress responses. Technological developments in whole genome and exome sequencing have accelerated the discovery of disease-causing mutations and are changing approaches for diagnosis. The discovery that non-coding variants and disorganization of the 3D genome are associated with limb patterning disorders has revealed an additional level of complexity in the regulatory framework of skeletal development and disease mechanisms. This chapter focuses on a selection of human skeletal pathologies which illustrate how new findings about the coding and noncoding genome, combined with functional modeling, are contributing to deeper understanding of skeletal development, mechanisms of disease, with therapeutic potential for chondrodysplasias.

1. Introduction

Congenital skeletal dysplasia arises when genetic alterations result in the disruption of the pattern, structure and growth of the skeleton. Such perturbations manifest as one or more phenotypes affecting the shape and size of individual skeletal elements, such as short, stubby fingers, duplications of fingers or toes, clubfeet, missing bones, fragile bones or curved spines. Normal skeletogenesis requires spatial and temporal control and integration of various transcription factors and signaling pathways to coordinate precisely the initial condensation of mesenchymal cells, specification of osteo-chondroprogenitors and the sequential phases of chondrocyte differentiation, proliferation, cell cycle exit and maturation, hypertrophy and the transition to the osteoblast lineage. Several of these factors are described in detail in other chapters. Many genes and pathways were discovered through identifying causative mutations associated with human skeletal syndromes. Advances and affordable technologies for sequencing whole genomes have accelerated the discovery of new genes and genetic loci in skeletal dysplasias (Bonafe et al., 2015; Geister & Camper, 2015). In this chapter, we highlight the contribution of recent discoveries of causative mutations in human skeletal dysplasias, combined with functional genomics, to the identification of key genes and pathways and gene regulatory mechanisms that govern different phases of skeletal development (Fig. 1). We also briefly illustrate how knowledge of the underlying molecular pathogenesis is being exploited for clinical translation, leading to human trials on chondrodysplasias.

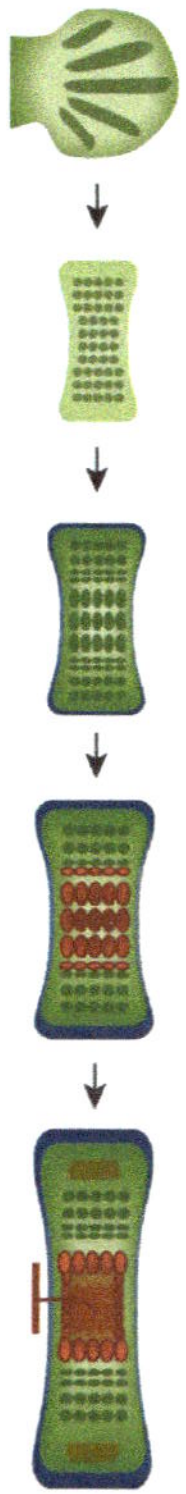

Skeletal disorders and affected processes	Affected genes	Physiological function of affected genes	Molecular pathomechanism
Limb bud patterning			
Preaxial Polydactyly (OMIM#174500)	*SHH* *GLI3*	SHH-Gli3 counterbalance patterns AP axisin limb bud	Ectopic anterior expression of *SHH* perturbs GLI3 repressor gradient across the limb field, leading to digit duplication
Synpolydactyly (OMIM#186000)	*HOXD13*	Activates RA signaling to suppress chondrogenesis	Mutant HOXD13 with expanded Ala repeats fail to upregulate *Raldh2* and therefore RA signaling at interdigital mesenchyme
Bardet-Biedl syndrome (OMIM#209900) Joubert syndrome (OMIM#213300)	*BBS* *TALPID3*	Ciliary components that modulate HH signal transduction	Ciliary dysfunction disrupts GLI protein processing to compromise HH transduction.
Chondrocyte differentiation			
Brachydactyly type A-1 (OMIM#112500)	*IHH*	Regulates chondrocyte proliferation and hypertrophy via interaction with PTHrP signaling pathway	Alteration in IHH signaling capacity and range leads to reduced chondrogenic templates for the future middle phalange
Brachydactyly type B-1 (OMIM#113000)	*ROR2*	Major receptor to transduce WNT/PCP activity	Truncated ROR2 operates as dominant negative mutants to interfere wild type ROR2 activity at cell membrane
Achondroplasia (OMIM#100800)	*FGFR3*	Inhibits both chondrocyte proliferation and hypertrophy	Constitutive FGFR3 activation induces STAT1 and MAPK pathway to arrest chondrocyte proliferation and maturation
Blomstrand chondrodysplasia (OMIM#215045) Jansen metaphyseal dysplasia (OMIM#156400)	*PTHR1*	Suppresses chondrocyte hypertrophy	Mutant PTHR1 alters the pace of chondrocyte hypertrophy, and therefore endochondral bone formation
Robinow syndrome (AD; OMIM#180700) Robinow syndrome (AR; OMIM#268310)	*WNT5A* *DVL1/3* *ROR2*	Transduce WNT/PCP signaling for directional growth of chondrocytes for proximal-distal elongation of limb	Impaired ability to induce VANGL2 activity distrubs chondrocyte polarization that is required for limb elongation
Campomelic dysplasia (OMIM#114290) Acampomelic dysplasia (OMIM#114290) Pierre Robin Sequence (OMIM#261800)	*SOX9*	Regulates chondrocyte differentiation, proliferation and entry to hypertrophy	Impaired precartilaginous mesenchymal condensation and premature hypertrophy and osteogenesis
Secretion of matrix proteins			
Multiple epiphyseal dysplasias (OMIM#132400)	*MATN3*	Organisation of collagen and other cartilage proteins	Retention of mutant MATN3 in ER induces ER stress and therefore premature chondrocyte apoptosis
Pseudoachondroplasia (OMIM#177170)	*COMP*	Binding of extracellular matrix proteins to chondrocytes	Presence of mutant COMP elicits novel ER stress respsone independent of UPR, promoting chondrocyte apoptosis
Metaphyseal chondrodysplasia, type Schmid (OMIM#156500)	*COL10A1*	Regulators of mineralisation	Intracellular accumulation of mutant COL10A1 activates ER stress signaling pathways causing a reversal of hypertrophic chondrocyte differentiated state and thereby a delay in endochondral ossification
Osteoblast differentiation			
Progressive osseous heteroplasia (OMIM#166350) Fibrous dysplasia (OMIM#174800)	*GNAS*	Regulates osteoblastogenesis via maintaining a balance between WNT and HH signaling pathways	Distinct *GNAS* mutations instigate extreme WNT or HH sinaling to inhibit bone formation or cause heterotpic ossification

Fig. 1 Mechanistic insights into skeletal development gained from skeletal dysplasias. See the main text for detailed information.

2. Genetic control of patterning the appendicular skeleton

Human limbs consist of bones and soft tissues of particular size and shape arranged in a precise pattern. Structural abnormalities are often unique and diagnostic. One of the most recognizable limb phenotypes is polydactyly (Greek for "many fingers"). Digits in human hands and feet are formed in a highly conserved pentadactyl pattern, but individuals with polydactyly have additional digits arising on the side of the thumb (preaxial), the little finger (postaxial), or the central fingers (central). Polydactyly can occur by itself or as part of a congenital syndrome. Its prevalence is estimated to be 0.3–3.6/1000 of live births, with preaxial and postaxial abnormalities being more common than central polydactyly (Malik, 2014).

The emergence of extra digits is sometimes associated with webbing of the adjacent fingers, resulting in synpolydactyly (SPD, OMIM#186000). One type of SPD is caused by polyalanine expansions and frameshift mutations in homeobox d13 (*HOXD13*), the most 5′ gene in the HOXD gene cluster (Muragaki, Mundlos, Upton, & Olsen, 1996). The pathogenic mechanism of Ala repeat expansions is of particular interest as the length of the Ala expansion correlates with the severity and penetrance of the SPD phenotype. Moreover, inactivation of *Hoxd13* in mice does not fully phenocopy SPD, indicating that the malformation is not simply due to loss of function of this homeobox gene (Zakany & Duboule, 1996). *In vitro*, poly-Ala expansions in HOXD13 induced the formation of cytoplasmic aggregates that sequestered the wild-type HOXD13, preventing the latter from entering the nucleus (Albrecht et al., 2004). Mutant HOXD13 with expanded Ala repeats fails to upregulate *Raldh2*, the rate-limiting enzyme for retinoic acid (RA) synthesis in the limb, which suppresses chondrogenesis in the interdigital space (Kuss et al., 2009). These findings suggest that HOXD13, which is normally expressed in interdigital mesenchyme, functions as a repressor of chondrogenesis via activating RA signaling to prevent the formation of additional cartilage condensations.

The important roles of TGF beta and BMP signaling in skeletal morphogenesis and patterning and in skeletal dysplasias are well established (Baldridge, Shchelochkov, Kelley, & Lee, 2010; Wu, Chen, & Li, 2016). Their role in limb patterning is highlighted by the association of mutations in the BMP subfamily member *CDMP-1* gene (*Gdf5*, in mice) with Acromesomelic dysplasia of the Hunter Thompson (OMIM#201250) and Grebe type chondrodysplasias (OMIM#200700), DuPan syndrome (missing or hypoplastic fibula and reduced distal bones) (OMIM#228900) and brachydactyly type C (OMIM#113100) (Faiyaz-Ul-Haque, Ahmad, Wahab, et al., 2002; Faiyaz-Ul-Haque, Ahmad, Zaidi, et al., 2002 Polinkovsky et al., 1997; Thomas et al., 1997). In Hunter-Thompson chondrodysplasia, a frameshift mutation in *CDMP-1* causes loss of function. In Grebe-type chondrodysplasia, *CDMP-1* mutation results in the synthesis of an inactive peptide that is not secreted and may act dominant negatively; in brachydactyly type C mutant *CDMP-1* peptides cannot dimerize (Everman et al., 2002).

Initial genetic screens of families with preaxial polydactyly (PPD, OMIM#174500) mapped the critical locus to a 450-kb region on chromosome 7q36, and this was subsequently refined to intron 5 of the *LMBR1* gene by FISH and sequencing of cosmid clones. The corresponding region is perturbed in the Sasquatch (*Ssq*) and hemimelic extra toe (*Hx*) mouse models that display PPD syndrome (Clark, Marker, & Kingsley, 2000; Lettice et al., 2002; Zguricas et al., 1999). However, analysis of cell lines from PPD patients found no transcript truncation or pathogenic mutations within the *LMBR1* coding region, and the gene itself has no role in limb development (Lettice et al., 2002). Sonic hedgehog (*Shh*), which is 1 Mb from the PPD-associated region, is mis-expressed in an ectopic anterior site in the limb bud of *Ssq* mice, thus linking *Shh* dysregulation to the generation of extra digits in PPD (Masuya, Sagai, Wakana, Moriwaki, & Shiroishi, 1995; Sharpe et al., 1999). SHH is the major signaling molecule produced by a posterior limb bud domain called the zone of polarizing activity (ZPA), which induces mirror-image duplications when grafted to the anterior region of a recipient limb. Ectopic implantation of *Shh*-expressing fibroblasts into the anterior limb bud domain is sufficient to cause digit duplications, confirming a crucial role of SHH in anteroposterior patterning in the limb (Riddle, Johnson, Laufer, & Tabin, 1993).

How does *Shh* misexpression lead to supernumerary digits in PPD? SHH is a member of the vertebrate HH protein family that also includes Desert Hedgehog (DHH) and Indian Hedgehog (IHH). HH proteins are classical morphogens because of their ability to act over a long range to influence activities of the responding cells in a time- and dose-dependent manner. Mammals have three hedgehog-regulated GLI transcription factors: GLI1-GLI3. GLI1 functions exclusively as a transcriptional activator, whereas GLI2 and GLI3 can be proteolytically processed into a repressor. In the absence of HH, GLI2 and GLI3 are trafficked in a complex with

SUFU and KIF7 to the tip of primary cilia where activated protein kinase A (PKA) phosphorylates sites in GLI3, leading to partial proteolysis and generation of a Gli3 repressor (GLI3R). Binding of HH ligands to the transmembrane receptor Patched (PTCH1) releases Smoothened (SMO), which is phosphorylated and translocated into the primary cilium where it counteracts the inhibitory effect of SUFU and enables the full-length GLI3 activator (Gli3A) to translocate to the cell nucleus where it activates target genes (Yang, Andre, Ye, & Yang, 2015). During embryonic development, GLI3 is expressed across the limb bud; however, in the posterior part of the bud where SHH concentration is high, GLI3A concentration is high and GLI3R concentration is low, whereas the opposite is true in the anterior region that is further away from the SHH-producing ZPA (Wang, Fallon, & Beachy, 2000). This SHH-mediated counteraction of GLI3 thereby sets up a GLI3A:GLI3R gradient along the anterior-posterior axis of the limb field. This gradient controls digit number and, to a lesser extent, digit identity (Zhu & Mackem, 2017). In the case of PPD, ectopic anterior *Shh* expression creates an unusually high GLI3A concentration at the anterior margin of the limb bud, leading to transformation of anterior digits into more posterior digit types (i.e., triphangial thumb) and, depending on the concentration of ectopic SHH, generation of extra digits (Anderson, Peluso, Lettice, & Hill, 2012). Of note, genetic experiments in mice have identified *Irx3*/5-dependent anterior progenitors as the "cells-of-origin" for the extra digits in PPD because of their function in promoting anterior *Gli3* expression in the initiating limb bud (Li et al., 2014). This finding suggests that PPD can arise from cellular differentiation defects at the prepatterning stage prior to *Shh* activation.

Another HH family member implicated in skeletal patterning is *IHH*. Heterozygous mutations in the human *IHH* gene cause brachydactyly A1 (BDA1, OMIM#112500), the first described human Mendelian autosomal dominant disorder which features shortening of or missing middle phalanges (Gao et al., 2001; Hellemans et al., 2003). Structural analyses of mutant IHH^{E95K} protein and analyses of a Ihh^{E95K} knock-in mouse model showed that this BDA1 mutation in IHH impairs its interaction with the receptor PTCH1 and antagonist HIP1, allowing it to travel further through the phalangeal growth plates than what happens in wild-type mice. As a consequence, the cells neighboring prehypertrophic chondrocytes receive suboptimal IHH stimulation, while cells in the periarticular and perichondrial regions receive too much, upregulating genes such as parathyroid hormone (PTH)-related peptide (PTHrP) (Gao et al., 2009; McLellan et al., 2008).

Because PTHrP forms a negative feedback loop with IHH, the increased dosage of PTHrP prevents induction of IHH signaling in the distal condensed mesenchyme (Vortkamp et al., 1996). This combination of events leads to impaired chondrogenesis and a smaller chondrogenic template for the future middle phalanges, and eventually a small or absent middle phalange. Disturbing the signaling capacity and range of a morphogen is therefore as detrimental as the absence of the morphogen. In summary, discovery of mutations causing limb patterning defects and functional studies have revealed that integration of a complex network of regulatory feedback loops, fields of repression and activation that are mediated by morphogen gradients and interaction between HH, PTHrP, and BMP pathways are critical for proper digit formation and patterning of the human limb.

3. Skeletal morphogenesis: Integrated control of chondrocyte differentiation

The cascade of differentiation steps in endochondral bone development is controlled by a combination of key transcription factors and the integrated action of signaling pathways, as exemplified by a number of skeletal dysplasias. The discovery of SOX9 as a master regulator of chondrogenesis came from the identification of its causative role in Campomelic dysplasia (CD, OMIM#114290), a rare, semi-lethal autosomal dominant congenital skeletal disorder that affects approximately 1 in 40,000 to 80,000 newborns (Scherer, Zabel, & Nishimura, 2013). CD was first reported by the Maroteaux group in 1971, who proposed the term campomelic to describe the characteristic bowing of the long bones in CD patients (Maroteaux et al., 1971). Other radiological features of CD include hypoplastic scapulae and pelvis, bilateral clubfeet, scoliosis, underdeveloped vertebrae and Pierre Robin Sequence (PRS, OMIM#261800), a craniofacial condition consisting of micrognathia, cleft palate and retroglossoptosis. Early characterization of three independent *de novo* translocations in CD patients and mutation analysis in clinically confirmed CD patients established *SOX9* as the gene responsible for both CD and autosomal sex reversal (Foster et al., 1994; Wagner et al., 1994). Since then, loss of function studies in mice have revealed the multiple roles of SOX9 in regulating mesenchymal condensation of chondroprogenitors and the steps of chondrocyte differentiation and entry into hypertrophy (reviewed by Lefebvre in this book).

How is CD caused by heterozygous mutations in *SOX9*? Because most CD patients carry heterozygous mutations in the coding region of *SOX9* and these mutations appear to cause loss of SOX9 function, SOX9 haploinsufficiency was proposed to be the cause of CD. In support of this, *Sox9*$^{+/-}$ mouse mutants recapitulate most clinical features of the CD syndrome, including bowing of long bones, which is most likely a consequence of defective precartilaginous mesenchymal condensations and premature chondrocyte hypertrophy and mineralization (Bi et al., 2001). However, this mouse model does not address the molecular consequences of all genetic events that cause CD since they can be: (1) Amino acid substitutions (missense mutations); (2) Truncations and insertions/deletions of amino acid residues (nonsense and frameshift mutations); (3) Splice site mutations in the coding region of *SOX9*; or (4) Chromosomal rearrangements and translocations surrounding the *SOX9* locus.

Point mutations carried by CD patients have been mapped to different positions in the *SOX9* gene, affecting various protein domains. Nonsense and frameshift mutations that introduce a premature termination codon are the most prevalent class of mutations, contributing approximately 45% of all the modifications reported. These premature termination codons are scattered at different positions of the *SOX9* gene, resulting in RNA transcripts of different lengths. In most cases, the nonsense-mediated mRNA decay pathway is activated to degrade these mutant transcripts. However, some nonsense mutations escape the pathway and are translated into truncated SOX9 proteins, causing distinct phenotypes. For instance, the W86X, Q375X and E400X nonsense mutations result in SOX9 proteins that either lack the high mobility group (HMG) DNA-binding domain or possess disrupted PQA or transactivation domains. Most patients with such severely impaired SOX9 proteins died in the neonatal period. However, a patient carrying the Q117X nonsense mutation that deleted part of the HMG domain survived for 12 years (Meyer et al., 1997). It therefore appears that disease severity is not correlated with the type and position of a mutation within *SOX9* that would be consistent with a haploinsufficiency mechanism. However, the case of the Y440X mutation that truncates the C-terminal transactivation domain is an exception. This nonsense mutation is the most frequent mutation identified in CD patients, and affected individuals usually have a milder phenotype than patients with other CD mutations. In cell transfection experiments, SOX9^{Y440X} protein behaves as a hypomorph with residual transactivation ability, and might be causally related to the longer survival time of patients carrying this mutation (Meyer et al., 1997; Pop, Zaragoza, Gaudette, Dohrmann, & Scherer, 2005). Truncated polypeptides could also act as dominant negatives to disrupt the functions of wild-type proteins by competing for binding to target genes or partner factors. Furthermore, truncations might create novel protein interfaces that could recruit cofactors to transactivate a new set of target genes. Determining whether hypomorphic and/or dominant negative mechanisms act in CD would yield important insights into transcription complexes and gene regulatory networks controlling chondrogenesis.

While heterozygous mutations in human *IHH* cause BDA1, homozygous mutations cause acrocapitofemoral dysplasia (OMIM#607778), characterized by short stature with cone-shaped epiphyses. Thus, unlike SHH, which functions primarily during the mesenchymal condensation stage of limb patterning, IHH acts later to drive chondrocyte differentiation and proliferation. Indeed, homozygous inactivation of *Ihh* in mice results in a severe and lethal chondrodysplasia characterized by profound reduction in chondrocyte proliferation, premature hypertrophy and absence of mature osteoblasts (St-Jacques, Hammerschmidt, & McMahon, 1999). By contrast, overexpression of *Ihh* in chick embryos prevents proliferating chondrocytes from initiating hypertrophic differentiation. This inhibitory effect of IHH on chondrocyte hypertrophy is mediated largely by its ability to induce parathyroid hormone-related peptide (PTHrP) in periarticular chondrocytes (Vortkamp et al., 1996). PTHrP is a paracrine hormone that acts on a G-protein-coupled receptor, PTH1R, which is highly expressed in prehypertrophic cells. Null and gain-of-function mutations in the human *PTH1R* gene cause Blomstrand chondrodysplasia (BLC, OMIM#215045) and Jansen metaphyseal dysplasia (MCDJ, OMIM#156400), respectively. Blomstrand chondrodysplasia is an extremely rare condition characterized by advanced endochondral bone maturation and very short limbs. MCDJ is characterized by progressive growth plate changes that are caused by abnormal chondrocyte growth and differentiation, eventually leading to short and bowed legs. Ablation of PTHrP or PTH1R in mice accelerates chondrocyte maturation and thus premature mineralization of the entire skeleton, while mice expressing a constitutively active mutation in *PTH1R* display impaired chondrocyte differentiation and delayed mineralization (Lanske et al., 1999; Schipani et al., 1996).

The downstream molecular mechanisms through which PTHrP inhibits chondrocyte differentiation are examples of coordinated and interdependent action involving the key regulators of chondrogenesis and osteogenesis: IHH pathway, SOX9 and RUNX2. *RUNX2* is the master regulator of osteoblast differentiation and mutations lead to Cleidocranial dysplasia (CCD) in humans (OMIM#119600). PTHrP can act through cAMP-dependent PKA to phosphorylate SOX9, thereby increasing its transcriptional activity and delaying maturation of prehypertrophic chondrocytes (Huang, Chung, Kronenberg, & de Crombrugghe, 2001). Induction of PTHrP signaling also activates protein phosphatase 2A to dephosphorylate HDAC4, which enhances nuclear translocation of HDAC4 and consequently repression of the activity of the pro-hypertrophy transcription factor MEF2C (Kozhemyakina, Cohen, Yao, & Lassar, 2009). Furthermore, PTHrP signals block RUNX2, whose activity accelerates chondrocyte hypertrophy, by at least three parallel pathways. PTHrP drives the expression of the transcriptional repressor *Nkx3.2/Bapx1*, which directly represses *Runx2* expression (Provot et al., 2006). PTHrP also stimulates phosphorylation and degradation of RUNX2 and RUNX3 proteins in a cyclin-D1-dependent manner (Zhang et al., 2009). Lastly, PTHrP acts through the cAMP/PKA signaling pathway to induce the expression of *Zfp521*, a zinc finger transcriptional co-regulator which antagonizes RUNX2 transcriptional activity in a HDAC4-dependent manner (Correa et al., 2010). Ablation of *Zfp521* from chondrocytes rescues the chondrodysplasia phenotype in a mouse model of Jansen metaphyseal dysplasia, supporting the conclusion that this factor is a key effector of PTHrP signaling (Seriwatanachai et al., 2011). Expression of a constitutively active PTHrP receptor in *Ihh*-deficient mice prevents premature chondrocyte hypertrophy but fails to restore normal chondrocyte proliferation, indicating that IHH regulates chondrocyte proliferation and maturation by both a PTHrP-dependent and a PTHrP-independent manner (Karp et al., 2000). Indeed, IHH stimulates chondrocyte proliferation directly by modulating cyclin D1 expression (Long, Zhang, Karp, Yang, & McMahon, 2001). Through conditional activation and inactivation of HH signaling in *PTHrP*-null mutants, Mak et al. demonstrated that IHH also promotes chondrocyte hypertrophy in a PTHrP-independent manner (Mak, Kronenberg, Chuang, Mackem, & Yang, 2008). This aspect of IHH function is necessary for postnatal cartilage homeostasis and is likely to be masked by a more potent PTHrP-dependent activity of IHH, which is to suppress chondrocyte hypertrophy in embryonic development. Taken together, these genetic experiments support a model in which IHH and PTHrP interact in a negative feedback loop regulating the onset of hypertrophic differentiation. Through genetic and functional studies, a complex gene regulatory network emerges that controls chondrocyte differentiation via cooperative and antagonistic interactions between transcription factors such as SOX9, RUNX2, MEF2C and the HH-PTHrP feedback loops.

Targeting their downstream effectors, *Zfp521*, for example, might represent a potential therapeutic strategy in diseases caused by the collapse of the feedback loop.

4. Integrated signaling control of osteoblast differentiation and activity

The mechanisms underlying several bone disorders highlight the integration of signaling pathways in controlling bone formation. Progressive osseous heteroplasia (POH, OMIM#166350) is an autosomal dominant disorder characterized by widespread and disabling heterotopic ossification of skeletal muscle and deep connective tissues. It is caused by a null mutation of *GNAS*, which encodes $G\alpha_s$, a protein that transduces signals from G protein-coupled receptors (Shore et al., 2002). In contrast, activating mutations in *GNAS* cause fibrous dysplasia (FD, OMIM#174800), in which osteoblastic differentiation of mesenchymal progenitors is impaired (Riminucci, Robey, Saggio, & Bianco, 2010). These distinct phenotypic outcomes occur because GNAS modulates two key signaling pathways: HH and WNT/β-catenin. On one hand, $G\alpha_s$ inhibits HH signaling by promoting generation of the GLI3 repressor and restraining GLI2 activation through PKA (Tuson, He, & Anderson, 2011). On the other hand, it enhances WNT/β-catenin signaling at the level of β-catenin destruction-complex assembly by binding AXIN (Regard et al., 2011). In a mouse model of POH in which *Gnas* was conditionally inactivated in limb mesenchymal progenitor cells, Jean et al. found that suppression of HH signaling by $G\alpha_s$ safeguarded spatial restriction of bone formation at appropriate sites, and in the absence of $G\alpha_s$, increased HH signaling drove ectopic bone formation (Regard et al., 2013). Recent analyses of a mouse model of FD, carrying the corresponding human FD mutation (R201H), revealed that target genes of HH signaling were down-regulated, but there was significant upregulation of WNT/β-catenin target genes. $Gnas^{R201H}$ expression was found to impair osteoblastic differentiation of bone marrow stromal cells, and reduction of WNT/β-catenin signaling by removing one copy of *Lrp6* in the mutant line rescued the FD phenotype (Khan et al., 2018). Therefore, $G\alpha_s$ maintains a critical balance between the WNT/β-catenin and HH pathways so that osteogenesis occurs to the correct extent and at the right place. This finding also implies that activation of the HH pathway is sufficient to initiate osteoblastogenesis, but excess WNT activity can severely hamper this process.

The stage-specific roles of WNT and HH signaling during osteoblast differentiation were corroborated in several mouse mutant models. In *Ihh*-null mice, perichondrial progenitors fail to express *Runx2*, an indispensable osteoblast regulator (Chung, Schipani, McMahon, & Kronenberg, 2001). However, forced expression of *Runx2* in skeletogenic cells did not restore osteoblast differentiation in *Ihh*-null embryos, suggesting that other effectors are required to induce osteoblastogenesis (Tu, Joeng, & Long, 2012). In fact, the simultaneous activation of GLI2 and removal of GLI3 are needed to revive the osteogenic program in *Ihh*-null mice (Joeng & Long, 2009). Regarding the WNT pathway, genetic inactivation of β-catenin in mesenchymal progenitors using *Prx1-Cre* and *Dermo1-Cre* greatly impaired mature osteoblast formation in both endochondral and intramembranous bones. The osteogenic program in these mice was arrested at the RUNX2+ preosteoblast stage, resulting in ectopic cartilage formation in place of bones (Day, Guo, Garrett-Beal, & Yang, 2005; Hill, Spater, Taketo, Birchmeier, & Hartmann, 2005; Hu et al., 2005). Furthermore, loss of canonical WNT activity in Osterix (OSX)-expressing osteoblast precursors abolished their progression into mature Osteocalcin (OC)-expressing osteoblasts, and strikingly caused their conversion into chondrocytes. In contrast, persistent stabilization of β-catenin in OSX-expressing cells resulted in their dramatic proliferation and accelerated progression toward osteoblasts. However, their terminal maturation from low to high OC-expressing osteoblasts was impaired, indicating that the cessation of canonical WNT signaling is necessary for this transition (Rodda & McMahon, 2006). Taken together, clinical and genetic studies of mice demonstrate that β-catenin is dispensable for the expression of *Runx2* and *Osx*, which are downstream targets of IHH signaling. This observation places β-catenin downstream of the IHH signaling pathway and of OSX during early osteoblast differentiation in endochondral bones.

The power of exome sequencing led to the identification of biallelic truncating mutations in the secreted frizzled related protein 4 gene, *SFRP4*, a WNT inhibitor, in Pyle disease (OMIM# 265900) (Kiper et al., 2016). This metaphyseal dysplasia is characterized by thinned and fragile cortical-bone and limb deformity. Mouse models recapitulating the disease displayed increased trabecular bone and thinned cortical bone linked to dysregulated WNT and BMP signaling. Notably, the defects in SFRP4-deficient mice could be corrected by treatment with a soluble BMP2 receptor (RAP-661) or with anti-Sclerostin antibodies, reinforcing the need to consider the integrated actions of pathways.

5. Ciliopathies and the primary cilia in skeletal development

In the past decade, primary cilia emerged as important modulators of vertebrate HH signaling and their dysfunction has been linked to a spectrum of human diseases, collectively termed ciliopathies (Huber & Cormier-Daire, 2012). As cilia are a component of almost all cells, ciliary dysfunction often affects multiple organs and the phenotypic outcome is characteristic of aberrant HH signaling (Waters & Beales, 2011). For example, Meckel's syndrome (MES, OMIM#249000), Bardet-Biedl syndrome (BBS, OMIM#209900) and Joubert syndrome (JBTS, OMIM#21330), all present polydactyly as part of their highly distinct clinical phenotypes. These recessive pleiotropic disorders are genetically heterogeneous; BBS has been associated with mutations in at least 15 genes, which are involved in ciliary, basal body, centrosomal or intraflagellar transport function (Priya, Nampoothiri, Sen, & Sripriya, 2016). The molecular pathogenesis of polydactyly in BBS has been elucidated only in part. *In vitro* studies demonstrated that Smo and PTCH1 are endogenous cargos of the BBSome, a complex involved in ciliary membrane biogenesis. Loss of *BBS* genes causes accumulation of SMO and PTCH1 in cilia and leads to a decrease in SHH response which might result in polydactyly (Zhang, Seo, Bugge, Stone, & Sheffield, 2012). Similarly, JBTS is associated with mutations in any one of at least 10 different genes, all of which play a role in the formation or function of sensory cilia. Amongst them, KIF7 regulates vertebrate SHH signaling through organizing the cilium architecture (Dafinger et al., 2011; He et al., 2014). The *TALPID3* gene, encoding a centrosomal protein required for formation of the primary cilium, is also mutated in individuals with JBTS (Alby et al., 2015; Stephen et al., 2015). Conditional deletion of the equivalent gene in mice causes severe polydactyly (Bangs et al., 2011). Loss of TALPID3 activity disrupts GLI protein processing and phosphorylation through interaction with PKA (Li et al., 2017). The consequential loss of HH signal transduction and perturbation of the GLI3A:GLI3R ratio in the limb bud is the probable cause of digit abnormalities in skeletal ciliopathies.

6. Planar cell polarity in the development of growth plate

The Planar Cell Polarity (PCP) pathway controls the process of convergent extension and collective cell migration and thereby the elongation of the body axis and shapes of many organs (Henderson, Long, & Dean, 2018).

In endochondral ossification, the growth plate architecture of organized columns of cells requires PCP activity and its defects predispose humans to various skeletal dysplasias (Wang, Sinha, Jiao, Serra, & Wang, 2011). Robinow syndrome (RS) is a genetically and phenotypically heterogeneous skeletal condition characterized by short stature, mesomelic limb shortening, brachydactyly, genital hypoplasia, and craniofacial abnormalities. Two forms of RS have been described: Autosomal recessive Robinow syndrome (RRS, OMIM#268310) and the milder autosomal dominant Robinow syndrome (DRS, OMIM#180700). Although they share highly similar clinical features, patients with DRS rarely show rib fusions and their short stature is less pronounced (Mazzeu et al., 2007). A subtype of DRS, termed DRS-OS, was recently reported and includes osteosclerosis in addition to the phenotypes listed above (Bunn et al., 2015). RRS results from loss-of-function mutations in the gene encoding the tyrosine kinase-like orphan receptor 2, *ROR2* (Afzal et al., 2000; van Bokhoven et al., 2000). By contrast, DRS can be caused by mutations in several PCP gens, including *WNT5A*, *DVL1*, and *DVL3* (Person et al., 2010; White et al., 2015, 2016). WNT5A is a classical non-canonical WNT ligand that binds and signals through ROR2 to activate the WNT-planar cell polarity pathway, hence controlling directional growth of chondrocytes along the proximal-distal axis for limb elongation (Gao et al., 2011). DVL1 and DVL3 play important roles in the transduction of both canonical and non-canonical WNT signaling (Gentzel & Schambony, 2017).

In the mouse, *Ror2* is expressed in the reserve and proliferative zones of the growth plate and perichondrium, but not in the hypertrophic zone. Mice lacking ROR2 display widespread skeletal abnormalities in endochondral but not intramembranous bones, and intriguingly distal long bones in the limb are more affected than proximal long bones. These anomalies are attributed to the initially shortened and disorganized cartilage anlagen in $Ror2^{-/-}$ embryos, suggesting that ROR2 is required for proper expansion of chondrocytes and patterning of the cartilaginous templates (DeChiara et al., 2000; Takeuchi et al., 2000). Later analyses confirmed that ROR2 is needed for chondrocyte maturation and differentiation, in part via positive regulation of the IHH signaling pathway (Witte, Chan, Economides, Mundlos, & Stricker, 2010). Shortly after, Gao et al. found that ROR2 and VANGL2 act together to transduce the WNT5A signal and establish planar cell polarity, hence directional growth of chondrocytes along the proximal-distal axis for limb elongation (Gao et al., 2011). Collectively, DRS and RRS-associated mutations in PCP genes impair chondrocyte polarization and arrangement that are necessary for the proximal-distal elongation of the long bone cartilage. Because distal limb chondrogenesis requires higher WNT-PCP activity, this might explain the more severe phenotype presented by distal long bones in RS patients (Gao et al., 2011).

Mutations in *ROR2* are also associated with a phenotypically different condition: Dominant brachydactyly type B1 (BDB1, OMIM#113000) (Schwabe et al., 2000). BDB1 is characterized by hypoplasia and/or aplasia of distal phalanges, often accompanied by nail dysplasia resulting in an amputation-like phenotype. The genetic lesions found in BDB1 patients consist of frame shift and nonsense mutations that cluster in two hotspots immediately upstream and downstream of the tyrosine kinase domain. They are predicted to generate truncated proteins that operate as a gain-of-function or dominant negative mutants (Afzal & Jeffery, 2003). By contrast, RRS-associated biallelic mutations are scattered throughout *ROR2* and, as supported by gene inactivation experiments in mice, these changes most likely lead to a complete loss of ROR2 activity (Afzal & Jeffery, 2003). *In vitro* studies have suggested that RRS ROR2 mutant proteins are retained within the endoplasmic reticulum (ER), whereas the BDB1 mutants should be able to reach the cell membrane and interfere with normal signaling (Chen, Bellamy, Seabra, Field, & Ali, 2005). This hypothesis is supported by a rare finding of a nonsense mutation in ROR2 (p.R441X) in patients with RRS and severe recessive brachydactyly. This mutation is located in the same position as a previously described frame shift mutation (p.R441fsX15) that causes dominant BDB1, arguing for a complex pathological mechanism rather than simple loss- or gain-of-function. The R441X mutant protein is not completely retained in the ER; a significant fraction is able to reach the cell membrane in a similar manner to R441fsX15, albeit at a lower total protein level (Schwarzer, Witte, Rajab, Mundlos, & Stricker, 2009). These findings provide a quantitative biochemical explanation for the recessive nature of the BDB1 phenotype in patients harboring the p.R441X mutation, as the amount of truncated ROR2 at the cell membrane is insufficient to elicit a dominant BDB1 phenotype. Thus the RRS versus BDB1 phenotype is determined by the relative degrees of intracellular retention and membrane localization of mutant ROR2, illustrating how minimal sequence changes in a polypeptide can generate overlap of two discrete phenotypes as well as a dosage requirement for ROR2 function in regulating both chondrocyte differentiation and convergent extension in the growth plate.

7. The impact of ER stress signaling on chondrocyte differentiation

The different cell types in the mammalian skeleton are embedded in tissue-characteristic complex extracellular matrix (ECM) networks, composed of collagens, proteoglycans, glycosaminoglycans, and glycoproteins. The ECM is important not only in providing structural support but by influencing cell adhesion, proliferation, migration, survival, differentiation and control of cell fate and morphogenesis. The pivotal role of the ECM is reflected in the major contribution of disruption in genes encoding ECM components to diverse skeletal disorders (Bonafe et al., 2015; Geister & Camper, 2015). Mounting evidence over the past decade suggests that instead of causing peptide truncation, some mutations affect conserved residues that are structurally important for protein assembly and secretion. Accumulation of the resulting misfolded or unfolded protein in the ER can induce ER stress signaling (ERSS) pathways and the unfolded protein response (UPR) to enhance protein folding, attenuate translation and degrade misfolded protein. Luminal distension and/or dilation of the ER, a hallmark of cells undergoing ER stress, is commonly observed in several heritable human skeletal dysplasias caused by mutations in ECM components (Hughes, Oxford, Tawara, Jorcyk, & Oxford, 2017; Tsang, Chan, Bateman, & Cheah, 2010).

The connection between ERSS and skeletal dysplasias is exemplified by mutations in Matrilin-3 (*MATN3*), cartilage oligomeric matrix protein (COMP), type II collagen and type X collagen, resulting in multiple epiphyseal dysplasias (MED, OMIM#132400), pseudoachondroplasia (PSACH, OMIM#177170), Type II collagenopathies and metaphyseal chondrodysplasia type Schmid (MCDS, OMIM#156500), respectively. MED and pseudoachondroplasia are relatively common skeletal conditions resulting in short-limbed dwarfism, joint pain and stiffness. They can be caused by mutations in various genes. Mutations in *COMP*, *MATN3* and type IX collagen genes account for the vast majority of classical autosomal dominant MED, while *SLC26A2* is the only gene known to be associated with autosomal recessive MED (Jackson et al., 2012; Posey, Coustry, & Hecht, 2018). Despite the strong clinical association of Matrilin-3 mutations with MED, *Matn3*-null mice show no appreciable skeletal malformations, suggesting that the MED phenotype is caused by a dominant negative effect of the mutant matrilin-3 rather than by the lack of matrilin-3 in cartilage ECM

(Ko et al., 2004; van der Weyden et al., 2006). In support of this view, a knock-in mouse model of MED, expressing the equivalent of the human p.Val194Asp mutation in mouse matrilin-3, exhibited retention of mutant matrilin-3 in the distended ER, elevated ER stress and activation of the UPR. This resulted in reduced proliferation, spatially dysregulated apoptosis and misalignment of mutant chondrocytes, collectively causing the short-limb dwarfism that is characteristic of MED (Leighton et al., 2007). Mutations in cartilage oligomeric matrix protein (COMP), clustering in the TSP3 repeat region of COMP that compromise its processing, result in Pseudoachondroplasia (PSACH). COMP-deficient mice do not phenocopy PSACH or MED, but a targeted mutation (equivalent to human p. T585M) in the C-terminal domain of mouse COMP causes mild pseudoachondroplasia, including short-limbed dwarfism with disordered growth plate morphology and mis-localized ECM protein (Pirog-Garcia et al., 2007; Svensson et al., 2002). Although the mutant COMP is not retained within the rough ER in this case, a mild ER stress is activated in chondrocytes as evidenced by the upregulation of several UPR markers such as BiP, eIF2-alpha and ATF6. This cellular response appears sufficient for the correct folding and secretion of $COMP^{T585M}$, but at the expense of reduced chondrocyte proliferation and increased apoptosis (Pirog-Garcia et al., 2007). Surprisingly, a knock-in mutation in the TSP3 region of *COMP* (p.Asp469del) results in the most severe form of pseudoachondroplasia and ER retention of the mutant COMP along with other ECM molecules. However, this does not induce a conventional UPR, but a novel form of ERSS that involves transcriptional changes in genes involved in oxidative stress, cell survival and apoptosis (Suleman et al., 2012). A possible cause of these discrepancies in the impact of ERSS is the different molecular organization of mutant protein aggregates. $MATN3^{V194D}$ forms non-native disulfide-bonded aggregates that render it inaccessible to degradation (Bell et al., 2012). By contrast, D469del COMP is retained in the ER in its native state of either tetramers or pentamers (Bell et al., 2013). These findings highlight variation in impact of activating ER stress in skeletal cells depending on their proliferation or differentiation state and in differences of their deployment of ERSS to restore ER homeostasis, with survival or apoptosis as the ultimate outcome. However, both outcomes result in skeletal dysplasia.

An example of how chondrocytes evade apoptosis and adapt to pathological ER stress is exemplified in MCDS, which is caused by heterozygous mutations in the *COL10A1* gene impairing collagen type X folding and trimeric assembly. Collagen type X misfolding induces ER stress and

UPR in transfected cells, and this was confirmed in transgenic mouse models expressing mutant collagen X in hypertrophic chondrocytes (HCs) (Ho et al., 2007; Tsang et al., 2007; Wilson, Freddi, Chan, Cheah, & Bateman, 2005). These mutant mice display characteristics of MCDS, disproportionate dwarfism with short limbs and a significant expansion of the hypertrophic zone. Importantly, the same phenotypes are present in mice that ectopically express an ER stress-inducing protein (*cog* mutant of thyroglobulin) in HCs, thereby confirming ER stress as a direct pathogenic factor in MCDS (Rajpar et al., 2009). In the transgenic mouse MCDS model (13del) which expressed the mouse equivalent of a human 13-base pair deletion in *COL10A1*, intracellular accumulation of mutant proteins, ER distention, and UPR activation were detected in HCs and the severity correlated with transgene dosage. Instead of differentiating to late HCs as normal HCs would, the 13del mutant HCs undergo cell-autonomous reversion to a pre-HC like state by re-expressing pre-hypertrophic markers and re-entering the cell cycle. Concomitantly, expression of both *Col10a1* and the 13del transgene is downregulated, thereby alleviating the load of mutant protein and ER stress exerted on the 13del HCs (Tsang et al., 2007). Similar findings were observed in a collagen type X (p.Asn617Lys) knock-in mouse model of MCDS in which suppression of vascular endothelial growth factor-mediated vascular invasion and osteoclast recruitment was proposed to have contributed to the reduced bone growth (Rajpar et al., 2009). Subsequent transcriptomic analysis identified CHOP-mediated transcriptional repression of C/EBP-β as responsible for the developmental arrest of ER-stressed $ColX^{N617K}$ chondrocytes in these mice (Cameron et al., 2015). A recent study in the 13del mouse model has provided further molecular insights into the mechanism by which ER stress causes MCDS pathology. Transcriptomic analysis of fractionated 13del growth plates revealed that, of the three sensing arms of the UPR, upregulation of PERK-p-eIF2 signaling (mediating preferential translation of transcription factors such as ATF4 and CHOP) plays a dominant role in controlling gene expression changes in ER-stressed HCs. It was shown that ATF4 directly activated *Sox9* in 13del HCs, reverting their differentiation to a more juvenile state, and moreover ATF4 worked cooperatively with CHOP to induce *Fgf21* and protect HCs from apoptosis in a cell autonomous manner (Wang et al., 2018). Taken together, these findings implicate cell fate change as an adaptive strategy to facilitate survival and recovery of chondrocytes in the face of ER stress. However, it comes with a debilitating cost; interruption of the differentiation program causes delayed endochondral ossification and chondrodysplasia.

Mutations that inactivate components of the ER sensors and stress transducers can also lead to skeletal defects. A prime example is *BBF2H7*, an ER-resident stress transducer. *Bbf2h7*-null mice show severe chondrodysplasia due to perturbed chondrocyte proliferation and differentiation, and importantly impaired intracellular trafficking of secreted proteins (Saito et al., 2009). Part of these defects can be explained by the ability of BBF2H7 to promote chondrocyte proliferation through direct regulation of the IHH-PTHrP pathway and to activate Sec23a for efficient ER-to-Golgi protein trafficking (Saito et al., 2014). Recently, whole exome sequencing identified compound heterozygous mutations in MBTPS1, encoding the serine protease Site-1 protease (S1P), in patients displaying spondyloepiphyseal dysplasia, dysmorphic facial features and retarded skeletal growth. S1P functions in the UPR by cleaving membrane bound transcription factors ATF6 and BBF2H7, key transducers of the UPR. The consequence of the *MBTPS1* mutations is ER stress caused by the accumulation of collagen in the ER, an impaired UPR caused by S1P deficiency and apoptosis (Kondo et al., 2018).

Other examples include XBP1 and ATF4, which are critical stress transducers in the UPR. Cartilage-specific inactivation of *Xbp1* and global ablation of *Atf4* each causes a dramatic dwarfism phenotype characterized by dysregulated chondrocyte proliferation and hypertrophy (Cameron et al., 2015; Wang et al., 2009). There are clear phenotypic differences between these mouse models. The chondrodysplasia and growth plate abnormalities in mice lacking chondrocytic XBP1 signaling are much milder than those observed in either *Bbf2h7* or *Atf4* null mice. This implies that components of the ER stress-response pathways regulate endochondral ossification through a variety of mechanisms. These studies using genetically manipulated mouse models clearly establish a causative role of ERSS in several chondrodysplasias caused by the synthesis of misfolded or unfolded secreted proteins.

The pathogenic impact of ERSS in skeletal dysplasia is not restricted to chondrocyte differentiation. Osteogenesis imperfecta (OI, OMIM#166200) is another example of a disease associated with the negative impact of ERSS on osteoblast differentiation and maturation. In OI many heterozygous mutations in the type I collagen genes *COL1A1* and *COL1A2* can perturb folding and secretion and trigger ERSS (Bateman, Boot-Handford, & Lamande, 2009). Recent studies showed that activation of the UPR in osteoblasts caused increased WNT signaling that delays their maturation resulting in prolonged active bone synthesis and generalized hyperostosis. The phenotypic similarity to craniodiaphyseal dysplasia (CDD, OMIM#122860) caused by heterozygous mutations in *SOST* that result in impaired protein folding implicate ERSS as a possible disease mechanism (Chan et al., 2017).

8. Non-coding mutations and regulatory control of skeletal development

The discovery of mutations in non-coding genomic regions that cause skeletal dysplasia has brought a new dimension to our understanding of the regulatory control of skeletogenesis (Fig. 2). An outstanding example is the identification of a highly conserved *cis*-regulatory element within the preaxial polydactyly (PPD) transcription-associated region, called the ZPA regulatory sequence (ZRS). This enhancer is responsible for the initiation and spatially restricted expression of *Shh* in the ZPA, despite its extreme distance ($\sim$1 Mb) from the *Shh* promoter, in mice and humans (Lettice et al., 2003). When tested in transgenic mice, all human ZRS mutations cause ectopic anterior expression of *Shh* in the forming limb bud and the degree of misexpression is directly related to the severity of the polydactylous phenotype in PPD patients (Lettice, Hill, Devenney, & Hill, 2008). In accordance, tandem duplications of ZRS that presumably result in greater dysregulation of SHH segregate with severe forms of type Haas polysyndactyly (OMIM#186200) and triphalangeal thumb-polysyndactyly syndrome (OMIM#174500) (Sun et al., 2008; Wieczorek et al., 2010).

Factors whose binding to ZRS might be disrupted or enhanced by the mutations can be predicted bioinformatically. Through this approach and by chromatin immunoprecipitation (ChIP), members of the ETS transcription factor family were shown to act directly at the ZRS in defining *Shh* spatial expression pattern. Binding of GABPα and ETS1 to the ZRS delineate the boundary of the ZPA, whereas ETV4/ETV5 occupancy restricts *Shh* expression outside the ZPA (Lettice, Devenney, De Angelis, & Hill, 2017). When ZRS mutation creates an additional GABPα/ETS binding site, it overrides the ETV4/ETV5-mediated spatial constraint, leading to ectopic anterior SHH signaling and PPD (Lettice et al., 2012). Alternatively, the Werner mesomelic syndrome (OMIM#188740) that also has PPD as part of its symptoms is caused by point mutations that abrogate binding of a repressor at ZRS, and thus permit ectopic *Shh* expression and PPD (Lettice et al., 2017). Therefore, non-coding mutations in the ZRS have two modes of action, gain or loss of transcription factor interaction, but both result in dominant manifestation of PPD.

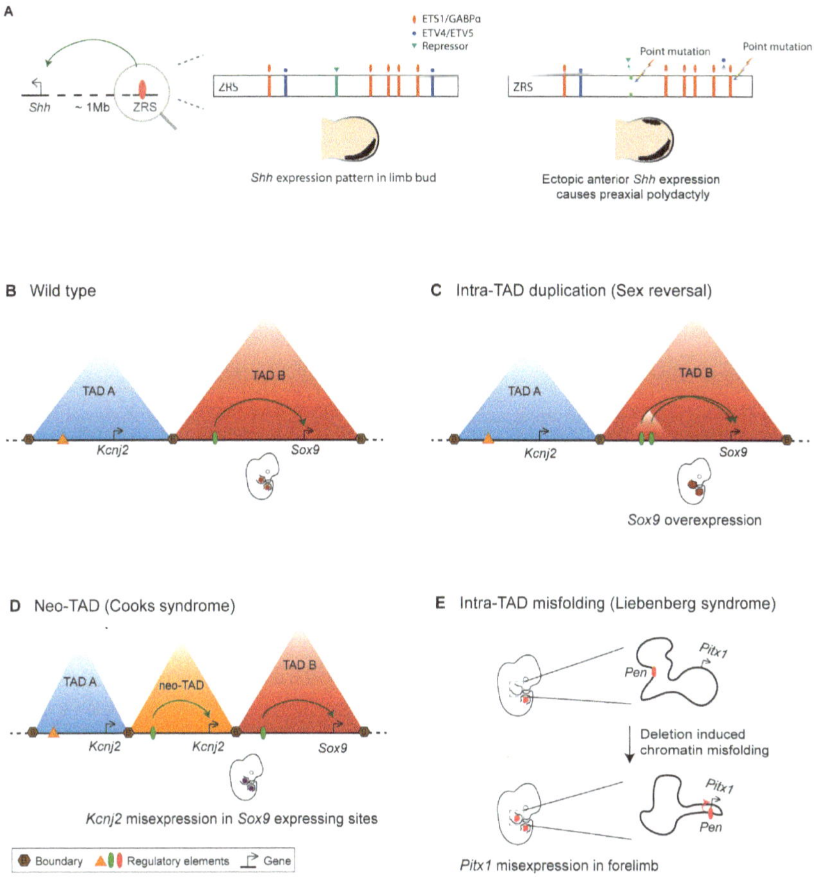

Fig. 2 Disrupting non-coding regulatory elements and 3D genome folding causes skeletal dysplasia. (A) Multiple ETS family members and an undefined repressor protein bind directly to ZRS enhancer to define *Shh* spatial expression in limb bud. Point mutations that change their binding profile drive ectopic anterior expression of *Shh* leading to preaxial polydactyl (Lettice et al., 2012). (B) In the wild-type genomic locus, *Sox9* and *Kcnj2* are compartmentalized into two different TADs separated by boundary elements. This spatial organization restricts chromatin interactions of a limb-specific regulatory element with its cognate target, *SOX9* (Franke et al., 2016). (C) An intra-TAD duplication does not change the overall TAD configuration but can result in increased interaction of duplicated regulatory elements with *Sox9*, which may underlie female to male sex reversal in humans (Franke et al., 2016). (D) Duplications of the boundary and the flanking sequence can create a new chromatin domain (neo-TAD). In the neo-TAD, the duplicated *Kcnj2* is regulated by the duplicated regulatory element that originally belonged to the *Sox9* TAD, driving ectopic *Kcnj2* expression in the developing limbs that underlies Cooks syndrome (Franke et al., 2016). (E) *Pitx1* is regulated by a pan-limb enhancer (*Pen*) that shows activity in both forelimbs and hindlimbs. However, *Pen* and *Pitx1* are physically separated in forelimbs because of chromatin folding. Liebenberg syndrome-associated deletion can convert the three-dimensional chromatin conformation of this locus, thereby permitting contact between *Pitx1* and *Pen*, and *Pitx1* misexpression in forelimbs, causing partial arm-to-leg transformation in humans (Kragesteen et al., 2018).

Another notable skeletal condition caused by disruption of non-coding elements is CD. As discussed earlier, CD can be caused by sequence changes within and surrounding the *SOX9* gene. Individuals carrying genomic lesion outside the *SOX9* gene, particularly balanced chromosomal rearrangements, tend to have better clinical outcomes and longer life expectancy (Leipoldt et al., 2007; Pop et al., 2004; Velagaleti et al., 2005). One example is the eponymous feature of CD, campomelia, which is detected in over 90% of the *SOX9* coding mutation cases but in only 50% of the translocation cases, classifying the latter as acampomelia CD (ACD) patients (Pfeifer et al., 1999). Non-coding chromosomal aberration cases are over-represented in long-term survivors of CD, ACD and isolated PRS individuals (Leipoldt et al., 2007). These observations indicate that chromosomal aberrations that presumptively alter *SOX9* expression do not affect embryonic development to the same extent as the heterozygous *SOX9* point mutations. Adding to this complexity, a range of chromosomal translocations, inversions and deletions have been described to cause CD of varying severity. These breakpoints are scattered over a ~1 Mb region upstream of *SOX9*, and occasionally downstream, and fall into two major clusters: A proximal cluster at 50–375 kb and a distal cluster at 789–932 kb upstream of *SOX9* (Leipoldt et al., 2007). Intriguingly, a striking correlation exists between symptom severity and the location of the breakpoint. Patients with translocation breakpoints within the proximal cluster tend to have severe or moderate bowing of long bones, while all lesions at the distal cluster result in straight long bones—that is, ACD. Additionally, a balanced translocation breakpoint at ~1.3 Mb downstream of *SOX9* has been reported in a sex-reversed ACD patient (Velagaleti et al., 2005). Two cases of sex-reversed ACD presented large-scale deletions distal to *SOX9* (Lecointre et al., 2009; Pop et al., 2004). These deletion cases support a causative relationship between loss of crucial *SOX9* regulatory elements and CD pathology, because in translocation cases the altered *SOX9* expression might be due to a positional effect invoked by an ectopic chromosomal environment.

Another notable skeletal disorder caused by structural variation in the genomic regions flanking *SOX9* is Pierre Robin sequence (PRS). This craniofacial birth defect was previously classified as part of CD and ACD, but genetic studies have identified translocations and deletions far up- and down-stream of *SOX9* in several isolated PRS cases (Benko et al., 2009; Xu et al., 2016). These PRS-associated sequence variants are clustered in a region around 1.2 Mb upstream of *SOX9*, which forms a third hotspot of genomic lesions, even further upstream from *SOX9* than the proximal and distal breakpoint clusters, that is associated with PRS only (Gordon et al., 2014). Therefore, PRS represents the mildest phenotype amongst the spectrum of disorders caused by *SOX9* perturbations. This notion of CD-ACD-PRS being a phenotypic continuum with descending disease severity can be explained by the deletion or disruption of variable amounts of regulatory elements, resulting in different degrees of *SOX9* dysregulation and thus different CD endophenotypes. Most recently, a 97-kb deletion at ~2.3 Mb upstream of *SOX9* was reported in a patient with a classical CD phenotype. Although this lesion site is more distal than any previously described breakpoints, the patient presented a severe disease phenotype, suggesting the presence of critical *SOX9* regulatory elements beyond the PRS cluster (Antwi et al., 2018). *SOX9* enhancers work in concert rather than singly. Therefore, the identification of the combination of enhancers that control stage- and cell-type specific expression of *SOX9 in vivo* will provide insight into the regulatory network controlling chondrogenesis. These loci will also be important for genome wide association studies in providing functional information for variants.

9. Impacting 3D genome folding in skeletal disorders

Genomic DNA in the nucleus is organized as topologically associating domains (TADs), which are fundamental structural units that guide the physical interaction between *cis*-regulatory elements and promoters while insulating adjacent domains from inappropriate contacts (Dixon et al., 2012; Rao et al., 2014). Structural variations caused by chromosomal rearrangements or insertions or deletions affecting TADs can have profound effects on gene regulation and thereby disease outcomes (Kragesteen et al., 2018) (Fig. 2B–D).

Evidence that TADs' configuration and integrity is crucial for gene regulation in skeletal development comes from two pioneering studies that used CRISPR/Cas genome editing to generate mice with structural variations recapitulating human skeletal disease alleles. By combining clinical genetics and genome-wide chromosome conformation capture assay (Hi-C), Lupianez et al. first identified a series of structural variants that disrupt the *EPHA4*-containing TAD in patients with brachydactyly, polydactyly and F-syndrome (OMIM#102510), a limb malformation disorder characterized by index/thumb syndactyly. Adopting the same Hi-C technique in mice that were engineered to phenocopy the diseases, the team demonstrated that these disease-associated structural changes cause TAD reorganization and *de novo* interaction between distal limb enhancers and target genes such as *Pax3*, *Wnt6* and *Ihh*, causing their misexpression at ectopic sites and, consequently, limb malformations (Lupianez et al., 2015). Regarding the *SOX9* locus, duplications within the regulatory domain of *Sox9* (intra-TAD structural variations) promote contacts between the duplicated enhancer elements and the *Sox9* promoter, likely causing *Sox9* overexpression that leads to female-to-male sex reversal in the patient (Fig. 2C). By contrast, a duplication that spans two TADs (inter-TAD structural variation) creates new chromatin domains (neo-TAD) that can elicit different effects, depending on their size and content. For instance, a large duplication of non-coding elements 5′ of *Sox9* causes formation of neo-TAD with the inclusion of the flanking gene, *Kcnj2*. The resulting ectopic interaction of *Kcnj2* with regulatory elements that originally belonged to the *Sox9* TAD leads to misexpression of *Kcnj2* in a *Sox9*-like pattern, and this produces a digit and joint malformation phenotype associated with Cooks syndrome (Fig. 2D). A slightly smaller structural change that does not incorporate the *Kcnj2* gene in the neo-TAD has no phenotypic consequences as no gene is positioned in a novel regulatory landscape (Franke et al., 2016).

Genetic manipulation of ZRS at the endogenous locus has defined two discrete domains that encode the enhancer activity of this sequence. The 5′ half of ZRS is crucial in directing spatiotemporal activity at a short range, whereas the 3′ half mediates long-range enhancer-promoter interactions to ensure robust *Shh* activation for normal limb patterning (Lettice et al., 2014). Deletion of the 3′ end of ZRS reduces the ability of ZRS to adopt the appropriate chromosomal conformation that is necessary for full *Shh* activation, resulting in a limb phenotype similar to those caused by *Shh* or ZRS knockout. The importance of chromatin organization in *Shh* regulation is further illustrated in a recent report that profiled the *cis*-interactome at the *Shh* locus by chromosome conformation capture sequencing. By analyzing a series of mouse strains carrying a structurally rearranged *Shh* allele, the authors found that ZRS can act pervasively across the entire TAD in which they occur. Remarkably, changing enhancer-promoter distance within the TAD has little effect on *Shh* expression. In contrast, a TAD-breaking inversion event abrogates physical and functional interactions between ZRS and *Shh*, unless the genomic distance between the two is substantially reduced. In this case, partial restoration of ZRS-*Shh* communication permits sporadic *Shh* activation producing less impaired limb structures (Symmons et al., 2016). Finally, several studies have found that genetic changes outside of the ZRS can cause *SHH* misexpression, leading to limb anomalies. Petit et al. reported that a 2 kb deletion located ~240 kb upstream from the *SHH* promoter is associated with autosomal dominant PPD, and the deleted sequence contains a silencer element capable of repressing *SHH* transcription when tested *in vitro* (Petit et al., 2016). Additionally, mutations in a newly defined, evolutionarily conserved sequence 500 bp upstream of ZRS, termed the pre-ZRS (pZRS), have been associated with PPD and triphalangeal thumb-polydactyly syndrome in humans. Although the mode of action of pZRS remains unknown, it is predicted to function as an independent enhancer of *SHH* as its mutation leads to ectopic activity throughout the entire limb bud (Potuijt et al., 2018; Xiang et al., 2017).

Another disease that is caused by disorganization of TADs is Liebenberg syndrome (OMIM#186550). It is a very rare autosomal dominant condition that involves homeotic transformation of the upper extremities, in which the arms acquire morphological characteristics of the legs (Spielmann et al., 2012). Genetic screening has documented heterozygous deletions and translocations 5′ of the *PITX1* gene as the cause of Liebenberg syndrome (Al-Qattan, Al-Thunayan, Alabdulkareem, & Al Balwi, 2013; Spielmann et al., 2012). PITX1 has an established role in driving hindlimb morphology. *Pitx1* is exclusively expressed in murine hindlimb bud mesenchyme and its inactivation abrogates hindlimb patterning and outgrowth (Duboc & Logan, 2011; Marcil, Dumontier, Chamberland, Camper, & Drouin, 2003). Conversely, misexpression of *Pitx1* in mouse forelimb results in the transformation and translocation of musculoskeletal components so that they acquire a hindlimb-like morphology (DeLaurier, Schweitzer, & Logan, 2006; Logan & Tabin, 1999).

How do genomic rearrangements cause ectopic *PITX1* expression in the forelimb? Examination of the *cis*-regulatory landscape surrounding the deletion and translocation sites discovered a forelimb-specific enhancer, which is able to drive *Pitx1* misexpression and produce a Liebenberg-like phenotype in transgenic assays (Spielmann et al., 2012). Recently, Kragesteen et al. reported that *Pitx1* is regulated by a pan-limb enhancer (*Pen*) which displays strong activity in both forelimbs and hindlimbs. The hindlimb-specific activation of *Pitx1* is achieved by a switch in 3D chromatin architecture such that *Pitx1* is disconnected from *Pen* in the forelimb, but is close to its enhancer in the hindlimb. By reengineering a human Liebenberg deletion in mice, Kragesteen et al. demonstrated that the lesion causes misfolding of the locus and permits ectopic interactions between *Pitx1* and *Pen*. The resulting regulatory endo-activation of *Pitx1* in the forelimbs induces partial transformed morphogenesis in which the arms are partially transformed into legs, resembling the Liebenberg syndrome (Kragesteen et al., 2018) (Fig. 2E). This example, together with the *SOX9* and *SHH* cases, shows how structural variations elicit pathogenic gene expression through a change in chromatin conformation. Given that genome organization is a fundamental property in every cell, chromatin misfolding is likely linked to the onset and progression of a broad range of human diseases.

10. Mechanistic insights from skeletal disorders: Impacting the path to therapy

The ultimate hope for skeletal dysplasia patients is the availability of therapies that can ameliorate or prevent the dysmorphology. In recent years some progress has been made toward the development of therapeutic approaches for certain types of congenital dwarfism, based on the underlying mechanistic insights gained from fundamental research. Here, we highlight two examples for which mechanistic insights have been exploited and have entered human clinical trials: Achondrodysplasia and Schmid metaphyseal chondrodysplasia.

Achondroplasia (ACH, OMIM#100800) is the most common type of short-limbed dwarfism in humans, occurring in 1 out of every 15,000 to 40,000 live births. Rhizomelic shortening of the limbs, macrocephaly with frontal bossing, trident hands, and pronounced lumbar lordosis are typical features of this genetic disorder (Unger, Bonafe, & Gouze, 2017). Genetic linkage studies revealed that almost all ACH patients have one of two mutations at nucleotide 1138, c.1138G>A or c.1138G>C, in the Fibroblast Growth Factor Receptor 3 gene (*FGFR3*), causing a glycine-to-arginine substitution at position 380 (G380R) in its transmembrane domain (Rousseau et al., 1994; Shiang et al., 1994). Although the G380R mutant FGFR3 remains ligand-dependent for its dimerization and activation, the mutation stabilizes the ligand/receptor complex at the cell membrane and impairs receptor internalization and degradation, culminating in excessive kinase activity of the receptor (Cho et al., 2004; Monsonego-Ornan, Adar, Feferman, Segev, & Yayon, 2000). Mutations affecting the extracellular and intracellular domains result in ligand-independent constitutive activation of the receptor, leading to the more severe forms of dwarfism, Thanatophoric Dysplasia Types I and II (OMIM#187600) (Naski, Wang, Xu, & Ornitz, 1996; Tavormina et al., 1995). In support of an inhibitory role for FGFR3 on endochondral bone growth, FGFR3 loss-of-function mutations cause skeletal overgrowth in patients (Makrythanasis et al., 2014).

A myriad of *in vitro* and mouse genetic studies have demonstrated that FGFR3 signaling uses two independent pathways to stunt bone growth: It acts through STAT1 to inhibit chondrocyte proliferation and the MAPK pathway to suppress chondrocyte differentiation (Krejci et al., 2004; Murakami et al., 2004; Parafioriti et al., 2009). Another key player that participates in FGFR3 signaling in skeletal dysplasia is SNAIL1. This transcription factor has been implicated in chondrocyte differentiation as it represses *Col2a1* and *Aggrecan* transcription (Seki et al., 2003). Aberrant Snail1 activation in mice causes an ACH-like phenotype, while Snail1 inhibition abrogates FGFR3 activity in cultured chondrocytes. The growth-inhibitory function of Snail1 was attributed to its ability to induce STAT1 nuclear translocation, which upregulates the cell cycle inhibitor *p21* to arrest chondrocyte proliferation (de Frutos et al., 2007). Snail1 also promotes activation and nuclear translocation of phosphorylated Erk1/2, which has been found to be responsible for retarded skeletal growth (Sebastian et al., 2011; Smith et al., 2014).

The molecular knowledge gained from the above mouse models is pivotal for designing rational therapy to treat FGFR3-associated skeletal defects. Besides the traditional approaches aimed at directly blocking FGFR3 activation using FGFR3-binding peptide or soluble FGFR3 decoy receptor, drugs that disrupt the intracellular signaling events appear to be a promising therapeutic approach for ACH. The C-type natriuretic peptide (CNP) analog, BMN111. CNP antagonizes FGFR3 signaling by inhibiting the MAPK pathway, particularly the ERK arm of this pathway at the level of Raf-1 (Krejci et al., 2005; Ozasa et al., 2005). Targeted overexpression of CNP in chondrocytes or continuous intravenous delivery of CNP rescues the impaired bone growth phenotype of ACH mice (Yasoda et al., 2004, 2009). Similarly, BMN111 administration alleviates the dwarfism phenotype in two *Fgfr3* ACH mouse models and in juvenile cynomolgus monkeys in a dose-dependent manner (Lorget et al., 2012; Wendt et al., 2015). The Phase 1 clinical trial of BMN111 (Vosoritide) for ACH treatment was completed in 2012 and the results indicated that the drug was generally well tolerated with no major toxicities. Phase 2 data also revealed a favorable safety profile and efficacy of BMN111 in children with ACH, demonstrating a 50% increase in annualized growth velocity compared with their pretreatment growth rate (Klag & Horton, 2016). Based on these encouraging results, BioMarin Pharmaceuticals has initiated a Phase 3 trial to evaluate the efficacy and safety of BMN111 in over 100 subjects (www.clinicaltrials.gov identifier NCT03424018).

Exploiting the insights gained into the impact of the UPR on chondrocyte differentiation for therapy is another example. As described in Section 7, the impact of the UPR and ER stress is implicated in a number of skeletal disorders. The use of chemical chaperones such as 4-PBA in OI models has been tried with varying success (Besio et al., 2018; Gioia et al., 2017). Major insights have been gained into the underlying mechanisms of the impact of ER stress on hypertrophic chondrocytes and the development of the growth plate in MCDS. One important lesson learnt from these findings is that pathological ER stress does not necessarily result in apoptotic cell death; in contrast, it can alter the cell differentiation trajectory and disrupt normal development culminating in disease. Thus, ERSS components are viable therapeutic targets, as illustrated by evidence that their inhibition has beneficial effects in MCDS mouse models. Carbamazepine (CBZ), an FDA-approved drug used to treat epilepsy, stimulates both autophagy and proteasomal degradation pathways. Administration of CBZ to a mouse model of MCDS reduced the level of ER stress caused by the accumulation of the mutant collagen type X protein and partially ameliorated the dwarfism phenotype (Mullan et al., 2017). These promising results have now led to the initiation of clinical trials in 2018 to test the efficacy of CBZ in MCDS patients (https://mcds-therapy.eu/).

However, in CBZ treated MCDS mice, the disrupted hypertrophic differentiation and enlarged hypertrophic zone defects were not completely rescued, implying that stimulation of mutant protein degradation *per se* is not sufficient to restore normal chondrocyte differentiation in MCDS. Another small molecule, named Integrated Stress Response InhiBitor (ISRIB), that desensitizes cells to eIF2α phosphorylation and thereby the preferential translation of *Atf4* and *Chop* transcripts, was recently shown to prevent or ameliorate the growth plate defect and dwarfism in a mouse model of 13del MCDS mice. Treatment of these mice with ISRIB was shown to dampen the PERK-mediated increased synthesis of ATF4 and CHOP. As a consequence, ATF4 induction of ectopic *Sox9* expression in the hypertrophic chondrocytes was prevented, chondrocyte differentiation was normalized and linear skeletal growth was restored to 93–97% (Wang et al., 2018). Although the limited solubility of ISRIB restricts its use in humans, this experiment provides a proof of concept that targeting both the cause (misfolded proteins) by enhanced degradation and the consequence (impact of inappropriate expression of signaling effectors) of the

UPR may have clinical efficacy. The phosphorylation of the translation initiation factor eIF2, triggered by the UPR to modulate protein synthesis, is high in the PERK control pathway and is common to diverse cellular stress responses, e.g., hypoxia, oxidative stress and inflammation. MCDS is an example of how this common mechanism, termed the integrated stress response (ISR), results in a reprogramming of gene expression and differentiation. These findings raise the possibility of a causative role for increased expression of the ISR in other skeletal syndromes and identifies eIF2B as a potential therapeutic target. Unfortunately, small molecules such as ISRIB and the recently described eIF2B activator 2Bact are not suitable for use in humans (Wong et al., 2019). However, the promising results obtained in ameliorating chondrodysplasia in mouse models reinforce the concept of targeting eIF2B and the ISR for treatment and provide an impetus for developing improved compounds that are suitable for use in humans.

11. Future directions and perspectives

Development of therapeutic approaches for congenital skeletal dysplasias requires knowledge of the lineage origins of skeletal cells and the properties of resident stem/progenitor populations in development, growth and disease. In recent years, lineage tracing experiments have revealed that a substantial fraction of hypertrophic chondrocytes survive and differentiate into osteoblastic cells during endochondral bone development and repair (Park et al., 2015; Yang, Tsang, Tang, Chan, & Cheah, 2014; Yang, Zhu, et al., 2014; Zhou et al., 2014). The latest discovery that skeletal stem cells in the resting zone of the growth plate contribute to trabecular osteoblasts is further evidence of the plasticity of chondrocytes (Mizuhashi et al., 2018). This existence of a lineage continuum from chondrocytes to osteoblasts raises questions about the origin of cells involved in reduced or excessive osteogenesis in bone disorders (Tsang, Chan, & Cheah, 2015). Thus, mutations in genes that are expressed in cartilage may contribute to bone pathologies.

Although the contribution of dysregulated chondrocyte to osteoblast lineage progression to human skeletal dysplasias has not yet been demonstrated, a candidate disorder is heterotopic ossification, characterized by ectopic bone formation in soft tissues, most commonly in muscles, tendons and ligaments. While non-hereditary forms arise following severe musculoskeletal trauma, the inherited form known as fibrodysplasia ossificans progressive

(FOP, OMIM#135100) is caused by mutation in the Activin-like kinase 2 (*ALK2*; also known as *ACVR1*) gene that encodes a transforming growth factor-beta (TGFβ)/bone morphogenetic protein (BMP) type I receptor (Shore et al., 2006). The current working model proposes that the *ACVR1* mutation in FOP hypersensitizes responding cells to osteogenic BMP signaling and promotes their osteoblastic differentiation. However, the cellular origin of the responding cells remains unclear. Circulating osteogenic precursor cells of hematopoietic origin have been detected in preosseous fibroproliferative FOP lesions, but they are not sufficient to trigger ectopic bone formation alone as tested by marrow transplant assays (Kaplan et al., 2007; Suda et al., 2009). Lineage tracing in Tie2-Cre reporter mice suggested that approximately 50% of the heterotopic chondrogenic and osteogenic cells may be of endothelial origin. Skeletal muscle cells and vascular smooth muscle cells as marked by MyoD-Cre and SMMHC-Cre activity showed a minimal (<5%) and no contribution to the heterotopic skeletogenesis, respectively (Lounev et al., 2009; Medici et al., 2010). Given the fact that the extraskeletal bone occurs through an endochondral process with a cartilage intermediary, the question of whether hypertrophic chondrocytes may represent a source of osteoblasts in FOP merits further study.

Another candidate disorder that may involve chondrocyte contributions to a bone phenotype is the newly identified PTHrP-salt-inducible kinase 3 (SIK3) syndrome, a disorder that shares radiographic similarities to the Jansen type of metaphyseal chondrodysplasia (MCDJ, OMIM#168468). While MCDJ is caused by constitutively active heterozygous mutations in *PTH1R*, the SIK3 syndrome results from homozygous missense mutation in the catalytic domain of SIK3 (Csukasi et al., 2018). *SIK3* is mainly expressed in proliferating and hypertrophic chondrocytes and loss of function in mice causes severe skeletal abnormality and loss of hypertrophic chondrocytes (Sasagawa et al., 2012). Human *SIK3* mutations cause disorganization and reduction of proliferating and hypertrophic chondrocytes and bone abnormalities. Functional studies of the impact of SIK3 mutations revealed a new signaling pathway, linking PTH/PTHrP, SIK3 and mTOR, that regulates endochondral bone and membranous bone development (Csukasi et al., 2018). Lineage tracing in mouse models of *SIK3* MCDJ would be important for addressing whether the bone defects may be due to disruption of the transition of chondrocytes to osteoblasts. Some common complex disorders, such as osteoarthritis, that involve excessive formation of osteophytes may also be seen as including the consequence of aberrant reactivation of chondrocyte-to-osteoblast transition. The contributions of multipotent skeletal stem cells are another possibility worthy of further investigation (Chan et al., 2018).

Affordable and rapid whole genome and exome sequencing is accelerating the discovery of skeletal dysplasia-causing mutations. The delineation of the underlying molecular pathogenetic mechanisms has seen great progress. This has been aided by enabling technologies for analyzing whole genomes, transcriptomes and epigenomes down to single cell resolution and the development of CRISPR-Cas9 mediated mutagenesis for creating disease models. The advances are changing genetic diagnostic approaches applied in the clinic (Adams & Eng, 2018). Data obtained on the genetic architecture of each affected individual include information on their mutation burden and structural and sequence variants. Mining such information, in concert with data available in various large consortium projects, such as the Human Phenotype Ontology (Kohler et al., 2019), the United Kingdom 100,000 Genomes (Moss & Wernham, 2018), FinnGen (https://www.finngen.fi/en) and the Human Cell Atlas (Ponting, 2019), will be invaluable for future discoveries of disease modifying variants, and the understanding of disease phenotypes and penetrance. At the same time, the wealth of "Big Data" highlights the challenges of functional interpretations and the need to integrate knowledge of gene regulatory networks, genetic variants, phenotypes and proteomes into a system level understanding of the molecular basis of skeletal dysplasias. Key genetic regulatory information is carried by the non-coding regions and 3D architecture of the genome. Disruption of their integrity has been shown to be associated with a broad range of skeletal disorders. A challenge ahead is to identify and categorize additional regulatory elements in disease-associated regions, to demonstrate their functional attributes and determine their link to diseases. Other challenges are to be able to apply systems level knowledge of the molecular base of dysplasias for genetic diagnosis, prognosis and treatment beyond rare single gene skeletal disorders. The acquired insights and their impacts are broad, relevant to understanding the genetic basis of common disorders such as osteoarthritis, osteoporosis and intervertebral disc degeneration.

Acknowledgments

K.C. and D.C. are supported by the Hong Kong Research Grants Council T12-708/12N. We thank Tiong Tan for valuable comments on the manuscript and Wilson Chan for assistance with figure drawing.

References

Adams, D. R., & Eng, C. M. (2018). Next-generation sequencing to diagnose suspected genetic disorders. *The New England Journal of Medicine, 379*, 1353–1362.

Afzal, A. R., & Jeffery, S. (2003). One gene, two phenotypes: ROR2 mutations in autosomal recessive Robinow syndrome and autosomal dominant brachydactyly type B. *Human Mutation, 22*, 1–11.

Afzal, A. R., Rajab, A., Fenske, C. D., Oldridge, M., Elanko, N., Ternes-Pereira, E., et al. (2000). Recessive Robinow syndrome, allelic to dominant brachydactyly type B, is caused by mutation of ROR2. *Nature Genetics, 25*, 419–422.

Albrecht, A. N., Kornak, U., Boddrich, A., Suring, K., Robinson, P. N., Stiege, A. C., et al. (2004). A molecular pathogenesis for transcription factor associated poly-alanine tract expansions. *Human Molecular Genetics, 13*, 2351–2359.

Alby, C., Piquand, K., Huber, C., Megarbane, A., Ichkou, A., Legendre, M., et al. (2015). Mutations in KIAA0586 cause lethal ciliopathies ranging from a hydrolethalus phenotype to short-rib polydactyly syndrome. *American Journal of Human Genetics, 97*, 311–318.

Al-Qattan, M. M., Al-Thunayan, A., Alabdulkareem, I., & Al Balwi, M. (2013). Liebenberg syndrome is caused by a deletion upstream to the PITX1 gene resulting in transformation of the upper limbs to reflect lower limb characteristics. *Gene, 524*, 65–71.

Anderson, E., Peluso, S., Lettice, L. A., & Hill, R. E. (2012). Human limb abnormalities caused by disruption of hedgehog signaling. *Trends in Genetics, 28*, 364–373.

Antwi, P., Hong, C. S., Duran, D., Jin, S. C., Dong, W., DiLuna, M., et al. (2018). A novel association of campomelic dysplasia and hydrocephalus with an unbalanced chromosomal translocation upstream of SOX9. *Cold Spring Harbor Molecular Case Studies, 4*, 1–13.

Baldridge, D., Shchelochkov, O., Kelley, B., & Lee, B. (2010). Signaling pathways in human skeletal dysplasias. *Annual Review of Genomics and Human Genetics, 11*, 189–217.

Bangs, F., Antonio, N., Thongnuek, P., Welten, M., Davey, M. G., Briscoe, J., et al. (2011). Generation of mice with functional inactivation of talpid3, a gene first identified in chicken. *Development, 138*, 3261–3272.

Bateman, J. F., Boot-Handford, R. P., & Lamande, S. R. (2009). Genetic diseases of connective tissues: Cellular and extracellular effects of ECM mutations. *Nature Reviews. Genetics, 10*, 173–183.

Bell, P. A., Pirog, K. A., Fresquet, M., Thornton, D. J., Boot-Handford, R. P., & Briggs, M. D. (2012). Loss of matrilin 1 does not exacerbate the skeletal phenotype in a mouse model of multiple epiphyseal dysplasia caused by a Matn3 V194D mutation. *Arthritis and Rheumatism, 64*, 1529–1539.

Bell, P. A., Wagener, R., Zaucke, F., Koch, M., Selley, J., Warwood, S., et al. (2013). Analysis of the cartilage proteome from three different mouse models of genetic skeletal diseases reveals common and discrete disease signatures. *Biology Open, 2*, 802–811.

Benko, S., Fantes, J. A., Amiel, J., Kleinjan, D. J., Thomas, S., Ramsay, J., et al. (2009). Highly conserved non-coding elements on either side of SOX9 associated with Pierre Robin sequence. *Nature Genetics, 41*, 359–364.

Besio, R., Iula, G., Garibaldi, N., Cipolla, L., Sabbioneda, S., Biggiogera, M., et al. (2018). 4-PBA ameliorates cellular homeostasis in fibroblasts from osteogenesis imperfecta patients by enhancing autophagy and stimulating protein secretion. *Biochimica et Biophysica Acta, Molecular Basis of Disease, 1864*, 1642–1652.

Bi, W., Huang, W., Whitworth, D. J., Deng, J. M., Zhang, Z., Behringer, R. R., et al. (2001). Haploinsufficiency of Sox9 results in defective cartilage primordia and premature skeletal mineralization. *Proceedings of the National Academy of Sciences of the United States of America, 98*, 6698–6703.

Bonafe, L., Cormier-Daire, V., Hall, C., Lachman, R., Mortier, G., Mundlos, S., et al. (2015). Nosology and classification of genetic skeletal disorders: 2015 revision. *American Journal of Medical Genetics. Part A, 167A*, 2869–2892.

Bunn, K. J., Daniel, P., Rosken, H. S., O'Neill, A. C., Cameron-Christie, S. R., Morgan, T., et al. (2015). Mutations in DVL1 cause an osteosclerotic form of Robinow syndrome. *American Journal of Human Genetics*, *96*, 623–630.
Cameron, T. L., Bell, K. M., Gresshoff, I. L., Sampurno, L., Mullan, L., Ermann, J., et al. (2015). XBP1-independent UPR pathways suppress C/EBP-beta mediated chondrocyte differentiation in ER-stress related skeletal disease. *PLoS Genetics*, *11*, e1005505.
Cameron, T. L., Gresshoff, I. L., Bell, K. M., Pirog, K. A., Sampurno, L., Hartley, C. L., et al. (2015). Cartilage-specific ablation of XBP1 signaling in mouse results in a chondrodysplasia characterized by reduced chondrocyte proliferation and delayed cartilage maturation and mineralization. *Osteoarthritis and Cartilage*, *23*, 661–670.
Chan, C. K. F., Gulati, G. S., Sinha, R., Tompkins, J. V., Lopez, M., Carter, A. C., et al. (2018). Identification of the human skeletal stem cell. *Cell*, *175*, 43–56 [e21].
Chan, W. C. W., Tsang, K. Y., Cheng, Y. W., Ng, V. C. W., Chik, H., Tan, Z. J., et al. (2017). Activating the unfolded protein response in osteocytes causes hyperostosis consistent with craniodiaphyseal dysplasia. *Human Molecular Genetics*, *26*, 4572–4587.
Chen, Y., Bellamy, W. P., Seabra, M. C., Field, M. C., & Ali, B. R. (2005). ER-associated protein degradation is a common mechanism underpinning numerous monogenic diseases including Robinow syndrome. *Human Molecular Genetics*, *14*, 2559–2569.
Cho, J. Y., Guo, C., Torello, M., Lunstrum, G. P., Iwata, T., Deng, C., et al. (2004). Defective lysosomal targeting of activated fibroblast growth factor receptor 3 in achondroplasia. *Proceedings of the National Academy of Sciences of the United States of America*, *101*, 609–614.
Chung, U. I., Schipani, E., McMahon, A. P., & Kronenberg, H. M. (2001). Indian hedgehog couples chondrogenesis to osteogenesis in endochondral bone development. *The Journal of Clinical Investigation*, *107*, 295–304.
Clark, R. M., Marker, P. C., & Kingsley, D. M. (2000). A novel candidate gene for mouse and human preaxial polydactyly with altered expression in limbs of hemimelic extra-toes mutant mice. *Genomics*, *67*, 19–27.
Correa, D., Hesse, E., Seriwatanachai, D., Kiviranta, R., Saito, H., Yamana, K., et al. (2010). Zfp521 is a target gene and key effector of parathyroid hormone-related peptide signaling in growth plate chondrocytes. *Developmental Cell*, *19*, 533–546.
Csukasi, F., Duran, I., Barad, M., Barta, T., Gudernova, I., Trantirek, L., et al. (2018). The PTH/PTHrP-SIK3 pathway affects skeletogenesis through altered mTOR signaling. *Science Translational Medicine*, *10*, eaat9356.
Dafinger, C., Liebau, M. C., Elsayed, S. M., Hellenbroich, Y., Boltshauser, E., Korenke, G. C., et al. (2011). Mutations in KIF7 link Joubert syndrome with sonic hedgehog signaling and microtubule dynamics. *The Journal of Clinical Investigation*, *121*, 2662–2667.
Day, T. F., Guo, X., Garrett-Beal, L., & Yang, Y. (2005). Wnt/beta-catenin signaling in mesenchymal progenitors controls osteoblast and chondrocyte differentiation during vertebrate skeletogenesis. *Developmental Cell*, *8*, 739–750.
de Frutos, C. A., Vega, S., Manzanares, M., Flores, J. M., Huertas, H., Martinez-Frias, M. L., et al. (2007). Snail1 is a transcriptional effector of FGFR3 signaling during chondrogenesis and achondroplasias. *Developmental Cell*, *13*, 872–883.
DeChiara, T. M., Kimble, R. B., Poueymirou, W. T., Rojas, J., Masiakowski, P., Valenzuela, D. M., et al. (2000). Ror2, encoding a receptor-like tyrosine kinase, is required for cartilage and growth plate development. *Nature Genetics*, *24*, 271–274.
DeLaurier, A., Schweitzer, R., & Logan, M. (2006). Pitx1 determines the morphology of muscle, tendon, and bones of the hindlimb. *Developmental Biology*, *299*, 22–34.
Dixon, J. R., Selvaraj, S., Yue, F., Kim, A., Li, Y., Shen, Y., et al. (2012). Topological domains in mammalian genomes identified by analysis of chromatin interactions. *Nature*, *485*, 376–380.

Duboc, V., & Logan, M. P. (2011). Pitx1 is necessary for normal initiation of hindlimb outgrowth through regulation of Tbx4 expression and shapes hindlimb morphologies via targeted growth control. *Development*, *138*, 5301–5309.

Everman, D. B., Bartels, C. F., Yang, Y., Yanamandra, N., Goodman, F. R., Mendoza-Londono, J. R., et al. (2002). The mutational spectrum of brachydactyly type C. *American Journal of Medical Genetics*, *112*, 291–296.

Faiyaz-Ul-Haque, M., Ahmad, W., Wahab, A., Haque, S., Azim, A. C., Zaidi, S. H., et al. (2002). Frameshift mutation in the cartilage-derived morphogenetic protein 1 (CDMP1) gene and severe acromesomelic chondrodysplasia resembling Grebe-type chondrodysplasia. *American Journal of Medical Genetics*, *111*, 31–37.

Faiyaz-Ul-Haque, M., Ahmad, W., Zaidi, S. H., Haque, S., Teebi, A. S., Ahmad, M., et al. (2002). Mutation in the cartilage-derived morphogenetic protein-1 (CDMP1) gene in a kindred affected with fibular hypoplasia and complex brachydactyly (DuPan syndrome). *Clinical Genetics*, *61*, 454–458.

Foster, J. W., Dominguez-Steglich, M. A., Guioli, S., Kwok, C., Weller, P. A., Stevanovic, M., et al. (1994). Campomelic dysplasia and autosomal sex reversal caused by mutations in an SRY-related gene. *Nature*, *372*, 525–530.

Franke, M., Ibrahim, D. M., Andrey, G., Schwarzer, W., Heinrich, V., Schopflin, R., et al. (2016). Formation of new chromatin domains determines pathogenicity of genomic duplications. *Nature*, *538*, 265–269.

Gao, B., Guo, J., She, C., Shu, A., Yang, M., Tan, Z., et al. (2001). Mutations in IHH, encoding Indian hedgehog, cause brachydactyly type A-1. *Nature Genetics*, *28*, 386–388.

Gao, B., Hu, J., Stricker, S., Cheung, M., Ma, G., Law, K. F., et al. (2009). A mutation in Ihh that causes digit abnormalities alters its signalling capacity and range. *Nature*, *458*, 1196–1200.

Gao, B., Song, H., Bishop, K., Elliot, G., Garrett, L., English, M. A., et al. (2011). Wnt signaling gradients establish planar cell polarity by inducing Vangl2 phosphorylation through Ror2. *Developmental Cell*, *20*, 163–176.

Geister, K. A., & Camper, S. A. (2015). Advances in skeletal dysplasia genetics. *Annual Review of Genomics and Human Genetics*, *16*, 199–227.

Gentzel, M., & Schambony, A. (2017). Dishevelled paralogs in vertebrate development: Redundant or distinct? *Frontiers in Cell and Development Biology*, *5*, 59.

Gioia, R., Tonelli, F., Ceppi, I., Biggiogera, M., Leikin, S., Fisher, S., et al. (2017). The chaperone activity of 4PBA ameliorates the skeletal phenotype of Chihuahua, a zebrafish model for dominant osteogenesis imperfecta. *Human Molecular Genetics*, *26*, 2897–2911.

Gordon, C. T., Attanasio, C., Bhatia, S., Benko, S., Ansari, M., Tan, T. Y., et al. (2014). Identification of novel craniofacial regulatory domains located far upstream of SOX9 and disrupted in Pierre Robin sequence. *Human Mutation*, *35*, 1011–1020.

He, M., Subramanian, R., Bangs, F., Omelchenko, T., Liem, K. F., Jr., Kapoor, T. M., et al. (2014). The kinesin-4 protein Kif7 regulates mammalian Hedgehog signalling by organizing the cilium tip compartment. *Nature Cell Biology*, *16*, 663–672.

Hellemans, J., Coucke, P. J., Giedion, A., De Paepe, A., Kramer, P., Beemer, F., et al. (2003). Homozygous mutations in IHH cause acrocapitofemoral dysplasia, an autosomal recessive disorder with cone-shaped epiphyses in hands and hips. *American Journal of Human Genetics*, *72*, 1040–1046.

Henderson, D. J., Long, D. A., & Dean, C. H. (2018). Planar cell polarity in organ formation. *Current Opinion in Cell Biology*, *55*, 96–103.

Hill, T. P., Spater, D., Taketo, M. M., Birchmeier, W., & Hartmann, C. (2005). Canonical Wnt/beta-catenin signaling prevents osteoblasts from differentiating into chondrocytes. *Developmental Cell*, *8*, 727–738.

Ho, M. S., Tsang, K. Y., Lo, R. L., Susic, M., Makitie, O., Chan, T. W., et al. (2007). COL10A1 nonsense and frame-shift mutations have a gain-of-function effect on the growth plate in human and mouse metaphyseal chondrodysplasia type Schmid. *Human Molecular Genetics, 16*, 1201–1215.

Hu, H., Hilton, M. J., Tu, X., Yu, K., Ornitz, D. M., & Long, F. (2005). Sequential roles of Hedgehog and Wnt signaling in osteoblast development. *Development, 132*, 49–60.

Huang, W., Chung, U. I., Kronenberg, H. M., & de Crombrugghe, B. (2001). The chondrogenic transcription factor Sox9 is a target of signaling by the parathyroid hormone-related peptide in the growth plate of endochondral bones. *Proceedings of the National Academy of Sciences of the United States of America, 98*, 160–165.

Huber, C., & Cormier-Daire, V. (2012). Ciliary disorder of the skeleton. *American Journal of Medical Genetics. Part C, Seminars in Medical Genetics, 160C*, 165–174.

Hughes, A., Oxford, A. E., Tawara, K., Jorcyk, C. L., & Oxford, J. T. (2017). Endoplasmic reticulum stress and unfolded protein response in cartilage pathophysiology; contributing factors to apoptosis and osteoarthritis. *International Journal of Molecular Sciences, 18*, 665–683.

Jackson, G. C., Mittaz-Crettol, L., Taylor, J. A., Mortier, G. R., Spranger, J., Zabel, B., et al. (2012). Pseudoachondroplasia and multiple epiphyseal dysplasia: A 7-year comprehensive analysis of the known disease genes identify novel and recurrent mutations and provides an accurate assessment of their relative contribution. *Human Mutation, 33*, 144–157.

Joeng, K. S., & Long, F. (2009). The Gli2 transcriptional activator is a crucial effector for Ihh signaling in osteoblast development and cartilage vascularization. *Development, 136*, 4177–4185.

Kaplan, F. S., Glaser, D. L., Shore, E. M., Pignolo, R. J., Xu, M., Zhang, Y., et al. (2007). Hematopoietic stem-cell contribution to ectopic skeletogenesis. *The Journal of Bone and Joint Surgery. American Volume, 89*, 347–357.

Karp, S. J., Schipani, E., St-Jacques, B., Hunzelman, J., Kronenberg, H., & McMahon, A. P. (2000). Indian hedgehog coordinates endochondral bone growth and morphogenesis via parathyroid hormone related-protein-dependent and -independent pathways. *Development, 127*, 543–548.

Khan, S. K., Yadav, P. S., Elliott, G., Hu, D. Z., Xu, R., & Yang, Y. (2018). Induced Gnas(R201H) expression from the endogenous Gnas locus causes fibrous dysplasia by up-regulating Wnt/beta-catenin signaling. *Proceedings of the National Academy of Sciences of the United States of America, 115*, E418–E427.

Kiper, P. O. S., Saito, H., Gori, F., Unger, S., Hesse, E., Yamana, K., et al. (2016). Cortical-bone fragility–insights from sFRP4 deficiency in Pyle's disease. *The New England Journal of Medicine, 374*, 2553–2562.

Klag, K. A., & Horton, W. A. (2016). Advances in treatment of achondroplasia and osteoarthritis. *Human Molecular Genetics, 25*, R2–R8.

Ko, Y., Kobbe, B., Nicolae, C., Miosge, N., Paulsson, M., Wagener, R., et al. (2004). Matrilin-3 is dispensable for mouse skeletal growth and development. *Molecular and Cellular Biology, 24*, 1691–1699.

Kohler, S., Carmody, L., Vasilevsky, N., Jacobsen, J. O. B., Danis, D., Gourdine, J. P., et al. (2019). Expansion of the human phenotype ontology (HPO) knowledge base and resources. *Nucleic Acids Research, 47*, D1018–D1027.

Kondo, Y., Fu, J., Wang, H., Hoover, C., McDaniel, J. M., Steet, R., et al. (2018). Site-1 protease deficiency causes human skeletal dysplasia due to defective inter-organelle protein trafficking. *JCI Insight, 3*, e121596.

Kozhemyakina, E., Cohen, T., Yao, T. P., & Lassar, A. B. (2009). Parathyroid hormone-related peptide represses chondrocyte hypertrophy through a protein phosphatase 2A/histone deacetylase 4/MEF2 pathway. *Molecular and Cellular Biology, 29*, 5751–5762.

Kragesteen, B. K., Spielmann, M., Paliou, C., Heinrich, V., Schopflin, R., Esposito, A., et al. (2018). Dynamic 3D chromatin architecture contributes to enhancer specificity and limb morphogenesis. *Nature Genetics*, *50*, 1463–1473.

Krejci, P., Bryja, V., Pachernik, J., Hampl, A., Pogue, R., Mekikian, P., et al. (2004). FGF2 inhibits proliferation and alters the cartilage-like phenotype of RCS cells. *Experimental Cell Research*, *297*, 152–164.

Krejci, P., Masri, B., Fontaine, V., Mekikian, P. B., Weis, M., Prats, H., et al. (2005). Interaction of fibroblast growth factor and C-natriuretic peptide signaling in regulation of chondrocyte proliferation and extracellular matrix homeostasis. *Journal of Cell Science*, *118*, 5089–5100.

Kuss, P., Villavicencio-Lorini, P., Witte, F., Klose, J., Albrecht, A. N., Seemann, P., et al. (2009). Mutant Hoxd13 induces extra digits in a mouse model of synpolydactyly directly and by decreasing retinoic acid synthesis. *The Journal of Clinical Investigation*, *119*, 146–156.

Lanske, B., Amling, M., Neff, L., Guiducci, J., Baron, R., & Kronenberg, H. M. (1999). Ablation of the PTHrP gene or the PTH/PTHrP receptor gene leads to distinct abnormalities in bone development. *The Journal of Clinical Investigation*, *104*, 399–407.

Lecointre, C., Pichon, O., Hamel, A., Heloury, Y., Michel-Calemard, L., Morel, Y., et al. (2009). Familial acampomelic form of campomelic dysplasia caused by a 960 kb deletion upstream of SOX9. *American Journal of Medical Genetics. Part A*, *149A*, 1183–1189.

Leighton, M. P., Nundlall, S., Starborg, T., Meadows, R. S., Suleman, F., Knowles, L., et al. (2007). Decreased chondrocyte proliferation and dysregulated apoptosis in the cartilage growth plate are key features of a murine model of epiphyseal dysplasia caused by a matn3 mutation. *Human Molecular Genetics*, *16*, 1728–1741.

Leipoldt, M., Erdel, M., Bien-Willner, G. A., Smyk, M., Theurl, M., Yatsenko, S. A., et al. (2007). Two novel translocation breakpoints upstream of SOX9 define borders of the proximal and distal breakpoint cluster region in campomelic dysplasia. *Clinical Genetics*, *71*, 67–75.

Lettice, L. A., Devenney, P., De Angelis, C., & Hill, R. E. (2017). The conserved sonic hedgehog limb enhancer consists of discrete functional elements that regulate Precise spatial expression. *Cell Reports*, *20*, 1396–1408.

Lettice, L. A., Heaney, S. J., Purdie, L. A., Li, L., de Beer, P., Oostra, B. A., et al. (2003). A long-range Shh enhancer regulates expression in the developing limb and fin and is associated with preaxial polydactyly. *Human Molecular Genetics*, *12*, 1725–1735.

Lettice, L. A., Hill, A. E., Devenney, P. S., & Hill, R. E. (2008). Point mutations in a distant sonic hedgehog cis-regulator generate a variable regulatory output responsible for preaxial polydactyly. *Human Molecular Genetics*, *17*, 978–985.

Lettice, L. A., Horikoshi, T., Heaney, S. J., van Baren, M. J., van der Linde, H. C., Breedveld, G. J., et al. (2002). Disruption of a long-range cis-acting regulator for Shh causes preaxial polydactyly. *Proceedings of the National Academy of Sciences of the United States of America*, *99*, 7548–7553.

Lettice, L. A., Williamson, I., Devenney, P. S., Kilanowski, F., Dorin, J., & Hill, R. E. (2014). Development of five digits is controlled by a bipartite long-range cis-regulator. *Development*, *141*, 1715–1725.

Lettice, L. A., Williamson, I., Wiltshire, J. H., Peluso, S., Devenney, P. S., Hill, A. E., et al. (2012). Opposing functions of the ETS factor family define Shh spatial expression in limb buds and underlie polydactyly. *Developmental Cell*, *22*, 459–467.

Li, D., Sakuma, R., Vakili, N. A., Mo, R., Puviindran, V., Deimling, S., et al. (2014). Formation of proximal and anterior limb skeleton requires early function of Irx3 and Irx5 and is negatively regulated by Shh signaling. *Developmental Cell*, *29*, 233–240.

Li, J., Wang, C., Wu, C., Cao, T., Xu, G., Meng, Q., et al. (2017). PKA-mediated Gli2 and Gli3 phosphorylation is inhibited by hedgehog signaling in cilia and reduced in Talpid3 mutant. *Developmental Biology*, *429*, 147–157.

Logan, M., & Tabin, C. J. (1999). Role of Pitx1 upstream of Tbx4 in specification of hindlimb identity. *Science*, *283*, 1736–1739.
Long, F., Zhang, X. M., Karp, S., Yang, Y., & McMahon, A. P. (2001). Genetic manipulation of hedgehog signaling in the endochondral skeleton reveals a direct role in the regulation of chondrocyte proliferation. *Development*, *128*, 5099–5108.
Lorget, F., Kaci, N., Peng, J., Benoist-Lasselin, C., Mugniery, E., Oppeneer, T., et al. (2012). Evaluation of the therapeutic potential of a CNP analog in a Fgfr3 mouse model recapitulating achondroplasia. *American Journal of Human Genetics*, *91*, 1108–1114.
Lounev, V. Y., Ramachandran, R., Wosczyna, M. N., Yamamoto, M., Maidment, A. D., Shore, E. M., et al. (2009). Identification of progenitor cells that contribute to heterotopic skeletogenesis. *The Journal of Bone and Joint Surgery. American Volume*, *91*, 652–663.
Lupianez, D. G., Kraft, K., Heinrich, V., Krawitz, P., Brancati, F., Klopocki, E., et al. (2015). Disruptions of topological chromatin domains cause pathogenic rewiring of gene-enhancer interactions. *Cell*, *161*, 1012–1025.
Mak, K. K., Kronenberg, H. M., Chuang, P. T., Mackem, S., & Yang, Y. (2008). Indian hedgehog signals independently of PTHrP to promote chondrocyte hypertrophy. *Development*, *135*, 1947–1956.
Makrythanasis, P., Temtamy, S., Aglan, M. S., Otaify, G. A., Hamamy, H., & Antonarakis, S. E. (2014). A novel homozygous mutation in FGFR3 causes tall stature, severe lateral tibial deviation, scoliosis, hearing impairment, camptodactyly, and arachnodactyly. *Human Mutation*, *35*, 959–963.
Malik, S. (2014). Polydactyly: Phenotypes, genetics and classification. *Clinical Genetics*, *85*, 203–212.
Marcil, A., Dumontier, E., Chamberland, M., Camper, S. A., & Drouin, J. (2003). Pitx1 and Pitx2 are required for development of hindlimb buds. *Development*, *130*, 45–55.
Maroteaux, P., Spranger, J., Opitz, J. M., Kucera, J., Lowry, R. B., Schimke, R. N., et al. (1971). The campomelic syndrome. *Presse Médicale*, *79*, 1157–1162.
Masuya, H., Sagai, T., Wakana, S., Moriwaki, K., & Shiroishi, T. (1995). A duplicated zone of polarizing activity in polydactylous mouse mutants. *Genes & Development*, *9*, 1645–1653.
Mazzeu, J. F., Pardono, E., Vianna-Morgante, A. M., Richieri-Costa, A., Ae Kim, C., Brunoni, D., et al. (2007). Clinical characterization of autosomal dominant and recessive variants of Robinow syndrome. *American Journal of Medical Genetics. Part A*, *143*, 320–325.
McLellan, J. S., Zheng, X., Hauk, G., Ghirlando, R., Beachy, P. A., & Leahy, D. J. (2008). The mode of Hedgehog binding to Ihog homologues is not conserved across different phyla. *Nature*, *455*, 979–983.
Medici, D., Shore, E. M., Lounev, V. Y., Kaplan, F. S., Kalluri, R., & Olsen, B. R. (2010). Conversion of vascular endothelial cells into multipotent stem-like cells. *Nature Medicine*, *16*, 1400–1406.
Meyer, J., Sudbeck, P., Held, M., Wagner, T., Schmitz, M. L., Bricarelli, F. D., et al. (1997). Mutational analysis of the SOX9 gene in campomelic dysplasia and autosomal sex reversal: Lack of genotype/phenotype correlations. *Human Molecular Genetics*, *6*, 91–98.
Mizuhashi, K., Ono, W., Matsushita, Y., Sakagami, N., Takahashi, A., Saunders, T. L., et al. (2018). Resting zone of the growth plate houses a unique class of skeletal stem cells. *Nature*, *563*, 254–258.
Monsonego-Ornan, E., Adar, R., Feferman, T., Segev, O., & Yayon, A. (2000). The transmembrane mutation G380R in fibroblast growth factor receptor 3 uncouples ligand-mediated receptor activation from down-regulation. *Molecular and Cellular Biology*, *20*, 516–522.
Moss, C., & Wernham, A. (2018). The 100 000 genomes project: Feeding back to patients. *BMJ*, *361*, k2441.

Mullan, L. A., Mularczyk, E. J., Kung, L. H., Forouhan, M., Wragg, J. M., Goodacre, R., et al. (2017). Increased intracellular proteolysis reduces disease severity in an ER stress-associated dwarfism. *The Journal of Clinical Investigation*, *127*, 3861–3865.

Muragaki, Y., Mundlos, S., Upton, J., & Olsen, B. R. (1996). Altered growth and branching patterns in synpolydactyly caused by mutations in HOXD13. *Science*, *272*, 548–551.

Murakami, S., Balmes, G., McKinney, S., Zhang, Z., Givol, D., & de Crombrugghe, B. (2004). Constitutive activation of MEK1 in chondrocytes causes Stat1-independent achondroplasia-like dwarfism and rescues the Fgfr3-deficient mouse phenotype. *Genes & Development*, *18*, 290–305.

Naski, M. C., Wang, Q., Xu, J., & Ornitz, D. M. (1996). Graded activation of fibroblast growth factor receptor 3 by mutations causing achondroplasia and thanatophoric dysplasia. *Nature Genetics*, *13*, 233–237.

Ozasa, A., Komatsu, Y., Yasoda, A., Miura, M., Sakuma, Y., Nakatsuru, Y., et al. (2005). Complementary antagonistic actions between C-type natriuretic peptide and the MAPK pathway through FGFR-3 in ATDC5 cells. *Bone*, *36*, 1056–1064.

Parafioriti, A., del Bianco, S., Barisani, D., Armiraglio, E., Peretti, G., & Albisetti, W. (2009). Increased p21 expression in chondrocytes of achondroplasic children independently from the presence of the G380R FGFR3 mutation. *Journal of Orthopaedic Science*, *14*, 623–630.

Park, J., Gebhardt, M., Golovchenko, S., Perez-Branguli, F., Hattori, T., Hartmann, C., et al. (2015). Dual pathways to endochondral osteoblasts: A novel chondrocyte-derived osteoprogenitor cell identified in hypertrophic cartilage. *Biology Open*, *4*, 608–621.

Person, A. D., Beiraghi, S., Sieben, C. M., Hermanson, S., Neumann, A. N., Robu, M. E., et al. (2010). WNT5A mutations in patients with autosomal dominant Robinow syndrome. *Developmental Dynamics*, *239*, 327–337.

Petit, F., Jourdain, A. S., Holder-Espinasse, M., Keren, B., Andrieux, J., Duterque-Coquillaud, M., et al. (2016). The disruption of a novel limb cis-regulatory element of SHH is associated with autosomal dominant preaxial polydactyly-hypertrichosis. *European Journal of Human Genetics*, *24*, 37–43.

Pfeifer, D., Kist, R., Dewar, K., Devon, K., Lander, E. S., Birren, B., et al. (1999). Campomelic dysplasia translocation breakpoints are scattered over 1 Mb proximal to SOX9: Evidence for an extended control region. *American Journal of Human Genetics*, *65*, 111–124.

Pirog-Garcia, K. A., Meadows, R. S., Knowles, L., Heinegard, D., Thornton, D. J., Kadler, K. E., et al. (2007). Reduced cell proliferation and increased apoptosis are significant pathological mechanisms in a murine model of mild pseudoachondroplasia resulting from a mutation in the C-terminal domain of COMP. *Human Molecular Genetics*, *16*, 2072–2088.

Polinkovsky, A., Robin, N. H., Thomas, J. T., Irons, M., Lynn, A., Goodman, F. R., et al. (1997). Mutations in CDMP1 cause autosomal dominant brachydactyly type C. *Nature Genetics*, *17*, 18–19.

Ponting, C. P. (2019). The human cell atlas: Making 'cell space' for disease. *Disease Models & Mechanisms*, dmm037622.

Pop, R., Conz, C., Lindenberg, K. S., Blesson, S., Schmalenberger, B., Briault, S., et al. (2004). Screening of the 1 Mb SOX9 5′ control region by array CGH identifies a large deletion in a case of campomelic dysplasia with XY sex reversal. *Journal of Medical Genetics*, *41*, e47.

Pop, R., Zaragoza, M. V., Gaudette, M., Dohrmann, U., & Scherer, G. (2005). A homozygous nonsense mutation in SOX9 in the dominant disorder campomelic dysplasia: A case of mitotic gene conversion. *Human Genetics*, *117*, 43–53.

Posey, K. L., Coustry, F., & Hecht, J. T. (2018). Cartilage oligomeric matrix protein: COMPopathies and beyond. *Matrix Biology*, *71–72*, 161–173.

Potuijt, J. W. P., Baas, M., Sukenik-Halevy, R., Douben, H., Nguyen, P., Venter, D. J., et al. (2018). A point mutation in the pre-ZRS disrupts sonic hedgehog expression in the limb bud and results in triphalangeal thumb-polysyndactyly syndrome. *Genetics in Medicine*, *20*(11), 1405–1413.

Priya, S., Nampoothiri, S., Sen, P., & Sripriya, S. (2016). Bardet-Biedl syndrome: Genetics, molecular pathophysiology, and disease management. *Indian Journal of Ophthalmology*, *64*, 620–627.

Provot, S., Kempf, H., Murtaugh, L. C., Chung, U. I., Kim, D. W., Chyung, J., et al. (2006). Nkx3.2/Bapx1 acts as a negative regulator of chondrocyte maturation. *Development*, *133*, 651–662.

Rajpar, M. H., McDermott, B., Kung, L., Eardley, R., Knowles, L., Heeran, M., et al. (2009). Targeted induction of endoplasmic reticulum stress induces cartilage pathology. *PLoS Genetics*, *5*, e1000691.

Rao, S. S., Huntley, M. H., Durand, N. C., Stamenova, E. K., Bochkov, I. D., Robinson, J. T., et al. (2014). A 3D map of the human genome at kilobase resolution reveals principles of chromatin looping. *Cell*, *159*, 1665–1680.

Regard, J. B., Cherman, N., Palmer, D., Kuznetsov, S. A., Celi, F. S., Guettier, J. M., et al. (2011). Wnt/beta-catenin signaling is differentially regulated by Galpha proteins and contributes to fibrous dysplasia. *Proceedings of the National Academy of Sciences of the United States of America*, *108*, 20101–20106.

Regard, J. B., Malhotra, D., Gvozdenovic-Jeremic, J., Josey, M., Chen, M., Weinstein, L. S., et al. (2013). Activation of hedgehog signaling by loss of GNAS causes heterotopic ossification. *Nature Medicine*, *19*, 1505–1512.

Riddle, R. D., Johnson, R. L., Laufer, E., & Tabin, C. (1993). Sonic hedgehog mediates the polarizing activity of the ZPA. *Cell*, *75*, 1401–1416.

Riminucci, M., Robey, P. G., Saggio, I., & Bianco, P. (2010). Skeletal progenitors and the GNAS gene: Fibrous dysplasia of bone read through stem cells. *Journal of Molecular Endocrinology*, *45*, 355–364.

Rodda, S. J., & McMahon, A. P. (2006). Distinct roles for hedgehog and canonical Wnt signaling in specification, differentiation and maintenance of osteoblast progenitors. *Development*, *133*, 3231–3244.

Rousseau, F., Bonaventure, J., Legeai-Mallet, L., Pelet, A., Rozet, J. M., Maroteaux, P., et al. (1994). Mutations in the gene encoding fibroblast growth factor receptor-3 in achondroplasia. *Nature*, *371*, 252–254.

Saito, A., Hino, S., Murakami, T., Kanemoto, S., Kondo, S., Saitoh, M., et al. (2009). Regulation of endoplasmic reticulum stress response by a BBF2H7-mediated Sec23a pathway is essential for chondrogenesis. *Nature Cell Biology*, *11*, 1197–1204.

Saito, A., Kanemoto, S., Zhang, Y., Asada, R., Hino, K., & Imaizumi, K. (2014). Chondrocyte proliferation regulated by secreted luminal domain of ER stress transducer BBF2H7/CREB3L2. *Molecular Cell*, *53*, 127–139.

Sasagawa, S., Takemori, H., Uebi, T., Ikegami, D., Hiramatsu, K., Ikegawa, S., et al. (2012). SIK3 is essential for chondrocyte hypertrophy during skeletal development in mice. *Development*, *139*, 1153–1163.

Scherer, G., Zabel, B., & Nishimura, G. (2013). Clinical utility gene card for: Campomelic dysplasia. *European Journal of Human Genetics*, *21*, 792–794.

Schipani, E., Langman, C. B., Parfitt, A. M., Jensen, G. S., Kikuchi, S., Kooh, S. W., et al. (1996). Constitutively activated receptors for parathyroid hormone and parathyroid hormone-related peptide in Jansen's metaphyseal chondrodysplasia. *The New England Journal of Medicine*, *335*, 708–714.

Schwabe, G. C., Tinschert, S., Buschow, C., Meinecke, P., Wolff, G., Gillessen-Kaesbach,-G., et al. (2000). Distinct mutations in the receptor tyrosine kinase gene ROR2 cause brachydactyly type B. *American Journal of Human Genetics*, *67*, 822–831.

Schwarzer, W., Witte, F., Rajab, A., Mundlos, S., & Stricker, S. (2009). A gradient of ROR2 protein stability and membrane localization confers brachydactyly type B or Robinow syndrome phenotypes. *Human Molecular Genetics, 18*, 4013–4021.

Sebastian, A., Matsushita, T., Kawanami, A., Mackem, S., Landreth, G. E., & Murakami, S. (2011). Genetic inactivation of ERK1 and ERK2 in chondrocytes promotes bone growth and enlarges the spinal canal. *Journal of Orthopaedic Research, 29*, 375–379.

Seki, K., Fujimori, T., Savagner, P., Hata, A., Aikawa, T., Ogata, N., et al. (2003). Mouse Snail family transcription repressors regulate chondrocyte, extracellular matrix, type II collagen, and aggrecan. *The Journal of Biological Chemistry, 278*, 41862–41870.

Seriwatanachai, D., Densmore, M. J., Sato, T., Correa, D., Neff, L., Baron, R., et al. (2011). Deletion of Zfp521 rescues the growth plate phenotype in a mouse model of Jansen metaphyseal chondrodysplasia. *The FASEB Journal, 25*, 3057–3067.

Sharpe, J., Lettice, L., Hecksher-Sorensen, J., Fox, M., Hill, R., & Krumlauf, R. (1999). Identification of sonic hedgehog as a candidate gene responsible for the polydactylous mouse mutant Sasquatch. *Current Biology, 9*, 97–100.

Shiang, R., Thompson, L. M., Zhu, Y. Z., Church, D. M., Fielder, T. J., Bocian, M., et al. (1994). Mutations in the transmembrane domain of FGFR3 cause the most common genetic form of dwarfism, achondroplasia. *Cell, 78*, 335–342.

Shore, E. M., Ahn, J., Jan de Beur, S., Li, M., Xu, M., Gardner, R. J., et al. (2002). Paternally inherited inactivating mutations of the GNAS1 gene in progressive osseous heteroplasia. *The New England Journal of Medicine, 346*, 99–106.

Shore, E. M., Xu, M., Feldman, G. J., Fenstermacher, D. A., Cho, T. J., Choi, I. H., et al. (2006). A recurrent mutation in the BMP type I receptor ACVR1 causes inherited and sporadic fibrodysplasia ossificans progressiva. *Nature Genetics, 38*, 525–527.

Smith, B. N., Burton, L. J., Henderson, V., Randle, D. D., Morton, D. J., Smith, B. A., et al. (2014). Snail promotes epithelial mesenchymal transition in breast cancer cells in part via activation of nuclear ERK2. *PLoS One, 9*, e104987.

Spielmann, M., Brancati, F., Krawitz, P. M., Robinson, P. N., Ibrahim, D. M., Franke, M., et al. (2012). Homeotic arm-to-leg transformation associated with genomic rearrangements at the PITX1 locus. *American Journal of Human Genetics, 91*, 629–635.

Stephen, L. A., Tawamie, H., Davis, G. M., Tebbe, L., Nurnberg, P., Nurnberg, G., et al. (2015). TALPID3 controls centrosome and cell polarity and the human ortholog KIAA0586 is mutated in Joubert syndrome (JBTS23). *eLife, 4*, e08077.

St-Jacques, B., Hammerschmidt, M., & McMahon, A. P. (1999). Indian hedgehog signaling regulates proliferation and differentiation of chondrocytes and is essential for bone formation. *Genes & Development, 13*, 2072–2086.

Suda, R. K., Billings, P. C., Egan, K. P., Kim, J. H., McCarrick-Walmsley, R., Glaser, D. L., et al. (2009). Circulating osteogenic precursor cells in heterotopic bone formation. *Stem Cells, 27*, 2209–2219.

Suleman, F., Gualeni, B., Gregson, H. J., Leighton, M. P., Pirog, K. A., Edwards, S., et al. (2012). A novel form of chondrocyte stress is triggered by a COMP mutation causing pseudoachondroplasia. *Human Mutation, 33*, 218–231.

Sun, M., Ma, F., Zeng, X., Liu, Q., Zhao, X. L., Wu, F. X., et al. (2008). Triphalangeal thumb-polysyndactyly syndrome and syndactyly type IV are caused by genomic duplications involving the long range, limb-specific SHH enhancer. *Journal of Medical Genetics, 45*, 589–595.

Svensson, L., Aszodi, A., Heinegard, D., Hunziker, E. B., Reinholt, F. P., Fassler, R., et al. (2002). Cartilage oligomeric matrix protein-deficient mice have normal skeletal development. *Molecular and Cellular Biology, 22*, 4366–4371.

Symmons, O., Pan, L., Remeseiro, S., Aktas, T., Klein, F., Huber, W., et al. (2016). The Shh topological domain facilitates the action of remote enhancers by reducing the effects of genomic distances. *Developmental Cell, 39*, 529–543.

Takeuchi, S., Takeda, K., Oishi, I., Nomi, M., Ikeya, M., Itoh, K., et al. (2000). Mouse Ror2 receptor tyrosine kinase is required for the heart development and limb formation. *Genes to Cells*, *5*, 71–78.

Tavormina, P. L., Shiang, R., Thompson, L. M., Zhu, Y. Z., Wilkin, D. J., Lachman, R. S., et al. (1995). Thanatophoric dysplasia (types I and II) caused by distinct mutations in fibroblast growth factor receptor 3. *Nature Genetics*, *9*, 321–328.

Thomas, J. T., Kilpatrick, M. W., Lin, K., Erlacher, L., Lembessis, P., Costa, T., et al. (1997). Disruption of human limb morphogenesis by a dominant negative mutation in CDMP1. *Nature Genetics*, *17*, 58–64.

Tsang, K. Y., Chan, D., Bateman, J. F., & Cheah, K. S. (2010). In vivo cellular adaptation to ER stress: Survival strategies with double-edged consequences. *Journal of Cell Science*, *123*, 2145–2154.

Tsang, K. Y., Chan, D., & Cheah, K. S. (2015). Fate of growth plate hypertrophic chondrocytes: Death or lineage extension? *Development, Growth & Differentiation*, *57*, 179–192.

Tsang, K. Y., Chan, D., Cheslett, D., Chan, W. C., So, C. L., Melhado, I. G., et al. (2007). Surviving endoplasmic reticulum stress is coupled to altered chondrocyte differentiation and function. *PLoS Biology*, *5*, e44.

Tu, X., Joeng, K. S., & Long, F. (2012). Indian hedgehog requires additional effectors besides Runx2 to induce osteoblast differentiation. *Developmental Biology*, *362*, 76–82.

Tuson, M., He, M., & Anderson, K. V. (2011). Protein kinase A acts at the basal body of the primary cilium to prevent Gli2 activation and ventralization of the mouse neural tube. *Development*, *138*, 4921–4930.

Unger, S., Bonafe, L., & Gouze, E. (2017). Current care and investigational therapies in achondroplasia. *Current Osteoporosis Reports*, *15*, 53–60.

van Bokhoven, H., Celli, J., Kayserili, H., van Beusekom, E., Balci, S., Brussel, W., et al. (2000). Mutation of the gene encoding the ROR2 tyrosine kinase causes autosomal recessive Robinow syndrome. *Nature Genetics*, *25*, 423–426.

van der Weyden, L., Wei, L., Luo, J., Yang, X., Birk, D. E., Adams, D. J., et al. (2006). Functional knockout of the matrilin-3 gene causes premature chondrocyte maturation to hypertrophy and increases bone mineral density and osteoarthritis. *The American Journal of Pathology*, *169*, 515–527.

Velagaleti, G. V., Bien-Willner, G. A., Northup, J. K., Lockhart, L. H., Hawkins, J. C., Jalal, S. M., et al. (2005). Position effects due to chromosome breakpoints that map approximately 900 Kb upstream and approximately 1.3 Mb downstream of SOX9 in two patients with campomelic dysplasia. *American Journal of Human Genetics*, *76*, 652–662.

Vortkamp, A., Lee, K., Lanske, B., Segre, G. V., Kronenberg, H. M., & Tabin, C. J. (1996). Regulation of rate of cartilage differentiation by Indian hedgehog and PTH-related protein. *Science*, *273*, 613–622.

Wagner, T., Wirth, J., Meyer, J., Zabel, B., Held, M., Zimmer, J., et al. (1994). Autosomal sex reversal and campomelic dysplasia are caused by mutations in and around the SRY-related gene SOX9. *Cell*, *79*, 1111–1120.

Wang, B., Fallon, J. F., & Beachy, P. A. (2000). Hedgehog-regulated processing of Gli3 produces an anterior/posterior repressor gradient in the developing vertebrate limb. *Cell*, *100*, 423–434.

Wang, W., Lian, N., Li, L., Moss, H. E., Wang, W., Perrien, D. S., et al. (2009). Atf4 regulates chondrocyte proliferation and differentiation during endochondral ossification by activating Ihh transcription. *Development*, *136*, 4143–4153.

Wang, B., Sinha, T., Jiao, K., Serra, R., & Wang, J. (2011). Disruption of PCP signaling causes limb morphogenesis and skeletal defects and may underlie Robinow syndrome and brachydactyly type B. *Human Molecular Genetics*, *20*, 271–285.

Wang, C., Tan, Z., Niu, B., Tsang, K. Y., Tai, A., Chan, W. C. W., et al. (2018). Inhibiting the integrated stress response pathway prevents aberrant chondrocyte differentiation thereby alleviating chondrodysplasia. *eLife*, *7*, e37673.

Waters, A. M., & Beales, P. L. (2011). Ciliopathies: An expanding disease spectrum. *Pediatric Nephrology*, *26*, 1039–1056.

Wendt, D. J., Dvorak-Ewell, M., Bullens, S., Lorget, F., Bell, S. M., Peng, J., et al. (2015). Neutral endopeptidase-resistant C-type natriuretic peptide variant represents a new therapeutic approach for treatment of fibroblast growth factor receptor 3-related dwarfism. *The Journal of Pharmacology and Experimental Therapeutics*, *353*, 132–149.

White, J. J., Mazzeu, J. F., Hoischen, A., Bayram, Y., Withers, M., Gezdirici, A., et al. (2016). DVL3 alleles resulting in a −1 frameshift of the last exon mediate autosomal-dominant Robinow syndrome. *American Journal of Human Genetics*, *98*, 553–561.

White, J., Mazzeu, J. F., Hoischen, A., Jhangiani, S. N., Gambin, T., Alcino, M. C., et al. (2015). DVL1 frameshift mutations clustering in the penultimate exon cause autosomal-dominant Robinow syndrome. *American Journal of Human Genetics*, *96*, 612–622.

Wieczorek, D., Pawlik, B., Li, Y., Akarsu, N. A., Caliebe, A., May, K. J., et al. (2010). A specific mutation in the distant sonic hedgehog (SHH) cis-regulator (ZRS) causes Werner mesomelic syndrome (WMS) while complete ZRS duplications underlie Haas type polysyndactyly and preaxial polydactyly (PPD) with or without triphalangeal thumb. *Human Mutation*, *31*, 81–89.

Wilson, R., Freddi, S., Chan, D., Cheah, K. S., & Bateman, J. F. (2005). Misfolding of collagen X chains harboring Schmid metaphyseal chondrodysplasia mutations results in aberrant disulfide bond formation, intracellular retention, and activation of the unfolded protein response. *The Journal of Biological Chemistry*, *280*, 15544–15552.

Witte, F., Chan, D., Economides, A. N., Mundlos, S., & Stricker, S. (2010). Receptor tyrosine kinase-like orphan receptor 2 (ROR2) and Indian hedgehog regulate digit outgrowth mediated by the phalanx-forming region. *Proceedings of the National Academy of Sciences of the United States of America*, *107*, 14211–14216.

Wong, Y. L., LeBon, L., Basso, A. M., Kohlhaas, K. L., Nikkel, A. L., Robb, H. M., et al. (2019). eIF2B activator prevents neurological defects caused by a chronic integrated stress response. *eLife*, *8*, e42940.

Wu, M., Chen, G., & Li, Y. P. (2016). TGF-beta and BMP signaling in osteoblast, skeletal development, and bone formation, homeostasis and disease. *Bone Research*, *4*, 16009.

Xiang, Y., Jiang, L., Wang, B., Xu, Y., Cai, H., & Fu, Q. (2017). Mutational screening of GLI3, SHH, preZRS, and ZRS in 102 Chinese children with nonsyndromic polydactyly. *Developmental Dynamics*, *246*, 392–402.

Xu, J. X., Kilpatrick, N., Baker, N. L., Penington, A., Farlie, P. G., & Tan, T. Y. (2016). Clinical and molecular characterisation of children with Pierre Robin sequence and additional anomalies. *Molecular Syndromology*, *7*, 322–328.

Yang, J., Andre, P., Ye, L., & Yang, Y. Z. (2015). The hedgehog signalling pathway in bone formation. *International Journal of Oral Science*, *7*, 73–79.

Yang, L., Tsang, K. Y., Tang, H. C., Chan, D., & Cheah, K. S. (2014). Hypertrophic chondrocytes can become osteoblasts and osteocytes in endochondral bone formation. *Proceedings of the National Academy of Sciences of the United States of America*, *111*, 12097–12102.

Yang, G., Zhu, L., Hou, N., Lan, Y., Wu, X. M., Zhou, B., et al. (2014). Osteogenic fate of hypertrophic chondrocytes. *Cell Research*, *24*, 1266–1269.

Yasoda, A., Kitamura, H., Fujii, T., Kondo, E., Murao, N., Miura, M., et al. (2009). Systemic administration of C-type natriuretic peptide as a novel therapeutic strategy for skeletal dysplasias. *Endocrinology*, *150*, 3138–3144.

Yasoda, A., Komatsu, Y., Chusho, H., Miyazawa, T., Ozasa, A., Miura, M., et al. (2004). Overexpression of CNP in chondrocytes rescues achondroplasia through a MAPK-dependent pathway. *Nature Medicine, 10*, 80–86.

Zakany, J., & Duboule, D. (1996). Synpolydactyly in mice with a targeted deficiency in the HoxD complex. *Nature, 384*, 69–71.

Zguricas, J., Heus, H., Morales-Peralta, E., Breedveld, G., Kuyt, B., Mumcu, E. F., et al. (1999). Clinical and genetic studies on 12 preaxial polydactyly families and refinement of the localisation of the gene responsible to a 1.9 cM region on chromosome 7q36. *Journal of Medical Genetics, 36*, 32–40.

Zhang, Q., Seo, S., Bugge, K., Stone, E. M., & Sheffield, V. C. (2012). BBS proteins interact genetically with the IFT pathway to influence SHH-related phenotypes. *Human Molecular Genetics, 21*, 1945–1953.

Zhang, M., Xie, R., Hou, W., Wang, B., Shen, R., Wang, X., et al. (2009). PTHrP prevents chondrocyte premature hypertrophy by inducing cyclin-D1-dependent Runx2 and Runx3 phosphorylation, ubiquitylation and proteasomal degradation. *Journal of Cell Science, 122*, 1382–1389.

Zhou, X., von der Mark, K., Henry, S., Norton, W., Adams, H., & de Crombrugghe, B. (2014). Chondrocytes transdifferentiate into osteoblasts in endochondral bone during development, postnatal growth and fracture healing in mice. *PLoS Genetics, 10*, e1004820.

Zhu, J., & Mackem, S. (2017). John Saunders' ZPA, Sonic hedgehog and digit identity—How does it really all work? *Developmental Biology, 429*, 391–400.

Kolibri 360 GmbH
Ferdinand-Jühlke-Straße 7
99095 Erfurt, Deutschland
produktsicherheit@kolibri360.de